Roderic P. Hunt

April 2002

Thermoplastic Elastomers

Thermoplastic Elastomers

2nd Edition

Edited by

Geoffrey Holden, Norman R. Legge, Roderic P. Quirk, and Herman E. Schroeder

With contributions by

R.K. Adams, F.S. Bates, P. Bayard, W.J. Brittain, A.Y. Coran, S. Davison,
R. Fayt, G.H. Fredrickson, K.D. Gagnon, W.P. Gergen, W. Goyert,
P.T. Hammond, T. Hashimoto, G.K. Hoeschele, G.H. Hofmann, G. Holden,
C. Jacobs, R. Jerome, J.P. Kennedy, E.N. Kresge, N.R. Legge,
R.D. Lundberg, R.G. Lutz, W.J. MacKnight, W. Meckel, M. Morton,
R.P. Patel, R.P. Quirk, R.W. Rees, M.F. Rubner, H.E. Schroeder,
G.O. Schulz, P. Teyssie, S.K. Varshnay, W. Wieder, W.K. Witsiepe

Hanser Publishers, Munich Vienna New York

Hanser/Gardner Publications, Inc., Cincinnati

The Editors:
Dr. Geoffrey Holden, 1385 Iron Springs Road, #106, Prescott, AZ 86301, USA; *Dr. Norman R. Legge*, 19 Barkentine Road, Rancho Palos Verdes, CA 90274, USA; *Professor Roderic P. Quirk*, College of Polymer Science and Engineering, University of Akron, Akron, OH 44325-3909, USA; *Dr. Herman E. Schroeder*, 74 Stonegates, 4031 Kennett Pike, Wilmington, DE 19807, USA

Distributed in the USA and in Canada by
Hanser/Gardner Publications, Inc.
6600 Clough Pike, Cincinnati, Ohio 45244-4090, USA
Fax: (513) 527-8950
Phone: (513) 527-8977 or 1-800-950-8977

Distributed in all other countries by
Carl Hanser Verlag
Postfach 86 04 20, 81631 München, Germany
Fax: +49 (89) 98 12 64

The use of general descriptive names, trademarks, etc., in this publication, even if the former are not especially identified, is not to be taken as a sign that such names, as understood by the Trade Marks and Merchandise Marks Act, may accordingly be used freely by anyone.

Library of Congress Cataloging-in-Publication Data
Thermoplastic elastomers / edited by Geoffrey Holden ... [et al.].
2nd. ed.
    p.   cm.
Includes bibliographical references and index.
ISBN 1-56990-205-4
1. Elastomers.   2. Thermoplastics.   I. Holden, G.
TS1925. T445   1996
678–dc20   96-14538

Die Deutsche Bibliothek – CIP-Einheitsaufnahme
Thermoplastic elastomers / ed. by Geoffrey Holden ... -2. ed.
- Munich; Vienna; New York: Hanser; Cincinnati:
Hanser/Gardner, 1996
    ISBN 3-446-17593-8
NE: Holden, Geoffrey [Hrsg.]

© Carl Hanser Verlag, Munich Vienna New York, 1996
Typeset in England by Techset Composition Ltd., Salisbury
Printed and bound in Germany by Schoder Druck GmbH & Co KG, Gersthofen

Preface to Second Edition

In the preface to the first edition of this work, the editors stated that they planned to cover the entire field of Thermoplastic Elastomers – history, chemistry, polymer structure, morphology, rheology, physical properties and typical applications. Looking back, we believe that this rather ambitious goal was met, and that the book has been a valuable source of information to workers in this field. The time has now come to update and revise the original work and this Second Edition is the result. Once again the authors of the chapters include both technical experts from the various companies that produce Thermoplastic Elastomers and academic researchers who deal with some of the newer developments.

We again express our appreciation to the authors, their secretarial staffs and in many cases their spouses for their dedicated efforts in the production of this work. We similarly acknowledge the cooperation of the companies who provided the time and support for their employees to write many of these chapters. These are Advanced Elastomer Systems, L.P., Bayer A.G., Dow Chemical Company, E.I. du Pont de Nemours and Company, Exxon Research and Engineering Company, Goodyear Tire and Rubber Company and Olin Corporation.

In the original preface, the editors, particularly NRL, gratefully acknowledged the assistance of Jean Legge in the many tasks involved in editing a work of this complexity. On this occasion, GH similarly acknowledges the assistance that he has received from Brenda Holden in these matters. The editors also thank the staff of Chernow Editorial Services, New York, USA, Dr. E.H. Immergut and Mr Jon Glover of Hanser Publishers, New York, USA, and Ms. Martha Kürzl of Carl Hanser Verlag, Munich, Germany, for their meticulous attention to the details of copy-editing, printing and producing this book.

February 1996

G. Holden
N.R. Legge
R.P. Quirk
H.E. Schroeder

Contents

1 Introduction and Plan 1

G. Holden, N.G. Legge, H.E. Schroeder, and R.P. Quirk

2 Thermoplastic Polyurethane Elastomers 15

W. Meckel, W. Goyert, and W. Wieder

2.1 Introduction 16
2.2 Raw Materials 17
2.3 Synthesis 24
2.4 Morphology 25
2.5 Properties 32
2.6 Processing 37
2.7 Blends 37
2.8 Applications 39
2.9 Recycling 41
2.10 Future Trends 41

3 Styrenic Thermoplastic Elastomers 47

G. Holden and N.R. Legge

3.1 Introduction 48
3.2 Historical Review 48
3.3 Structure 49
3.4 Synthesis 51
3.5 Properties 54

4 Research on Anionic Triblock Copolymers 71

R.P. Quirk and M. Morton

4.1 Introduction 72
4.2 Methods and Problems in the Laboratory Synthesis of Triblock Copolymers 74
4.3 Structure–Property Relationships of Triblock Copolymers 78
4.4 Research on Other Anionic Triblock Copolymers 89
4.5 Incompatibility and Processability in Triblock Copolymers 93
4.6 Interphase Adhesion and Tensile Strength 95
4.7 Triblock Copolymers with Crystallizable End Blocks 96

5 Polyolefin-Based Thermoplastic Elastomers 101

E.N. Kresge

5.1 Introduction 102
5.2 Random Block Polymers 103
5.3 Block Copolymers 112
5.4 Graft Copolymers 117
5.5 Polyolefin Blend Thermoplastic Elastomers 119

6 Thermoplastic Elastomers Based on Halogen-Containing Polyolefins 129

G.H. Hofmann

6.1 Introduction 130
6.2 Melt Processable Rubber 130
6.3 Poly(vinyl chloride)–Nitrile Rubber Blends 143
6.4 Poly(vinyl chloride)–Copolyester Elastomer Blends 148
6.5 Pol(vinyl chloride)–Polyurethane Elastomer Blends 150

**7 Thermoplastic Elastomers Based on Dynamically Vulcanized Elastomer–Thermoplastic
Blends** 153

A.Y. Coran and R.P. Patel

7.1 Introduction 154
7.2 Preparation of Elastomer–Plastic Blends by Dynamic Vulcanization 156
7.3 Properties of Blends Prepared by Dynamic Vulcanization 157
7.4 Technological Applications 185

8 Thermoplastic Polyether Ester Elastomers 191

R.K. Adams, G.K. Hoeschele, and W.K. Witsiepe

8.1 Introduction 192
8.2 Early Fiber Research 192
8.3 Synthetic Methods 195
8.4 Polymer Structure and Morphology 197
8.5 Properties of Commercial Copolyether Ester Elastomers 211
8.6 Structural Variations 213
8.7 Polymer Blends with Polyether Ester Elastomers 219
8.8 Commercial Aspects 221

9 Thermoplastic Elastomers Based on Polyamides 227

R.G. Nelb and A.T. Chen

9.1 Introduction 228
9.2 Segmented Block Copolymers 228
9.3 Polyamide Thermoplastic Elastomers 230
9.4 Structure–Property Relationships 239
9.5 Physical Properties 241
9.6 Processing Conditions 252
9.7 Applications 254
9.8 Summary 254

**10A Ionomeric Thermoplastic Elastomers: Early Research–Surlyn® and Related
Polymers** 257

R.W. Rees

10A.1 Introduction 258
10A.2 Ionomer Discovery 258
10A.3 Development of Ionomer Technology and Applications 259

10A.4 Extension of Ionic Crosslinking to Other Polymer Types 264
10A.5 Diamine Ionomers. 267
10A.6 Structural Studies on Ionomers 268
10A.7 Ionomers in Blends and Alloys 269
10A.8 Product Development 269

10B Research and Ionomeric Systems

10B Research and Ionomeric Systems 271

W.J. MacKnight and R.D. Lundberg

10B.1   Introduction 272
10B.2   Theory 272
10B.3   Morphological Experiments 274
10B.4   Recent Developments: Synthesis 280
10B.5   Recent Developments: Properties 285
10B.6   Preferential Plasticization 288
10B.7   Ionic Interactions in Polymer Blends 288
10B.8   Applications of Ionomeric Elastomers 291
10B.9   Conclusions 292
10B.10 Future Developments 293

11 Hydrogenated Block Copolymers in Thermoplastic Elastomer Interpenetrating Polymer Networks

11 Hydrogenated Block Copolymers in Thermoplastic Elastomer Interpenetrating
    Polymer Networks 297

W.P. Gergen, R.G. Lutz, and S. Davison

11.1 Introduction 298
11.2 Hydrogenated Diene Block Copolymers 299
11.3 Thermoplastic Interpenetrating Polymer Network Formation and Properties 312
11.4 Experimental 329

12 Block Copolymer Thermodynamics: Theory and Experiment

12 Block Copolymer Thermodynamics: Theory and Experiment 335

Frank S. Bates and Glenn H. Fredrickson

12.1 Introduction 336
12.2 Strong Segregation Limit 339
12.3 Weak Segregation Limit 345
12.4 Surface Behavior 355
12.5 Discussion and Outlook 357
12.6 Update 358

13 Thermoplastic Elastomers by Carbocationic Polymerization

13 Thermoplastic Elastomers by Carbocationic Polymerization 365

J.P. Kennedy

13.1 Introduction 366
13.2 Thermoplastic Elastomer Graft Copolymers 367
13.3 Thermoplstic Elastomer Block Copolymers 380
13.4 Thermoplastic Elastomer Ionomers 389
13.5 Summary 391

14 Macromonomers as Precursors for Thermoplastic Elastomers 395

R.P. Quirk, W.J. Brittain, and G.O. Schulz

14.1 Introduction 396
14.2 Synthesis of Macromonomers 398
14.3 Homopolymerization of Macromonomers 410
14.4 Copolymerization of Macromonomers 411
14.5 Summary 425

15A Order–Disorder Transition in Block Polymers 429

Takeji Hashimoto

15A.1 Introduction 430
15A.2 Nature of the Order–Disorder Transition of Block Copolymers 431
15A.3 Equilibrium Aspects of the Order-Disorder Transition 434
15A.4 Characterization of the Order–Disorder Transition by Scattering Techniques 437
15A.5 Changes of Spatial Concentration Fluctuations Accompanied by Order–Disorder
        Transitions 442
15A.6 Kinetic Aspects 0of the Order–Disorder Transition 456
15A.7 Changes of Properties Accompanied by the Order–Disorder Transition 460

15B Thermoplastic Elastomers Produced by Bacteria 465

Karla D. Gagnon

15B.1   Introduction 466
15B.2   The Structure–Property Relationship 467
15B.3   Polymer Biosynthesis and Characterization 469
15B.4   Characterization 472
15B.5   Morphology 473
15B.6   Crystallization 475
15B.7   Mechanical and Elastic Properties 478
15B.8   Biodegradation 483
15B.9   Summary 484
15B.10 The Future 485

**15C Polymer Blends Containing Styrene/Hydrogenated Butadiene Block Copolymers:
        Solubilization and Compatibilization** 489

D.R. Paul

15C.1 Introduction 490
15C.2 Solubilization 491
15C.3 Compatibilization 496
15C.4 Summary 517

15D Polyacrylate-Based Thermoplastic Elastomers 521

R. Jerome, P. Bayard, R. Fayt, C. Jacobs, S.K. Varshney, and P. Teyssie

15D.1 Introduction 522
15D.2 Synthesis of Poly(MMA-*b*-*t*BA-*b*-MMA) Precursors 523

15D.3 Derivatization of Poly(MMA-*b*-Alkyl Acrylate-*b*-MMA).   524
15D.4 Synthesis of Star-Shaped Diblock Copolymers.   526
15D.5 Mechanical Properties of Block and Star-Branched Block Copolymers   529
15D.6 Conclusions   535

15E Novel Optical and Mechanical Properties of Diacetylene-Containing Segmented
Polyurethanes   537

P.T. Hammond and M.F. Rubner

15E.1 Introduction.   538
15E.2 Characterization of Polyurethane–Diacetylene Segmented Elastomers   543
15E.3 Mechanical Properties of Polyurethane–Diacetylenes   547
15E.4 Linear Optical Properties of Segmented Polyurethane–Diacetylenes   553
15E.5 More Recent and Future Developments   566
15E.6 Conclusions.   569

16   Applications of Thermoplastic Elastomers   573

G. Holden

16.1 Introduction.   574
16.2 Composition   576
16.3 Commercial End-Uses of Thermoplastic Elastomers   580
16.4 Economics and Summary   599

17   Future Trends   603

G. Holden, N.R. Legge, H.E. Schroeder, and R.P. Quirk

Index   611
Biographies   619

Contributors

R.K. Adams, 35 Linden Street, Claymont, DE 19803, USA

Frank S. Bates, Department of Chemical Engineering and Materials Science, University of Minnesota, Minneapolis, MN 55455, USA

Philippe Bayard, Center for Education and Research on Macromolecules, 56-4000 Liège, Belgium

William J. Brittain, Maurice Morton Institute of Polymer Science, University of Akron, Akron, OH 14325-3903, USA

A.Y. Coran, The Institute of Polymer Engineering, University of Akron, Akron, OH 44325-0301, USA

S. Davison, P.O. Box 82069, Houston, TX 77282, USA

Roger Fayt, Center for Education and Research on Macromolecules, 86-4000 Liège, Belgium

Glenn H. Fredrickson, Department of Chemical and Nuclear Engineering, University of California, Santa Barbara, CA 93106, USA

Karla D. Gagnon, University of Massachusetts, Amherst, MA 01003, USA. Present address: Advanced Elastomer Systems, Akron, OH 44334-0354, USA

W.P. Gergen, 11311 Hylander, Houston, TX 77070, USA

W. Goyert, Bayer AG, K-A Forschung, 41538 Dormagen, Germany

Paula T. Hammond, Department of Chemical Engineering, Massachusetts Institute of Technology, Cambridge, MA 02139, USA

Takeji Hashimoto, Department of Polymer Chemistry, Faculty of Engineering, Kyoto University, Kyoto 606, Japan

G.K. Hoeschele, 2007 Dogwood Lane, Faulk Woods, Wilmington, DE 19810, USA

George H. Hofmann, DuPont Packaging and Industrial Polymers, Wilmington, DE 19880-0269, USA

G. Holden, Holden Polymer Consulting Incorporated, Prescott, AZ 86301, USA

Christian Jacobs, Center for Education and Research on Macromolecules, 86-4000 Liège, Belgium

Robert Jerome, Center for Education and Research on Macromolecules, 86-4000 Liège, Belgium

Joseph P. Kennedy, Institute of Polymer Science, University of Akron, Akron, OH 44325-3909, USA

E.N. Kresge (retired), Exxon Chemical Company, Watchung, NJ 07060, USA

N.R. Legge, Consultant, Rancho Palos Verdes, CA 90274, USA

R.D. Lundberg, Exxon Research and Engineering Company, Annandale, NJ 08801, USA

R.G. Lutz, 2959 Joy Road, Occidental, CA 98465, USA

W.J. MacKnight, University of Massachusetts, Amherst, MA 01003, USA

W. Meckel, Bayer AG, K-A Forschung, 41538 Dormagen, Germany

M. Morton (deceased)

R.P. Patel, Advanced Elastomer Systems, Akron, OH 44334, USA

Roderic P. Quirk, Maurice Morton Institute of Polymer Science, University of Akron, Akron, OH 44325-3903, USA

Richard W. Rees, 105 Sharply Road, Wilmington, DE 19803, USA

Michael F. Rubner, Department of Materials Science and Engineering, Massachusetts Institute of Technology, Cambridge, MA 02139, USA

Herman E. Schroeder, 74 Stonegates, 4031 Kennett Pike, Wilmington, DE 19807, USA

Gerald O. Schulz, The Goodyear Tire & Rubber Company, Akron, OH 44316, USA

Philippe Teyssie, Center for Education and Research on Macromolecules, 86-4000 Liège, Belgium

Sunil K. Varshnay, Center for Education and Research on Macromolecules, 86-4000 Liège, Belgium

W. Wieder, Bayer AG, KA-P/S-Ökologie, Geb M16 Ost, 41538 Dormagen, Germany

W.K. Witsiepe (deceased)

1 Introduction and Plan

G. Holden, N.R. Legge, H.E. Schroeder, and R.P. Quirk

The editors had a number of reasons for publishing this second edition of *Thermoplastic Elastomers—A Comprehensive Review*. The growth of the thermoplastic elastomer (TPE) industry has now reached a high level of commercial importance, involving many new products and industrial participants, and a large number of strong academic research groups. From the discovery of the elastomeric character of plasticized poly(vinyl chloride) (PVC) in 1926 by Waldo Semon; and the first academic work on block co-polymers by Bolland and Melville in 1938; through the discovery of the polyurethanes by Otto Bayer; and all the practical industrial work on thermoplastic polyamides, copolyesters, and polyester amides during the 1930s and 1940s; to the contributions of Flory, Mark, Tobolsky, and many others, there has been an interaction between academic and industrial research enlivened by serendipitous discoveries that resulted in great advances. Thus we note that since 1967, the date of the first symposium on block copolymers [1], which was largely devoted to the polystyrene–polydiene type, numerous texts [2–21], symposia [22–32], and review articles [33–37] have been devoted either to block copolymers or to the related subject of TPEs.

The majority of TPEs are block or graft copolymers, and here we include a note on their nomenclature. The most commonly applied terminology for block polymers uses A to represent a block of A mer units, and similarly B, C, and so forth represent blocks of B mer and C mer units and so forth.

The following representations are widely used:

$$A–B \qquad A–B–C \qquad A–B–A$$

$$(A–B)_n \qquad (A–B)_n x$$

Here A–B represents a diblock polymer, A–B–C represents a triblock polymer in which all three segments are polymerized from different monomers, and A–B–A represents a block copolymer with two terminal A blocks and a B center block. $(A–B)_n$ represents an alternating block copolymer, A–B–A–B–A– ... etc., whereas $(A–B)_n x$ represents a branched block copolymer with n branches ($n = 2, 3, 4 \ldots$) and a junction point x.

Similarly, graft copolymers may be represented as

$$B \ldots B–B \left(B \atop {\overset{|}{A}} \right)_n B– \ldots B$$

This represents a polymer where each B block has (on average) n random grafts of A blocks.

It is common to use the first letter of the monomer unit to denote the polymer block. For example, a three-block copolymer, poly(styrene–*block*-butadiene–*block*-styrene) is written as S–B–S. If one of the blocks is itself a copolymer (e.g., ethylene–propylene rubber), the block copolymer poly(styrene–*block*-ethylene–*co*-propylene–*block*-styrene) is written as S–EP–S.

The main objective of this book is to record in one volume the research and development status of the TPE systems from the beginning up to the most recent academic research. To place these in the proper time frame, the order of the chapters correspond to the chronological sequence of developments insofar as this is possible.

A second objective is to stimulate further research and development in these systems, as it is now apparent that there are very many ways of arriving at TPEs. The consequences of this continued forward movement will be important technically and economically.

A brief survey of the events leading up to the present status of TPEs, as described in this book, will aid in orienting the reader.

In early research the term "thermoplastic elastomer" is seldom found. Nevertheless, polymeric products with both thermoplastic and elastomeric properties were discovered and utilized. For example, the plasticization of PVC, by high boiling point liquids, to give a flexible material resembling rubber or a leather was recognized by Waldo Semon [38] of B.F. Goodrich in 1926. Plasticized PVC was later marketed by Goodrich under the tradename Koroseal®.

A composition of matter patent covering vulcanized and unvulcanized blends of poly(acrylonitrile–co-butadiene) rubber (NBR) and PVC was applied for in 1940 by D.E. Henderson [39] of Goodrich (see Chapter 6). In 1947 Goodrich offered NBR–PVC blends under the tradename Geon Polyblend®. A typical blend having a composition of 45% NBR/55% PVC had a tensile strength of 13 MPa (1800 psi), and elongation at break of 450%.

The basic diisocyanate polyaddition reaction was discovered by Professor Otto Bayer [40] in the I.G. Farben laboratories in Leverkusen in 1937 and used to make a polyurethane fiber trademarked as Perlon®. Later, the elastomeric properties of some polyurethanes were recognized independently by chemists at DuPont [41, 42] and ICI [43]. Many of the polymers they obtained were thermoplastic but lacked adequate melt stability to function as practical TPEs. In the early 1950s chemists at Bayer, DuPont, Goodyear, and other companies were making and offering to their customers thermoplastic gum elastomers, but as we now realize, these did not have a sufficiently high concentration of hard segments to have good properties unless vulcanized.

In the late 1940s Coleman of ICI was attempting to impart dyeability to polyethylene terephthalate by copolymerizing small amounts of poly(oxyalkylene) glycols. Snyder of DuPont was using a somewhat similar approach in a program aimed at elastic fibers. In 1952 Snyder received a U.S. Patent [44], applied for in 1950, on elastic linear copolyesters prepared by melt copolymerizing, for example, terephthalic acid, suberic acid, and trimethylene glycol; melt copolymerizing separately a copolyester of terephthalic acid and ethylene glycol, and finally performing a carefully controlled melt-ester interchange reaction between the two polymers. The resulting linear copolyester had a higher strength and higher stretch modulus than any natural rubber threads. These copolyester elastic fibers had that essential property of natural rubber threads, lacking in most synthetic materials, of a very quick elastic recovery (snap!). The fibers could be extruded from the melt or spun from solvents. The former is a thermoplastic processing step and the latter indicates solubility, and therefore the absence of crosslinks. Thus, these elastic fibers can be classified as TPEs, possibly the first multiblock, (A–B)$_n$ copolymers in which the relationship between structure and property was clearly delineated.

Moncrief [45] notes a 1954 patent by DuPont [46] that describes an elastomeric polyurethane fiber based on copolymerizing polyethylene glycol and toluene-2,4-diisocyanate in the presence of water and a small amount of acid chloride. It had a tensile strength of 13.8 MPa (2000 psi) and elongation at break of 500%.

In 1958 DuPont [47] introduced to the trade an experimental segmented polyurethane identified as Fiber K. This fiber was commercialized in 1959 under the Lycra® trademark and was later revealed to be a segmented polyether urethane prepared from methylene bis(4-phenyl isocyanate).

In 1957 Schollenberger [48] presented an article on Polyurethan VC, a "virtually crosslinked elastomer." This was a linear polymer prepared from methylene bis(4-phenyl isocyanate), adipic acid, and 1,4-butanediol. The polymer was completely soluble and showed high elasticity, high extensibility, and excellent resistance to tear and abrasion. There was no explanation of the virtual crosslinking mechanism.

Charch and Shivers [49] in 1959 published an article on elastomeric condensation block copolymers that gave a very useful discussion of the stress–decay and tensile recovery properties of condensation block copolymers with different amounts of hard crystalline and soft low-melting copolymer blocks. The authors viewed these as prototypes of a very large family of condensation elastomers that combine stiff and flexible segments in the same polymer chain. More detailed discussions of this article are presented in Chapters 2 and 8. It is interesting to note that the article, which came close to describing the essentials of elastomeric, multisegmented TPEs, was oriented to elastic fibers and published in the *Textile Research Journal*.

However, it was not until 1966 that Cooper and Tobolsky [50] compared the properties of Polyurethan VC with those of an anionically polymerized, triblock poly(styrene–*block*-butadiene–*block*-styrene) copolymer (S–B–S). They concluded that the presence of the segregated hard and soft blocks in the polyurethane, rather than the hydrogen bonding, was the source of the TPE behavior.

In addition to Goodrich, at least five other companies have manufactured and sold thermoplastic polyurethane elastomer in the United States. These products have processed well in thermoplastic equipment and have shown very good elastomeric properties. Abrasion and tear resistance are outstanding and the materials have been commercially successful.

Professor Otto Bayer and his co-workers at Bayer AG in Leverkusen have accomplished major developments in polyurethane elastomers and thermoplastic polyurethane elastomers. The editors were delighted when Dr. W. Meckel, Dr. W. Goyert, and Dr. W. Wieder of Bayer AG agreed to write Chapter 2, Thermoplastic Polyurethane Elastomers.

Other developments in block copolymers were also occurring at about this time. Thus the first academic reference to block polymer formation that we have noted was reported in 1938 by Bolland and Melville [51], who found that a film of poly(methyl methacrylate), deposited on the walls of an evacuated tube, could initiate the polymerization of chloroprene. Melville later concluded [52] that trapped free radicals in the film initiated the second polymerization.

In 1942 (A–B)$_n$ type block copolymers were made by Hanford and Holmes [42] by reaction of diisocyanate with difunctional polymer. Bayer's Vulcolan® elastomer, produced in the early 1940s, was a diisocyanate–alkylene adipate condensation polymer (see Chapter 2).

An (A–B)$_n$ block polymer was reported in 1946 by Baxendale et al. [53], who linked low molecular weight polystyrene and poly(methyl acrylate) with diisocyanate to form a high molecular weight, multiblock polymer.

Nonionic detergents based on a triblock A–B–A type polymer with polyethylene oxide end blocks and a polypropylene oxide center block were introduced in 1951 by Lundsted [54].

In 1953 H.F. Mark [55] published an article on multiblock and multigraft polymers in which the blocks were of the order of 50 monomer units in length. Potential applications of these block polymers were considered mainly from the point of view of detergency and of surface treatment of fibers and films. The preparation of the multiblock and multigraft polymers was also discussed.

A comprehensive review of graft and block copolymers and their syntheses was published by E.H. Immergut and H.F. Mark [56] in 1956. At the British Rubber Producers Research Association Bateman [57] showed that graft copolymers of methyl methacrylate (PMMA) on natural rubber (NR) could exist in two physical forms depending on the precipitation method used. If precipitated by a nonsolvent for NR, the NR polymer chains were collapsed and the PMMA chains extended; the resulting material was hard, stiff, and nontacky. If precipitated by a nonsolvent for PMMA, the NR chains were extended and the PMMA chains collapsed; the material was soft and flabby. Both of these forms were stable under heavy milling. Merrett [58] discussed these observations in terms of the microseparation of phases. He pointed out that once this occurred any attempt to approach homogeneity would be improbable. He postulated that either form of the dry polymer would consist of domains of collapsed chains as a discrete phase in a continuous phase of extended chains. Merrett's comments constitute a precursor event to the domain theory of two-phase TPEs, which was not postulated as such until about 10 years later.

It is interesting to note Ceresa's comment on our state of knowledge of block and graft polymers in 1962 [59]. His book covered the syntheses of 1400 block or graft polymers. He stated that fewer than 5% of the block and graft copolymers described had been isolated with any reasonable purity and that fewer than 20 species had been analyzed and characterized fully. He attributed this to the contamination of the product block or graft copolymers with homopolymers. In many cases the block or graft copolymers comprised only a small percentage of the final mixture.

Anionic polymerization systems have been widely used in block polymerizations. However, as Halasa [60] has pointed out, Ziegler and his co-workers [61–64] in the late 1920s and early 1930s laid the foundation for living polymerization through their research on alkali metals and their organic derivatives (lithium and alkyllithium), with butadiene, isoprene, and piperylene in polar media. These were shown to be "nontermination" systems. The polymeric products were low in molecular weight and high in 1,2 addition polybutadiene, or 3,4 addition polyisoprene. Thus, the polymers had high T_gs and were resinous in nature.

An economic polymerization system that would produce a polyisoprene with a high *cis*-1,4 content (the structure of natural rubber) had been a long-term objective of the U.S. Synthetic Rubber Research Program. In 1955, Firestone Research reported to the Office of Synthetic Rubber U.S. Federal Facilities Corporation their discovery of a lithium metal catalyst polymerization system that produced high ($>90\%$) *cis*-1,4-polyisoprene. The following publication [65], as well as the Goodrich announcement [66] of a high *cis*-1,4-polyisoprene prepared via a Ziegler coordination catalyst system, aroused great interest in other rubber research groups. Because the Firestone report was distributed to the operators of the U.S. Government-owned synthetic rubber plants there is little doubt that those with adequate research facilities quickly turned to an examination of lithium metal initiated polymerizations of isoprene during the years 1955–56.

During this period the unique role of lithium as a counterion in hydrocarbon media was further elucidated by the work of Tobolsky and his co-workers. In 1957 they reported that the same high *cis*-1,4-polyisoprene microstructure was obtained either by using bulk polymerization initiated by a lithium dispersion or by using solution polymerization in heptane or benzene initiated by *n*-butyl lithium [67, 68]. Furthermore, the loss of the high *cis*-1,4 stereospecificity was found for analogous polymerizations initiated by lithium counterions in the presence of either tetrahydrofuran or diethyl ether. They also reported that solution polymerizations in heptane initiated by either sodium or potassium counterions gave high 3,4 and 1,2 enchainment [69].

Shell Development started research on lithium metal initiators for isoprene polymerization in 1956. It is apparent that they and others did not go back to the early work of Ziegler. However, it was soon discovered that the initiating species was an alkyllithium and that the system, if sufficiently pure, had no termination step. In 1957 Porter [70], working in a research department with both elastomers and plastics responsibility, described a process for the polymerization of styrene–diene block polymers using alkyllithium initiators. At this point the experimental polymers were not recognized as TPEs but the research contributed importantly to in-house background knowledge.

The alkyllithium polymerization of high *cis*-1,4-polyisoprene was taken through bench scale process development at Shell Development and then directly to plant scale in a modified styrene–butadiene–rubber (SBR) plant. This resulted in the first commercial production of high *cis*-polyisoprene in 1959 [71].

Crouch and Short [72] of Phillips discussed the use of S–B block copolymers, although not identified as such, in 1961. The polymers were produced commercially in 1962 and identified [73] as S–B polymers in 1964.

In 1956, Szwarc, Levy, and Milkovich [74–76] rediscovered the anionic living polymer systems, using sodium naphthalene diinitiators in tetrahydrofuran (THF) to prepare S–I–S block copolymers. By the time the Szwarc articles appeared, most of the major research groups working with isoprene polymerizations based on alkyllithium initiators were aware of the living polymer nature of these systems. Sodium naphthalene diinitiators required polar solvents such as THF. The use of these polar solvents resulted in a polyisoprene segment with very high 3,4 addition and this was not a very good elastomer. Thus, the Szwarc results were not directly applicable in the ongoing elastomer research projects.

In 1961, the staff of the Synthetic Rubber Research Laboratory of Shell Chemical Company were examining solutions to problems of excessive flow and poor green strength of high *cis*-1,4-polyisoprene (which was in commercial production at that time) and also of high 1,4-polybutadiene. Both these polymers were produced using alkyllithium initiation. As a part of this experimental program, low molecular weight polyisoprene and polybutadiene were prepared with very short polystyrene segments to give S–I and S–B diblock copolymers. These copolymers showed virtually no cold flow and appeared to have solved this part of the problem. Shortly thereafter S–I–S and S–B–S triblock copolymers were also prepared. All these polymers were heated and then pressed into flat sheets for determination of green strength. When tested, the samples prepared from S–I–S and S–B–S were found to have very high tensile strengths, high elongations at break, and very rapid elastic returns (snap). These properties were similar to those of conventional vulcanized rubbers, even though they were achieved without any vulcanization step. The sheets were soluble in toluene, thus demonstrating that no chemical crosslinks had been formed.

Fortunately at that time the Research Laboratory also housed a polystyrene research group fully equipped with thermoplastic forming processes. The combined research groups were quick to recognize the potential of TPEs and to evolve an explanation—the domain theory. This is the theory that in the solid state, the polystyrene end segments agglomerate to form a separate phase (the domains). These domains act as physical crosslinks to form a rubber network, similar in many ways to that of a vulcanized rubber. Physical properties and thermo–plastic processability of the S–I–S and S–B–S samples were so outstanding that the incentives to carry through the commercialization were clear.

The discovery of the S–I–S and S–B–S triblock TPEs was clearly serendipitous and drew on much research background [71]. Four factors of importance leading up to this discovery can be identified:

1. Alkyllithium polymerization of high *cis*-1,4-polyisoprene.
2. Early scouting research, also in-house, on a styrene–diene block polymer process.
3. Existence at the discovery location of an R&D staff with long experience in both elastomers and plastics.
4. The domain theory.

The triblock styrene–diene TPEs were announced in October 1965 [77] and the domain theory described and extended to other block copolymers in an article published in 1967 [78].

Chapter 3, Styrenic Thermoplastic Elastomers, is written by Geoffrey Holden and N.R. Legge, both members of the research team that discovered these materials in the 1960s.

Following the announcement [77] of these styrene–diene TPEs and the publication of articles on the domain theory [79, 80], there was a tremendous surge of interest in TPEs and in the two-phase systems. In this respect, the domain theory served as a paradigm [81], in that its simple explanation combined with the obvious commercial importance of the styrene–diene TPEs has spurred the development of a new field of polymer science. One of the most active of the academic research groups that studied the anionic polymerization systems and polymer properties of the block copolymers was at the University of Akron. The editors invited Professor Maurice Morton, Regents Professor Emeritus of Polymer Science, and Retired Director of the Institute of Polymer Science of the University of Akron to join with Professor Roderic P. Quirk, Professor of Polymer Science at the University of Akron, to write Chapter 4, Research on Anionic Triblock Copolymers. Sadly, while this was in progress, Professor Morton passed away. Professor Quirk completed the chapter and the other editors owe him a debt of gratitude for fitting this project into his busy schedule. In Chapter 4, Professor Morton and Professor Quirk have covered most of the research done on this subject at what is now the Maurice Morton Institute of Polymer Science.

A.V. Tobolsky in 1958 [82] predicted that "new block polymers might be synthesized, one block being composed of linear polyethylene, or isotactic polypropylene and the other block being a random copolymer of ethylene and propylene. Such a polymer would ... have high melting crystalline regions, and amorphous regions of low T_g. Blends of different varieties of ethylene propylene polymers may also prove interesting." This was a prophetic statement. Blends of ethylene–propylene–diene rubber (EPDM) or ethylene–propylene rubber (EPR) with polypropylene have proved to be of major commercial importance (see Chapter 16). E.G. Kontos and his co-workers at Uniroyal have done some thought provoking research on "living α-olefin polymerizations" [83–86]. These polymers were composed of random amorphous ethylene–propylene blocks and linear homopolymer polyethylene blocks, or isotactic polypropylene blocks. The semicrystalline stereoblock copolymers were said to have "plastic–rubber properties" [85]. Although these particular polymer systems were not commercialized, the editors believed that it would be useful to include them, along with the commercial products that were later introduced. We were very fortunate in persuading Dr. Edward N. Kresge, formerly of Exxon Chemical, to cover all these developments. He is one of the acknowledged authorities in this field and reviews it for us in Chapter 5, Polyolefin-Based Thermoplastic Elastomers.

Another important group of TPEs is based on halogen-containing polymers. This is very broad field and can potentially cover plasticized PVC [38], PVC–NBR blends [39], and the single phase melt processable rubber developed by DuPont and described by Wallace [87]. The editors felt that most of the work on plasticized PVC fell outside the intended scope of this book. We were pleased to have Dr. G.W. Hoffman of DuPont review all the other developments in this area in Chapter 6, Thermoplastic Elastomers Based on Halogen-Containing Polyolefins.

The next item of our agenda was a discussion of the developments in producing TPEs by dynamic vulcanization. Here the editors were attracted by the work done at Monsanto. This was described in an excellent series of articles, commencing in 1978 and continuing through 1985 [88–97], presented by A.Y. Coran, R.P. Patel, et al. These describe dynamically vulcanized blends of a wide variety of thermoplastics and elastomers. The editors are grateful to Dr. Coran (now with the University of Akron) and to Dr. Patel (now with Advanced Elastomer Systems) for agreeing to write Chapter 7, Thermoplastic Elastomers Based on Dynamically Vulcanized Elastomer–Thermoplastic Blends, and for including also a discussion of the earlier work by others. Their research is outstanding in coverage of the polymers examined, the breadth of the physical evaluations, the application of the results to guidelines for these systems, and in the commercialization of several very useful new products.

From the time of their early work on segmented liquid and solid elastomeric polyurethane gums (1950–60), and throughout the studies of the high-modulus elastomeric fibers such as Lycra, DuPont scientists had been seeking a unique species of TPE to round out their line of specialty elastomeric polymers. Their research showed that the polyurethanes had relatively poor melt stability because of a tendency to revert to macromonomer segments, and that this deficiency could not be corrected through use of economical structural variations. They then turned back to the early work on segmented condensation polymers of either polyamides or polyesters. These had produced materials of great interest as elastic fibers and had given indications that high-modulus rubbers could be made from segmented polymers with hard and soft segments.

In 1968 at DuPont, Dr. W.K. Witsiepe [98] discovered an outstanding new variant of this class in the form of a segmented copolyester based on poly(tetramethylene terephthalate) and poly(oxytetramethylene) glycol. In Chapter 8, Thermoplastic Elastomers Having Polyester Hard Segments, Dr. Witsiepe, Dr. R.K. Adams and Dr. G.K. Hoeschle (who are all now retired from DuPont) describe the background for this discovery, and then review the structure, morphology, and the physical and mechanical properties of these materials. The morphology has been of special interest because these block copolymers (together with the polyurethanes and polyamides described in Chapters 2 and 9) appear to have an entirely different structure from the styrenic triblock copolymers. In the first three block polymers there seem to be two interpenetrating, more or less co-continuous phases, made up of a network of crystalline domains in an elastomer matrix. Because of the unusual structure they have extraordinary physical properties. We are sad to say that Dr. Witsiepe died after revising this chapter.

When the polyurethanes and copolyesters revealed the virtues of crystalline hard segments in TPEs, others were quick to search for similar new polymers that might show special advantages [99, 100]. Polyamides were an obvious possibility and ATOCHEM and Dow Chemical Company have been especially active in this area. Block copolymers containing hard segments of high melting point alternating with elastomeric segments can be produced through reaction of either diisocyanate or diol macromonomers with dicarboxylic acids. A wide variety of block copolymers can be produced in this way, and many have good high temperature mechanical properties. Both authors are very experienced in this field. In Chapter 9, Thermoplastic Elastomers Based on Polyamides, Dr. R.G. Nelb II and Dr. A.T. Chen describe the synthesis and properties of copolyetheramides and copolyesteramides produced by condensation reactions. Dr. Nelb is a member of the Research Staff of Dow Chemical Company. Dr. Chen was formerly with the same organization; he is now a member of the Olin Research Center.

In a review [101] of research and patent activity in TPEs, ionomeric systems were prominent in both areas. The editors decided that this was another case in which some of the early work should be set forth as a preamble. We asked R.W. Rees, who is now retired from DuPont, to describe the early research on these products, including Surlyn®. The result is Chapter 10A, Ionomeric Thermoplastic Elastomers: Early Research—Surlyn and Related Products. R.W. Rees is the discoverer of the Surlyn polymers and the first to postulate, in 1964 [102], the clustering of the ionic structures in these systems. He and his collaborators noted the differences in crosslinks produced by different ion pairs, for example, the ionomers based on Na, K, and NH_4 were thermally reversible; similar materials based on Zn and Mg had very high melting points and were intractable.

For a more general review of ionomeric TPEs we invited Professor W.P. MacKnight and Dr. R.E. Lundberg to contribute Chapter 10B, Research on Ionomeric Systems. Both have much experience in this field. Professor MacKnight is Head of the Department of Polymer Science and Engineering of the University of Massachusetts at Amherst and Dr. R.E. Lundberg is a consultant in the Corporate Research Laboratories of the Exxon Research and Engineering Company.

In Chapter 11, S. Davison, W.P. Gergen, and R.G. Lutz discuss Hydrogenated Block Copolymers in Thermoplastic Interpenetrating Networks. The authors were engaged in Shell Development Company TPE research for over 20 years, specializing in rheology, morphology, and physical properties of the polymers and blends. They describe the structure and properties of hydrogenated diene block copolymers, poly(styrene–*block*-ethylene–*co*-butylene–*block*-styrene), S–EB–S, compared to those of poly(styrene–*block*-butadiene–*block*-styrene), S–B–S, of comparable block lengths. They then discuss the morphology and properties of interpenetrating network blends of S–EB–S with polypropylene, polybutylene, nylon, polybutylene terephthalate, polycarbonate, and other thermoplastics. Mr. Gergen is presently a consultant with Shell Development Company and the other two authors are retired from the same organization.

At this point in the book we have described the research and development of the major types of TPEs that are presently commercially available. However, the subject of TPEs is unusual in that in addition to its commercial importance it also has considerable theoretical interest. We now turn to this element. It is described in Chapter 12, Block Copolymer Thermodynamics: Theory and Experiment, by Professor Frank S. Bates of the Department of Chemical Engineering and Materials Science, University of Minnesota and Professor Glen H.

Fredrickson of the Department of Chemical and Nuclear Engineering, University of California at Santa Barbara. Dr. Bates and Dr. Fredrickson have updated an article from the *Annual Reviews of Physical Chemistry* that extensively covers the work in this field.

Since the first edition of this book appeared in 1987, there have been many attempts to expand the range of available TPEs. It is clear that the underlying feature of most existing TPEs is a system consisting of hard and soft phases, often chemically linked. One way to obtain such chemically linked phases is by block copolymerization. Another is by producing graft or comb type copolymers, in which one polymer species is attached pendant to the other. Our next two chapters explore these two approaches.

In Chapter 13, Professor Joseph Kennedy, Distinguished Professor of Polymer Science and Chemistry at the University of Akron, describes the synthesis and properties of novel block copolymers prepared by carbocationic polymerization. Structurally, these resemble the styrenic block copolymers discussed in Chapter 3. However, they show two important differences: first, a wide range of substituted styrenes can be polymerized as the end segments and second, the elastomer segment is polyisobutylene. These changes permit the direct synthesis of styrenic block copolymers with improved high-temperature properties and resistance to oxidative degradation.

Another approach is described in Chapter 14. In this chapter, Professor Roderic Quirk and Professor W.J. Britain (both of the University of Akron) and Dr. Gerald Schultz of the Goodyear Tire and Rubber Company describe the synthesis of "comb" graft copolymers using macromonomers (i.e., linear macromolecules with reactive groups at one or both ends of the chain). This synthesis technique opens up many possibilities for the production of TPEs based on new hard and soft phases.

In the first edition of this book, the editors invited a number of outstanding research workers to describe recent research in their laboratories. Their contributions were gathered into a multisection chapter entitled Research on Thermoplastic Elastomers. The editors did not request specific topics but rather suggested that the research leaders might describe a single project, or summarize work done by the laboratory in recent years. The choice was theirs. In this edition we have elected to continue this practice and the result is Chapter 15.

It was difficult to decide on the order of the sections in Chapter 15. Finally the editors selected an order, secure in the knowledge that most readers would approach this chapter with their own specific interest firmly in mind, and decide their own order of reading by title and author. Following are some comments.

Section A is Order–Disorder Transition in Block Copolymers by T. Hashimoto. The nature of the order–disorder transition in block copolymers has been explored extensively in recent years. Professor Hashimoto and his co-workers at the University of Kyoto have been active contributors. He has pointed out that this problem bears on such industrial applications of block copolymers as TPEs, pressure-sensitive hot-melt adhesives, viscosity stabilizers for oils, and so forth. For example, if the block copolymers are in the disordered state at processing temperatures their viscosities are low—an advantage in processability. On the other hand, if they are in the ordered state at processing temperatures they exhibit high viscosity and remarkable non-Newtonian behavior.

The order–disorder transition of block copolymers is also of great theoretical importance. It is related to structure and structure evolution (ordering) and dissolution (disordering) in a cooperative system—a fundamental problem in equilibrium and nonequilibrium statistical physics in the condensed state.

Section B is Thermoplastic Elastomers Produced by Bacteria by K.D. Gagnon. All the other TPE systems described in both the previous and the present edition of this book were produced by chemical synthesis or, in one case, the mixing of a synthetic polymer (polypropylene) with natural rubber. In this chapter, a different and novel approach has been taken by Dr. Gagnon. She and her fellow workers at the University of Massachusetts, Amherst [103, 104] have extensively studied polyesters produced by bacteria.

These polymers contain pendant groups that are arranged stereoregularly. Normally, this gives crystalline structures that are hard thermoplastics. However, when the polymers are grown under suitable conditions, the pendant groups can be mixed alkyls. This gives amorphous as well as crystalline regions in the polymers, which results in A–B–A–B–A–B ... structures, similar in some ways to those of the TPEs described in Chapters 2, 5, 8, and 9.

Section C is Polymer Blends Containing Styrene–Hydrogenated Butadiene Block Copolymers: Solubilization and Compatibilization by D.R. Paul. In 1978 [105] Professor Paul of the University of Texas presented a comprehensive review of the expected action of block copolymer molecules as interfacial agents in immiscible polymer mixtures. In Section C he provides a more focused review of results that have been published since then. The discussion here is limited to the use of styrene–hydrogenated butadiene block copolymers (S–EB–S) as compatibilizing agents. The systems considered are those in which these block copolymers are added to mixtures of polystyrene (PS) with low-density polyethylene (LDPE), high-density polyethylene (HDPE), and polypropylene (PP). One example is included in which the block copolymer segments are not miscible with either homopolymer—HDPE and poly(ethylene terephthalate) (PET). In addition to results from his own laboratory, Professor Paul has included results from other workers in this field. Throughout the discussions of these systems Professor Paul attempts to reach conclusions concerning mechanisms and the optimum structure for compatibilizing block copolymers.

Section D is Polyacrylate Based Elastomers Blends by R. Jérôme, Ph. Bayard, R. Fayt, C. Jacobs, S.K. Varshney, and Ph. Teyssié. In the first edition of this book, three workers from the Laboratoire de Chemie Macromoleculaire et de Catalyse Organique of the Université de Liège, (Professor R. Jérôme, Professor R. Fayt, and Professor Ph. Teyssié) described the preparation and properties of some novel block copolymers and blends. In this edition, the original authors have been joined by three collaborators (Dr. Ph. Bayard, Dr. C. Jacobs, and Dr. S.K. Varshney). Together, they describe some further work in a more specific field, that of A–B–A TPEs in which the elastomeric B segments (and in many cases the hard A segments also) are based on polyacrylates. Their properties are compared to those of analogous S–B–S TPEs of the type described in Chapters 3 and 4 of this edition.

Section E is Novel Optical and Mechanical Properties of Diacetylene Containing Segmented Polyurethanes by P.T. Hammond and M.F. Rubner. This section covers an interesting and novel extension of the thermoplastic polyurethane elastomers described in Chapter 2. In the polyurethanes described here, the hard domains contain a polydiacetylene network. This results in dramatic color changes in these polymers with changes in either temperature or strain. In their work at the Massachusetts Institute of Technology Dr. Hammond and Professor Rubner have extensively studied these polymers and their properties. Here they discuss both the theoretical background and also the potential applications of these novel materials.

Of course, there would not be much interest in the subject of TPEs (and not much demand for a second edition of this book!) if these materials did not have extensive commercial applications. In Chapter 16, Applications of Thermoplastic Elastomers, Dr. Geoffrey Holden covers applications of most of the commercially available TPEs and compares their physical properties. He also gives an extensive list of some of the tradenames under which these polymers are sold. Dr. Holden, who is now retired from Shell Development Company, has been active in the field of TPE research since the early 1960s, and has written many articles on the rheology, physical properties, compounding, and applications of TPEs.

In the first edition of this book, two of the editors with many years of experience in directing research on elastomers (Dr. N.R. Legge and Dr. H.E. Schroeder) reviewed in detail the contributions to the book and then examined further potential progress in TPE research. In Chapter 17, Future Trends, the other two editors of the present edition (Professor R.P. Quirk and Dr. G. Holden) have updated and revised these forecasts in the light of the many recent developments in this field.

References

1. J. Moacanin, G. Holden, and N.W. Tschoegl (Eds.), Symposium sponsored by the California Institute of Technology and the American Chemical Society, Pasadena, CA, 1967. *J. Polym. Sci. Pt. C 26* (1969)
2. D.C. Allport and W.H. Janes (Eds.), *Block Copolymers* (1972) John Wiley & Sons, New York
3. R.J. Ceresa (Ed.), *Block and Graft Polymerization* (1972) John Wiley & Sons, New York
4. J.J. Burke and V. Weiss (Eds.), *Block and Graft Copolymers* (1973) Syracuse Univ Press, New York
5. G.E. Molau (Ed.), *Colloidal and Morphological Behavior of Block and Graft Copolymers* (1970) Plenum Press, New York
6. S.L. Cooper and G.M. Estes (Eds.), *Multiphase Polymers, Advances in Chemistry Series, No. 176* (1979) American Chemical Society, Washington, D.C.
7. A. Eisenberg and M. King, *Ion Containing Polymers* (1977) Academic Press, New York
8. A. Noshay and J.E. McGrath, *Block Copolymers—Overview and Critical Survey* (1977) Academic Press, New York
9. D. Klempner and K.C. Frisch (Eds.), *Polymer Alloys—Blends, Blocks, Grafts, and Interpenetrating Networks* (1977) Plenum Press, New York
10. D. Klempner and K.C. Frisch (Eds.), *Polymer Alloys II* (1980) Plenum Press, New York
11. O. Olabisi, L.M. Robeson, and M.T. Shaw, *Polymer–Polymer Miscibility* (1979) Academic Press, New York
12. D.R. Paul and S. Newman (Eds.), *Polymer Blends, Vols. 1 and 2* (1978) Academic Press, New York
13. N.A.J. Platzer, *Copolymer, Polyblends, and Composites. Advances in Chemistry Series, No. 142* (1975) American Chemical Society, Washington, D.C.
14. L.H. Sperling, *Interpenetrating Polymer Networks and Related Materials* (1981) Plenum Press, New York
15. Yu. S. Lipatou and L.M. Sergeena, *Interpenetrating Polymeric Networks* (1979) Naukova Dumka, Kiev
16. B.M. Walker and C.P. Rader (Eds.), *Handbook of Thermoplastic Elastomers* (1988) Van Nostrand, New York
17. A.D. Thorn, *Thermoplastic Elastomers—A Review of Current Information* (1980) Rubber and Plastics Research Association of Great Britain, Shawbury, Shrewsbury, Shropshire SY4 4NR, Great Britain
18. S.L. Aggarwal (Ed.), *Block Polymers, Proc. Am. Chem. Soc. Symposium*, New York, 1969; Plenum Press, New York (1970)
19. L.H. Sperling (Ed.), *Recent Advances in Polymer Blends, Grafts, and Blocks, Proc. Am. Chem. Soc. Symposium*, Chicago, 1973; Plenum Press New York (1974)
20. C.E. Rogers and A. Skoulios (Eds.), *Polym. Eng. Sci. 17* (1977)
21. R.J. Ambrose and S.L. Aggarwal (Eds.), *J. Polym. Sci. Polym. Symp. 60*, (1977)
22. ACS Rubber Division Symposium on Thermoplastic Elastomers, Los Angeles, CA, April, 1985
23. ACS Rubber Division Symposium on Thermoplastic Elastomers, Cincinnati, OH, October 1988

24. ACS Rubber Division Symposium on Thermoplastic Elastomers, Nashville, TN, November 1992

25. ACS Symposium on Advances in Elastomers and Rubber Elasticity, Chicago, IL, September 10, 1985

26. First International Conference on Thermoplastic Elastomer Markets and Products, sponsored by Schotland Business Research, Orlando, FL, March, 1988

27. Second International Conference on Thermoplastic Elastomer Markets and Products, sponsored by Schotland Business Research, Orlando, FL, March, 1989

28. Third International Conference on Thermoplastic Elastomer Markets and Products, sponsored by Schotland Business Research, Dearborn, MI, March, 1990

29. Fourth International Conference on Thermoplastic Elastomer Markets and Products, sponsored by Schotland Business Research, Orlando, FL, February, 1991

30. Sixth International Conference on Thermoplastic Elastomer Markets and Products, sponsored by Schotland Business Research, Orlando, FL, January, 1992

31. Seventh International Conference on Thermoplastic Elastomer Markets and Products, sponsored by Schotland Business Research, Orlando, FL, February, 1993

32. Thermoplastic Elastomers Regional Technical Conference of the Society of Plastics Engineers, Fort Mitchell, KY, October, 1993

33. N.R. Legge, S. Davison, H.E. DeLaMare, G. Holden, and M.K. Martin, In *Applied Polymer Science*, 2nd edit. R.W. Tess and G.W. Poehlein (Eds.), *ACS Symposium Series No. 285* (1985_ American Chemical Society, Washington, D.C.

34. P. Dreyfuss, L.J. Fetters and D.R. Hansen, *Rubber Chem. Technol. 53*, 728 (1980)

35. M. Morton, *Rubber Chem. Technol. 56*, 1069 (1983)

36. G. Holden, In *Encyclopedia of Polymer Science and Engineering, Vol. 15*, 2nd edit. J.I. Kroschwitz (Ed.) (1986) John Wiley & Sons, New York, p. 216

37. G. Holden, In *Kirk-Othmer Encyclopedia of Chemical Technology*, 4th edit. J.I. Kroschwitz (Ed.) (1994) John Wiley & Sons, New York, p. 15

38. W.L. Semon (to B.F. Goodrich Co.), U.S. Patent 1,929,453 (October 10, 1933); W.L. Semon, *Ann. Tech. Conf. Soc. Plast. Eng. 30*, 693 (1972)

39. D.E. Henderson (to B.F. Goodrich Co.), U.S. Patent 2,330,353 (September 28, 1943)

40. O. Bayer, et al. (to I.G. Farben), German Patent 728,981 (1937)

41. A.E. Christ and W.E. Hanford (to DuPont), U.S. Patent 2,333,639 (1940)

42. W.E. Hanford and D.F. Holmes (to DuPont), U.S. Patent 2,284,896 (June 2, 1942)

43. (to ICI), British Patents 580,524 (1941) and 574,134 (1942)

44. M.D. Snyder (to DuPont), U.S. Patent 2,632,031 (1952)

45. R.W. Moncrief, *Man-Made Fibers*, 6th edit. (1975) Halsted Press, John Wiley & Sons, New York, p. 489

46. Ibid., p. 489 (to DuPont), U.S. Patent 2,692,873 (1954)

47. B.P. Corbman, *Textiles: Fibers to Fabric*, 4th edit. (1975) McGraw-Hill, New York, p. 462

48. C.S. Schollenberger (to B.F. Goodrich), U.S. Patent 2,871,218 (1955); C.S. Schollenberger, H. Scott, and G.R. Moore, Paper presented at the ACS Rubber Division Meeting, September 13, 1957, *Rubber World 137*, 549 (1958), *Rubber Chem. Technol. 35*, 742 (1962)

49. W.H. Charch and J.C. Shivers, *Textile Res. 36*, 536 (1959)

50. S.L. Cooper and A.V. Tobolsky, *Textile Res. 36*, 800 (1966)

51. J.H. Bolland and H.W. Melville, *Proc. First Rubber Technology Conference*, London, W. Heffer, London (1938), p. 239

52. H.W. Melville, *J. Chem. Soc.* 414 (1946)

53. J.H. Baxendale, M.G. Evans, and G.S. Parks, *Trans. Faraday Soc. 42*, 155 (1946)

54. L.G. Lundsted, *J. Am. Oil Chem. Soc. 28*, 294 (1951)

55. H.F. Mark, *Textile Res. 23*, 294 (1953)

56. E.H. Immergut and H. F. Mark, *Makromolekular Chem. 18/18*, 322 (1956)

57. L.C. Bateman, *Ind. Eng. Chem. 49*, 704 (1957)

58. F.M. Merrett, *J. Polym. Sci. 24*, 462 (1957)

59. R.J. Ceresa (Ed.), *Block and Graft Copolymers* (1962) Butterworths, Washington, D.C.

60. A.F. Halasa, *Rubber Chem. Technol. 54*, 627 (1981)

61. K. Ziegler and K. Bahr, *Chem. Ber. 61*, 253 (1928)

62. K. Ziegler, H. Colonius, and O. Schater, *Ann. Chem. 473*, 36 (1929)

63. K. Ziegler and O. Schater, *Ann. Chem. 479*, 150 (1930)

64. K. Ziegler, et al., *Ann Chem. 511*, 64 (1934)
65. F.E. Stavely, et al., Paper presented to the ACS Rubber Division, Philadelphia, PA, November 1955, *Ind. Eng. Chem. 48*, 778 (1956)
66. S.E. Horne, et al., *Ind. Eng. Chem. 48*, 784 (1956)
67. H. Hsieh and A.V. Tobolsky, *J. Polym. Sci. 26*, 245 (1957)
68. H. Hsieh, D.J. Kelley, and A.V. Tobolsky, *J. Polym. Sci. 26*, 240 (1957)
69. H. Morita and A.V. Tobolsky, *J. Am. Chem. Soc. 79*, 5853 (1957)
70. L.M. Porter (to Shell Oil Co.), U.S. Patent 3,149,185, 1964, filed October 28, 1957
71. N.R. Legge, Presented at a meeting of the ACS Rubber Division, May 4, 1985; Chemtech 13, 630–639 (1983)
72. W.W. Crouch and J.N. Short, *Rubber Plast. Age 42*, 276 (1961)
73. H.E. Railsback, C.C. Beard, and J.R. Haws, *Rubber Age 94*, 583 (1964)
74. M. Szwarc, M. Levy, and R. Milkovich, *J. Am. Chem. Soc. 78*, 2656 (1956)
75. M. Szwarc, *Nature 178*, 1168 (1956)
76. M. Szwarc, Polym. Preprints 26, (1) 198 (1985)
77. J.T. Bailey, et al., Presented at a meeting of the ACS Rubber Division, October 22, 1965; *Rubber Age 1966*, October, p. 69
78. G. Holden, E.T. Bishop, and N.R. Legge, *Proceedings of the International Rubber Conference*, 1967; Maclaren and Sons, London (1968), pp. 287–309; *J. Polym. Sci. Pt. C 26*, 37 (1969)
79. E.T. Bishop and S. Davison, *J. Polym. Sci. Pt. C 26*, 54 (1969)
80. D.J. Meier, *J. Polym. Sci. Pt. C 26*, 81 (1969)
81. T.S. Kuhn, *The Structure of Scientific Revolutions* (1970) University of Chicago Press, Chicago
82. A.V. Tobolsky, *Rubber World 138*, 857 (1959)
83. E.G. Kontos, E.K. Easterbrook, and R.D. Gilbert, *J. Polym. Sci. 61*, 69 (1962)
84. E.G. Kontos (to Uniroyal), U.S. Patent 3,378,606 (1968)
85. E.G. Kontos (to Uniroyal), U.S. Patent 3,853,969 (1974)
86. D. Puett, K.J. Smith, A. Ciferri, and E.G. Kontos, *J. Chem. Phys. 40*, (1) 253 (1964)
87. J.G. Wallace, Chapter 5 in Ref. 16
88. A.Y. Coran and R.P. Patel, Paper presented at the International Rubber Conference, Kiev, USSR, October, 1978
89. A.Y. Coran and R.P. Patel, *Rubber Chem. Technol. 53*, 141 (1980)
90. A.Y. Coran and R.P. Patel, *Rubber Chem. Technol. 53*, 781 (1980)
91. A.Y. Coran and R.P. Patel, *Rubber Chem. Technol. 54*, 91 (1981)
92. A.Y. Coran and R.P. Patel, *Rubber Chem. Technol. 54*, 892 (1981)
93. A.Y. Coran, R.P. Patel, and D. Williams, *Rubber Chem. Technol. 55*, 116 (1982)
94. A.Y. Coran, R.P. Patel, and D. Williams, *Rubber Chem. Technol. 55*, 1063 (1982)
95. A.Y. Coran and R.P. Patel, *Rubber Chem. Technol. 56*, 210 (1983)
96. A.Y. Coran and R.P. Patel, *Rubber Chem. Technol. 56*, 1045 (1983)
97. A.Y. Coran, R.P. Patel, and D. Williams-Headd, *Rubber Chem. Technol. 58*, 1014 (1985)
98. W.K. Witsiepe (to DuPont), U.S. Patent 3,651,014 (March 21, 1972)
99. A.T. Chen, et al. (to Dow Chemical Co.), U.S. Patent 4,129,715 (December 12, 1978)
100. R.G. Nelb II, et al. *SPE Ann. Tech. Conf. (ANTEC)*, Boston, May 4–7, 1981, p. 421
101. N.R. Legge, *Elastomerics 117*, (10) 17 (1985)
102. R.W. Rees, *Mod. Plastics 42*, 209 (1964)
103. K.D Gagnon, R.C. Fuller, R.W. Lenz, and R.J. Farris, *Rubber Chem. Technol. 65*, 4 (1992)
104. K.D Gagnon, R.C. Fuller, R.W. Lenz, and R.J. Farris, *Rubber World 207*, 32 November (1992)
105. D.R. Paul, In *Polymer Blends, Vol. 2*. D.R. Paul and S. Newman (Eds.) (1978) Academic Press, New York

2   Thermoplastic Polyurethane Elastomers

W. Meckel, W. Goyert, and W. Wieder

2.1	Introduction	16
2.2	Raw Materials	17
	2.2.1 Soft Segments	18
	2.2.1.1 Polyesters	18
	2.2.1.2 Polyethers	18
	2.2.2 Hard Segments	21
	2.2.2.1 Polyisocyanates	21
	2.2.2.2 Chain Extenders	21
	2.2.3 Additives	22
2.3	Synthesis	24
2.4	Morphology	25
	2.4.1 Structure of Hard Segments	25
	2.4.2 Thermal Transitions	27
	2.4.3 Dynamic Mechanical Property Measurements	30
	2.4.4 Stress–Strain and Ultimate Properties	31
2.5	Properties	32
	2.5.1 Mechanical Properties	33
	2.5.2 Thermal Properties	34
	2.5.3 Hydrolytic Stability	35
	2.5.4 Oil, Grease, and Solvent Resistance	36
	2.5.5 Resistance to Microorganisms	36
	2.5.6 UV Resistance	36
2.6	Processing	37
	2.6.1 Welding–Bonding	37
2.7	Blends	37
	2.7.1 TPU as the Minor Component	37
	2.7.2 TPU as an Equal Component with Other Resins	38
	2.7.3 TPU as the Major Component	38
2.8	Applications	39
	2.8.1 Film and Sheet	39
	2.8.2 Hose	39
	2.8.3 Shoes	39
	2.8.4 Automotive	40
	2.8.5 Mechanical Goods and Other Applications	40
	2.8.6 Medical	41
2.9	Recycling	41
2.10	Future Trends	41
	Acknowledgements	41
	References	42

2.1   Introduction

Polyurethane chemistry opened the way to a new class of high-performance materials such as coatings, adhesives, elastomers, fibers, and foams. Formed via a simple polyaddition reaction, the polyurethanes proved to be very versatile polymers. Materials with tailormade properties can be produced from the broad variety of chemicals used. Thermoplastic polyurethane elastomers (TPUs) were the first, homogeneous, thermoplastically processable elastomers, and today they play an important role within the rapidly growing family of thermoplastic elastomers. We first review the history leading to the discovery and development of TPUs.

Pioneering polyurethane work was done by Otto Bayer and his co-workers at I.G. Farbenindustrie at Leverkusen, Germany (now Bayer AG) in 1937 [1]. Their original goal was to duplicate or improve the properties of synthetic polyamide fibers. Subsequently, the elastomeric properties of polyurethanes were recognized by DuPont [2] and by ICI [3]. By the 1940s polyurethanes were produced on an industrial scale [4]. The first so-called "I-rubber," however, had very poor properties. To overcome the deficiencies, which were supposed to stem from an irregular elastomeric network, a polyurethane elastomer was synthesized that consisted of linear polyesters and 2-nitro-4,4'-diisocyanato-biphenyl [5]. In a second step the nitro groups were to be reduced to form azo linkages. Surprisingly, this latter step proved unnecessary, because the original polymer, which was chain extended by water, already showed interesting elastomeric behavior. A similar result was obtained when the nitrodiisocyanate was replaced by naphthalene-1,5-diisocyanate [6]. These unexpected results were explained in terms of the following reaction sequence [7]: First the strictly linear hydroxyl terminated polyester reacts with excess diisocyanate to form an α, ω-diisocyanate prepolymer. This prepolymer is subsequently chain extended by water, leading to urea linkages. The urea linkages react with additional diisocyanate to build an elastomeric network. Chain extension by short-chain diols proved to be the breakthrough to polyurethane elastomers, which were tradenamed Vulkollan® by Bayer. In the United States, Chemigum SL® was an early Vulkollan-type polyester urethane elastomer that was developed by Seeger et al. [8] of the Goodyear Tire and Rubber Co. DuPont marketed Adiprene®, a polyether [9] urethane.

Early polyurethane elastomers consisted of three basic components:

1. A polyester-or polyether macrodiol.
2. A chain extender such as water, a short-chain diol, or a diamine.
3. A bulky diisocyanate, for example, naphthalene-1,5-diisocyanate (NDI).

However, these polyurethane elastomers were not TPUs in the proper sense of the term, because their melting temperature was higher than the decomposition temperature of the urethane linkages.

Great progress was achieved when NDI was replaced by diphenylmethane-4,4'-diisocyanate (MDI) in the above mentioned three-component system. Schollenberger [10] of B.F. Goodrich described a TPU in 1958. Somewhat earlier DuPont announced an elastic fiber called Lycra®, a polyurethane based on MDI. In the United States, by the early 1960s Goodrich marketed Estane®, Mobay Texin®, and Upjohn Pellethane®. In Europe Bayer and Elastogran marketed Desmopan® and Elastollan®.

In the following years much effort was spent to elucidate the nature of the structure–property relationship, and so forth. Today it is well established that TPUs owe their

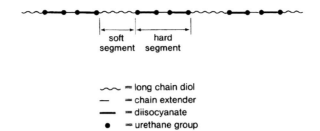

Figure 2.1   Schematic representation of a TPU composed of diisocyanate, long-chain diol, and chain extender

elastomeric properties to a domain structure that is achieved by the phase-separated systems of these multiblock polymers.

One type of block, the hard segment, is formed by addition of the chain extender, for example, butanediol, to the diisocyanate, for example, MDI. The other type is the soft segment and consists of the long flexible polyether or polyester chain that interconnects two hard segments (Fig. 2.1).

At room temperature, the low melting soft segments are incompatible with the polar and high melting point hard segments, which leads to a microphase separation. A part of the driving force for phase separation is the development of crystallinity in the hard segments. On heating above the melting temperature of the hard segments, the polymer forms a homogeneous viscous melt that can be processed by thermoplastic techniques such as injection molding, extrusion, blow molding, and so forth. Subsequent cooling leads again to segregation of hard and soft segments and recovery of the elastic properties.

Usually, the soft segments form an elastomer matrix which gives the elastic properties of TPU, with the hard segments acting as multifunctional tie points that function both as physical crosslinks and reinforcing fillers. However, these crosslinks can be removed by heat or by solvation. On subsequent cooling or desolvation the TPU network is reformed. Thus, the TPU network was described as "virtually crosslinked" [10]. To obtain thermoplasticity, the average functionalities of the starting materials should be close to 2.00. That is, each prepolymer or monomer unit should have 2 terminal reactive groups. This ensures formation of high molecular weight linear chains with no or only very few branch points [11–13].

2.2   Raw Materials

Thermoplastic polyurethanes are generally made from long-chain polyols with an average molecular weight of 600 to 4000, chain extenders with a molecular weight of 61 to 400, and polyisocyanates.

Among the broad variety of possible starting materials, only a limited number are of practical interest. However, as a result of the wide range of hard to soft segment variations possible, TPUs can be formulated that range from soft flexible, elastomeric materials to more brittle, high-modulus plastics.

The most common raw materials used and their most typical effects on the properties of TPUs are discussed in the following sections.

2.2.1   Soft Segments

The long flexible soft segment largely controls the low-temperature properties, the resistance to solvents, and the weather-resistant properties of TPUs.

There are two types of flexible segments of importance: the hydroxyl terminated polyesters and the hydroxyl terminated polyethers.

2.2.1.1   *Polyesters*

The typical hydroxyl terminated polyester is made from adipic acid and an excess of glycol such as ethylene glycol, 1,4-butanediol, 1,6-hexanediol, neopentyl glycol, or mixtures of these diols [14].

The reaction is carried out at temperatures of up to 200 °C and the resulting polyester should have an acid number of less than two. As in all polymeric structures, the polyesters are composed of all possible oligomers ranging from the monomeric glycol to high molecular weight species; the distribution follows a Flory probability [15]. The properties of the elastomer are governed mainly by the overall molecular weight of the polyester and only to a minor degree by the molecular weight distribution [16].

Starting from adipic acid and straight-chain diols, the resulting polyesters are crystalline products with melting points up to about 60 °C. The crystallinity can be reduced easily either by using mixtures of diols (e.g., 1,4-butanediol with ethylene glycol or 1,6-hexanediol and 1,6-hexanediol with neopentyl glycol) or by using mixtures of polyesters.

The use of other acids such as azelaic acid or ortho-or terephthalic acid has also been reported, either alone or in mixtures with adipic acid. Generally, the presence of aromatic or cycloaliphatic rings in the acid or in the diol increases the glass transition temperature of the polyester.

There are two special classes of polyesters of commercial interest, the polycaprolactones and the aliphatic polycarbonates.

The polycaprolactones are made from ε-caprolactone and a bifunctional initiator, for example, 1,6-hexanediol [17]. The properties of these polyesters are very similar to those of poly(1,4-butanediol, 1,6-hexanediol adipate) glycols.

The polycarbonates offer excellent hydrolytic stability [18]. They are made from diols, for example, 1,6-hexanediol, and phosgene or by transesterification with low molecular weight carbonates such as diethyl or diphenyl carbonate.

2.2.1.2   *Polyethers*

There are two classes of polyethers of technical importance: the poly(oxypropylene) glycols and the poly(oxytetramethylene) glycols.

The poly(oxypropylene) glycols are made by the base catalyzed addition of propylene oxide to bifunctional initiators, for example, propylene glycol or water [19]. By using propylene oxide, the resulting polyethers have predominantly secondary hydroxyl groups at the ends. Primary hydroxyl groups are introduced by using higher proportions of ethylene oxide, especially at the end of the reaction ("tipping"). Owing to side reactions, the functionality of poly(oxypropylene) glycols is always lower than the functionality of the initiator. Increasing amounts of allylic and isopropylidene end groups are generated with increasing molecular weight of the polyether. For example, a poly(oxypropylene) glycol with a molecular weight of 2000 has a functionality of about 1.96 instead of 2.00. Special catalysts are used to produce polyethers with almost no loss in functionality, and these give high-performance TPUs [19a].

Poly(oxytetramethylene) glycols are made by cationic polymerization of tetrahydrofuran [20]. The functionality is about 2.00.

The molecular weight distribution of the polyether oligomers follows the Poisson probability equation, yielding a less broad distribution compared to the thermodynamically controlled Flory probability [15] of analogous polyesters.

In the patent literature special polyethers have been claimed to have advantages; however, they have made no major breakthrough in the market for TPUs [21–27, 30].

Mixtures of polyethers and polyesters are of economic interest and give TPUs with very useful combinations of properties [28, 29].

As mentioned before, the character of the soft segment mainly governs the low-temperature flexibility and the long-term aging properties of TPUs. Table 2.1 shows the general trends in properties produced by the more important soft segments.

Despite the fact that the melting points of the polyols are mostly above room temperature, the soft segment in the TPU is normally in the amorphous state. The soft segments crystallize only at very low levels of hard segments or on prolonged cooling. The crystallization is noticeable as an increase in hardness. However, the crystallization tendency of the soft

Table 2.1 Important Polyols[a] and Corresponding Thermoplastic Polyurethane Elastomers[b]

Polyol nomenclature	Polyols		Elastomers	Hydrolytic stability
	T_e	T_m	T_e	
Poly(ethylene adipate) glycol	−46	52	−25	Fair
Poly(butylene-1,4 adipate) glycol	−71	56	−40	Good
Poly(ethylene butylene-1,4 adipate) glycol	−60	17	−30	Fair/good
Poly(hexamethylene 2,2-dimethylpropylene adipate) glycol	−57	27	−30	Good
Polycaprolactone glycol	−72	59	−40	Good
Poly(diethylene glycol adipate) glycol	−53	/	−30	Poor
Poly(1,6-hexanediol carbonate) glycol	−62	49	−30	Very good
Poly(oxytetramethylene) glycol	−100	32	−80	Very good

[a] Molecular weight 2000
[b] ca. 85 Shore hardness
T_e, Lower end of glass transition range
T_m, melting point
After Ref. [31], with permission

segment is very well observed during low and medium elongations of polyurethanes (Fig. 2.2). This induced crystallization leads to a self-reinforcing effect, which is noticed as a higher modulus compared to elastomers with noncrystalline soft segments. This effect disappears at temperatures above the melting points of the soft segment [32, 33]. On the other hand, the induced crystallization leads to a higher permanent set and compression set.

Thus, the character of the soft segment must be carefully adjusted to give the required property profile for the final application.

The low-temperature properties of TPUs are governed by the broadness and the location of the glass transition range, defined by starting with the very first softening of the glassy soft segment at T_e and ending at the temperature where the complete soft segment is molten. In low to medium hardness TPUs the lower end, T_e, of the glass transition temperature range is normally 20 °C to 30 °C above the corresponding temperature of pure soft segment, which has a rather narrow glass transition range.

The broadness of the glass transition range depends on the amount of hard segment and the separation of hard and soft segments, increasing with the concentration of the hard segments and their resulting intrusions into the soft segment phase. This leads to poorer low-temperature properties. Improved low-temperature flexibility [34, 35], characterized by a T_e well below room temperature and a narrow glass transition temperature range, is obtained through the use of soft segments such as polyethers that are less compatible with the hard segment. The incompatibility is also increased by increasing the molecular weight of the soft segment [34, 36, 37] or by annealing the elastomer [38].

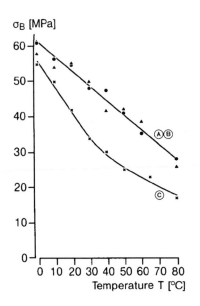
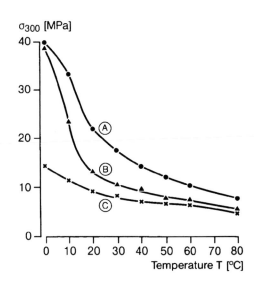

Figure 2.2   Effect of the soft segment crystallinity on the temperature dependence of the tensile strength at 300% elongation and the ultimate tensile strength. (From Ref. [31], with permission) (A) Poly(1,4-butylene adipate), (B) polycaprolactone, (C) poly(diethylene glycol adipate)

2.2.2   Hard Segments

2.2.2.1   *Polyisocyanates*

Among the commercially available polyisocyanates [39] only a very few are suitable for TPUs [35]. The most important diisocyanate is 4,4′-diphenylmethane diisocyanate (MDI).

$$OCN-\langle\bigcirc\rangle-CH_2-\langle\bigcirc\rangle-NCO$$

Other diisocyanates being used or having attracted some interest are the following:

Hexamethylene diisocyanate (HDI).

$$OCN-(CH_2)_6-NCO$$

4,4′-Dicyclohexylmethane diisocyanate (H_{12}-MDI) normally consists of an isomer mixture. Recently, interesting physical properties have been reported for TPUs made from the highly crystalline *trans, trans*-4,4′-dicyclohexylmethane diisocyanate [39a, b].

3,3′-Dimethyl-4,4′-biphenyl diisocyanate (TODI) [40]

1,4-Benzene diisocyanate and *trans*-cyclohexane-1,4-diisocyanate [42]

1,5-Naphthalene diisocyanate (NDI) [43]

2.2.2.2   *Chain Extenders*

The choice of chain extender and diisocyanate determines the characteristics of the hard segment and to a large extent the physical properties of TPU.

The most important chain extenders for TPU are linear diols such as ethylene glycol, 1,4-butanediol, 1,6-hexanediol, and hydroquinone bis(2-hydroxyethyl) ether. With diisocyanates these glycols form urethanes that are mostly well crystallized and melt without decomposition on thermoplastic processing. The use of ethylene glycol as a chain extender should be avoided for a TPU with a high hard segment content, because of thermal instability at higher temperature [44, 45].

1,4-Butanediol and hydroquinone bis(2-hydroxyethyl) ether are the most suitable diols for TPUs. The latter gives better high-and low-temperature properties and reduced compression set. Table 2.2 gives a survey of the most important hard segments.

Nonlinear diols [48] are usually not suitable for use in TPUs because the urethanes do not form well-crystallized hard segments and therefore exhibit poor low-and high-temperature properties. A chain extender mixture of straight-chain diols is sometimes recommended to produce a hard segment of lower order. This is especially valuable in extrusion grade TPUs to provide a broader processing range [49].

With increasing hard segment content, the polymers generally show an increase in hardness accompanied with an increase in the glass transition temperature [50] of the elastomeric phase. At levels of 60% to 70% by weight of hard segment, a phase transition occurs, which leads to a change in the overall behavior from that of an elastomeric polymer to a more brittle, high-modulus plastic [51, 52].

Although diamines are excellent chain extenders, they normally cannot be used for TPUs, because the urea groups melt well above the processing range of TPUs and also undergo some decomposition on melting. Sterically hindered diamines, for example, 1-amino-3-aminomethyl-3,5,5-trimethylcyclohexane (isophoronediamine), can be used in combination with aromatic and aliphatic diisocyanates [53, 54]. The use of a sterically hindered amine and a glycol as cochain extender produces a hard segment that is not well crystallized. The polymer exhibits poor elastomeric properties but this makes it especially suitable for energy-absorbing applications [54].

Water was the very first chain extender used for making elastomeric polyurethanes. One urea group is formed from two isocyanate groups. This reaction has been studied recently [30, 55], and the resulting polymers may allow melt processing because this reaction leads to hard blocks with a lower proportion of urea groups as compared to the use of diamines.

2.2.3   Additives

The following compounds are the most widely used additives for TPUs.

Mold release agents are the most commonly employed additives and are necessary for fast and economical cycle times. Chemically they belong to the class of fatty acid derivatives (e.g., esters and amides), and also silicones or fluoroplastics. The amount of mold release agent is about 0.1% to 2% by weight.

Polyester-based TPUs are stabilized against hydrolytic degradation by adding 1% to 2% by weight of sterically hindered aromatic carbodiimides. The carbodiimide group reacts with acid residues, generated by the hydrolysis of ester groups, which otherwise would catalyze further hydrolysis [56, 57, 57a].

Table 2.2 Melting Temperatures T_m of Diisocyanates, Chain Extenders, and Corresponding Hard Segments Ascertained by Differential Scanning Calorimetry (DSC) Measurements

Diisocyanate	T_m (°C)	Chain extender	T_m (°C)	X-ray[a]	Hard segments Melting point determined by DSC measurement Curve profile	T_m (°C)	Visual evaluation in the melting tube
Naphthalene-1,5-diisocyanate (NDI)	131	1,4-butanediol	19.5	C	Thermal effects as from 190 °C, no real melting up to 320 °C		Sintering as from about 260 °C, no melting up to 320 °C
Diphenylmethane-4,4'-diisocyanate (MDI)	42	1,4-butanediol	19.5	C		230	Initial softening at about 120 °C, melting at 230 °C to 237 °C[b]
Hexamethylene diisocyanate (HDI)	11	1,4-butanediol	19.5	C	Step at 75 °C to 100 °C	165	Initial softening up to about 100 °C and 148 °C, melting at 166 °C to 172 °C
2,4-TDI	24	1,4-butanediol	19.5	A	Step at 70 °C to 110 °C	217[b]	Initial softening at 78 °C to 120 °C, melting at 220 °C[b]
NDI	131	1,4-bis(β-Hydroxy-ethoxy) benzene	104	C	Step at 133 °C and 210 °C	302	Brown discoloration as from 288 °C, melting at 298 °C to 302 °C
MDI	42	1,4-bis(β-Hydroxy-ethoxy) benzene	104	C		252	Initial softening at 243 °C, melting at 247 °C to 260 °C[b]
HDI	11	1,4-bis(β-Hydroxy-ethoxy) benzene	104	C		214	Initial softening at 140 °C to 150 °C, melting at 212 °C to 226 °C[b]
2,4-TDI	24	1,4-bis(β-Hydroxy-ethoxy) benzene	104	C	Step at 140 °C	197	Initial softening at 140 °C, melting at 200 °C to 216 °C, breakdown as from 250 °C

[a] X-ray wide angle study
[b] Melts with breakdown
C, crystalline; A, amorphous
After Ref. [31], with permission

$$\text{Ar}-\text{N}=\text{C}=\text{N}-\text{Ar} + \text{R}-\text{C} \overset{\displaystyle \nearrow \text{O}}{\searrow \text{OH}} \longrightarrow \left[\text{Ar}-\text{N}=\overset{\displaystyle \overset{\text{H}}{|}}{\underset{\displaystyle \underset{\text{R}}{\underset{|}{\text{C}=\text{O}}}}{\text{C}}}-\text{N}-\text{Ar}\right] \longrightarrow \text{Ar}-\overset{\text{H}}{\underset{\displaystyle \underset{\text{R}}{\underset{|}{\text{C}=\text{O}}}}{\text{N}}-\overset{\text{O}}{\underset{}{\text{C}}}-\overset{\text{H}}{\underset{}{\text{N}}}-\text{Ar}$$

Sterically hindered phenols and certain amines are the preferred stabilizers used to reduce degradation by light, oxidation, and higher temperatures [45, 58, 59].

Inorganic materials (e.g., calcium carbonate, talc, or silicates) are added for better release properties in molding operations or in film production. They act either as crystallization promoters or surface rougheners. Certain minerals (e.g., mica, organic fibers [60]) and especially glass fibers [61] are used to reinforce TPUs.

The friction coefficient of TPUs can be markedly reduced by adding small amounts of graphite, molybdenum sulfide, fluorinated hydrocarbon, or silicone oil as well as mixtures thereof [62].

Soft grades of TPUs may be obtained by adding small amounts of plasticizers. Such grades are increasingly found in the market.

2.3   Synthesis

In general TPUs are made by reacting the ingredients together at temperatures above 80 °C. For optimum results, the ratio of isocyanate groups to the sum of isocyanate reactive groups should be close to 1.0. Polymers with insufficient molecular weight are obtained if this ratio is less than 0.96, whereas thermoplastic processing becomes increasingly difficult at ratios above 1.1 owing to crosslinking reactions. An average molecular weight M_n of 40,000 is sufficient for satisfactory property development [11]. This M_n is easily obtained at ratios of 0.98 and higher.

The reaction can be carried out in different ways. The so-called "one-shot method" involves mixing all the ingredients together. In the "prepolymer method" the polyol is reacted first with the diisocyanate to give an isocyanate containing prepolymer, which is then reacted with the chain extender. The reaction can be done batchwise [63] or continuously in a mixing chamber or reaction extruder [64–66].

For large-scale industrial production two methods are normally used, the belt process and the reaction extruder process. In the belt process all ingredients are mixed together. The liquid mixture is then poured onto a belt, where it is allowed to solidify. The slab produced is then granulated. The granulated material can be used as such but it is most often blended and extruded into more uniform pellets. In contrast, if a reaction extruder is used, the urethane reaction is almost complete at the end of the extruder and uniform pellets are obtained immediately.

The heat history during production is of extreme importance because the phase separation of the hard and soft segment, (and hence the properties of) TPUs is temperature dependent. Thus, starting from the same raw materials, the physical properties of the resulting polymers can be very different.

Various methods of affecting the properties of TPU are given in the literature [67–73, 73a].

2.4   Morphology

It is evident that the morphology of multiphase systems will play an important role in determining the final properties of a product. By controlled variation of the morphology, desired properties of a material can be obtained. Hence, a profound knowledge of the morphology is essential for understanding structure–property relationships. However, this has proved to be a formidable challenge, as the morphology of urethane block polymers is complicated by physical phenomena such as crystallization, interphase mixing, hydrogen bonding in both segments, dependence of properties on thermal history, and so forth.

Theoretically, phase separation occurs because of thermodynamic incompatibility of the phases. A model to predict phase separation phenomena based on a thermodynamic approach has been developed by Krause [74, 75]. In a copolymer molecule of given length phase separation becomes more difficult as the number of blocks increases. On the other hand, at fixed copolymer composition, an increase in molecular weight and number of blocks per molecule would favor phase separation. Generally a higher degree of phase separation is predicted for a copolymer system where one component is crystallizable. Both the soft and the hard phases in polyurethanes can be amorphous or partially crystalline.

2.4.1   Structure of Hard Segments

Considerable efforts have been made to elucidate the nature of TPU hard-segment domains [76–80]. Hard segments which are formed by linear glycols and MDI would be expected to be crystalline. Under normal conditions, however, crystallinity, as evidenced by wide-angle X-ray scattering (WAXS) techniques, appears to be inhibited [76]. Initial proposals for the structure of hard segments, based on small-angle X-ray scattering (SAXS) results, were made by Bonart and co-workers [77–79], and by Wilkes and Yusek [80]. Clough et al. [81, 82] also used SAXS to study domain structures in polyurethane block polymers. The ordered state of TPU hard segment domains has been referred to as paracrystalline [77]. Bonart [78, 79, 83] developed two-dimensional and three-dimensional models of MDI–butanediol hard segment crystals. Arrangements were postulated that provided optimum hydrogen bonding. Suitable heat treatment transformed the structure from paracrystalline into crystalline [78]. Relatively high temperatures and long annealing times (190 °C, 12 h) are reportedly required to produce significant hard segment crystallinity [84].

Blackwell et al. [85, 86] extensively studied the structure of hard segments in MDI–diol–polytetramethylene adipate polyurethanes by X-ray diffraction. They used butanediol (BDO), propanediol (PDO), and ethylene glycol (EDO) as chain extenders. Poly(MDI–BDO) was found to be the most crystalline hard segment. Based on conformational analysis and model compound considerations, it was concluded that poly(MDI–BDO) existing in its fully extended chain conformation can form a hydrogen-bonded network in the two dimensions perpendicular to the chain axis. In contrast, poly(MDI–PDO) and poly(MDI–EDO) crystallize in higher energy contracted conformations, which are necessary for a nonstaggered packing of the chains. These findings were confirmed by the recent results of Eisenbach and Guenther [87], who reported on the synthesis and analysis of a series of well-defined soft and hard segments under strictly controlled conditions. Oligomers from MDI and BDO were synthesized and endcapped. By SAXS measurements it was shown that MDI–BDO-based oligourethanes crystallize in an extended chain without chain folding. Recently, Blackwell and Lee [88] found hard segment polymorphism in MDI–diol-based polyurethane elastomers. In a MDI–polyester urethane, the authors varied the chain length of the extender from EDO to hexanediol (HDO) and observed crystallization phenomena by X-ray analyses. The HDO-and BDO-extended polyurethanes clearly showed polymorphism of the crystalline state. These findings were supported by DSC traces. The appearance of a second crystalline structure was accompanied by another melting point. The HDO-extended TPU was not very crystalline in the melt-pressed film, but developed high crystallinity on stretching and annealing. In contrast, EDO-based polyurethane seemed to show no polymorphism at all. An X-ray analysis on single crystals formed by MDI–methanol bisurethanes furnished a model for the arrangement of MDI–BDO polyurethane chains within the hard segment domains [89] (Fig. 2.3). This model compound was proven to exist in two crystal modifications [90].

Analogous studies were performed on model compounds which were obtained by reaction of diphenylmethane-4-monoisocyanate and glycols of the $HO-(CH_2)_n-OH$ structure, with $n = 2$ to 6 [91]. X-ray analyses on single crystals of these compounds revealed a more stable arrangement of hydrogen bonds between neighboring molecules in urethanes with "even" chain extenders, while urethanes containing "odd" chain extenders exhibited significant strains which reduce the stability of the physical crosslinking system. In a more recent paper, Born [92] published additional crystal structure data on some of these model compounds. Similarly, Born and Hespe [93] studied a bisurea produced by reacting diphenylmethane-4-isocyanate with 1,4-butanediamine. They found that the hydrogen

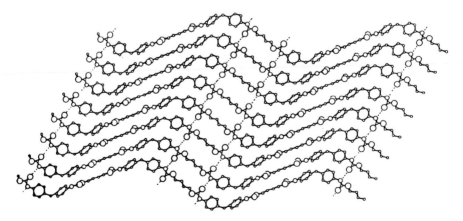

Figure 2.3   Chain arrangement of MDI–BDO hard segments derived from X-ray studies of a model compound. (From Ref. [89], with permission)

bonds in this system were bifurcated, which accounts for the greater heat stability of the bisurea crystals compared with those of the corresponding bisurethane. Based on single-crystal X-ray data, Blackwell et al. [94] derived structural parameters, such as bond lengths, bond angles, and bond torsion angles, for the prediction of polyurethane structures.

2.4.2   Thermal Transitions

Differential scanning calorimetry (DSC) is a common tool to determine changes, for example, phase segregation, glass transitions, and melting, in the state of organization of the molecules in a sample. The practical use of DSC in analyzing thermal response of a TPU with respect to engineering properties has been illustrated by Goyert and Hespe [31]. The effect of hard segment content on thermal response is shown in Fig. 2.4. The specimens of this series were prepared from the same components, but with differing molar ratios of polyesterdiol, BDO, and MDI (see Table 2.3). With increasing hard segment content, the glass transition is broadened and shifted toward higher temperatures. The observed glass transition shift may be explained by an increasing amount of hard segments that are "dissolved" in the soft matrix. Consequently, it seems plausible that a concentration gradient of hard segments exists near the phase interface. This assumption agrees with earlier explanations, in which the observed phenomena were attributed to irregular structures at the phase interfaces [95].

   This structure can be represented schematically by the model shown in Fig. 2.5 (left). As a result of these irregularities, which have also been observed by SAXS [34, 79], parts of the soft segments are heavily restricted in mobility and therefore appear as flexible components only at elevated temperatures. The increasing content of chain extender results not only in more but also in larger hard segments (see Table 2.3).

   The length of the hard segment blocks forms the upper limit to the size of the hard segment crystals in the chain direction, which, in turn, determines the melting point, and thus

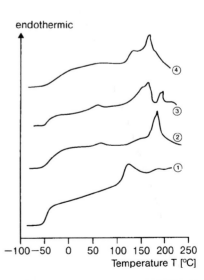

Figure 2.4   DSC scans of TPU samples with different hard segment content. (From Ref. [31], with permission)

Table 2.3 TPU Samples with Various Hard Segment Content

Sample No.	Molar ratio			Hard segment content (%)	Mean hard segment length (calculated) (μm)
	Polyester[a]	BDO	MDI		
1	1	: 1.77	: 2.8	31	5.0
2	1	: 3.55	: 4.6	40	8.6
3	1	: 5.55	: 6.6	52	12.7
4	1	: 10	: 11	65	21.7

[a] Soft segment: poly(ethylene butylene adipate) glycol $M_n = 2000$
From Ref. [31], with permission

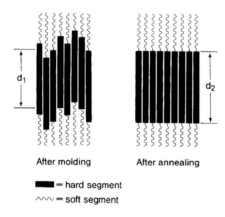

After molding   After annealing

■■■ = hard segment
∿∿ = soft segment

Figure 2.5 Model illustrating schematically the arrangement of the hard segments (left) after molding/cleavage of interfaces and (right) after additional annealing. d_1, d_2, crystal thickness. (From Ref. [31], with permission)

the thermal stability. In Fig. 2.4 the temperature range in which most hard segment crystals melt (maximum of the DSC curve) is seen to shift toward higher temperature as the proportion of hard segment increases. The maximum is achieved at ca. 190 °C (specimen No. 2 was prepared under different conditions). Moreover, the graphs show transitions over a broad temperature range, which suggests a wide distribution of the hard segment crystal thickness [96].

The crystallite size distribution, which results when a TPU is cooled from the melt, is governed not only by the length distribution of the hard segment but also by the crystallization kinetics. Figure 2.6 (curve 1) shows the DSC trace of a TPU based on MDI, BDO, and ethanediol–butanediol–polyadipate, prepared at the initial molar ratio of about 6 : 5 : 1. The material was molded into a panel. Several melting maxima are recognized, of which the most pronounced occurs at 205 °C. By thermal treatment at 118, 135, 180, and 205 °C for 5 min each (curves 2 to 5), the primary melting region can be shifted to a limit of ca. 230 °C.

The melting range narrows at the same time. This effect is attributable to a partial melting of the hard segment crystals as well as to recrystallization toward larger crystals of a better order. This process can be imagined as follows: On rapid cooling, the hard segment blocks, corresponding to their arrangement in the melt, form crystals, which are statistically displaced with respect to one another. The thermodynamically effective crystal thickness (d_1) is small,

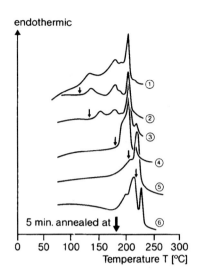

Figure 2.6 Effect of annealing on DSC transitions. Curves 2 to 5 illustrate the effect of thermal treatment at 118, 135, 180, and 205 °C for 5 min each. (From Ref. [31], with permission)

while the phase interface is large (Fig. 2.5, left). On annealing, the structure rearranges, resulting in a thermodynamically more favorable arrangement (Fig. 2.5, right). The melting point subsequently rises because of the increased crystal thickness (d_2). Of course, additional improvements of the crystalline order in the lateral direction cannot be excluded. The uniform block length shown in Fig. 2.5 is also a simplification. When the annealing temperatures exceed a limit of ca. 210 °C, the melting temperatures decrease again (curve 6 in Fig. 2.6). In this case the existing crystals are largely molten and the hard segments cannot form new crystals within the available time. Crystallization will occur only on subsequent cooling to form lower melting aggregates of relatively poor order.

The kinetics of phase separation have been investigated by Wilkes et al. [97, 98]. Polyester TPUs were quickly heated, then rapidly quenched to room temperature. Changes in phase separation were monitored by SAXS and DSC measurements as a function of time. It was shown that, as a result of kinetic and viscous effects, as much as several days were required to regain equilibrium.

Interphase mixing phenomena, as studied by thermal analysis and SAXS measurements, have been explained in terms of soft segment hydrogen bonding ability [81, 82, 99]. For the same molecular weight polyol ($M_n = 1000$) a 1 : 2 : 1 polyester–MDI–BDO system was single phased (compatible) whereas a 1 : 2 : 1 polyether–MDI–BDO system was phase separated [14]. The former material was much more transparent, in agreement with the formation of smaller crystallites and/or a much lower degree of crystallinity. A DSC transition at 60 °C to 80 °C was ascribed to urethane–polyester or urethane–polyether hydrogen bond disruption. More recent studies have shown, however, that the intermediate DSC transitions are not attributable to hydrogen bond dissociation [100]. The observed transition could alternatively be ascribed to the glass transition of the hard phase for the particular block length employed, combined with a plasticizer effect of the soft segments. Infrared studies indicate that it is the chain mobility, or T_g of the hard blocks, that controls hydrogen bond dissociation [96, 101–103]. Thus, hydrogen bonds serve to increase the cohesion of the hard domains. Within the hard domains, hydrogen bonds appear to determine the arrangement of urethane groups [77–79]. However, the importance of hydrogen bonds in TPUs should not be overemphasized. This is demonstrated in the example of polyether-ester elastomers, where no hydrogen bonding exists. Witsiepe [160] prepared them by condensation of chloroformates of short-chain glycols and of polytetramethylene or other long-chain ether diols with piperazine. The resulting "secondary" urethane elastomers have no urethane hydrogen for bonding, yet have physical properties fully comparable with the isocyanate-derived materials. In addition, since the reversion reaction to isocyanate and alcohol is not possible, they are much more stable thermally and have a wider processing "window." An important influence may be π-electron interactions between parallel oriented rings.

Schneider and co-workers [102–105] have extensively studied TDI-based polyurethanes. Block polyurethanes were prepared from 2,4-and 2,6-TDI isomers. Thermomechanical studies, DSC, X-ray scattering, and infrared analysis were applied to elucidate the structural organization and thermal transitions in these products. It was found that the degree of phase segregation strongly depended on the TDI symmetry and the soft segment molecular weight. The T_g of 2,4-TDI polyurethanes with polyether soft segment of M_n ca. 1000 showed a strong dependence on composition, indicating extensive phase mixing. In contrast, comparable 2,6-TDI-based polyurethanes displayed a highly ordered structure as revealed by a concentration-independent T_g and a high temperature transition attributed to melting of the crystallizable hard segment. A change of soft segment molecular weight from 1000 to 2000 led to phase segregation in the 2,4-TDI series and further improved phase segregation in the 2,6-TDI series. The results are in agreement with similar results obtained with a polycaprolactone diol-based TPU [35, 50].

It can be concluded that the soft segment T_g is a sensitive measure of the degree of phase separation. Recently, the above described findings of Schneider et al. were confirmed by Senich and MacKnight [106]. From a technical standpoint, as hardness is increased by increasing the hard block content, the extent of phase mixing also increases. This shows up as a decrease of low temperature flexibility. This drawback can be reduced by incorporation of soft segments with higher molecular weight or by the use of preextended polyols.

2.4.3   Dynamic Mechanical Property Measurements

The dynamic mechanical behavior of segmented polyurethanes has been studied by many research groups [35, 50, 95, 99, 107, 108]. An example is illustrated in Fig. 2.7, which represents the effect of a differing content of hard segments on the dynamic mechanical behavior. The curves for the samples designated in Table 2.3 clearly show the glass transitions of the soft segments, which start at ca. −40 °C and account for the pronounced decrease of the shear modulus as the temperature increases. A severe drop of the modulus occurs for specimen No. 1. With increasing hard segment content, this drop becomes less pronounced, demonstrating the function of hard segment domains as reinforcing fillers. Sample 1 maintains a rubbery plateau over a reasonable temperature range. A second drop in modulus occurs when the temperature is increased to the hard segment transition point. The two transitions depend strongly on block length, composition, phase mixing, and thermal history.

Cooper [84, 109] and Wilkes [107] reported on extensive studies of dynamic mechanical properties on hydrogen-bonded and non-hydrogen-bonded TPUs. In addition to the hard

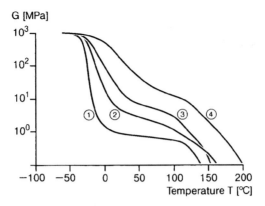

Figure 2.7   Shear modulus curves of TPU with various hard segment contents. (From Ref. [31], with permission) For sample designation see Table 2.3

segment transition and the soft segment T_g, several relaxations were found. These could be assigned to low-temperature transitions of the soft segment. A comparison between hydrogen-bonded and non-hydrogen-bonded [110] polyurethanes revealed a pronounced effect of hydrogen bonds on soft segment mobility. Low-temperature dynamic mechanical properties of MDI–BDO–polyether-based polyurethanes with various soft segments were studied by Schneider and co-workers [99]. They found that with increasing urethane concentration, the soft segment glass transition remained almost constant, but the crystallization of the soft segment was inhibited above a certain urethane content. Koleske et al. [35, 50] studied the dynamic mechanical properties of a polycaprolactonediol (PCP)–MDI–BDO polyurethane system. The hard segment concentration and the PCP molecular weight were varied. With a PCP M_n lower than 3000 a compatible noncrystalline system was observed. A two-phase morphology with soft segment crystallization was found for higher M_n PCP. The low molecular weight PCP system was very sensitive to hard segment content. In contrast the T_g of the higher M_n PCP system was much less affected over the same range of hard segment concentration, that is, about 40% to 60% hard segment content.

2.4.4   Stress–Strain and Ultimate Properties

The thermomechanical behavior of linear block copolyurethanes is basically different from the behavior of chemically crosslinked products. Applied mechanical stress causes orientation phenomena, which in turn lead to disruption and recombination of hydrogen bonds in an energetically more favorable position. These phenomena may explain typical features of TPUs such as high tensile strength, tear strength, and elongation, and also high permanent set. Studies of orientation using infrared dichroism [113, 114] showed that the soft segments may be readily oriented by an applied stress, but return to the unoriented state when the stress is removed. The hard blocks, however, show a more complex behavior of orientation and relaxation. This behavior is dependent on the magnitude of the applied stress, the molecular weight of the soft blocks, and the crystallinity of the hard blocks. Harrell [110] systematically studied the effect of segment size distribution on mechanical properties in a non-hydrogen-bonded TPU based on piperazine. Hard and soft segment molecular weight distributions were varied and their combined effects on stress–strain and permanent set were described. It was observed that the ultimate stress–strain properties were significantly better in the case of a narrow size distribution, in particular for the hard segment. In contrast, permanent set is better when the size distribution is broad. Apparently, a more regular physical network and a higher degree of hard segment domain perfection [109], accompanied by strain-induced crystallization, was responsible for the better ultimate properties. Strain-induced crystallization, however, accounts for an increase in permanent set [115], if the sample temperature was below the melting point of the soft segment. Cooper and co-workers [116] examined the effect of hard segment lengths and hard segment distribution on orientation phenomena using infrared dichroism experiments. In contrast to soft segment distribution, hard segment size distribution was found to exert a pronounced effect on the orientation tendency. At high strains, TPUs based on MDI oriented to a much lower degree than those based on piperazine. This is in contrast to the orientation behavior of non-hydrogen-bonded TPU containing monodisperse hard segment lengths [117].

2.5   Properties

TPUs were the first polymeric materials to combine both rubber elasticity and thermoplastic characteristics. The comparison of the Young's moduli of TPUs to those of other materials is

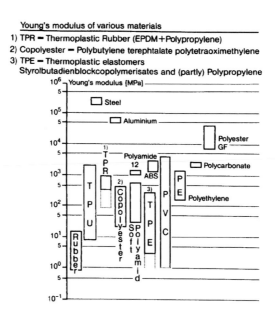

Figure 2.8   Comparison of the Young's moduli of TPUs with other similar plastic materials. (From Ref. [118], with permission)

shown in Fig. 2.8 [118]. The Young's moduli of TPUs range from 8 MPa up to about 2000 MPa (about 300,000 psi) depending on hard segment concentration [119, 120]. The stiffness of TPUs can also be increased by inorganic and organic fillers, especially glass fibers [61].

2.5.1   Mechanical Properties

TPUs offer excellent physical properties, for example, high tensile strength and elongation (see Fig. 2.9). Depending on their chemical structures and Shore hardness, the tensile strengths of TPU vary from 25 to 70 MPa (3600 to 10,000 psi). Softer grades (70 to 85 Shore A) show lower tensile strength, whereas harder grades (50 to 83 Shore D) exhibit higher values (Fig. 2.9). One of the main advantages of TPU is its high abrasion resistance. Therefore, polyurethanes, especially ester-or ether-ester polyurethanes, are the materials of choice for shoe soles and cable jacketing.

The resistance to tear propagation is so high that even skiboot buckles don't tear. TPUs exhibit flexibility over a wide temperature range and good resistance to many oils and greases. TPUs do not contain any plasticizers and are thus preferred for sandwich constructions with other materials such as polycarbonate and ABS.

The soft segment (polyol) determines the behavior at low temperatures and is responsible for many other properties as shown in Table 2.4. Most commercial TPUs are ester based. Ester TPUs show better abrasion and cut resistance and tensile and tear strength than ether TPUs. Ester TPUs also swell less in oil, grease, and water than ether TPUs (Table 2.4). Ether

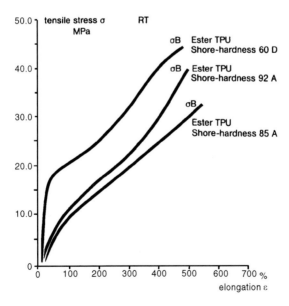

Figure 2.9   Stress–strain diagram of three TPUs with different Young's moduli (Shore D hardness). (From Ref. [118], with permission)

Table 2.4 Comparison of Properties of TPU Prepared from Different Polyols

	Ester (adipate or caprolactone)	Ester with carbodiimide	Ether	Ether-special ester
Tensile strength	+ +	+ +	0	+
Abrasion resistance	+ +	+ +	0	+
Low swelling in oil, grease, water	+	+	0	+
Weathering	+	+	0	+
Stability to energic radiation	+	+	0	+
Tear strength, Graves	+ +	+ +	0	+
Rebound resilience	+ + → +	+ + → +	+ + → +	+
Microbe and fungus resistance	+ → −	0 → −	+ +	+ +
Low-temperature impact resistance	+ → 0	+ → 0	+ + → +	+
Hydrolysis resistance	0	+ /0	+	+

+ +, excellent, favorable
+, good
0, indifferent
−, poor, unfavorable

TPUs are recommended if special properties such as resistance to hydrolysis and micro-biological degradation or good low-temperature flexibility are required. Further advantages often can be expected when using ether-ester TPUs produced by a special synthesis procedure [29]. Ether-ester TPUs are often a good compromise with respect to many properties. Therefore they have been used in applications such as fire hoses, cable jacketing, and films.

2.5.2   Thermal Properties

TPUs can be used over a wide range of temperatures [121]. The majority of articles made from TPUs can be used from −40 °C up to 80 °C for both long-and short-term applications. TPUs have a short-term resistance to temperatures up to 120 °C, although in some cases even higher temperatures can be tolerated as in secondary urethanes based on piperazine. The hard segment is the main contributor to high service temperature performance [122]; the harder the product (more isocyanate and chain extender) the higher the service temperature (Fig. 2.7). Besides being a function of the amount of chain extender, the high temperature performance is also affected by the type of chain extender (Fig. 2.7). Hydroquinone bis(2-hydroxyethyl) ether yields TPUs with higher service temperature than 1,4-butanediol or 1,6-hexanediol.

The type of diisocyanate also affects the high-temperature performance [122]. As is seen in Table 2.2 the hard segments produced from different diisocyanates and chain extenders show different melting points.

Mechanical properties such as stiffness and elasticity are dependent on temperature. This is demonstrated by plotting the shear moduli of four TPUs at a wide range of temperatures, Fig. 2.7. The Shore hardness of two TPUs show a similar dependence on temperature (Fig. 2.10).

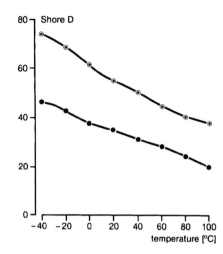

Figure 2.10   Dependence of the Shore hardness on temperature in case of two different TPUs. (From Ref. [118], with permission)

2.5.3   Hydrolytic Stability

At room temperature TPU can be used in pure water over a period of several years without any significant changes of properties. At 80 °C, however, the mechanical properties are affected after exposure for some weeks or months (see Fig. 2.11). The hydrolytic stability is dependent on the structure of the polyol. Ester TPUs are less resistant than ester TPUs protected by carbodiimide [123, 124]. The highest resistance to water at elevated temperatures is shown by ether-ester TPUs or pure ether TPUs (Fig. 2.11).

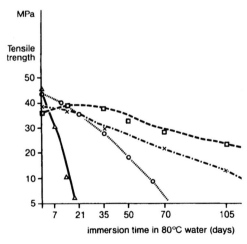

Figure 2.11   Effect of immersion in water at 80 °C on the tensile strength of different TPUs of about 87 Shore A hardness. △, Ester TPU without carbodiimide; ○, ester TPU with carbodiimide; X, ether TPU (tetramethyleneoxide); □, propylenglycol ether-carbonate TPU

With increasing hardness, TPUs become more hydrolytically stable owing to the hydrophobic character of the hard segment.

Polyurethanes are sensitive to acids and bases. At room temperature they are attacked slowly by diluted acids and bases. At higher temperatures they cannot withstand concentrated acids and alkalies. Different acids behave differently in contact with various urethanes. Immersion tests under simulated conditions are therefore recommended for each specific application.

2.5.4   Oil, Grease, and Solvent Resistance

Nonpolar solvents such as hexane, heptane, and paraffin oil have almost no effect on the polar polyurethanes. Even at high temperatures only a slight swelling is observed.

TPUs exhibit excellent resistance to pure mineral oils, diesel oils and greases. Some technical oils and greases can attack TPUs at elevated temperatures owing to the additives they may contain. Therefore it is advisable to test the effect of the oil on the TPU under service conditions.

Chlorinated hydrocarbons or aromatic liquids such as toluene cause a very severe swelling of TPU [125]. The degree of swelling is dependent on the structure of the polyurethane. Ester types swell less than ether types and hard polyurethanes swell less than soft ones.

Some polar solvents (e.g., tetrahydrofuran, methylethylketone, or dimethylformamide) are capable of partially or completely dissolving TPUs. For example, soft linear polyurethanes can be dissolved in methyl ethyl ketone–acetone mixtures and applied as adhesives, whereas harder linear polyurethanes are dissolved and applied as textile and leather coatings. These topics are not covered in this book but further information is available in a publication by Oertel [126].

TPUs are generally stable in contact with petroleum hydrocarbons if they do not contain alcohol. Fuels that contain aromatics or alcohols cause a reversible swelling of the polymer; the extent of the swelling depends on the amount of such ingredients.

2.5.5   Resistance to Microorganisms

Soft ester TPUs can be attacked by microorganisms after long contact with moist earth. Soft and hard ether TPUs, ether-ester TPUs, or hard TPUs are normally resistant to such attack. A slight discoloration at the surface of the articles can be caused by fungus, but it does not indicate a mechanical damage of the material.

2.5.6   UV Resistance

TPUs based on aromatic isocyanates exhibit yellowing, although that only minimally affects the properties of the polymer. Special UV absorbers can be used to minimize the yellowing.

2.6   Processing

TPUs are normally supplied as granules in moistureproof containers. They can be processed by the usual methods for thermoplastic materials such as injection molding, extrusion, blow molding, and calendering without any pretreatment.

When stored incorrectly the TPU, as well as any additives, may become moist and must be dried prior to processing. Moist granules should be dried at 100 °C to 110 °C in a circulating air oven or in a flash drier for 1 to 2 h. The moisture content of the granules should be below 0.1%.

After processing the articles achieve nearly their final properties after storage at room temperature for 4 to 6 weeks. If some special properties (e.g., low compression set at elevated temperatures) are desired, annealing at 100 °C to 130 °C for 12 h is recommended. Detailed information is given in the literature [127].

2.6.1   Welding–Bonding

The following techniques [127] are used to weld TPUs with other TPUs:

- heated mirror welding,
- hot air or nitrogen welding,
- heating tool and thermal impulse welding,
- high-frequency welding, and
- friction welding.

A TPU can be bonded to itself using solvents such as *N*-methylpyrrolidone or dimethylformamide as long as the surface is small. Two-component polyurethane adhesives are suitable for bonding TPU with itself and other polar plastic materials, metals, wood, leather, and so forth [126]. Epoxy resins may also be applied.

2.7   Blends

The literature dealing with blends of TPUs with other polymers is quite large [128–130].

Except for very nonpolar resins such as polyethylene or polypropylene, TPUs can be blended with many other resins in wide ratios, as long as the processing temperatures are below ca. 280 °C. Depending on the portion of TPU, the blends can be formally divided into three areas, as discussed in the following sections.

2.7.1   TPU as the Minor Component

In this case TPU is acting as a modifier. In particular, softer grades of TPUs are used to modify higher modulus plastics, for example, unsaturated polyester resins, epoxy resins, poly(oxymethylene) [131] or poly(butylene terephthalate) [132]. The main improvement gained by adding small amounts of TPU is an increase in impact strength and low-temperature flexibility without impairing the other properties of the plastic. TPU is also used as a nonmigrating, nonvolatile plasticizer for PVC and can be blended in any ratio with PVC [133].

2.7.2   TPU as an Equal Component with Other Resins

Blending equal amounts of TPU with other thermoplastics usually results in additive mechanical properties. Examples of such blends are:

TPU–Polycarbonate [34]: The addition of polycarbonate results in a higher modulus and the blends exhibit excellent processing properties, making them useful for automotive applications.

TPU–ABS: These two resins can be blended in any ratio. Increasing the ABS proportion results in a higher modulus and lower abrasion resistance and tear strength. The blends may offer cost advantages because of the lower cost of ABS [135]. Blends of copolymers such as styrene and maleic anhydride [136] or styrene and maleimides [137] with TPUs exhibit improved impact strength while retaining high Vicat softening temperatures.

The blends of TPU with the following compounds are basically comparable to TPU–ABS blends: copolymers of styrene, methacrylic acid, and alkadienes [138]; copolymers of styrene with acrylonitrile or methacrylic acid ester [139]; and rubbery copolymers of styrene, butadiene, and either acrylonitrile or methyl methacrylate [140].

Examples of ternary blends are:

TPU–polycarbonate–ABS: These blends are claimed to have better processing properties [141] and better fuel resistance [142]. TPU–polycarbonate–polybutylene terephthalate: These blends show less stress cracking with solvents [143].

2.7.3   TPU as the Major Component

The use of ABS as an impact modifier for higher modulus TPUs or a phase compatibilizer for polyether-based TPUs has been described in the patent literature and proved to be successful in the market [119, 144, 145].

Certain acrylic polymers can serve as processing aids for TPUs [146]. Neutralized salts of ethylene acrylic acid copolymer have been used to improve the processing of TPUs in blow molding operations [147]. Generally, ionic groups in nonpolar resins are acting as compatibilizers [148].

Besides blending TPU with other polymers, the mixing of different TPUs is often practiced.

Blends of hard and soft grades have been used to obtain TPUs with medium hardness or to achieve TPUs with better processing properties. These blends are especially useful for filled TPUs. Polyester and polyether grades have been blended to obtain special properties [149].

Blends of TPUs having different intrinsic melt indexes and hardnesses are claimed to give better demolding properties and less blocking in blow molding operations [150].

2.8   Applications

2.8.1   Film and Sheet

Films and sheets of TPUs can be extruded in thicknesses ranging from a few micrometers to several millimeters. They may be pigmented in any color. These films have outstanding properties: resistance to abrasion and puncture and tear propagation resistance, in combination with high elasticity, bondability, and weldability. Films of TPU are therefore being used for conveyor belts, and, depending on the grade of TPU used, these conveyor belts may be used for contact with dry, aqueous, and fatty foods. Special TPU types meet the requirements for food contact of the Federal German Health office (BGA) [151]. Some special grades of TPU meet the corresponding regulations of the United States (FDA) [152]. Recently, the use of TPU powders in a powder slush process has attracted interest as a replacement for PVC in automotive applications [152a].

TPU has low permeability to air and may be used for many other applications where these properties are required. Other applications for TPU films include:

- welded hollow bodies,
- textile lamination,
- protective coverings,
- sealing of foams, and
- abrasion resistant coatings.

2.8.2   Hose

Many different types of hoses are made from TPUs. The high elongation, good resistance to hydrolysis and microbes, the excellent mechanical properties (e.g., tensile strength and resistance to tear propagation) of ether-ester or ether TPUs provide an ideal combination for the inner liner of fire hoses. The high tensile strength of TPUs allows a thinner wall thickness to be used in comparison to conventional hoses [153]. Fire hoses made with an inner lining of TPU are lighter than conventional hoses and thus allow a fireman to operate more effectively.

Tubes with an inner layer of TPU are very useful in the transport of sand and stone slurries. The excellent abrasion and cut resistance of polyurethanes gives such tubes a relatively long service life.

2.8.3   Shoes

The outer materials of ski boots are made primarily from TPUs. Properties such as good abrasion and cut resistance, permanent buckling strength, and the high impact strength at low temperatures are particularly useful in this application. No other plastic material has this high resistance to tear and tear propagation. TPUs of varying stiffnesses are used for ski boots, with E (Young's)-moduli ranging from 100 to 600 MPa (15,000 to 90,000 psi) (50 to 66 Shore D). Various TPUs may be used to meet the property requirements of the different shoe parts. Ice hockey boots, which are similar to ski boots, are often also made from thermoplastic polyurethanes. They must also show a very high impact strength at low temperatures.

Many sport shoe soles are manufactured from TPUs. The largest application is soccer shoes. They are made mostly from ester polyurethanes with a Shore hardness of 85 to 90 Shore A. The excellent abrasion and cut resistance of ester polyurethanes is the main reason for their use.

2.8.4   Automotive

Exterior automotive body parts can be injection molded from pure TPUs, TPU–PC blends, and, recently, glass-reinforced TPUs. Parts with E-modulus up to 2500 MPa (360,000 psi) can be produced. Their good recovery after deformation, cut resistance, good weathering properties, and resistance to oils and fuels are decisive factors contributing to the use of TPUs in this application. Other automotive applications include bearing bushings and gaskets for wheel components, tie rods and suspension link pivots, membranes for hydropneumatic suspension systems, bellows for steering assemblies and shock absorbers, tank bleeding tubes, and catches and seals for door locks [153].

2.8.5   Mechanical Goods and Other Applications

The high E-modulus (compared to conventional rubber) and the high dynamic load-bearing capacity of TPU are some useful properties for toothed belts. These can be produced by extrusion of ester TPU with Shore hardness in the range of 85 to 93 Shore A [153].

TPUs are used for couplings. Depending on the application and the size of the part, ester TPUs in the hardness range of 85 Shore A up to 60 Shore D are used. The high E-modulus of the polyurethane elastomer enables the transmission of high energies.

Precision cogged wheels are molded from ester TPUs. The vibration damping of TPU makes them useful in modern business machines.

Screens made of TPUs can be used to classify dry and wet materials (e.g., gravel, coal, and coke), or can be applied for sorting, washing, and separating. Because of their high abrasion resistance these screens exhibit a longer lifetime in comparison to rubber and steel elements. TPU sheathing for geophysical measuring cables also shows high abrasion, cut, and tear resistance. Other sheathing applications are connecting leads for electrical tools (in industry and in the home) and spiral leads.

Ear tags of TPU ensure a reliable identification of different animals. These ear tags must have a high tensile strength and tear propagation resistance as well as good weathering properties.

2.8.6   Medical

TPUs show a good compatibility to human skin. The good compatibility of ether TPUs with human blood and tissues [154] allows catheters and tubes for blood to be made from TPU [155]. Even a microporous, biodegradable, compliant, and blood compatible vascular prothesis was developed [156]. More information about this topic is given in the literature [157–159].

2.9   Recycling

Like all other thermoplastics TPUs can be recycled. This has been done since the very beginning by using processing scrap such as spruces, off grade parts, and so forth in a minor proportion in virgin material without any decrease in quality. However, TPU parts obtained from 100% recycled material show a decrease of some properties due to thermal and mechanical degradation under processing conditions [161]. Therefore, quality has to be carefully monitored, and in particular the regrind may need to be dried (see Section 2.6) before being reused.

2.10   Future Trends

In addition to the continuing need to improve quality, it will be necessary to broaden the range of applications for TPUs beyond the current limits. This means producing TPUs with lower hardness and greater long-term temperature resistance as well as TPUs with higher hardness that still retain good low-temperature properties.

To achieve these goals it will be necessary to have a deeper understanding of the morphology of TPUs and its effect on properties. Since the development of new and better building blocks is rather limited, major improvement will be achieved by a better understanding of morphology, resulting in further optimized production and processing equipment to obtain the ultimate properties from a given formulation.

A major object of research will be the development of more sophisticated blends with other resins to gain the best combination of needed properties.

Acknowledgments

The authors thank J.F. Dormish, H. Hespe, H.G. Hoppe, B. Krüger, B. Quiring, N. Schön, H. Wagner, H.G. Wussow, Bayer AG, and E.C. Ma, Miles Inc. for helpful discussions.

References

1. O. Bayer, H. Rinke, W. Siefken, L. Ortner, and H. Schild (to I.G. Farben), German Patent 728 981 (1937)
2. A.E. Christ and W.E. Hanford (to DuPont), U.S. Patent 2333 639 (1940)
3. British Patent 580 524 (1941); (to ICI), British Patent 574 134 (1942)
4. P. Pinten (to Dynamit AG), German Patent 932 633 (1943)
5. E. Müller, S. Petersen, and O. Bayer (to Bayer), German Patent 896 413 (1944)
6. S. Petersen, E. Müller, and O. Bayer (to Bayer), German Patent 883 347 (1944)
7. O. Bayer, E. Müller, S. Petersen, H. F. Piepenbrink, and E. Windemuth, *Angew. Chem. 62*, 57 (1950)
8. T.G. Mastin and N.V. Seeger (to Goodyear), U.S. Patent 2625535 (1953)
9. F.B .Hill, C.A. Young, J.A. Nelson, and R.G. Arnold, *Ind. Eng. Chem. 48*, 927 (1956)
10. C.S. Schollenberger, H. Scott, and G.R. Moore, *Rubber World 137*, 549 (1958); C.S. Schollenberger (to B.F. Goodrich), U.S. Patent 2871218 (1955)
11. C.S. Schollenberger and K. Dinbergs, *J. Elastoplastics 5*, 222 (1973); 7, 65 (1975)
12. C.S. Schollenberger and K. Dinbergs, *Polym. Preprints Am. Chem. Soc. Div. Polym. Chem. 20 (1)*, 532 (1979)
13. R. Becker and H.U. Schimpfle, *Plaste und Kautschuk 22*, 15 (1975); J.H. Saunders and K.C. Frisch, *High Polymers XVI: "Polyurethanes, Part I, Chemistry* (1962) Interscience, New York
14. J. Rohr, K. Koenig, H. Koepnick, and K.-H. Seemann, *Polyester*, Ullmanns Encyklopaedie der technischen Chemie, 4 (1980) Auflage, Verlag Chemie, Weinheim
15. P.J. Flory, *Principles of Polymer Chemistry (1953) Cornell University Press, Ithaca, NY*
16. G.L. Lunardon, Y. Sumida, and O. Vogl, *Angew. Makrom. Chem. 87*, 1 (1980)
17. F. Hostettler (to Union Carbide), U.S. Patent 2933477 (1956)
18. E. Mueller, *Angew. Makrom. Chem. 14*, 75 (1970)
19. E. Windemuth, H. Schnell, and O. Bayer (to Bayer), German Patent 974371 (1951); (a) C.P. Smith, J.W. Reisch and J.M. O'Connor, *J. Elast. Plast. 24*, 306 (1992)
20. F.B. Hill (to DuPont), U.S. Patent 2929800 (1953)
21. E. Pechold (to DuPont), U.S. Patent 4120850 (1977)
22. C.G. Seefried, Jr., R.D. Whitman, and R. van Cleve (to Union Carbide), German Patent Appl. 2550830 (1975)
23. F.X. O'Shea and C.L. Mao (to Uniroyal), U.S. Patent 4041105 (1976)
24. H.W. Bonk and T.M. Shah (to Upjohn), German Patent Appl. 2537775 (1975)
25. D.D. Russel and G. Shkapenko (to Samuel Moore), U.S. Patent 4010146 (1977)
26. H. Holtschmidt (to Bayer), German Patent 1039232 (1956); H.Schwarz, W. Kallert, C. Muehlhausen, and H. Holtschmidt (to Bayer), U.S. Patent 2844566 (1958)
27. E. Mueller and G. Braun (to Bayer), German Patent 1039744 (1955); W. Thoma, H. Rinke, H. Oertel, and E. Mueller (to Bayer), German Patent 1149520 (1961)
28. E.G. Kolycheck (to B.F. Goodrich), German Patent 1720843 (1967)
29. E. Meisert, A. Awater, C. Muehlhausen, and U.J. Doebereiner (to Bayer), German Patent 1940181 (1969)
30. G.L. Statton (to Atlantic Richfield), U.S. Patent 3987012 (1975)
31. W. Goyert and H. Hespe, *Kunststoffe 68*, 819 (1978)
32. L. Morbitzer and R. Bonart, *Kolloid Z. Z. Polymere 232*, 764 (1969)
33. L. Morbitzer and H. Hespe, *J. Appl. Polym. Sci. 27*, 2891 (1982)
34. R. Bonart and E.H. Mueller, *J. Macromol. Sci. 10*, 177 and 345 (1974)
35. C.G. Seefried, Jr., J.V. Koleske, and F.E. Critchfield, *J. Appl. Polym. Sci. 19*, 2493 and 3185 (1975)
36. R.J. Zdrahala, S.L. Hager, R.M. Gerkin, and F.E. Critchfield, *J. Elast. Plast. 12*, 225 (1980)
37. N.E. Rustad and R.G. Krawiec, *J. Appl. Polym. Sci. 18*, 4101 (1974)
38. T.K. Kwei, *J. Appl. Polym. Sci. 27*, 2891 (1982)
39. G. Becker/D. Braun, *Kunststoffhandbuch*, 2. Auflage, VII; p. 63 (1983) Hanser Verlag, Munich; (a) S.D. Seneker, L. Born, H.G. Schmelzer, C.D. Eisenbach, and K. Fischer, *Colloid Polym. Sci. 270*, 593 (1992); (b) S.D. Seneker (to Miles), U.S. Patent 5208315 (1991)
40. H.W. Bonk and T.M. Shah (to Upjohn), U.S. Patent 3899467 (1974)

41. R. Roberts (to Union Carbide), U.S. Patent 4055549 (1976)
42. H. Schulze, H. Zengel, W. Brodowski, F. Huntjems, J. Schutijer, and P. Hentschel (to AKZO), German Patent Appl. 2829199 (1978)
43. W. Goyert (to Bayer), German Patent Appl. 3329775 (1983)
44. K.J. Vorhees and R.P. Lattimer, *J. Polym. Sci. Polym. Chem. Ed. 20*, 1457 (1982)
45. H.J. Fabris, *Adv. Ureth. Chem. Technol. 6*, 173 (1978)
46. T.M. Shah (to Upjohn), U.S. Patent 3901852 (1974)
47. Y. Camberlin, J.P. Pascault, J.M. Letoffe, and P. Claude, *J. Polym. Sci. Polym. Chem. Ed. 20*, 1445 (1982)
48. H. Salzburg, H. Meyborg, W. Goyert, and J.M. Barnes (to Bayer), German Patent Appl. 3111093 (1981), 3302603 (1983)
49. B. Quiring, H.G. Niederdellmann, W. Goyert, and H. Wagner (to Bayer), European Patent Appl. 4393 (1978)
50. C.G. Seefried, Jr., J.V. Koleske, and F.E. Critchfield, *J. Appl. Polym. Sci. 19*, 2503 (1975)
51. R.J. Zdrahala, R.M. Gerkin, S.L. Hager, and F.E. Critchfield, *J. Appl. Polym. Sci. 24*, 2041 (1979)
52. S. Abouzahr, L.C. Wilkes, and Z. Ophir, *Polymer 23*, 1077 (1982)
53. B. Quiring, J. Wulff, and A. Eitel (to Bayer), German Patent Appl. 2423764 (1974)
54. H.P. Mueller, W. Oberkirch, K. Wagner, and B. Quiring (to Bayer), German Patent Appl. 2644434 (1976)
55. B. Quiring, W. Wenzel, H.G. Niederdellmann, H. Wagner, and W. Goyert (to Bayer), German Patent Appl. 2925944 (1979)
56. W. Neumann and P. Fischer, *Angew. Chem. 74*, 806 (1962)
57. W. Neumann, et al. (to Bayer), Belgium Patent 610969 and 612040 (1961); (a) B. Quiring, T. Muenzmay, W. Henning, E. Mayer, W. Meckel, and W. Goyert (to Bayer), German Patent Appl. 4018184 (1991)
58. C.S. Schollenberger and F.D. Stewart, *J. Elastoplastics 4*, 294 (1972)
59. J.E. Kresta, *Polymer Additives (1984) Plenum Press, New York and London, pp. 49 and 135*
60. S. Inoue, S. Shibata, Y. Kaneko, T. Nishi, and T. Matsunaga (to Bridgestone Tire), German Patent Appl. 2220306 (1972)
61. W. Goyert, W. Grimm, A. Awater, H. Wagner, and B. Krüger (to Bayer), German Patent Appl. 2854406 (1978)
62. J.W. Britain and G.J. Schexnayder (to Mobay), German Patent Appl. 2740711 (1976)
63. J.H. Saunders and K.A. Piggot (to Mobay), U.S. Patent 3214411 (1965)
64. B.F. Frye, K.A. Piggot, and J.H. Saunders (to Mobay), U.S. Patent 3233025 (1966); K.W. Rausch, Jr. and T.R. McClellan (to Upjohn), U.S. Patent 3642964 (1969)
65. E. Meisert, U. Knipp, B. Stelte, M. Hederich, A. Awater, and R. Erdmenger (to Bayer), German Patent 1964834 (1969)
66. R.M. Erdmenger, M. Ulrich, M. Hederich, E. Meisert, B. Stelte, A. Eitel, and R. Jacob (to Bayer), German Patent Appl. 2302564 (1973)
67. J.A. Obal and I.S. Megna (to American Cyanamid), German Patent Appl. 2648246 (1976)
68. K.H. Illers and H. Stutz (to BASF), German Patent Appl. 2547864 (1975)
69. K.H. Illers and H. Stutz (to BASF), German Patent Appl. 2547866 (1975)
70. S. Abouzahr and G.L. Wilkes, *J. Appl. Polym. Sci. 29*, 2695 (1984)
71. G. Heinz, H.-J. Maas, P. Herrmann, and H.-D. Schumann (to VEB Chemieanlagen), German Patent Appl. 2523987 (1975)
72. J.W. Britain and W. Meckel (to Mobay), German Patent Appl. 2323393 (1973)
73. H. Meisert, W. Goyert, A. Eitel, and W. Krohn (to Bayer), German Patent Appl. 2418075 (1974); (a) E. Orthmann, K. Wulff, P. Hoeltzenbein, H. Judat, H. Wagner, G. Zaby, and H. Heidingsfeld (to Bayer), European Patent Appl. 554718, 554719 (1992)
74. S. Krause, *Block and Graft Copolymers*, J.J. Burke and V. Weiss (Eds.) (1973) Syracuse University Press, Syracuse, NY
75. S. Krause and P.A. Reismiller, *J. Polym. Sci. 13*, 663 (1975)
76. S.L. Cooper and A.V. Tobolsky, *J. Appl. Polym. Sci. 10*, 1837 (1966)
77. R. Bonart, *J. Macromol. Sci. B2*, 115 (1968)
78. R. Bonart, L. Morbitzer, and G. Hentze, *J. Macromol. Sci. B3*, 337 (1969)
79. R. Bonart, L. Morbitzer, and E.H. Müller, *J. Macromol. Sci. B9*, 447 (1974)
80. C.W. Wilkes and C. Yusek, *J. Macromol. Sci. B7*, 157 (1973)

81. S.B. Clough and N.S. Schneider, *J. Macromol. Sci. B2*, 553 (1968)
82. S.B. Clough, N.S. Schneider, and A.O. King, *J. Macromol. Sci. B2*, 641 (1968)
83. R. Bonart, *Angew. Makromol. Chemie 58/59*, 259 (1977)
84. D.S. Huh and S.L. Cooper, *Polym. Eng. Sci. 11*, 369 (1971)
85. J. Blackwell and M.R. Nagarajan, *Polymer 22*, 202 (1981)
86. J. Blackwell, M.R. Nagarajan, and T.B. Hoitink, *ACS Symp. Ser. 172*, 179 (1981)
87. C.D. Eisenbach and C. Guenther, *Am. Chem. Soc. Org. Coat. Appl. Sci. Proc. 49*, 239 (1983)
88. J. Blackwell and C.D. Lee, *J. Polym. Sci. Phys. 22*, 759 (1984)
89. J. Blackwell and K.H. Gardner, *Polymer 20*, 13 (1979)
90. J. Hocker and L. Born, *J. Polym. Sci. Polym. Lett. Ed. 17*, 723 (1979)
91. L. Born, H. Hespe, J. Crone, and K.H. Wolf, *Colloid Polym. Sci. 260*, 819 (1982)
92. L. Born, *Z. Kristallographie 167*, 145 (1984)
93. L. Born and H. Hespe, *Colloid Polym. Sci. 263*, 335 (1985)
94. J. Blackwell, J.R. Quay, M.R. Nagarajan, L. Born, and H. Hespe, *J. Polym. Sci. Phys. Ed. 22*, 1247 (1984)
95. H. Hespe, E. Meisert, U. Eisele, L. Morbitzer, and W. Goyert, *Kolloid-Z. 250*, 797 (1972)
96. R.W. Seymour and S.L. Cooper, *Macromolecules 6*, 48 (1973)
97. G.L. Wilkes, S. Bagrodia, W. Humphries, and R. Wildnauer, *Polym. Lett. Ed. 13*, 321 (1975)
98. G.L. Wilkes and J.A. Emerson, *J. Appl. Phys. 47*, 4261 (1976)
99. J.L. Illinger, N.S. Schneider, and F.E. Karasz, *Polym. Eng. Sci. 12*, 25 (1972)
100. C.S. Schollenberger and L.E. Hewitt, *Polym. Preprints Am. Chem. Soc. Div. Polym. Chem. 19*, 17 (1978)
101. R.W. Seymour, G.M. Estes, and S.L. Cooper, *Macromolecules 3*, 579 (1970)
102. C.S. Paik Sung and N.S. Schneider, *Macromolecules 8*, 68 (1975)
103. C.S. Paik Sung and N.S. Schneider, *Macromolecules 10*, 452 (1977)
104. N.S. Schneider and C.S. Paik Sung, *Polym. Eng. Sci. 17*, 73 (1977)
105. N.S. Schneider, C.S. Paik Sung, R.W. Matton, and J.L. Illinger, *Macromolecules 8*, 62 (1975)
106. G.A. Senich and W.J. MacKnight, *Adv. Chem. Ser. 176*, 97 (1979)
107. S.L. Samuels and G.L. Wilkes, *J. Polym. Sci. Pt. C 43*, 149 (1973)
108. R.W. Seymour and S.L. Cooper, *Rubber Chem. Technol. 47*, 19 (1974)
109. H.N. Ng., A.E. Allegrezza, R.W. Seymour, and S.L. Cooper, *Polymer 14*, 255 (1973)
110. L.L. Harrell, *Macromolecules 2*, 607 (1969)
111. A.L. Chang and E.L. Thomas, *Adv. Chem. Ser. 176*, 31 (1979)
112. M.A. Vallance, J.L. Castles, and S.L. Cooper, *Polymer 25*, 1734 (1984)
113. R.W. Seymour, A.E. Allegrezza, and S.L. Cooper, *Macromolecules 6*, 896 (1973)
114. S.L. Cooper, G.M. Estes, and R.W. Seymour, *Macromolecules 4*, 452 (1971)
115. L. Morbitzer and H. Hespe, *J. Appl. Polym. Sci. 16*, 2697 (1972)
116. S.B. Lin, K.S. Hwang, S.Y. Tsay, and S.L. Cooper, *Colloid Polym. Sci. 263*, 128 (1985)
117. A.E. Allegrezza, R.W. Seymour, H.N. Ng, and S.L. Cooper, *Polymer 15*, 433 (1974)
118. W. Goyert, *Swiss Plastics 4*, 7 (1982)
119. W. Goyert, J. Winkler, H. Wagner, and H.G. Hoppe (to Bayer), European Patent Appl. 15049 (1984)
120. J. Goldwasser and K. Onder (to Upjohn), U.S. Patent 4376 834 (1981)
121. Desmopan, Thermoplastic polyurethane elastomer, Order No. Pu 52016 a/e, Edition 10/84, Bayer AG, 5090 Leverkusen, Germany
122. G. Oertel, *Polyurethane Handbook* (1985) Hanser, Munich, p. 412 ff.; W. Goyert, Thermoplastic PU-Elastomers, Properties
123. W. Neumann, H. Holtschmidt, J. Peter, and P. Fischer (to Bayer), U.S. Patent 3193522, (1965)
124. C.S. Schollenberger and F.D. Stewart, *Angew. Makromol. Chemie 29/30*, 413 (1973)
125. Technische Information; Beständigkeit von Elastollan-Typen (= TPU) gegenüber Chemikalien, Elastogran-Chemie, Lemförde, Germany
126. G. Oertel, *Polyurethane Handbook* (1985) Hanser, Munich, pp. 548 ff., 510 ff.
127. G. Oertel, *Polyurethane Handbook* (1985) Hanser, Munich, p. 408 and following pages by B. Krüger
128. C.G. Seefried, Jr., J.V. Koleske, and F.E. Critchfield, *Polym. Eng. Sci. 16*, 771 (1976)
129. R.D. Deanin, S.B. Driscoll, and J.T. Krowchun, Jr., *Org. Coat. Plast. Chem. Preprints 40*, 664 (1979)
130. J.M. Buist, *Developments in Polyurethanes-1*, 54 (1978) Applied Science, London
131. T.J. Dolce, F. Berardinelli, and D.E. Hudgin (to Celanese), U.S. Patent 3144431 (1959); G.W. Miller (to

Mobay), Canadian Patent 842325 (1969); P.N. Richardson (to DuPont), European Patent Appl. 117748 (1984); E.A. Flexman (to DuPont), European Patent Appl. 116456, 117667, 120711 and 121407 (1984); E. Reske and E. Wolters (to Hoechst), German Patent Appl. 330376 (1983)

132. M. Cramer and A.D. Wambach (to General Electric), U.S. Patent 4279801 (1975)
133. C.B. Wang and S.L. Cooper, *J. Appl. Polym. Sci. 26*, 2989 (1981)
134. K.B. Goldblum (to General Electric), U.S. Patent 3431224 (1962)
135. C.N. Georgacopoulos and A.A. Sardanopoli, *Mod. Plastic Intern.* May, 96 (1982); G. Demma, E. Martuscelli, A. Zanetti, and M. Zarzetto, *J. Mater. Sci. 18*, 89 (1983)
136. M. Freifeld, G.S. Mills, and R.J. Nelson (to GAF) German Patent Appl. 1694315 (1967)
137. R.A. Fava (to ARCO Polymers), U.S. Patent 4287314 (1980)
138. C.E. Chaney (to ARCO Polymers), U.S. Patent 4284734 (1980)
139. H. Sakano, F. Nakai, and Y. Tomari (to Sumitomo Naugatuck), U.S. Patent 4373063 (1981)
140. K.H. Tan and J.L. de Greef (to Borg Warner), U.S. Patent 4251642 (1979)
141. W.J. O'Connell (to General Electric), U.S. Patent 3813358 (1972)
142. E.J. Frencken, N.G.M. Hoen, and T.B.R. Drummen (to Stamicarbon), European Patent Appl. 104695 (1983)
143. A.L. Baron and J.V. Bailey (to Mobay), U.S. Patent 4034016 (1976)
144. T.S. Grabowski (to Borg Warner), U.S. Patent 3049505 (1962)
145. R. Roxburgh and D.M. Aitken (to ICI), British Patent 2021600 (1978)
146. R.P. Carter (to Mobay). U.S. Patent 4179479 (1978)
147. I.S. Megna (to American Cyanamid), U.S. Patent 4238574 (1979)
148. M. Rutkowska and A. Eisenberg, *J. Appl. Polym. Sci. 29*, 755 (1984)
149. R. Roxburgh, J.P. Aitken, and D.M. Brown (to ICI), British Patent Appl. 2021603 (1978)
150. G. Zeitler, F. Werner, G. Bittner, and H.M. Rombrecht (to BASF), European Patent Appl. 11682 (1983)
151. Empfehlung XXXIX der Kunststoffkommission of the Federal German Health Office (BGA)
152. Title 21, § 177.2600 of the FDA, USA: Rubber articles for repeated use; (a) G. Zeitler and G. Lehr, German Patent Appl. 3916874 (1989)
153. G. Oertel, *Polyurethane Handbook* (1985) Hanser, Munich, p. 424 and following pages: H.-G. Hoppe, Application for polyurethane elastomers
154. A.J. Coury, K.E. Cobian, P.T. Cahalan, and A.J. Jevne, In *Advances in Urethane Science and Technology, Vol. 9* K.C. Frisch and D. Klempner (Eds.) (1984) p. 130, Technomics Publishing, Lancaster, PA
155. H. Ulrich and H.W. Bonk, Presented at Proceedings of the SPI 27th Annual Technical/Marketing Conference
156. Gogolewski, S., Abstract and lecture presented at Internationales Kolloquium: Polyurethane in der Medizin-Technik, Stuttgart, Germany, January 27–29, 1983, Nr. 29
157. H.M. Leeper and R.W. Wright, *Rubber Chem. Technol. 56*, 523 (1983)
158. S. Gogolewski, *Colloid Polym. Sci. 267*, 757 (1989)
159. C.P. Sharma and M. Szycher, *Blood Compatible Materials and Devices* (1991) Technomics Publishing, Lancaster, PA (1991)
160. W.K. Witsiepe (to DuPont), U.S. Patent 3,377,322 April 9 (1968)
161. H.G. Hoppe, *Plastverarbeiter 44* (October), 40 (1993)

3 Styrenic Thermoplastic Elastomers

G. Holden and N. R. Legge

3.1 Introduction 48

3.2 Historical Review 48

3.3 Structure 49

3.4 Synthesis 51

3.5 Properties 54
    3.5.1 Tensile Properties 54
    3.5.2 Swelling 59
    3.5.3 Viscous and Viscoelastic Properties 59
    3.5.4 Solution Properties 62
    3.5.5 Morphology 63
    3.5.6 Stress-Optical Properties 65
    3.5.7 Critical Molecular Weights for Domain Formation 65
    3.5.8 Miscibility 66
    3.5.9 Structural Variations 67

References 68

3.1   Introduction

Of the various types of thermoplastic elastomers (TPEs) described in this book, those based on styrenic block copolymers were among the earliest to be investigated and are now commercially produced in the largest volume [1] (see also Chapter 17). Early work was concentrated on triblock copolymers of polystyrene and polydienes and their derivatives, but recently there have been interesting developments in similar polymers in which the elastomer segment is polyisobutylene (see Chapter 13). Because of their relatively simple molecular structure and the fact that this structure is uniform, unequivocal, and reproducible, the triblock copolymers of polystyrene and polydienes have served as model polymers from which the properties of other block copolymers could be deduced by analogy. In one of the earliest articles dealing with these polymers [2], the explanation of their properties, the domain theory (see later), was generalized to include "any block copolymer having or containing the structure A–B–A, where A represents a block which is glassy or crystalline at service temperatures but fluid at higher temperatures, and B represents a block which is elastomeric at service temperatures." In serving as model polymers for the whole spectrum of polymers covered by this definition, the polystyrene–polydiene block copolymers have acted as a paradigm, that is, their obvious technical and commercial importance combined with a reasonable and simple explanation of their properties has spurred the development of a new field of polymer science.

3.2   Historial Review

The roots from which the triblock polystyrene–polydiene TPEs grew are described in Chapter 1. Details of this innovation have been given in a article [3] and an address [4]. The various steps that led to it are described in the following paragraphs.

First, it is necessary to consider the state of research in the synthetic rubber industry in the early 1960s. For many years, one prime objective of this research was the economical polymerization of polyisoprene with a high cis-1,4 structure—that is, the production of a synthetic version of natural rubber. In the mid-1950s this work was stimulated by articles describing polymerization on a semicommercial scale using both Ziegler type catalysts [5] and lithium metal catalysts [6]. About this time, workers at the Shell Development Company investigated lithium metal initiators for isoprene polymerization and found that alkyllithiums were far more convenient than the metal itself, a conclusion duplicated at other industrial research laboratories [7]. With these lithium-based catalysts and with properly purified monomer, there were no chain termination or chain transfer steps. Thus when all the original monomer was consumed, the polymer chain still remained active—that is, it could initiate further polymerization if more monomer (either of the same or of a *different* species) was added. In 1957 [8], a process was described for the manufacture of polystyrene–polydiene block copolymers using alkyllithium initiators. Although at the time these polymers were not recognized as TPEs, this research contributed to our background knowledge. About this time triblock copolymers were also reported in which the polymerization initiator was difunctional

[9, 10]. These were produced under conditions that gave polydiene segments with relatively low 1,4 content and so the products had rather poor elastomeric properties.

In parallel with these developments, the melt rheology of polybutadiene [11] and polyisoprene [12] was being investigated. Both (especially polybutadiene) show Newtonian behavior, that is, the viscosities of the pure polymers approach constant values as the shear rate approaches zero. Thus even at room temperature, bales of these elastomers, although they appear to be solids, are in fact very viscous liquids. This gives serious problems in their commercial manufacture and subsequent storage. In the course of Shell Chemical research directed at this problem, polydiene elastomers of various molecular weights were polymerized. Later investigations included work on block copolymers with short blocks of polystyrene, first at only one end of the molecule and later at each end. In contrast to the diene homopolymers, these block copolymers showed non-Newtonian behavior, in that their viscosities tended toward infinity as the shear rate approached zero. In addition, their viscosities were anomalously high at other shear rates [2]. An even more striking anomaly was given by the physical properties of the triblock copolymers. When pressed into tensile sheets, they showed properties in the *unvulcanized* state that were very similar to those of conventional vulcanizates—that is, they had high tensile strength, high elongations, and rapid and almost complete recovery after elongation. These examples of anomalous behavior led to the development of the domain theory. This is the theory that in the bulk state, the polystyrene segments of these block copolymers agglomerate. At temperatures significantly below the glass transition temperature of polystyrene, these agglomerations (the "domains") act as strong, multifunctional junction points and so the block copolymers behave as though they are joined in a crosslinked network [13].

When this discovery was made, the laboratory also housed a polystyrene research group who were able to contribute much expertise on the processing of thermoplastics. The combined research groups were thus able to quickly recognize that the combination of properties—thermoplasticity and elasticity—possessed by these block polymers was so outstanding that the incentives to carry the work through to commercialization were very clear. Within days of the original discovery, potential applications in injection molded footwear, solution based adhesives, and injection molded mechanical goods were foreseen.

It is important to note that the discovery of these TPEs was serendipitous and drew on much research background. In retrospect we may identify four essential elements as contributing to this innovation. They are:

1. The in-house development and commercialization of alkyllithium polymerization systems for the manufacture of polyisoprene.
2. Early scouting research on a polystyrene–polydiene block copolymer process.
3. The existence at the discovery location of an R&D staff with much experience in both elastomers and plastics.
4. The domain theory.

3.3   Structure

Many of the polystyrene–polydiene block copolymers that are thermoplastic elastomers have the basic structure poly(styrene–*block*-butadiene–*block*-styrene) or poly(styrene–*block*-isoprene–*block*-styrene), using the nomenclature of Ceresa [14]. For convenience, they will be referred to here as S–B–S and S–I–S respectively. The most important result of this structure is that they are phase-separated systems, quite unlike the corresponding random copolymers. The two phases, polystyrene and polydiene, retain many of the properties of the respective homopolymers. For example, such block copolymers have two glass transition temperatures (T_g) [2, 15–19], characteristic of the respective homopolymers whereas the equivalent random copolymers have a single intermediate T_g, as shown in Fig. 3.1. This means that at room temperature, the polystyrene phase in S–B–S and S–I–S block copolymers is strong and rigid while the polydiene phase is soft and elastomeric. If the polystyrene phase is only a minor part of the total volume, it is reasonable to postulate a phase structure shown (in idealized form) in Fig. 3.2. In this structure, the polystyrene phase consists of separate spherical regions (domains). Since both ends of each polydiene chain are terminated by polystyrene segments, these rigid polystyrene domains act as multifunctional junction points to give a crosslinked elastomer network similar in many respects to that of a conventional vulcanized rubber. However, in this case the crosslinks are formed by a physical rather than a chemical process and so are labile. At room temperature, a block copolymer of this type has many of the properties of a vulcanized rubber. However, when it is heated, the domains soften, the network loses its strength, and eventually the block copolymer can flow. When the heated block copolymer is cooled down, the domains become hard again and original properties are regained. Similarly, such a block copolymer will be soluble in many solvents (generally, those that are solvents for both of the respective homopolymers) but it will regain its original properties when the solvent is evaporated.

This explanation has been given in terms of S–I–S and S–B–S block copolymers but it should apply (with similar restrictions on relative phase volumes) to other block copolymers such as S–I–S–I–S–I ... and (S–B)$_n$x (where x represents a multifunctional junction point), since these should be able to form continuous network structures similar to that shown in Fig. 3.2. On the other hand, block copolymers such as S–I and B–S–B cannot form these structures, since only one end of each polydiene chain is is terminated by a polystyrene

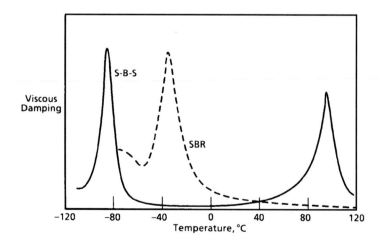

Figure 3.1   Viscous damping of S–B–S and SBR copolymers

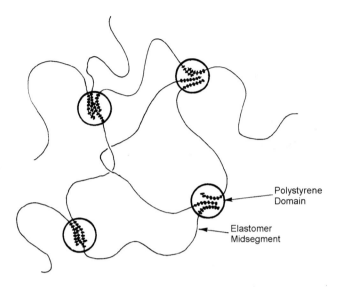

Polystyrene
Domain

Elastomer
Midsegment

Figure 3.2   Phase structure of S–B–S and similar block copolymers (diagrammatic)

segment. Because of this, S–I, B–S–B, and similar block copolymers are weak materials with no resemblance to conventional vulcanized rubbers [2].

When the above explanation was first postulated, it was generalized [2] to include all block copolymers with alternating hard and soft segments and specifically it was applied to polycarbonate–polyether and poly(dimethylsiloxane)–poly(silphenylenesiloxane) block copolymers. A similar explanation has been applied to segmented polyurethanes [20, 21] and it is now accepted as the underlying mechanism that gives most TPEs their valuable properties.

3.4   Synthesis

S–B–S, S–I–S, and similar block copolymers are made by anionic polymerization [22, 23]. This reaction can be used to polymerize three common monomers—styrene (including some substituted styrenes), butadiene, and isoprene. It is usually carried out in an inert hydrocarbon solvent such as cyclohexane or toluene and it is necessary to rigorously exclude oxygen, water, or any impurity that can react with the highly active propagating species. Under these conditions the polymeric molecular weights can be precisely controlled. This is in contrast to other block and graft copolymers, which generally have broad distributions both of the segmental molecular weights and also of their number in the polymer molecule. The preferred initiators for anionic polymerization are organo-lithiums (although others can be used [22]) and there are three basic methods for polymerization:

1. Sequential, that is, start polymerization at one end of the molecule and continue to the other.

2. Coupling, that is, start polymerization at each end of the molecule and then join the reactive chains together by a coupling or linking agent.
3. Multifunctional initiation, that is, start polymerization at the center of the molecule and continue to the ends, using initiators that have more than one active group.

In the first two polymerization methods, *sec*-butyllithium is the preferred initiator because it initiates the polymerization very readily [7]. The rate of the initiation reaction is high compared to that of the subsequent polymerization, which leads to a narrow molecular weight distribution. The initiator (R^-Li^+) first reacts with one molecule of styrene monomer and this is known as the initiation reaction.

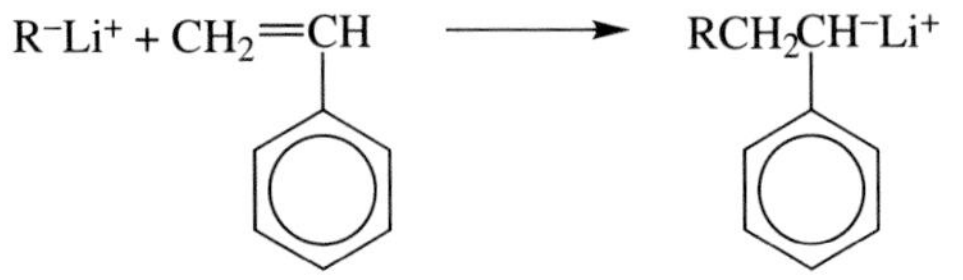

$$R^-Li^+ + CH_2{=}CH \longrightarrow RCH_2CH^-Li^+ \tag{3.1}$$

The product can then initiate polymerization of the styrene and this is known as the propagation reaction.

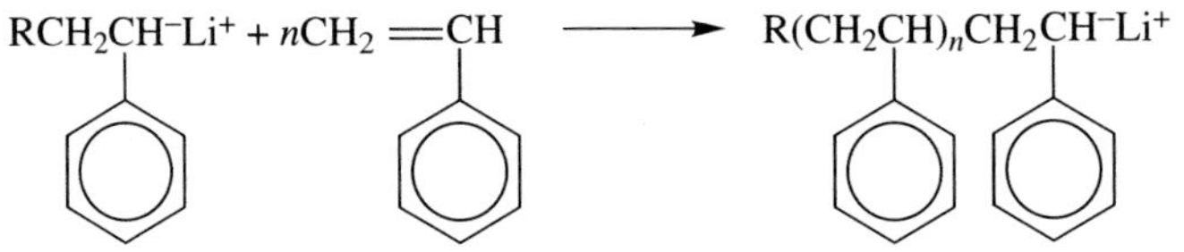

$$RCH_2CH^-Li^+ + nCH_2{=}CH \longrightarrow R(CH_2CH)_nCH_2CH^-Li^+ \tag{3.2}$$

The new end product is termed poly(styrl) lithium (the effects of the terminal radical are ignored) and we will denote it as S^-Li^+. If a diene (in this case butadiene) is added, the S^-Li^+ can initiate further polymerization:

$$S^-Li^+ + nCH_2{=}CHCH{=}CH_2 \longrightarrow S(CH_2CH{=}CHCH_2)_{n-1}CH_2CH{=}CHCH_2^-Li^+ \tag{3.3}$$

In the above example, the polymerization is shown to take place exclusively through the end or 1,4 carbon atoms. Polymerization in hydrocarbon solvents that are inert and nonpolar gives at least 90% of the polymer in this arrangement. The remaining material is polymerized through either the 1,2 carbon atoms (in the case of butadiene) or through the 3,4 carbon atoms (in the case of isoprene). For the above reaction, we will denote the product $S{-}B^-Li^+$. It also is an initiator, so that if more styrene monomer is now added, it will polymerize onto the "living" end of the polymer chain:

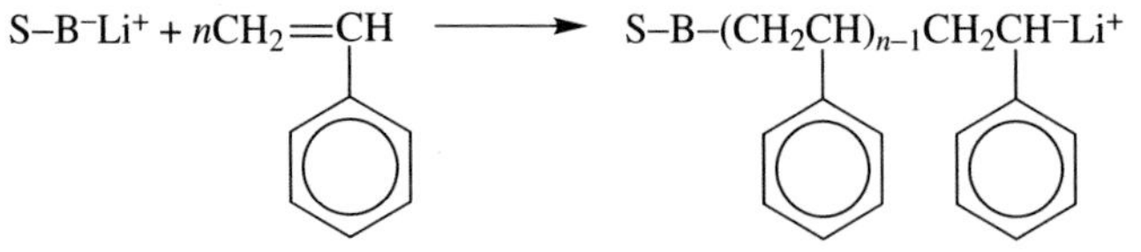

$$S{-}B^-Li^+ + nCH_2{=}CH \longrightarrow S{-}B{-}(CH_2CH)_{n-1}CH_2CH^-Li^+ \tag{3.4}$$

This can give difficulties, because the rate of initiation of styrene polymerization by the $S{-}B^-Li^+$ is slow compared to that of the subsequent propagation reaction. This effect broadens the molecular weight distribution of the second polystyrene segment. In extreme cases there may even be some $S{-}B^-Li^+$ molecules still unreacted when all the added styrene monomer is consumed. The problem can be avoided by adding solvating agents, such as ethers, just before the styrene is added. These increase the initiation rate and give a much narrower molecular weight distribution to the second polystyrene segment. It is important not to add solvating agents before this step, because they also alter the microstructure of the polydiene, reducing the amount of 1,4 enchainment.

When this last reaction is complete, the product (S–B–S⁻Li⁺) can be inactivated by the addition of a protonating species such as an alcohol. This terminates the reaction:

$$S-B-S^-Li^+ + ROH \longrightarrow S-B-SH + ROLi \tag{3.5}$$

If the polymer is to be made by coupling, the first three reactions shown above are unchanged, but instead of the S–B⁻Li⁺ initiating further polymerization of styrene, it is reacted with a coupling agent:

$$2S-B^-Li^+ + X-R-X \longrightarrow S-B-R-B-S + 2LiX \tag{3.6}$$

Many coupling agents have been described, including esters, organo-halogens, and silicon halides [24, 25]. The example above shows the reaction of a difunctional coupling agent, but those of higher functionality (e.g., $SiCl_4$) can also be used and these give branched or star-shaped molecules $(S-B)_n x$. If divinylbenzene is added at the end of the reaction the products are highly branched that is, the value of n is very large [24, 25].

The third method of producing these block copolymers uses multifunctional initiation. In this method a multifunctional initiator (Li⁺⁻R⁻Li⁺) is first reacted with the diene (in this case butadiene).

$$n CH_2 = CHCH = CH_2 + Li^{+-}R^-Li^+ \longrightarrow Li^{+-}B-R-B-Li^+ \tag{3.7}$$

The final two steps are similar to the corresponding steps in the sequential polymerization described above. When the reaction to produce the Li⁺⁻B–R–B⁻Li⁺ is completed, styrene monomer is added and the Li⁺⁻B–R–B⁻Li⁺ in turn initiates its polymerization onto the "living" chain ends to give Li⁺⁻S–B–R–B–S⁻Li⁺. A protonating species is then added to terminate the reaction and give the S–B–R–B–S polymer. This example shows the use of a difunctional initiator. There is no reason in principle why initiators of higher functionality could not be used but none appears to have been reported in the literature.

At first, multifunctional initiation received less attention than the preceding two methods, although early workers used a sodium naphthalene diinitiator [9, 10]. It has the serious deficiency that "living" chain ends associate when the polymerization is carried out in hydrocarbon solvent. This causes such multifunctionally initiated polymers to gel as soon as the reaction starts, which gives significant problems with heat removal, and so forth. Although this association can be prevented by the use of solvating agents such as ethers, as noted previously, these in turn alter the microstructure of the polydiene. More recently, there has been renewed interest in multifunctional initiation, including production of triblock copolymers in which the end segments are copolymers of styrene and α-methylstyrene [26, 27].

S–I–S and S–B–S block copolymers are the precursors of styrenic block copolymers with saturated elastomer center segments (see Chapter 11). If S–B–S polymers are used, they are polymerized in the presence of a structure modifier to give elastomer segments that are a mixture of 1,4 and 1,2 isomers and these are subsequently hydrogenated to give ethylene–butylene copolymers (EB):

$$—CH_2—CH{=}CH—CH_2{\text{-}}CH—CH_2— \xrightarrow{H_2} —CH_2—CH_2—CH_2—CH_2{\text{-}}CH—CH_2— \quad (3.8)$$

$$\begin{array}{ccc} & CH & & CH_2 \\ & \| & & | \\ & CH_2 & & CH_3 \\ 1,4 & 1,2 & E & B \end{array}$$

Similarly polyisoprene elastomer segments can be hydrogenated to give ethylene–propylene (EP) copolymers. The resultant S–EB–S and S–EP–S block copolymers have improved resistance to degradation.

Almost all the above anionically polymerized block copolymers have polystyrene end segments. This has been the preferred material for this type of block copolymer but substituted polystyrenes can be used also. Among those used are poly(α-methylstyrene) [28], copolymers of α-methylstyrene and styrene [26, 27, 29, 30] and poly(p-$tert$-butyl-styrene) [31]. The attractive feature of all three is their relatively high values of T_g, which should lead to improved upper service temperatures (see Chapter 16) in the corresponding block copolymers. However, α-methylstyrene is difficult to polymerize [25] because it has a slow reaction rate and a low ceiling temperature. (This is the temperature at which the rate of the depolymerization reaction is the same as the rate of the polymerization reaction). Poly(p-$tert$-butylstyrene) is apparently rather compatible with the polydienes and so forms a phase-separated system only at relatively high molecular weights [31].

The synthesis of poly(styrene–$block$-isobutylene–$block$-styrene) (S–IB–S) and similar materials by carbo-cationic polymerization is extensively covered in Chapter 13 and details will not be discussed here. However, it should be noted that carbo-cationic polymerization changes the available options for the type of polymers used both for the hard end segments and also for the elastomeric mid segments. Polyisobutylene is the preferred choice for the midsegment, but a wide range of polyaromatics is available for the end segments. There are important practical consequences. First, carbo-cationic polymerization of isobutylene allows the direct production of block copolymers having highly stable elastomer segments. This eliminates the hydrogenation step that must be used to produce similar polymers (e.g., S–EB–S and S–EP–S) by anionic polymerization. Second, the variety of aromatic monomers that can be polymerized to produce the end segments includes halogenated polystyrenes, as well as other substituted polystyrenes and similar polyaromatics. All these polyaromatics have higher values of T_g than polystyrene and so block copolymers based on them should show improved upper service temperatures.

3.5   Properties

3.5.1   Tensile Properties

From a commercial point of view, the most interesting property of these polymers is their resemblance, at least at room temperatures, to vulcanized rubbers. This was shown in very early work [32], where the stress–strain behavior of an S–B–S was compared to that of vulcanized natural rubber and vulcanized SBR (Fig. 3.3). This and similar block copolymers have tensile strengths up to about 30 MPa (about 4000 psi) and elongations of up to 800%. These values (particularly tensile strength) are much higher than those obtained from unreinforced vulcanizates of SBR or polybutadiene. This apparent anomaly has been explained by two possible mechanisms [2]. The first postulates that the hard polystyrene domains act similarly to reinforcing filler particles (e.g., carbon black) in conventional vulcanizates. This is supported by the fact that these domains are about the same size (about 300 Å) as typical particles of reinforcing fillers, and like them, are well dispersed and firmly bound to the elastomer phase. The second takes account of the increased tensile strength resulting from the slippage of entangled chains. It is quite possible that both apply. What is clear is that for materials of constant polystyrene content, the tensile moduli and tensile strengths of both S–B–S [2] and S–I–S [33] polymers are not molecular weight dependent, so long as the polystyrene molecular weight is high enough to cause the formation of strong, well-phase-separated domains under the conditions of the test.

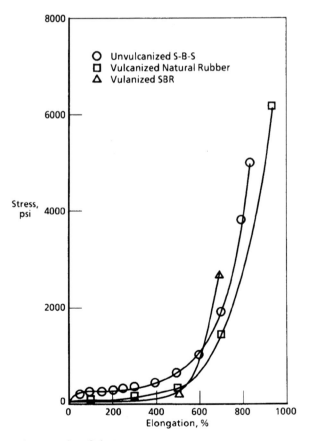

Figure 3.3   Stress–strain properties of elastomers

Another interesting point is the elastic modulus of these materials. It is anomalously high and does not vary with molecular weight. In a simple approach, this is attributed to trapped entanglements in the elastomer center segments acting as crosslinks. If this is the case, the molecular weight between entanglements is the critical parameter. Taking this approach a step further, it was also noted that the classical Mooney–Rivlin treatment of rubber elasticity [34] can be combined with the Guth–Smallwood treatment [35] of the effect of filler on the elastic modulus. When this combined treatment is applied to styrenic thermoplastic elastomers [2], it gives an equation:

$$f = (\rho RT/M_c + 2C_2/\lambda)(\lambda - 1/\lambda^2)(1 + 2.5\phi_s + 14.1\phi_s^2) \qquad (3.9)$$

where

f is the tensile stress applied to the sample,
λ is the extension ratio produced,
ρ is the density of the sample,
R is the gas constant,
T is the absolute temperature,
M_c is the molecular weight between chain entanglements in the elastomer phase,
C_2 is a constant that represents the deviation from ideal elastic behavior, and
ϕ_s is the volume fraction of the polystyrene domains.

The equation predicts that a plot of $f/(\lambda - 1/\lambda^2)(1 + 2.5\phi_s + 14.5\phi_s^2)$ against $1/\lambda$ should be linear. This was confirmed using an S–I–S block copolymer (see Fig. 3.4). Furthermore, the value of M_c obtained from this plot was about 20,000. This value is similar to the molecular weight between entanglements measured for polyisoprene using the

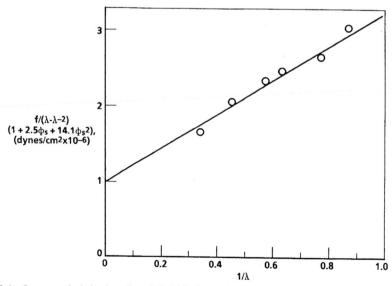

Figure 3.4   Stress–strain behavior of an S–I–S block copolymer (Mooney–Rivlin plot)

viscosity–molecular weight relationship (see later). The value of C_2 was similar to that reported for natural rubber vulcanizates [2].

Thus in these block copolymers, M_e, the molecular weight between entanglements in the elastomer phase (rather than the segmental molecular weight), is used to calculate both the elastic modulus [2] and the degree of swelling in solvents (see later). This approach gives much better agreement with observed behavior. The subject was covered in much more detail in an article in the first edition of this work [36] and the basic conclusion was that by far the largest contribution to the modulus derives from trapped entanglements. M_e has been determined using various methods. Values for polyisobutylene, polyisoprene, polybutadiene, and poly(ethylene–butylene) are given in Table 3.1. The result predicts that at constant styrene content, the elastic modulus of otherwise similar block copolymers should increase in the order S–IB–S, S–I–S, S–B–S, S–EB–S. A designed comparative experiment (i.e., one using well-characterized polymers of comparable molecular weight and polystyrene content) has not been reported for these four types of block copolymers. However, a comparison of S–B–S and S–EB–S block copolymers (see Chapter 11) gave a marked increase in elastic modulus in the expected order and recent results on an S–IB–S block copolymer showed it to be exceptionally soft [37].

These observations led to the interesting and still unresolved question of the mechanism of tensile failure in styrenic block copolymers. There appear to be three possible mechanisms:

1. Ductile failure in the styrenic domains.
2. Brittle fracture in the styrenic domains.
3. Elastic failure in the polydiene center segments.

The case for the first mechanism has been put very strongly [33] (see Chapter 4). It is supported by the fact that in otherwise similar polymers, when the end segments were changed from polystyrene to poly(α-methylstyrene), both the tensile strength and the tensile modulus were almost doubled [38]. This suggests that the glass transition temperature of the styrenic phase, which for poly(α-methylstyrene) is about 70 °C higher than for polystyrene [30], is the controlling parameter. This of course is exactly what would be expected if ductile flow of the domains is the mechanism leading to failure. Against this may be argued the fact that the tensile properties are not affected by molecular weight (see above), a factor that must

Table 3.1   Molecular Weights Between Entanglements (M_e)

Elastomer	M_e (dynamic)	M_e (viscous)
Poly(ethylene–butylene)	1660[a]	—
Polybutadiene	1700	5,600
Polyisoprene	6100[b]	14,000
Polyisobutylene	8900	15,200

[a] Taken from the value quoted for poly(ethylene– propylene)
[b] Taken from the value quoted for natural rubber
Data taken from Tables 13-I and 13-II of Ref. [48]. The dynamic values for M_e were obtained from integration of the loss compliance. The corresponding values obtained from measurements of the viscosity–molecular weight relationship are significantly larger but still in the same order.

strongly affect ductile flow. Both ductile flow and rupture of the domains have been observed in electron micrographs. The failure envelope (the locus of the point defining the tensile strength and elongation at break under various test conditions) has been measured for an S–B–S polymer (Fig. 3.5) and appeared to go through some kind of transition at about 40 °C [38]. It was suggested that failure took place in the domains and at about this temperature the mode changed from brittle to ductile. The case for failure taking place in the elastomer phase rests with the fact that as S–B–S, S–I–S or S–EB–S triblock polymers are diluted by S–B, S–I, or S–EB diblocks, tensile strength and tensile modulus both decrease.

It is possible that which of the mechanisms will apply in a particular case depends on the conditions. At high temperatures, as the domains soften, they will be the "weak link in the chain" and ductile failure will predominate. The same effect will apply when the time scale of the test is long. At lower temperatures or shorter times one of the other two mechanisms will take over. For relatively pure triblock polymers, it is hard to say which of these two mechanisms will predominate but as more diblock is added, this weakens the elastomer phase to the point where it becomes the site of failure.

The analogy between the domains in these block copolymers and the reinforcing fillers (such as carbon black) in conventional vulcanizates also extends to the stress-softening or Mullins effect (see later). Thus the block copolymers show a loss in modulus between the first and subsequent extensions [39, 40].

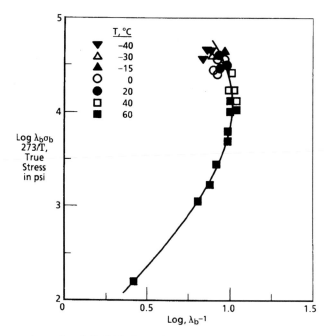

Figure 3.5   Failure envelope of an S–B–S block copolymer

3.5.2   Swelling

In conventional rubber vulcanizates, the degree of swelling in liquids may be used to estimate the molecular weight between effective crosslinks. According to the Flory–Rehner equation [41]:

$$M_c = \frac{\rho_2 V_1(\phi_2^{1/3} - \phi_2/2)}{\ln(1 - \phi_2) + \phi_2 + \chi_1 \phi_2^2} \tag{3.10}$$

where

M_c is the molecular weight between effective crosslinks,
ϕ_2 is the volume fraction of the elastomer in the swollen state,
ρ_2 is the density of the elastomer in the unswollen state,
V_1 is the molar volume of the swelling agent, and
χ_1 is the Flory–Huggins solvent interaction parameter.

This equation can be applied to styrenic block copolymers if a swelling agent is chosen that is very compatible with the rubber phase but that does not affect the polystyrene domains. This approach was applied to a series of S–I–S block copolymers that were swollen in isooctane and the M_c was found to be about 10,000 [42]. If the M_c is identified with the M_e (the molecular weight between entanglements), the value for polyisoprene is consistent with that obtained by other methods.

3.5.3   Viscous and Viscoelastic Properties

A very striking feature of S–B–S and S–I–S block copolymers is their melt viscosities [2, 43]. Under low shear conditions, these are much higher than those of either polybutadiene [11], polyisoprene [12], or random copolymers of styrene and butadiene [44] of equivalent molecular weights. Figure 3.6, for example, shows a comparison of the viscosities of an S–B–S and a polybutadiene each of about 75,000 molecular weight. Moreover, these block copolymers show non-Newtonian behavior, that is, their viscosities increase as the shear is decreased and apparently approach infinite values at zero shear. This behavior is shown both under steady-state [2, 43] and dynamic conditions [45, 46] (see Figs. 3.7 and 3.8). It is attributed [2, 45] to the persistence of a two-phase structure in the melt similar to that shown in Fig. 3.1. In such a structure, flow can take place only by the polystyrene segments at the ends of the elastomer chains being pulled out of their domains. Above a critical molecular weight (see later), the polystyrene segments are phase separated at all temperatures of practical importance and so even though the polystyrene is above its T_g (and therefore fluid), it requires an extra amount of energy to bring it into the elastomer phase. This energy is manifested as an increased viscosity. It should increase with the degree of incompatibility between the end and center segments and therefore the viscosity should also increase with the segmental incompatibility. This is seen to a very striking degree in similar S–EB–S block copolymers, which have very high (and very non-Newtonian) viscosities because of their extreme segmental incompatibility (see Chapter 11). Conversely, in similar polymers with

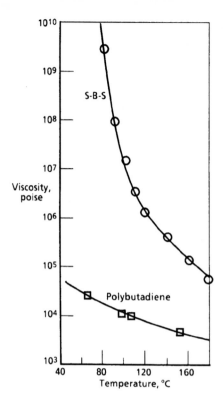

Figure 3.6   Viscosities of polymers at constant shear stress

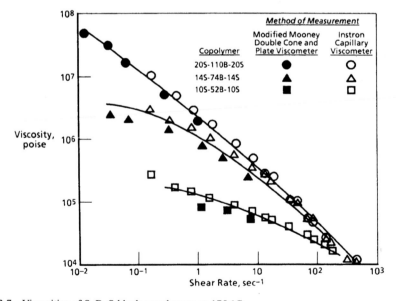

Figure 3.7   Viscosities of S–B–S block copolymers at 175 °C

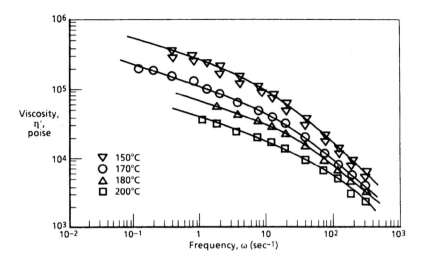

Figure 3.8   Dynamic viscosity of an S–B–S block copolymer at various temperatures

polyethylene end segments (i.e., E–EB–E) the end and center segments are sufficiently compatible so that there is apparently no phase separation in the melt. For this reason E–EB–E block copolymers have very low melt viscosities [47].

The dynamic mechanical behavior of these block copolymers is also unusual. For most polymers the results can be expressed by using the WLF [48] approach. In this method the dynamic mechanical properties are measured at various fixed temperatures and over a range of frequencies. The data are plotted on a master curve by applying appropriate shift factors, which depend on the difference between the temperature of measurement and a reference temperature. This reference temperature is related to the T_g of the polymer. As noted previously, these block copolymers have two T_gs [2, 15–19] and so this technique cannot be directly applied. Several modifications have been used. In one the shift factors for these block copolymers at low temperatures were calculated using a reference temperature appropriate for polybutadiene and at high temperatures using one appropriate for polystyrene. At other temperatures they were calculated using a "sliding" T_g intermediate between the two [49]. In another approach, an extra factor was added at higher temperatures to reflect the viscoelastic response of the polystyrene domains [50] and in later work it was concluded that shift factors dependent both on temperature and on frequency were necessary to describe the behavior of S–B–S block copolymers with both high and low 1,2 content in the polybutadiene segments [51].

From a practical point of view, one of the most interesting aspects of the viscoelastic behavior of block copolymers is in their application as pressure sensitive adhesives (see Chapter 16). Here, the concept of using resins to both soften the polymer by diluting the elastomer phase and also to modify its viscoelastic response has resulted in a most fruitful and elegant explanation of the adhesive behavior.

3.5.4   Solution Properties

The properties of dilute solutions of these polymers in relatively good solvents are quite normal. Of course, there are some difficulties applying theories of molecular behavior that apply to theta solvents, since no solvent can simultaneously provide theta conditions for both the elastomer and the polystyrene segments. Nevertheless, good approximations can be made to such dilute solution properties as intrinsic viscosity, which has been measured in a range of solvents (Fig. 3.9) [52]. It was maximized when the solubility parameter of the solvent was about 8.6 $(cal/cc)^{1/2}$. In contrast, the viscosities of concentrated (24.8% wt/vol) solutions were minimized in the same region (Fig. 3.10) [52]. Allowances must be made for the fact that there is some interference between the coils of the two segmental types and also that the environment surrounding each segmental chain is not just that of the solvent alone but also that of the other segments in the same molecule.

As the solutions become more concentrated phase separation begins and evidence of ordered structures is observed. The first indications were given by the iridescence of these solutions [53], and in later work small-angle X-ray scattering was used to establish domain sizes, interdomain distances, and morphologies [54–58]. The domain sizes depend on the polystyrene molecular weights and on the thermal histories of the solutions [59].

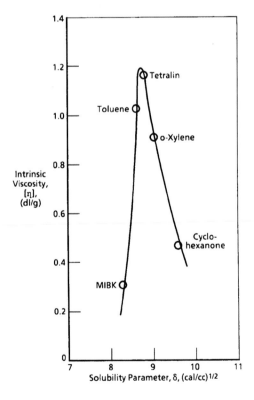

Figure 3.9   Effect of solvent solubility parameter on the intrinsic viscosity of an S–B–S block copolymer

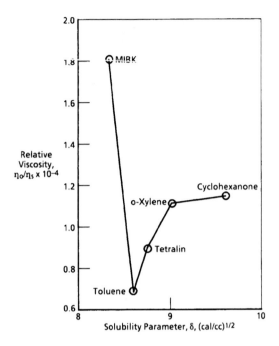

Figure 3.10   Effect of solvent solubility parameter on the solution viscosity of an S–B–S block copolymer

3.5.5   Morphology

The morphology of the systems is one of their most interesting features. The idealized structure shown in Fig. 3.1 was postulated from the mechanical and rheological behavior of S–I–S and S–B–S block copolymers—there were no direct observations to support it. At the same time changes in phase arrangement with the relative proportions of the two segmental types were also postulated—again without any direct observations [2]. One problem was that the domains are too small to be observed by visible light, which is why these block copolymers are transparent. However, the development of a staining technique using osmium tetroxide [60] allowed the morphologies to be established by electron microscopy [17] and this work confirmed that the postulated structures were at least conceptually correct. Later, a more detailed picture of the morphological changes with block copolymer composition was proposed [61] (see Fig. 3.11) in which, as the styrene content is increased, the morphology of the polystyrene phase changes from spheres to cylinders, both dispersed in a continuous elastomer phase. When the volume fractions of the elastomer and polystyrene phases are about equal, the two form alternating lamellae. With further increase in the styrene

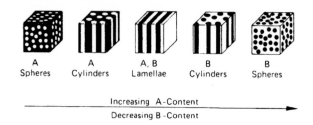

Figure 3.11   Changes in the morphology of an A–B–A block copolymer as a function of composition

content a continuous polystyrene phase forms in which either cylinders or spheres of the elastomer are dispersed. This picture was supported by electron micrographs showing regularly spaced lamellae in block copolymers of intermediate (i.e., about 50%) styrene content [62]. Later work showed that at lower (about 30%) styrene content, a remarkably regular hexagonal array of polystyrene domains dispersed in the elastomer matrix could be obtained, especially if the sample was slowly cast from the solvent [63, 64] (see Fig. 3.12). Rods containing hexagonally packed cylindrical polystyrene domains were produced by extrusion under carefully controlled conditions and shown by small-angle X-ray scattering to have almost "single crystal" perfection. Electron microscopy of ultrathin sections of these rods showed the dramatic regularity and effectively infinite length of these cylindrical polystyrene domains [65]. Application of an oscillating shear field was found to be even more effective in ordering block copolymer domains into either hexagonally packed cylindrical arrays or alternating lamellae [66, 67]. Which was formed in a particular case depended on the polystyrene content of the polymer, with the formation of lamellae being favored when the volume fractions of the polystyrene and the elastomer were about equal.

Figure 3.12   Electron micrograph of an $(S–I)_n$ block copolymer. The polyisoprene phase is stained black

The morphologies of solution cast films depend on the nature of the solvent. Good solvents for the polystyrene segments favor the formation of a continuous polystyrene phase. This gives products that are relatively stiff and inelastic. Conversely, good solvents for the elastomer segments favor the formation of a dispersed polystyrene phase and this gives softer, more elastic products [68, 69]. When these block copolymer films (particularly those with a continuous polystyrene phase) are extended, there is an obvious stress-softening, similar to the Mullins effect in conventional reinforced vulcanizates [69, 70]. Thus when the film is stretched to an elongation below its ultimate elongation, allowed to retract, and then restretched, it shows much higher modulus during the first extension than during the second and subsequent ones. This behavior appears to be caused by rupture of the continuous polystyrene phase during elongation to give discrete domains [71] and has been investigated by both electron microscopy and small-angle X-ray scattering.

3.5.6   Stress-Optical Properties

Birefringence measurements during elongation of S–B–S block copolymers at several temperatures supported the view that below temperatures of about 70 °C, the decreases in strength and modulus were not caused by flow of the polystyrene segments but rather by increased mobility of the polybutadiene chains [70]. In an unstrained S–B–S sample, the difference between the calculated and observed birefringence was attributed to residual stresses in the domains [72]. However, when the samples were stretched, clamped, and annealed below the polystyrene T_g, molecular orientation caused a very large increase in the birefringence [73]. Increasing the annealing temperature removed this effect.

3.5.7   Critical Molecular Weights for Domain Formation

In block copolymers complete miscibility of the segments will take place if the free energy of mixing (ΔG_m) is favorable, that is, negative. This free energy can be expressed as:

$$\Delta G_m = \Delta H_m - T\Delta S_m \tag{3.11}$$

where ΔH_m and ΔS_m are the enthalpy and entropy of mixing and T is the absolute temperature. Thus the condition for domain formation is that:

$$\Delta H_m > T\Delta S_m \tag{3.12}$$

For hydrocarbon polymers, ΔH_m is usually positive, since there are no strongly interacting groups, and will increase as the structures of the two polymers forming the segments become less alike. ΔS_m will always be positive but will approach zero as the molecular weights of the segments become large. Thus we can expect domain formation to be favored by several factors:

1. A high degree of structural difference between the segments.
2. High segmental molecular weight.
3. Low temperatures.

Using this approach, the theory of domain formation has been extensively developed and quantified (see Chapters 12 and 15, Section A). In experimental work, the effects of structural differences are shown by the fact that E–EB–E block copolymers are apparently not phase separated in the melt [47] whereas a strong separation exists for corresponding S–EB–S copolymers (see Chapter 11). The effects of molecular weight and temperature have been demonstrated by work on an experimental S–B–S with end segment molecular weight of 7000. Measurements of both the steady-state [74] and dynamic viscosities [75] showed that for this polymer the critical temperature for domain formation was about 150 °C. Another S–B–S block copolymer with end segment molecular weight of 8000 was reported to show Newtonian behavior at 160 °C [75] and so is apparently not phase separated at this temperature. On the other hand, similar block copolymers with end segment molecular weights of 10,000 and greater appeared to be phase separated at temperatures up to 200 °C [2, 45]. These values of the critical molecular weights and temperatures for domain formation agree quite well with the predictions of the theory. In n-tetradecane, a good solvent for polybutadiene but a poor one for polystyrene, an S–B–S copolymer with higher segmental molecular weights gave a phase-separated gel at lower temperatures but an apparently homogeneous solution at higher temperatures. The temperature at which the solution became homogeneous depended strongly on concentration [76].

3.5.8   Miscibility

Since styrenic block copolymers are two-phase systems, when another polymer or an oligomer is added, there are several possibilities as to the distribution of the added material between the phases. Since almost all end-uses of these polymers involve the addition of such polymers or oligomers, this has important practical consequences (see Chapter 16). Materials mix only if the free energy of mixing (ΔG_m in equation 3.11) is negative. Thus for mixing to take place

$$T\Delta S_m > \Delta H_m \tag{3.13}$$

For mixtures of hydrocarbons, ΔH_m is usually positive. ΔS_m is a function of the reciprocal molecular weight and thus for mixtures of polymers, $T\Delta S_m$ approaches zero and so is less than ΔH_m. For this reason most polymer pairs are immiscible. Of course, if the two polymers are the same, ΔH_m is zero and so the polymer is miscible with itself.

Block copolymers present a different case. When the effects of adding homopolymers to block copolymers were considered [77], it was shown that even if the homopolymer is structurally identical to one segment of the block copolymer, significant amounts will not be miscible unless the homopolymer molecular weight is much less than that of the corresponding segment in the block copolymer. For example, if a homopolystyrene with a molecular weight of about 100,000 is added to a styrenic block copolymer, the homo-polystyrene is not miscible with the polystyrene domains to any significant extent. Instead, it forms a separate phase. There are restrictions on the size of the polystyrene domains in these block copolymers. A very simple calculation [2] shows that because of these restrictions, the domains are too small to give much scattering of visible light—thus even though the styrenic block copolymers are two-phase systems, they are transparent. However, there are no restrictions on the size of the particles of the added polystyrene and so these particles are

large and do scatter light—thus mixtures of homopolystyrene and styrenic block copolymers
are opaque.

The amount of a homopolystyrene (H) able to dissolve in the polystyrene phase of an
S–B block copolymer has been calculated [77], with the results shown in Fig. 3.13. (This
subject is also discussed in Chapter 15, Section C.) If the molecular weight of the
homopolystyrene (M_H) is similar to the molecular weight of the polystyrene segment in
the block copolymer (M_S), very little of the added homopolystyrene can mix with the
polystyrene domains. Significant mixing between the homopolystyrene and the polystyrene
domains occurs only if $M_H \ll M_S$. The same considerations apply to the elastomer phase.
Thus as a rule only very low molecular weight resins and oils should be compatible with
either phase of a block copolymer. This has also been at least qualitatively confirmed [78].
An exception is poly(phenyleneether) (PPE). As discussed in Chapter 15, Section C,
because of a positive interaction between the PPE and polystyrene, in this case ΔH_m, is
negative and so added PPE of higher molecular weight is miscible with the polystyrene
domains in styrenic block copolymers [79].

3.5.9   Structural Variations

Almost all of the work described above has dealt with triblock copolymers having terminal
polystyrene segments and center elastomer segments. Commercially these are the most
important, but other types have been prepared. In early work linear multiblock copolymers
with the general structure $(S–I)_n$ and having a total of up to nine segments were reported
[80]. Later the rheological properties of a series of linear and branched polymers having
30% polystyrene content and with the structures S–B–S, $(S–B)_3x$, $(S–B)_4x$, B–S–B, and
$(B–S)_3x$ were compared [81]. At the same total molecular weight, polymers of the first

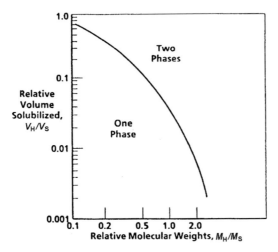

Figure 3.13   Compatibility of an S–B block copolymer with homopolystyrene. Subscript S refers to the
polystyrene segments in the S–B block copolymer; subscript H refers to the added homopolystyrene

three types (i.e., those with terminal polystyrene segments) showed much higher viscosities than those of the last two types (those with terminal polybutadiene segments). At equal *total* molecular weights, the linear polymers (S–B–S and B–S–B) were much more viscous than the branched equivalents. However, if the linear polymers were considered as dibranched materials (i.e., (S–B)$_2$x and (B–S)$_2$x, respectively), then at equal arm length there was little difference between the viscous or viscoelastic responses of equivalent linear and branched polymers. Solution viscosities and Melt Flows [82] of a series of S–B–S, (S–B)$_3$x and (S–B)$_4$x polymers with 25% styrene content also showed the same tendency [83], that is, the polymers behaved very similarly if they were compared at equal arm lengths. The behavior in the melt state supports the view that in this state these polymers are phase separated [2, 45]. In the case of an S–B–S or (S–B)$_n$x polymer, flow will take place by disruption of the disperse polystyrene phase and extra energy will be required for this. This extra energy will depend on the molecular weight of the polystyrene segments and branching in the polybutadiene phase will not have much effect on it. Conversely, for B–S–B and (B–S)$_n$x polymers, flow can take place in the continuous polybutadiene phase without disruption of the polystyrene domains. The energy required for this will be much less and so the viscosity will be lower.

References

1. M.S. Reisch, *C and E News* p. 30 (May 4, 1992)
2. G. Holden, E.T. Bishop, and N.R. Legge, *Proceedings of the International Rubber Conference, 1967*, MacLaren and Sons, London, p. 287 (1968); *J. Polym. Sci. Pt. C 26*, 37 (1969).
3. N.R. Legge, Presented at a Symposium on Innovation and Creativity in the Rubber Industry, ACS Rubber Division Meeting, Philadelphia, PA, May 4, 1982; *Chemtech 13*, 630 (1983)
4. N.R. Legge, Charles Goodyear Medal Address, presented at the ACS Rubber Division Meeting, Montreal, Quebec, May 26, 1987; *Rubber Chem. Technol. 60*, G83 (1987)
5. S.E. Horne Jr., et al., *Ind. Eng. Chem. 48*, 784 (1956)
6. F.W. Stavely, et al., *Ind. Eng. Chem. 48*, 778 (1956)
7. L.E. Foreman, *Polymer Chemistry of Synthetic Elastomers, Part II*. J.P. Kennedy and E. Tornquist (Eds) (1969) John Wiley & Sons, New York, p. 497
8. L.M. Porter (to Shell Oil Co.), U.S. Patent 3,149,182, filed 1957 (1964)
9. M. Szwarc, M. Levy, and R. Milkovich, *J. Am. Chem. Soc. 78*, 2656 (1956)
10. M. Szwarc, *Nature 178*, 1168 (1956). See also A.F. Halasa, *Rubber Chem. Technol. 54*, 627 (1981) and M. Szwarc, *Polym. Preprints 26(1)*, 198 (1985)
11. J.T. Gruver and G. Kraus, *J. Polym. Sci. Pt. A 2*, 797 (1964)
12. G. Holden, *J. Appl. Polym. Sci. 9*, 2911 (1965)
13. G. Holden and R. Milkovich (to Shell Oil Co.), U.S. Patent 3,265,765, filed January 1962 (1964)
14. R.J. Ceresa, *Block and Graft Copolymers* (1962) Butterworth, Washington, D.C.
15. G. Kraus, C.W. Childers, and J.T. Gruver, *J. Appl. Polym. Sci. 11*, 1581 (1967)
16. R.J. Angelo, R.M. Ikeda, and M.L. Wallach, *Polymer 6*, 141 (1965)
17. H. Hendus, K.H. Illers, and E. Ropte, *Kolloid Z.Z. Polymere 216–217*, 110 (1967)
18. J.F. Beecher, L. Marker, R.D. Bradford, and S.L. Aggarwal, *J. Polym. Sci. Pt. C 26*, 117 (1969)
19. D.G. Fesko and N.W. Tschoegl, *Int. J. Polym. Mater. 3*, 51 (1974)
20. (a) S.L. Cooper and A.V. Tobolsky, *J. Appl. Polym. Sci. 10*, 1837 (1966); (b) S.L. Cooper and A.V. Tobolsky, *Textile Res. J. 36*, 800 (1966)
21. W.H. Charch and J.C. Shivers, *Textile Res. J. 29*, 536 (1959)

22. M. Morton, *Anionic Polymerization: Principles and Practice* (1983) Academic Press, New York
23. J.E. McGrath (Ed.), *Anionic Polymerization. Kinetics Mechanics and Synthesis, ACS Symposium Series No. 166* (1981) American Chemical Society, Washington, D.C.
24. N.R. Legge, S. Davison, H.E. DeLaMare, G. Holden, and M.K. Martin, In *Applied Polymer Science*, 2nd edit. R.W. Tess and G.W. Poehlein (Eds.), *ACS Symposium Series No. 285* (1985) American Chemical Society, Washington, D.C.
25. P. Dreyfuss, L.J. Fetters, and D.R. Hansen, *Rubber Chem. Technol. 53*, 728 (1980)
26. (a) L.H. Tung and G.Y.-S. Lo, *Macromolecules 27*, 2219 (1994); (b) C.J. Bredeweg, A.L. Gatzke, G.Y.-S. Lo, and L.H. Tung, *Macromolecules 27*, 2225 (1994); (c) G.Y.-S. Lo, E.W. Otterbacher, A.L. Gatzke, and L.H. Tung, *Macromolecules 27*, 2233 (1994); (d) G.Y.-S. Lo, E.W. Otterbacher, R.G. Pews, and L.H. Tung, *Macromolecules 27*, 2241 (1994); (e) A.L. Gatzke and D.P. Green, *Macromolecules 27*, 2249 (1994)
27. (a) L.H. Tung, G.Y.-S. Lo, and D.E. Beyer (to Dow Chemical Co.), U.S. Patent 4,196,154 (1980); (b) L.H. Tung, G.Y.-S. Lo, J.W. Rakshys, and B.D. Beyer (to Dow Chemical Co.), U.S. Patent 4,201,729 (1980)
28. L.J. Fetters and M. Morton, *Macromolecules 2*, 190 (1969)
29. F.E. Neumann (to Shell Oil Co), British Patent 1,264,741
30. G.Y. Lo and L.H. Tung, Paper presented at ACS Symposium on Advances in Elastomers and Rubber Elasticity, Chicago, IL, September 10, 1985
31. R.E. Cunningham, *J. Appl. Polym. Sci. 22*, 2907 (1978)
32. J.T. Bailey, E.T. Bishop, W.R. Hendricks, G. Holden, and N.R. Legge, *Rubber Age 98(10)*, 69 (1966)
33. M. Morton, *Rubber Chem. Technol. 56*, 1069 (1983)
34. S.M. Gumbrell, L. Mullins, and R.S. Rivlin, *Trans. Faraday Soc. 49*, 1495 (1953)
35. E. Guth, *J. Appl. Phys. 16*, 20 (1945)
36. J.K. Bard and C.I. Chung, In *Thermoplastic Elastomers—A Comprehensive Review*. N.R. Legge, G. Holden, and H.E. Schroeder, (Eds.) (1987) Hanser, Munich and Oxford University Press, New York
37. M. Gyor, Zs. Fodor, H.-C. Wang, and R. Faust, *Polym. Preprints (2)*, 562 (1993); *J. Macromol. Sci.* (in press) *12* 2053 (1994)
38. T.L. Smith and R.A. Dickie, *J. Polym. Sci. Pt. C 26*, 163 (1969)
39. D. Puett, K.J. Smith, and A. Ciferri, *J. Phys. Chem. 69*, 141 (1965)
40. E. Fisher and J.F. Henderson, *Rubber Chem. Technol. 40*, 1313 (1967)
41. P.J. Flory and J. Rehner, *J. Chem. Phys. 18*, 108 (1943)
42. E.T. Bishop and S. Davison, *J. Polym. Sci. Pt. C 26*, 59 (1969)
43. C.W. Childers and G. Kraus, *Rubber Chem. Technol. 40*, 1183 (1967)
44. G. Kraus and G.T. Gruver, *Trans. Soc. Rheol. 13*, 15 (1969)
45. G. Kraus and G.T. Gruver, *J. Appl. Polym. Sci. 11*, 2121 (1967)
46. K.R. Arnold and D.J. Meier, *J. Appl. Polym. Sci. 14*, 427 (1970)
47. M. Morton, N.-C. Lee, and E.R. Terrill, *Polym. Preprints 2*, 136 (1981); *ACS Symposium Series No. 193* (1982) American Chemical Society, Washington, D.C., p. 101
48. J.D. Ferry, *Viscoelastic Properties of Polymers*, 2nd. edit. (1971) John Wiley & Sons, New York, p. 344
49. M. Shen and D.H. Kaelble, *Polym. Lett. 8*, 149 (1970)
50. C.K. Lim, R.E. Cohen, and N.W. Tschoegl, *Advances in Chemistry Series, No. 99* (1971) American Chemical Society, Washington, D.C.
51. R.E. Cohen and N.W. Tschoegl, *Trans. Soc. Rheol. 20*, 153 (1976)
52. D.R. Paul, J.E. St. Lawrence, and J.H. Troell, *Polym. Eng. Sci. 10*, 70 (1970)
53. E. Vanzo, *J. Polym. Sci. Al 4*, 1727 (1966)
54. M. Shibayama, T. Hashimoto, and H. Kawai, *Macromolecules 16*, 16 (1983)
55. T. Hashimoto, M. Shibayama, H. Kawai, H. Wanatabe, and T. Kotaka, *Macromolecules 16*, 361 (1983)
56. M. Shibayama, T. Hashimoto H. Hasegawa, and H. Kawai, *Macromolecules 16*, 1427 (1983)
57. T. Hashimoto, M. Shibayama, and H. Kawai, *Macromolecules 16*, 1093 (1983)
58. M. Shibayama, T. Hashimoto, and H. Kawai, *Macromolecules 16*, 1434 (1983)
59. C.J. Stacy and G. Kraus, *Polym. Eng. Sci. 17*, 627 (1977)
60. K. Kato, *Polym. Eng. Sci. 7*, 38 (1967)
61. G.E. Molau, *Block Polymers*, S.L. Aggarwal (Ed.) (1970) Plenum Press, New York, p. 79
62. E.B. Bradford and E. Vanzo, *J. Polym. Sci. Pt. Al 6*, 1661 (1968)
63. P.R. Lewis and C. Price, *Polymer 13*, 20 (1972)

64. L.K. Bi and L.J. Fetters, *Macromolecules 8*, 98 (1975)
65. J.A. Odell, J. Dlugosz, and A. Keller, *J. Polym. Sci. Polym. Phys. Ed. 14*, 861 (1976)
66. G. Hadziioannou, A. Mathis, and A. Skoulios, *Colloid Polym. Sci. 257*, 136 (1979)
67. G. Hadziioannou and A. Skoulios, *Macromolecules 15*, 258 (1982), *15*, 263 (1982), *15*, 267 (1982)
68. J.F. Beecher, L. Marker, R.D. Bradford, and S.L. Aggarwal, *J. Polym. Sci. Pt. C 26*, 117 (1969)
69. D.M. Brunwin, E. Fischer, and J.F. Henderson, *J. Polym. Sci. Pt. C 26*, 117 (1969)
70. E. Fischer and J.F. Henderson, *J. Polym. Sci. Pt. C 26*, 149 (1969)
71. M. Fujimura, T. Hashimoto, and H. Kawai, *Rubber Chem. Technol. 51*, 215 (1978)
72. T. Pakula, K. Saijo, H. Kawai, and T. Hashimoto, *Macromolecules 18*, 1294 (1985)
73. T. Pakula, K. Saijo, and T. Hashimoto, *Macromolecules 18*, 2037 (1985)
74. C.I. Chung and J.C. Gale, *J. Polym. Sci. Polym. Phys. Ed. 14*, 1149 (1976)
75. E.V. Gouinlock and R.S. Porter, *Polym. Eng. Sci. 17*, 535 (1977)
76. H. Wanatabe, S. Kuwahara, and T. Kotaka, *Trans. Soc. Rheol. 28*, 393 (1974)
77. D.J. Meier, *Polym. Preprints 18*, 340 (1977)
78. R.-J. Roe and W.-C. Zin, *Macromolecules 17*, 189 (1984)
79. P.S. Tucker, J.W. Barlow, and D.R. Paul, *Macromolecules 21*, 1678 and 2744 (1988)
80. S.Ye. Bresler, L.M. Pyrkov, S.Ya. Frenkel, L.A. Laius, and S.I. Klenin, *Vysokmolekul. Soedin. 4*, 250 (1962); *Polym. Sci. (USSR) 4*, 89 (1963)
81. G. Kraus, F.E. Naylor, and K.W. Rollman, *J. Polym. Sci. Pt. A2 9*, 1839 (1971)
82. ASTM Specification D 1238-57T. American Society for Testing Materials, Philadelphia, PA
83. O.L. Marrs, R.P. Zelinski, and R.C. Doss, Paper presented at ACS Rubber Division Meeting, Denver, CO. October 1973

4 Research on Anionic Triblock Copolymers

R.P. Quirk and M. Morton*

4.1 Introduction 72

4.2 Methods and Problems in the Laboratory Synthesis of Triblock Copolymers 74
    4.2.1 Three-Stage Process with Monofunctional Initiators. 74
    4.2.2 Two-Stage Process with Monofunctional Initiators. 75
    4.2.3 Two-Stage Process with Difunctional Initiators. 76
    4.2.4 "Star" Block Copolymers 78

4.3 Structure–Property Relationships of Triblock Copolymers. 78
    4.3.1 Effect of Sample Preparation Methods. 78
    4.3.2 Morphology. 80
    4.3.3 Mechanical Properties—Uniaxial Stress–Strain Behavior. 82
        4.3.3.1 S–I–S Triblock Copolymers 82
        4.3.3.2 S–B–S Triblock Copolymers. 85
        4.3.3.3 Substituting α-Methylstyrene for Styrene in S–I–S. 86

4.4 Research on Other Anionic Triblock Copolymers 89
    4.4.1 Triblock Copolymers of α-Methylstyrene and Propylene Sulfide 89
    4.4.2 Triblock Copolymers Based on Polysiloxanes 91

4.5 Incompatibility and Processability in Triblock Copolymers 93

4.6 Interphase Adhesion and Tensile Strength 95

4.7 Triblock Copolymers with Crystallizable End Blocks. 96

References 99

---

* Deceased March 23, 1994

4.1   Introduction

This chapter is essentially a review of the research carried out on these types of block copolymers over the past 20 years in our laboratories, much of it in collaboration with Professor L.J. Fetters and his group of investigators. Appropriate reference is, of course, made to work by other investigators that has some bearing on our results and conclusions. The original research project was centered on the polystyrene–*block*-polydiene–*block*-polystyrene (S–D–S) type of triblock copolymer, which first demonstrated the properties of "thermoplastic elastomers," as pointed out in the basic patent (and associated publications) [1] on this composition of matter, and many of the results presented here relate to these materials. However, as will be seen, other variations of triblock (and polyblock) copolymers were also studied by us, and are included herein.

Since the application of these block polymers as thermoplastic elastomers represented a breakthrough in rubber technology, it is appropriate to define more precisely the meaning of this term. I should first be recognized that, by definition, the term "elastomer" refers to a network of flexible polymer chains. This is because such chains do not exhibit the properties of an elastomer (high degree of retraction after deformation) unless they are crosslinked ("vulcanized") into a network. The latter, then, generally loses the ability to flow. However, the term "thermoplastic elastomer," which may at first seem to be a contradiction in terms, simply refers to a material that, at normal service conditions, behaves like a crosslinked network, but at elevated temperatures has the ability to flow (hence "thermoplastic"). Thus, thermoplastic elastomers can be reprocessed, in contrast to crosslinked networks, which are thermosets and not reprocessable.

The reason why the triblock copolymers, such as the polystyrene–*block*-polybutadiene–*block*-polystyrene type (S–B–S), exhibit the properties of thermoplastic elastomers can be understood by considering the morphology of triblock copolymers [2, 3]. Because of the basic thermodynamic incompatability of the polymer blocks, a two-phase morphology is obtained. The observed equilibrium morphologies generally correspond to domains of the minor component dispersed in a matrix of the major component; the critical parameters determining domain morphology are the volume fractions of the components, molecular weights and the degree of incompatibility between different blocks, as well as molecular architecture. Because the blocks are chemically linked to each other, the dimensions of the separatated phases are restricted; thus, domain sizes of the minor component are on the order of hundreds of angstroms. For a typical copolymer composition of 10,000 to 15,000 g/mol polystyrene end block molecular weights and 50,000 to 70,000 g/mol polybutadiene center block molecular weights, the equilibrium domain morphology corresponds to a fine dispersion of spheres of polystyrene thermoplastic, chemically bonded to the surrounding matrix of elastic polybutadiene chains (which are, of course, not chemically bonded to each other). Thus it can be concluded that, in this type of network, the elastic chains are held together only by the thermoplastic domains. The morphology described above is illustrated in Fig. 4.1, which shows a schematic for a typical triblock copolymer S–B–S. Two things should be noted about this representation: (1) the polystyrene is shown as the dispersed phase, since it is the minor constituent, and (2) the polystyrene domains are, presumably, spherical in shape. It turns out that both of these aspects are dependent on the relative proportions of the

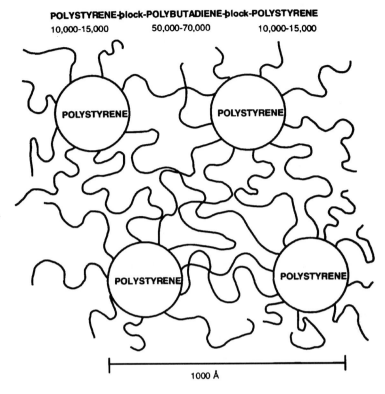

POLYSTYRENE-block-POLYBUTADIENE-block-POLYSTYRENE
10,000-15,000   50,000-70,000   10,000-15,000

Figure 4.1   Schematic of an S–B–S triblock copolymer

two constituents, as well as on the method of preparation of the specimens, and this is discussed in a later section.

Because these types of triblock copolymers exhibit the remarkable property of thermoplastic elastomers, the research at our laboratories had, as its prime objective, the elucidation of the structure–property relationships of these unusual "networks." The main questions requiring answers could be listed as follows:

1. What is the role of the polystyrene domains (other than as thermoplastic "crosslinks"), and do they also act as reinforcing "fillers"?
2. Does the molecular weight of the polydiene blocks represent the "molecular weight between crosslinks (M_c)" in these networks?
3. In this connection, to what extent do these networks behave like conventional cross-linked (and filled) elastomers, with regard to such parameters as modulus and tensile strength?

Finding the answers to such questions required the preparation of a series of S–D–S triblock copolymers. Furthermore, since the morphologies and properties of these networks are governed by the structures of the triblock copolymers from which they originate, it is obvious that delineation of the relationships between polymer structure and property measurements requires the preparation of well-defined polymers with low degrees of compositional heterogeneity. The synthesis of block copolymers by alkyllithium-initiated anionic polymerization involves sophisticated laboratory techniques, because of the "living" carbanionic character of the growing chains, so that a review of methods and problems of synthesis is most appropriate.

4.2   Methods and Problems in the Laboratory Synthesis of Triblock Copolymers

The high-vacuum techniques used to carry out carbanionic polymerization in our laboratories have been described in the literature [4–6]. The application of these techniques to the synthesis of triblock copolymers has also been discussed [7–9]. The requirements for the synthesis of S–D–S triblock copolymers having predictable molecular weights and narrow molecular weight distributions (M_w/M_n, MWD) can be listed as follows:

1. Any possible chain termination reactions must be reduced to a negligible level.
2. The polymerization system must be capable of producing a polydiene block of high 1,4 chain structure, that is, a low T_g elastomer.
3. The initiation rate for each block must be competitive with or faster than the propagation rate ($R_i \geq R_p$) in order to obtain a narrow molecular weight distribution [10].

Organolithium initiators, especially *sec*-butyllithium [11], can fulfill all of these requirements by suitable choice of solvents, for example, nonpolar, hydrocarbon solvents to ensure high 1,4 content [8].

In the special case of block copolymer synthesis, it is necessary not only to ensure the absence of impurities in the polymerization "reactor," but also to take special pains in purifying the monomers. Because monomers are added sequentially, each monomer addition may offer an opportunity for the introduction of impurities that may terminate some of the "living" chains. Bearing this in mind, it is possible to list three basic methods for synthesis of S–D–S triblock copolymers using organolithium-initiated polymerization.

4.2.1   Three-Stage Process with Monofunctional Initiators

This is the standard method, used in our laboratories. It requires a rapid initiation of the polystyrene chain in a nonpolar solvent using a reactive initiator such as *sec*-butyllithium. The "crossover" reaction of poly(styryl) lithium with the diene is known [12–14] to be very rapid, ensuring a fast initiation of the center polydiene block. Finally, to overcome the well-known slow crossover reaction [12–14] with the final styrene charge in nonpolar media, it is necessary to add a small amount of a Lewis base, for example, tetrahydrofuran (THF) [4, 6]. The excellent results that can be obtained by this method, with rigorous exclusion of impurities, is demonstrated by the size-exclusion chromatograms (SECs) [4] shown in Fig.

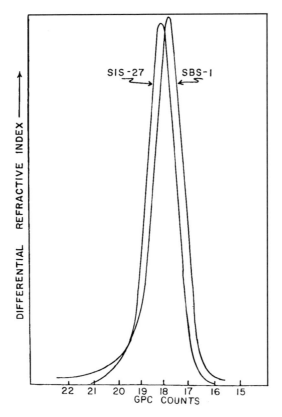

Figure 4.2   SEC curves of polystyrene–*block*-polyisoprene–*block*-polystyrene (S–I–S) and polystyrene–*block*-polybutadiene–*block*-polystyrene (S–B–S) triblock copolymers. (Reprinted from Ref. [4], with permission of John Wiley & Sons)

4.2 for both S–I–S and S–B–S block copolymers [11]. The polymer formed in each step of this sequential synthesis can be analyzed as shown in Fig. 4.3 for the synthesis of a polystyrene–*block*-poly(myrcene)–*block*-polystyrene (S–M–S) triblock copolymer.

4.2.2   Two-Stage Process with Monofunctional Initiators

This process also uses a monofunctional organolithium initiator, as in the above case, but the polymerization is carried only to the diblock (S–D) stage. The triblock copolymers are then formed by using a linking agent, for example, a dihalide or an ester, to join the lithium chain ends of the diblock [17]. This has the advantage of involving only two monomer additions, thus reducing the possibility of introduction of impurities. However, it has the serious disadvantage of requiring very careful adjustment of the ratio of linking agent to chain-end concentration, any deviation from exact stoichiometry leading to formation of free diblocks. The latter have been found [7, 15] to have a dramatic effect on the strength of the material

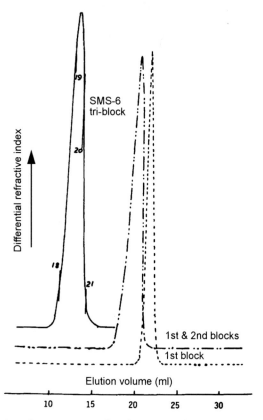

Figure 4.3   Size exclusion chromatograms of each segment for the synthesis of polystyrene–*block*-poly(myrcene)–*block*-polystyrene (S–M–S) triblock copolymer. (Reprinted with permission from Ref. [16]. Copyright 1984, Plenum Press)

(free chain ends in the network). It is interesting to note, in this connection, that a certain amount of free polystyrene (terminated monoblock) can be easily tolerated [7, 15] since it is apparently incorporated into the polystyrene domains. Another advantage of the two-step process is that it is more versatile with respect to the chemical composition of the center block. With this process, the center block can be a more reactive monomer that would not be capable of reinitiating polymerization of styrene because of the increased stability of the chain end. For example, poly(α-methylstyrene)–*block*-poly(propylene sulfide)–*block*-poly(α-methylstyrene) was prepared by the stepwise polymerization of (1) α-methylstyrene followed by (2) propylene sulfide and then coupling the active lithium thiolate chain ends with phosgene [18].

4.2.3   Two-Stage Process with Difunctional Initiators

This process requires a difunctional initiator, so that the center block (D) is formed first, by a dianionic polymerization, followed by addition of styrene to form the two end blocks. It, too, has the advantage of requiring only two monomer additions. However, it has several serious disadvantages. In the first place, it is difficult to obtain a dilithium initiator that is soluble in hydrocarbon media, as required for polymerization of dienes, because of the association of the chain ends [19] to form insoluble, networklike structures [20]; however, some success has been reported [8, 9]. Furthermore, any loss of difunctionality, either in the initiator or after the addition of the diene, leads to formation of undesirable diblocks. However, this method can be especially useful in the case of "unidirectional" block copolymerization, that is, where monomer A can initiate monomer B but not vice versa, for example, for polar B monomers. The dilithium initiator formed by the dimerization of 1,1-diphenylethylene with lithium in cyclohexane in the presence of anisole has been utilized for the synthesis of a poly(α-methylstyrene)–polyisoprene–poly(α-methylstyrene) triblock copolymer [21]. The dilithium initiator formed by the addition of 2 mol of sec-butyllithium with 1,3-bis(1-phenylethenyl) benzene forms a hydrocarbon-soluble, dilithium initiator that has been used in the presence of

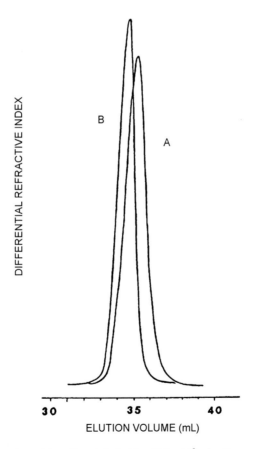

Figure 4.4   SEC curves for polybutadiene [M_n (calc) $= 58.9 \times 10^3$ g/mol] prepared with dilithium initiator from 1,3-bis(1-phenylethenyl) benzene (A) and polystyrene-$block$-polybutadiene–$block$-polystyrene triblock copolymer [M_n (calc) $= 98.8 \times 10^3$ g/mol] prepared by adding styrene to this α,ω-dilithiumpolybutadiene (B). (Reprinted with permission from Ref. [9]. Copyright 1991, Society of Chemical Industry)

lithium alkoxide to form well-defined S–B–S [9] (see Fig. 4.4) and poly(methylmethacry-late)–*block*-polyisoprene–*block*-poly(methylmethacrylate) triblock copolymers [22].

4.2.4   "Star" Block Copolymers

No discussion of the synthesis of triblock copolymers would be complete without mention of the "star" (or "radial") type of block copolymers. This can be considered a variant of Method 2 described previously, except that the original S–D diblocks are linked by means of a polyfunctional, instead of a difunctional, linking agent. This can be expressed by means of the following chemical equations for the case of a styrene (S)–isoprene (I) star block copolymer:

$$RLi + nS \rightarrow PSLi$$

$$PSLi + ml \rightarrow PS-b\text{-}PILi$$

$$PS-b\text{-}PILi + R'X_z \rightarrow (PS-b\text{-}PI)_z R' + zLiX$$

where $R'X_z$ is a linking agent with a functionality of z. This, of course, leads to a star-branched structure but the block sequence in all branches is still (S–D–S). The synthesis, morphology, and properties of these systems have been studied extensively by L.J. Fetters and associates in recent years, using first divinyl benzene [23] and later polychlorosilanes [24, 25] as polyfunctional linking systems.

Living linking reactions have been developed for the synthesis of heteroarm, star-branched block copolymers in which the arms differ in either molecular weight or composition [26, 27]. The living coupling reaction of poly(styryl) lithium with 1,3-bis(1-phenylethenyl) benzene (MDDPE) produces a difunctional, linked product, $(PS)_2(MDDPE)Li_2$ which can initiate the sequential growth of butadiene and styrene block segments to form a star-branched thermoplastic elastomer, $(PS)_2MDDPE(PBD\text{-}b\text{-}PS)_2$ [28, 29]:

$$2PSLi + MDDPE \rightarrow (PS)_2MDDPELi_2$$

$$(PS)_2MDDPELi_2 + mBD \rightarrow (PS)_2MDDPE(PBDLi)_2$$

$$(PS)_2MDDPE(PBDLi)_2 + nS \rightarrow (PS)_2MDDPE(PBD-b\text{-}PSLi)_2$$

$$(PS)_2MDDPE(PBD-b\text{-}PSLi)_2 + ROH \rightarrow (PS)_2MDDPE(PBD-b\text{-}PS)_2$$

In concluding the discussion of block copolymer synthesis, it should be emphasized that, as will be seen later, the experimental precision exercised during the synthesis (and sample preparation) of these polymers has a strong bearing on the morphology and properties of these materials.

4.3   Structure–Property Relationships of Triblock Copolymers

4.3.1   Effect of Sample Preparation Methods

The thermoplastic elastomeric behavior of the triblock copolymers depends on phase separation on cooling from the processing and forming temperatures. Therefore, it might

Table 4.1 Tensile Properties of Molded and Cast Films of S–I–S

(S–I–S) Block MW ($\times 10^{-3}$)	S (%)	Tensile strength (MPa)		Cast film[b]
		Molded sample[a]		
		0.13 cm	0.03 cm[c]	0.03 cm
13.7–63.4–13.7	30	29.0	25.0	32.0
21.1–63.4–21.1	40	29.0	27.0	34.0

[a] Compression molded for 10 min at 140 °C
[b] From soln. of 90% tetrahydrofuran/10% methyl ethyl ketone (MEK)
[c] Molded from 0.03 cm cast film
Source: Ref. [15]

be expected that their morphology and resulting mechanical behavior would be affected by their processing history.

In our earlier work, these polymers were processed like conventional elastomers, that is, by compression molding. However, it soon became apparent that these composite materials were far from their equilibrium phase separation under these conditions, and that film-casting from solvents allowed a closer approach to the equilibrium state. This is demonstrated in Table 4.1, which compares the tensile strength of molded and cast films of a typical S–I–S triblock copolymer. It is obvious from these results that films cast from solvent show superior properties, presumably as a result of better phase separation resulting from this technique, that is, higher purity of the polystyrene domains, which hold the network together. This effect was corroborated by the experiment in Table 4.1 where cast films were remolded, and thereby experienced an actual decrease in strength. It should be noted that proper care had to be taken in the use of film casting to select the best solvent system [6] and to remove the last traces of solvent.

The effect of solvent type and of traces of residual solvent is convincingly shown in Table 4.2. As can be seen, three different solvent systems were used [15]: 90% tetrahydro-

Table 4.2 Solution Casting of S–I–S Polymer Films (13.7– 41.1–13.7) $\times 10^3$ MW

Film treatment	90/10 THF/MEK		90/10 C_6H_6/C_7H_{16}		CCL_4	
	Wt% loss	Tensile strength (MPa)	Wt% loss	Tensile strength (MPa)	Wt% loss	Tensile strength (MPa)
Air dried	—	25.0	—	27.5	—	26.0
Vacuum dried[a]						
1 day	3.1	29.0	2.5	34.0	7.7	28.0
2 days	0.3	31.0	0.13	33.0	0.42	30.0
3 days	0.09	36.0	~0	35.0	~0	33.0
4 days	~0	36.0	~0	36.0	~0	33.5

[a] At 1 mm Hg, 25 °C
Source: Ref. [15]

furan/10% methyl ethyl ketone, 90% benzene/10% n-heptane, and carbon tetrachloride. These were chosen because electron microscopy studies [30] had shown that the best phase separation was obtained with the first and second of the above solvents, which represented specific solvents for polystyrene and polydienes, respectively, whereas the CCl_4 was a mutual solvent. The data in Table 4.2 corroborate this effect of the two specific solvents, that is, higher tensile strengths compared to the CCl_4-cast films. What is even more dramatic, however, is the effect of traces of residual solvent, and the time necessary to vacuum-dry the films. It should, therefore, be noted that all tensile data reported from our laboratories on solvent-cast films of triblock copolymers are based on samples rigorously vacuum-dried for several days. It is of interest also to note the stringent drying and annealing techniques used on solvent-cast films, as described in a recent article [25] on the equilibrium morphology of "star" block copolymers.

4.3.2   Morphology

The effect of various factors on the morphology of these two-phase systems has been studied extensively, both theoretically and experimentally [2, 3, 31–34], and a general picture has emerged. Thus, it is now accepted that, under equilibrium conditions, for example, in films cast from suitable solvents and well-annealed, the polystyrene domains will assume highly ordered but different morphologies as the polystyrene content is increased, starting from spheres, at less than 20% polystyrene, and changing to cylinders for 20% to 35% styrene and then lamellae as the polystyrene content rises to 50% (see Fig 3.11). Beyond that there is apparently a phase inversion with the polystyrene forming the continuous phase and the polybutadiene dispersion assuming the different shapes described above. Recent SAXS and TEM studies have revealed the existence of a new ordered bicontinuous double diamond morphology for block copolymers [2, 3, 35] (see also Chapter 12).

The remarkable order found in the morphology of these polymers is illustrated in the transmission electron microphotographs of ultrathin (~50 nm) films of a S–I–S triblock copolymer shown in Fig. 4.5, and of a $(S–I)_x$ "star" block copolymer, shown in Fig. 4.6.

The osmium tetroxide "staining" renders the polydiene matrix more opaque to the electron beam, and hence dark, while the polystyrene domains appear lighter. The circular shape of the latter does not necessarily reflect a spherical shape, since it could also represent a cylindrical cross-section.

The remarkable regularity shown in these figures is, of course, a natural outcome of the near-monodisperse MWD of the polymer blocks. Since this heterophase morphology results from the incompatibility of the two types of polymer blocks present, it might be instructive to consider the constraints and limitations on block molecular weight imposed by the material considerations, as follows:

1.   The styrene/diene ratio controls the modulus of the thermoplastic elastomer (poly-styrene domains acting as a "filler").
2.   The lower limit of polystyrene block size is set by incompatibility requirements.
3.   The upper limit of molecular weights is set by viscosity considerations, which affect both processability and efficiency of phase separation in the melt.

Figure 4.5   Transmission electron microphotograph of an
S–I–S triblock copolymer

Because of these considerations, the molecular weights of the polystyrene blocks are generally in the range of 10,000 to 15,000 whereas the polydiene molecular weights vary from 50,000 to 70,000, as indicated in Fig. 4.1.

In this connection, it might be of interest to consider the difference between phase separation of two amorphous phases, as a result of incompatibility, with the case where one of

Figure 4.6   Transmission electron microphotograph of an (S–I)$_x$ "radial" block copolymer. Ultrathin ($\sim$50 nm) section of cast film, stained with OsO$_4$. (Reprinted with permission from Ref. [23]. Copyright 1975, American Chemical Society)

the components separates by crystallization. In the first instance, that is, separation of two amorphous phases, it is obvious that this will be enhanced with increasing incompatibility, thus leading to less "phase mixing," and better mechanical properties. However, a greater incompatibility of the two phases would also lead to higher melt viscosity (and hence poorer processibility), since flow of such a two-phase system involves their free energy of mixing. In contrast, where phase "separation" occurs by crystallization of one of the components, no incompatibility is involved in the melt and the flow properties are similar to those of a homogeneous material, that is, lower melt viscosity.

4.3.3   Mechanical Properties–Uniaxial Stress–Strain Behavior

4.3.3.1   S–I–S Triblock Copolymers

To answer the questions raised in the introductory section about the elastic behavior of these polymers, a series of (S–I–S) triblock copolymers was prepared with varying polystyrene content and molecular weight of the polyisoprene center block. The stress–strain properties, up to break, are shown in Fig. 4.7, from which the following conclusions may be drawn:

1. The tensile modulus appears to be mainly dependent on the polystyrene content ("filler" effect) and independent of the molecular weight of the polyisoprene center block. Hence the latter cannot be considered as representing the "molecular weight between cross-links" (M_c) of the network. This is really not surprising, since the polyisoprene center block has a molecular weight of at least 40,000 while the "molecular weight between entanglements" (M_e) [36] is about 7000. Hence there are obviously a number of such chain entanglements between the polystyrene "crosslinks," and the "network chain" can really be considered as equivalent to the M_e value of polyisoprene (see also Chapter 3).
2. The tensile strength (end of each curve) of this series of S–I–S polymers appears to be largely independent of either polystyrene content or molecular weight, with the exception of the polymer having the lowest polystyrene molecular weight, where the strength is cut by almost one half. The most obvious conclusion is that, in the latter case,

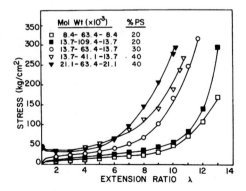

Figure 4.7   Effect of composition and block size on tensile properties of styrene–isoprene–styrene triblock copolymers ($1 \text{ kg} \cdot \text{cm}^{-2} = 0.1$ MPa). (Reprinted from Ref. [15], with permission of John Wiley & Sons)

Table 4.3 Effect of Polystyrene Molecular Weight on the Tensile Strength of S–I–S

Wt% styrene	Polymer MW (S–I–S) ($\times 10^{-3}$)	Tensile stress (MPa)	
		300% elong.	At break
20	13.7–100.4– 13.7	1.8	27.0
20	8.4–63.4–8.4	1.1	16.0
19	7.0–60.0–7.0	1.3	2.2
11	5.0–80.0–5.0	~0	~0

the polystyrene MW is too low for good phase separation, so that there is substantial plasticization of the polystyrene domains by the polyisoprene, decreasing their ability to withstand the stress. This is further illustrated in Table 4.3, where the molecular weight of the polystyrene end blocks has been further reduced (below 8400), showing the dramatic effect this has on the tensile strength. It can be concluded that, at a molecular weight of 5000 to 6000, the polystyrene is compatible with the polyisoprene, and no domains can be present, as predicted by Meier [37].

One further note should be made about Fig 4.7. These tensile curves represent the first stretch. The retraction curves exhibit a large hysteresis loop and considerable unrecovered deformation (set), which increases with degree of strain and polystyrene content. This phenomenon is apparently largely due to a distortion of the polystyrene domains, which has actually been observed by electron microscopy [30] of stretched specimens. The distorted domains can be restored to their original condition either by heating the sample at or above the T_g of the polystyrene, or by swelling in a specific solvent for the polystyrene, namely tetrahydrofurfuryl alcohol (THFA). This type of treatment also restores the tensile behavior of the prestretched sample, as convincingly demonstrated in Fig. 4.8 [15, 38]. Apparently it is the polystyrene phase which is wholly responsible for any unrecovered deformation, since swelling by a selective solvent for the polyisoprene (n-decane) has no effect.

These experiments on the behavior of these polymers under uniaxial stress seem to corroborate the fact that it is the polystyrene domains that are responsible for the integrity and strength of the network, and that the mechanism of tensile failure involves rupture of these domains. As a matter of fact, the exceptionally high strength shown by these polymers is undoubtedly a result of both the regularity of the network, which helps to distribute the stresses more evenly, and the fact that the polystyrene domains can act as an "energy sink" to absorb the elastic energy and thus delay failure.

This proposed strength mechanism is further supported by experiments involving the introduction of crosslinks [15] in the polyisoprene matrix, using dicumyl peroxide as crosslinking agent. This agent is known [39] to crosslink polyisoprene stoichiometrically without any observed chain scission. The results are shown in Table 4.4. The swelling data show that crosslinks were indeed introduced, yielding a tighter network. However, this had little effect on the modulus and a very noticeable deleterious effect on the tensile strength. Hence these experiments add further evidence pointing to the polystyrene domains as the key factor in the strength of these materials. The decrease in tensile strength after crosslinking is undoubtedly due to the action of the fixed crosslinks in preventing stress distribution in the entangled chain network.

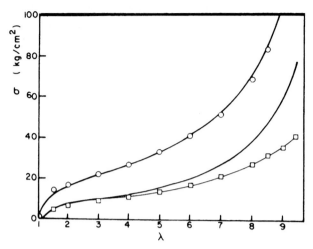

Figure 4.8   Effect of swelling on stress-softened S–I–S triblock copolymer (MW × 10^{-3} = 13.7–63.4–13.7). (From Ref. [38]). □, n-decane; ○, tetrahydrofurfuryl alcohol; —, first and second cycle unswollen (1 kg · cm^{-2} = 0.1 MPa)

Table 4.4 Crosslinking of S–I–S by Dicumyl Peroxide

	Swelling vol. ratio[a]	Stress at 300% elong. (MPa)	Tensile strength (MPa)
Original sample	9.8	5.0	22.5
After crosslinking	5.4	5.3	16.0

[a] In n-decane, 48 h

When dicumyl peroxide was used to introduce crosslinks into poly(p-methylstyrene–*block*-butadiene–*block*-p-methylstyrene) triblock copolymers, the tensile strength increased from 2.2 to 7.3 MPa (320 to 1060 psi); however, the elongation at break decreased from 197 to 12% [40]. In contrast, after the addition of ω-hydroperoxypolystyrene, a phase-specific crosslinking agent for the p-methylstyrene phase, the tensile strength increased to 10.4 MPa (1510 psi) and the elongation at break was 294%. Thus, crosslinking in the hard-phase domains increases the tensile strength and elongation, in contrast to crosslinking of the polydiene phase.

One additional feature of the tensile curves in Fig. 4.7 deserves mention. It can be seen that the curves representing the polymers containing 40% styrene show an initial yield point. It should be noted that this occurs only on the first stretch and not on any subsequent stretch, unless the sample is reheated, or allowed to rest (for several months at room temperature). This is believed to be due to the high concentration of polystyrene domains, which, at 40% content, are approaching the critical fraction for volume packing. Hence it can be expected that there would be some "connections" between the domains, for example, cylindrical polystyrene domain morphology [2, 3, 31–34], which would, of course, be broken on the first stretch. In a sense, at this stage, the morphology would be that of two interpenetrating continuous phases (cylindrical polystyrene domains in a continuous polybutadiene matrix). Needless to say, such yield points are also found at higher proportions of styrene.

4.3.3.2   S–B–S Triblock Copolymers

A similar series of S–B–S triblock copolymers was used to obtain the tensile curves shown in Fig. 4.9. It should be noted that these polymers were of an earlier vintage, from a time when compression molding was used in their preparation, rather than solution casting; this may account for the somewhat lower tensile strengths obtained. However, they also show some similarity in behavior to the S–I–S polymers, at least on two counts, that is, the styrene content does control the modulus, and the latter is independent of the molecular weights. However, the real difference from the S–I–S polymers lies in the fact that the tensile strength seems to depend greatly on the styrene content. (The difficulties of using compression molding for sample preparation are illustrated in the case of the 40% styrene-containing polymers, where "annealing," that is, allowing the sample to cool very slowly, at 1 °C per minute, from the molding temperature of 140 °C, increased the tensile strength substantially).

This unexpected effect of styrene content on the tensile strength of S–B–S is in all likelihood related to the question of incompatibility. This becomes apparent from an examination of the solubility parameters (δ) of polystyrene, polyisoprene, and polybutadiene, which are, respectively, 9.1, 8.1, and 8.4 $(\text{cal/cm}^3)^{1/2}$, as given in the literature [41]. Hence it appears that polybutadiene is more compatible than polyisoprene with polystyrene. Furthermore, the incompatibility of two polymers is dependent not only on their basic chemical structure but also on their molecular weights and volume fractions in the mixture, as given by the relationships:

$$\chi_{12} = M_1(\delta_1 - \delta_2)^2/\rho_1 RT \qquad (4.1a)[42]$$

and

$$\Delta E_m = V\Phi_1\Phi_2(\delta_1 - \delta_2)^2 \qquad (4.1b)[43]$$

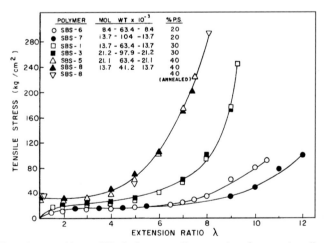

Figure 4.9   Effect of composition and block size on tensile properties of styrene–butadiene–styrene triblock copolymers (1 kg · cm^{-2} = 0.1 MPa). (Reprinted from Ref. [15], with permission of John Wiley & Sons)

where

> χ_{12} = interaction parameter of components 1 and 2,
> M_1 = molecular weight of component 1,
> δ_1 and δ_2 = solubility parameters of components 1 and 2,
> ρ_1 = density of component 1,
> ΔE_m = heat of mixing,
> Φ_1 and Φ_2 = volume fractions of components 1 and 2, and
> V = total volume of mixture.

Hence the increase in tensile strength of S–B–S with increasing styrene content shown in Fig. 4.9 is most likely related to a better phase separation of the polystyrene domains at higher volume fractions of this component, that is, less phase mixing and plasticization by the polybutadiene. In the case of the S–I–S polymers, it appears that the incompatibility of the two phases is already high enough not to be affected very much, at those levels, by the styrene content.

In this connection, it is interesting to note the experimental results we obtained [15] by solution blending polystyrene with S–I–S and S–B–S. For this purpose, a polystyrene was prepared by organolithium polymerization, having approximately the same molecular weight as the polystyrene end blocks in the S–I–S or S–B–S. When this polystyrene was blended into an S–B–S polymer, it was found [15] that the blend had the same tensile properties as an S–B–S with a similar styrene block content, that is, higher modulus and higher tensile strength than the original S–B–S used in the blend. However, when a similar blend was made of S–I–S and PS, only the modulus was raised, the tensile strength remaining unchanged. This behavior is similar to the results shown in Fig. 4.7. These experiments offered a gratifying confirmation of the results shown in Figs. 4.7 and 4.9, and of the proposed strength mechanism.

The effect of composition on properties has also been investigated for the heteroarm, star-branched thermoplastic elastomers [(PS)₂MDDPE((PB-*b*-PS)₂; S₂(BS)₂] (see Section 4.2.4) whose structures are schematically shown below in comparison with S–B–S [26–29].

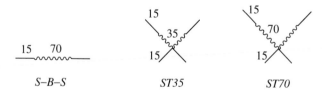

| S–B–S | ST35 | ST70 |

The stress–strain properties for these star polymers, in comparison to an S–B–S linear analog, are shown in Fig. 4.10. These results are comparable to the data for the corresponding linear S–B–S polymer shown in Fig. 4.9, that is, there is little effect of branching on tensile properties of these thermoplastic elastomers.

4.3.3.3   Substituting α-Methylstyrene for Styrene in S–I–S

In view of the temperature limitations imposed by the glass transition temperature of the polystyrene in the triblocks (these materials show a serious loss of strength even at 60 °C), it was of interest to substitute α-methylstyrene (mS) for styrene, since poly-α-methylstyrene has

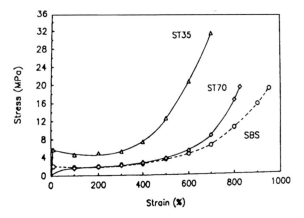

Figure 4.10   Stress–strain curves for heteroarm star-branched $S_2(B–S)_2$ thermoplastic elastomers. (From Ref. [28]). $ST70 = (15)_2–(70–15)_2$; $ST35 = (15)_2–(35–15)_2$; $SBS = 15–70–15$

a T_g of about 165 °C instead of the 105 °C for polystyrene. α-Methylstyrene is polymerizable by organolithiums. Because of its low "ceiling" temperature, the polymerization of the α-methylstyrene [21] had to be carried out at reduced temperature, using polar solvents to accelerate the rate. This also then required the use of a dilithium initiator [21] (Method 3 described previously), which yielded excellent results and a mS–I–mS triblock of narrow MWD and high purity.

The stress–strain properties of this polymer are shown in Fig. 4.10, compared with those of an S–I–S triblock of very similar composition and architecture. It is most interesting to note the substantially higher modulus of the mS–I–mS triblock, providing a strong confirmation of the idea that the plastic domains are the principal stress bearers, since poly(α-methylstyrene) has the higher tensile modulus. Furthermore, Fig. 4.11 shows the marked superiority of the

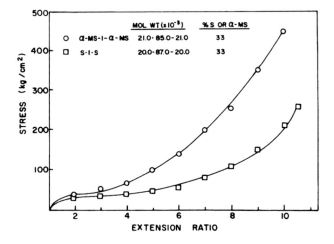

Figure 4.11   Tensile properties of mS–I–mS triblock copolymers ($1 \, kg \cdot cm^{-2} = 0.1$ MPa). (Reproduced from Ref. [15], with permission of John Wiley & Sons)

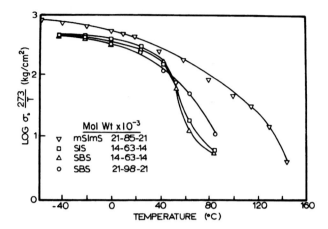

Figure 4.12   Effect of temperature on the tensile strength of triblock copolymers (1 kg · cm⁻² = 0.1 MPa). (Reproduced from Ref. [15], with permission of John Wiley & Sons)

mS–I–mS polymer at elevated temperatures, its tensile strength still being substantial (7 MPa (1000 psi)) even at 100 °C! The dependence of tensile strength on the temperature also points to the plastic domains as the key factor in the strength of the network.

It is interesting, in this connection, to compare the temperature dependence of tensile strength of conventional rubber networks with that of these triblock copolymers. According to the viscoelastic theory of tensile strength of elastomers [44], the latter is inversely

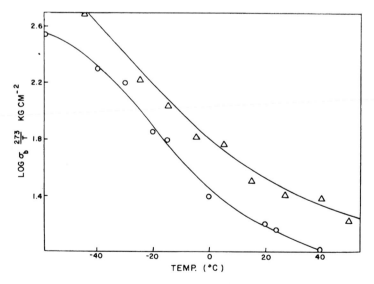

Figure 4.13   Effect of temperature on the tensile strength of gum vulcanizates of SBR (Δ) and polybutadiene (O) (1 kg · cm⁻² = 0.1 MPa). (From Ref. [46], with permission.)

dependent on the difference between the test temperature and the glass transition temperature (T_g) of the elastomer, that is, the S–I–S should be considerably stronger than the S–B–S based on a T_g of about $-65\ °C$ for the polyisoprene and a T_g of $-95\ °C$ for the polybutadiene. Yet, as Fig. 4.12 shows, both the S–B–S and S–I–S curves lie very close together. In contrast, the tensile strength vs. temperature curves for conventional, sulfur-vulcanized polybutadiene and SBR ($T_g \sim -50\ °C$), as shown [45] in Fig. 4.13, show the expected difference. All of these differences in tensile behavior illustrate quite unequivocally the strong dependence of the properties of these thermoplastic elastomers on the nature of the plastic domains that hold the network together.

4.4   Research on Other Anionic Triblock Copolymers

Since the advent of the S–D–S triblock copolymers as novel thermoplastic elastomers, it was realized that other heterophase polymers containing "hard" thermoplastic domains dispersed in, and chemically bonded to, an elastomer matrix should also behave similarly. In fact, it was soon recognized that some of the polyurethane elastomers, which were developed much earlier, represented similar heterophase systems with the polyester or polyether chains being the soft, continuous phase (see Chapter 2). Since then, of course, there has been a rapid development of several analogous thermoplastic elastomers, based on block or graft polymers such as polyesters, polyamides and polyolefins, among others (see Chapters 8, 9 and 5).

We were similarly interested, in our laboratories, in exploring other triblock copolymers that could be prepared by anionic polymerization under rigorous conditions and thus offer new systems for study of their structure–property relationships.

4.4.1   Triblock Copolymers of α-Methylstyrene and Propylene Sulfide

A successful synthesis [18] of a triblock copolymer having poly(α-methylstyrene) end blocks and a poly(propylene sulfide) center block (mS–PS–mS) was accomplished. The method involved initiation by ethyllithium, and sequential addition of α-methylstyrene and propylene sulfide, followed by coupling of the chain ends by phosgene. This use of Method 2, as described previously, was necessary since the lithium thiolate chain ends cannot initiate the polymerization of α-methylstyrene to form the second end block. The success of this method is demonstrated by the gel permeation chromatograms in Fig. 4.14, all of which show a satisfactory narrow MWD, but include very small shoulders corresponding to unlinked (mS–PS) diblock.

The uniaxial tensile properties of these triblock copolymers, containing 20, 30, and 40 wt% α-methylstyrene end blocks, are shown in Fig. 4.15, and indicate typical behavior similar to that of the S–I–S triblocks. However, their tensile strength is substantially lower, especially considering that the end blocks consist of poly(α-methylstyrene) (see Fig. 4.10). Possibly this is partly due to the small fraction of diblock, visible in Fig. 4.14. However, in the absence of any knowledge about the morphology of these polymers (these saturated polymers were not amenable to osmium tetroxide staining for electron microscopy), it is difficult to

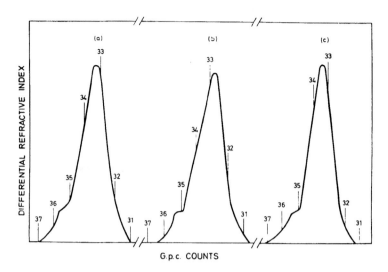

Figure 4.14   Gel permeation chromatograms of mS–PS–mS triblock copolymers. Wt% α-methylstyrene: (a) 20, (b) 30, (c) 40. (Reprinted with permission from Ref. [18]. Copyright 1971 American Chemical Society)

draw any definitive conclusions about these differences in tensile strength. Furthermore, other factors, such as interphase adhesion, which is discussed in a later section, may also play a role.

It should be noted here that the test samples used for the data in Fig. 4.15 were, as usual, cut from solution-cast films. However, it was also observed that these polymers could not be compression molded, in view of the high temperature required for flow of the mS–PS–mS

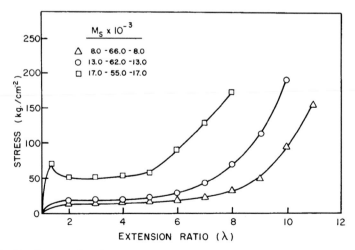

Figure 4.15   Tensile properties of mS–PS–mS triblock copolymers. Wt% α-methylstyrene: (Δ) 20, (○) 30, (□) 40 (1 kg·cm^{-2} = 0.1 MPa). (Reprinted with permission from Ref. [18]. Copyright 1971 American Chemical Society)

triblock copolymer (≈ 200 °C) which led to decomposition. Presumably, the thermally weak link was the central dithiocarbonyl group (–S–CO–S–) resulting from the coupling reaction. In this respect, these polymers behaved in analogous fashion to the mS–I–mS triblocks, which also could not be compression molded, due to decomposition of the polyisoprene center block.

4.4.2   Triblock Copolymers Based on Polysiloxanes

Even though it was found that poly(α-methylstyrene) end blocks gave triblock copolymers with good strength at elevated temperatures, this could not be put to practical use because the center blocks (e.g., polydienes) were not able to withstand the higher molding temperatures required. In this connection, a new candidate for the rubbery center block was considered, that is, a polysiloxane. The anionic polymerization of cyclic siloxanes has been known for some time, but this required the use of strong bases, for example, KOH, and was notorious for its bond interchange tendencies. However, the more recent discovery [47] that the use of the less well-known cyclic trimer, hexamethylcyclotrisiloxane, instead of the use of the corresponding cyclic tetramer, made it possible to use organolithium initiators and thus to obtain a "living" polymer with no noticeable bond interchange. In view of the known high-temperature stability of polysiloxanes, this method was considered as feasible for the synthesis of triblock copolymers based on polysiloxanes [48].

Because the block copolymerization of vinyl monomers and cyclic siloxanes can only be "unidirectional," Method 2, using *sec*-butyllithium as initiator, was used. Either styrene or α-methylstyrene was first added, followed by the cyclic trimer, hexamethylcyclo-trisiloxane, to form the diblocks, mS–D_x/2. The latter were then coupled by reacting with dimethyldichloro-silane, so that no heterogeneous chemical groups were introduced in the chain. Experimental details are described elsewhere [48]. A gel permeation chromatogram of a typical mS–D_x–mS triblock copolymer, containing 40 wt% α-methylstyrene, is shown in Fig. 4.16. The MWD is not quite as narrow as that of the S–D–S triblocks, but is still quite satisfactory. It is interesting to note the absence of any unlinked diblocks, which might ordinarily be expected (see Fig. 4.14), and instead to find the presence of a small amount of free poly-α-methylstyrene (at count 57), obviously due to some termination occurring during addition of the cyclic trisiloxane.

In this case, it was possible to obtain transmission electron microphotographs of thin films, since the silicon atoms are much denser than carbon atoms. A typical example is shown in Fig. 4.17 for a triblock containing 40 wt% poly-α-methylstyrene end blocks. Here the white domains represent the poly-α-methylstyrene. The same type of orderly phase separation can be noted, but the contrast in the photomicrograph is not as good as in the case of the S–D–S polymers. This either represents a less efficient phase separation, or is an artifact of the electron microscopy.

The tensile properties of solution-cast films of a series of these triblocks are shown in Fig. 4.18. The increase in modulus and tensile strength with increase in "hard block" content is again observed, but the tensile strength values are disappointingly low, compared to the values obtained for the mS–I–mS triblock (see Fig. 4.10). It is difficult to offer any definitive explanation for the low tensile strengths, since several possibilities may apply. One concerns

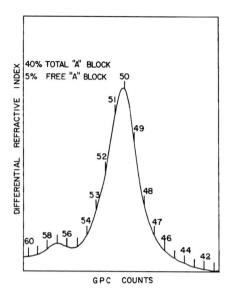

Figure 4.16   Gel permeation chromatogram of mS– D_x– mS–6 triblock copolymer (40 wt% α- methylstyrene). (Reproduced from Ref. [48], with permission of John Wiley & Sons)

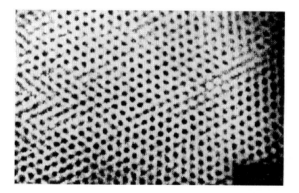

Figure 4.17   Transmission electron microphotograph of ultrathin (≈ 50 nm) section of cast film of mS–D_x– mS–6 (cut parallel to film surface). (Reproduced from Ref. [48], with permission of John Wiley & Sons)

the efficiency of phase separation, already mentioned previously, although the disparity in solubility parameters of poly-α-methylstyrene (8.8) [49, 50] and the polydimethylsiloxane (7.5) [41] is even greater than in the case of mS–I–mS triblocks. Another possibility may be based on the much lower viscoelasticity of polydimethylsiloxanes, which may make it impossible for the elastic chains to transfer the stress to the "hard domains," thus changing the mechanism of failure. An alternative explanation may invoke the interphase adhesion between the domains and the matrix, as discussed in a later section. In this connection, it should be noted that the tensile strength of these triblocks, containing 30% to 40% end blocks, compares very favorably with that of conventional silicone rubber vulcanizates, reinforced by fine silica fillers.

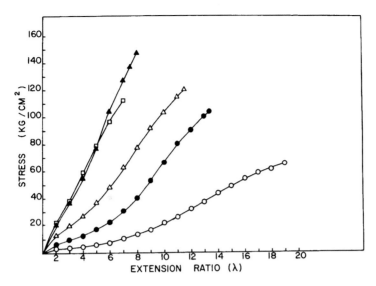

Figure 4.18   Tensile properties of mS–D$_x$–mS triblock copolymers. (Reproduced from Ref. [48], with permission of John Wiley & Sons)

MW ($\times 10^{-3}$)	% End block
◯ 12.5–141.3–12.5	15
● 11.9–80.4–11.9	22
△ 2.9–60.8–12.9	31
☐ 12.5–38.4–12.5	40
▲	36 (fractionated sample of 40% end block polymer, with no free poly-α-MS)

4.5   Incompatibility and Processability in Triblock Copolymers [46]

The various well-characterized A–B–A triblock copolymers described above, all prepared by anionic polymerization under rigorous experimental conditions, offer an opportunity to draw some conclusions about the relationship between heterophase incompatibility and processability of these materials. Thus far, only uniaxial tensile data have been discussed. However, it has also been postulated, in a previous section, that the degree of incompatibility between the two phases affects the flow properties, since this involves the heat of mixing of the two phases. In other words, although incompatibility between the blocks is desirable for phase separation, too high an incompatibility can introduce flow problems.

Because of the greater difference in solubility parameters between polydimethylsiloxanes and hydrocarbon polymers, triblock copolymers containing polydimethylsiloxane center blocks could conveniently be compared to triblocks containing, say, polyisoprene center blocks. For this purpose, the flow properties were determined by measuring the compliance rate of the polymers from the deformation of a sample compressed between parallel plates.

Table 4.5 Flow Properties of Triblock Copolymers

Polymer[a] rate	M_n ($\times 10^{-3}$)	$\Delta\delta$[b] ($cal^{1/2}$ $cm^{-3/2}$)	Temperature (°C)	Compliance (MPa^{-1} s^{-1})
S–I–S	102	1.1	130	0.25
S–D$_x$–S	53	1.7	130	0.04
mS–I–mS	132	0.7	200	0.50
mS–D$_x$–mS	87	1.3	200	0.06

[a] All polymers at 30 wt%, end block
[b] Difference between solubility parameters of blocks
Source: Ref. [45]

The experimental details are described elsewhere [51]. The results are shown in Table 4.5. In addition to the data shown in this table, it should be noted that the values of M_e (molecular weight between entanglements), which govern the melt viscosity of the elastic center blocks, are 7000 and 12,000, respectively, for polyisoprene [36] and polydimethylsiloxane [52].

The much poorer flow properties (lower compliance values) of the polydimethylsiloxane-containing triblocks are immediately apparent, despite the much higher M_e values and lower overall molecular weights (M_n) of the polydimethylsiloxane center blocks. In other words, because of its higher M_e and lower M_n, a polydimethylsiloxane of the same molecular weight as the center block in the above S–D$_x$–S and mS–D$_x$-mS polymers would have a much lower viscosity than a polyisoprene having the same molecular weight as the center block in the corresponding S–I–S and MS–I–mS copolymers. These data illustrate dramatically how the incompatibility dominates the flow properties of triblock copolymers.

The flow properties of block copolymers are also very sensitive to molecular architecture. The dynamic melt viscosities, η^*, were measured at 165 °C for linear S–B–S and two heteroarm, star-shaped triblock copolymers and the results are shown in Fig. 4.19 [28]. Thus, although branching had very little effect on the tensile properties of block copolymers as shown in Fig. 4.10, it gave a dramatic decrease in melt viscosity. The

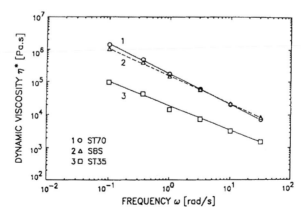

Figure 4.19   Dynamic melt viscosities of linear SBS and star-branched S$_2$(BS)$_2$ thermoplastic elastomers. (From Ref. [28]). ST70 = (15)$_2$–(70–15)$_2$; ST35 = (15)$_2$–(35–15)$_2$; SBS = 15–70–15

branched polymer ST70 ($M_n = 199 \times 10^3$ g/mol) exhibits almost the same melt viscosity as a linear S–B–S ($M_n = 103 \times 10^3$ g/mol) with one half of the molecular weight. Furthermore, ST35 ($M_n = 123 \times 10^3$ g/mol) exhibits a melt viscosity that is a factor of 10 lower than that of the linear S–B–S with approximately the same molecular weight. In considering the balance between properties and processability, it is noteworthy that these star-branched polymers also exhibit excellent tensile properties as shown in Fig. 4.10.

4.6   Interphase Adhesion and Tensile Strength [46]

It has already been pointed out in this discussion that the "hard phase" that separates from the soft, rubbery matrix forms domains that have the dimensions of reinforcing fillers in rubber (≈ 30 nm). Recent studies in our laboratories [53] have been concerned with the use of model polymeric fillers, for example, polystyrene, in rubber vulcanizates, and the effect of uniaxial strain on the extent of dewetting of such fillers, using sensitive density measurements to determine the volume dilation. Such measurements have shown that there is an inverse relationship between such volume dilation and tensile strength of the composite, indicating a strong effect of the filler–elastomer adhesion.

The poor tensile strength of the polysiloxane-based triblock copolymers described previously raises questions about the possible effect of the well-known low surface free energy of polysiloxane on the interphase adhesion. To determine whether this interphase adhesion does indeed play a role in the tensile strength, volume dilation measurements were carried out, as a function of strain, on a series of triblock copolymers [46]. Experimental details can be found elsewhere [51]. As before [53], the volume dilation ($\Delta V/V$) was found to increase linearly as a function of extension ratio (λ), so that the slopes of the $\Delta V/V$ vs. λ plot, that is, $\Delta V/V\lambda$, could be defined as the "specific dilation" parameter for any given sample. These can then be considered as an inverse measure of the adhesion between the two phases.

What is even more interesting is to relate the tensile strength, the strain dilation, and the calculated work of adhesion between the two phases. The latter, W_a, as defined by Shafrin and Zisman [54], is expressed as

$$W_a = (2 + b_d\gamma_d)\gamma_m - b_d(\gamma_m)^2 \tag{4.2}$$

where

λ_d and λ_m are the critical surface free energies of the dispersed phase and the matrix, respectively.
b_d is defined by the equation: $\cos \theta = \text{constant} - b_d\gamma_d$, and
$\theta = $ contact angle between phases.

The required values of W_a for the various block copolymers were calculated from the published values of critical surface free energy (γ) for the polymer blocks, as shown in Table 4.6.

The correlation between tensile strength, calculated interphase adhesion (W_a) values, and measured volume dilation ($dV/V\lambda$) values is shown in Table 4.7 for a series of triblock copolymers. All of the polymers, with the exception of the Kraton D1102, were prepared in

Table 4.6 Critical Surface Free Energies of Polymers

Polymer	γ (mN/m)
Polydimethylsiloxane [54]	24
Polybutadiene [54]	31
Polyisoprene [55]	31
Polystyrene [56, 57]	33
Poly(α-methylstyrene) [57, 58]	36

Table 4.7 Effect of Interphase Adhesion on Tensile Strength of Triblock Copolymers

Polymer	MW ($\times 10^{-3}$)	W_a (mJ/m^2)	Specific dilation ($\Delta V/V\lambda \times 10^3$)	Tensile strength (MPa)
S–D$_x$–S–1	8–37–8	54	6.90	0.6
S–D$_x$–S–2	16–76–16	54	—	1.8
mS–D$_x$–mS	13–61–13	57	4.79	7.4
Kraton D1102[a]	10–55–10[a]	64	3.57	26
mS–I–mS	21–85–21	67	2.86	45

[a] S–B–S; Ref. [59]

our laboratories, and were chosen because they all had a roughly equivalent "end block" content (hard phase) of about 30%.

It can be seen from Table 4.7 that, at least qualitatively, there is an excellent correlation between interphase adhesion, strain dilation, and tensile strength, that is, better adhesion leads to lower dilation and higher tensile strength. The much lower tensile strength of the S–D$_x$–S–1 sample as compared to the S–D$_x$–S–2 despite the same W_a values, can be easily explained by the much lower molecular weight of the end blocks of the S–D$_x$–S–1 which would lead to more phase mixing, as discussed previously (see Table 4.3). The same type of correlation between low strain dilation and high tensile strength was found in the studies of model fillers in rubber vulcanizates [53].

4.7   Triblock Copolymers with Crystallizable End Blocks

It has already been pointed out in a previous part of this chapter that triblock thermoplastic elastomers based on an amorphous two-phase system embody an inherent structural balance, because phase separation is caused by incompatibility, and an increase in the latter, while favoring good phase separation, decreases processability, that is, melt flow. Therefore, it was of interest to extend the study of structure–property relationships of anionic triblock copolymers to systems in which the two types of blocks might be compatible in the melt, but the end blocks could form crystalline domains on cooling. Some of this work has been reported recently [46, 60] and one of the systems studied will be discussed here, as illustrative of this type of approach.

This particular triblock involved the *sec*-butyllithium-initiated polymerization of a single monomer, butadiene, in such a way that the end blocks were high in 1,4 structure (> 90%), while the center block had about 45% of 1,2 chain units. This was accomplished by preparing the first polybutadiene end block in hydrocarbon solution using *sec*-butyllithium as initiator, then adding 5% diethyl ether before polymerizing the second charge of butadiene; the final triblock copolymer was formed by coupling the living diblock with dimethyldichloro-methane. Subsequent hydrogenation of the polymer solution with the Falk catalyst [61] converted the end blocks into a type of "linear low-density" polyethylene, while the center block became a random copolymer of ethylene and 1-butene. Hence the hydrogenated end blocks were crystallizable while the hydrogenated center block was an elastomer. The experimental details of synthesis and characterization are described elsewhere [60, 62], but suffice it to say that the original polybutadiene triblock had the desired narrow MWD, that hydrogenation efficiency was very close to 100%, and that no significant chain scission occurred with the special hydrogenation method used. Furthermore, the end blocks were found to crystallize to form spherulites, similar to those of conventional, low-density polyethylene, and showed a melting point of 107 °C, again very similar to that of commercial low density polyethylene.

A series of four such triblock copolymers was prepared, and their architecture is described in Table 4.8. It can be seen that the four polymers contained two levels of end block content ("hard phase") and two levels of molecular weight. The latter variation was thought to be especially important, in view of the known effect of molecular weight on crystallization; and optical microscopy indeed showed [60, 62] that the higher molecular weight end-blocks formed better spherulites.

The uniaxial tensile behavior of these four polymers is shown in Fig. 4.20. As expected, the two polymers containing 60% "hard block" showed a higher modulus than the 30% end block polymers, regardless of molecular weight, and they also showed more of a yield point at initial stretch. In contrast, however, the two polymers with the shorter end blocks (B–60–19 and B–30–18) both showed similar (and lower) tensile strengths. Since the two polymers containing 60% end blocks (B–60–19 and B–60–54) both exhibited more plastic than elastic behavior, that is, high unrecovered deformation on stretching, the only real thermoplastic elastomers were those triblocks containing 30 wt% end blocks; and of these, the B–30–55 (high MW) had excellent tensile strength, thus demonstrating the critical importance of molecular weight on formation of the crystalline domains. It should be noted that this effect

Table 4.8 Molecular Structure of Hydrogenated Polybutadiene Triblocks

Polymer	Wt% end blocks	MW ($\times 10^{-3}$)
B–60–19	60	19–25–19
B–30–18	30	18–58–18
B–60–54	60	54–72–54
B–30–55	30	55–257–55

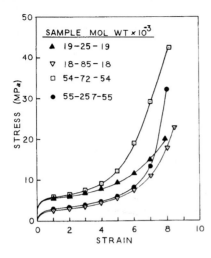

Figure 4.20   Tensile properties of hydrogenated polybutadiene triblock copolymers. (Reprinted with permission from Ref. [60]. Copyright 1982, American Chemical Society)

was not so critical in the amorphous, styrene-based triblocks, except at the low molecular weights, where phase separation was reduced.

Two other aspects about the behavior of these crystallizable triblocks are worthy of note. They seemed to show a somewhat higher degree of tensile "set" (unrecovered deformation on stretching) than their styrene-based counterparts. Although no direct morphological data were obtained, it appears that the crystalline domains presumably suffer a greater distortion than the amorphous, glassy polystyrene domains. This is perhaps not surprising, considering the tendency of fibers to "cold draw." However, it should also be noted that these hard phases

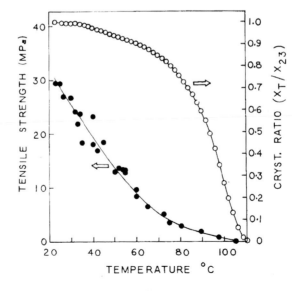

Figure 4.21   Comparative effect of temperature on tensile strength and crystallinity of B–30–55 triblock copolymers. (Reprinted with permission from Ref. [60]. Copyright 1982, American Chemical Society)

exhibit only about 50% crystallinity; thus, 50% of the hard phase is actually deformable, "elastomeric" polyethylene.

The second aspect concerns the tensile strength of these crystalline triblocks at elevated temperature, where they seem to be somewhat better than the S–I–S or S–B–S triblocks (but not the mS–I–mS type). Thus, at 60 °C, the tensile strength of the B–30–55 triblock is still at the respectable level of 10 MPa (1450 psi) (see Fig. 4.19), compared to only 2 MPa (290 psi) for a similar S–I–S triblock (see Fig. 4.11). An interesting aspect of the effect of temperature on these crystalline triblocks is shown in Fig. 4.19, where both the tensile strength and the degree of crystallinity (as measured by DTA) of the B–30–55 polymer are plotted against temperature. Thus it can be seen that, at 60 °C, where the sample has lost two thirds of its strength, it has lost only 10% of its crystallinity, while at 80 °C, where its strength is only 10% of the original, it still retains almost 80% of its original crystalline content. It would be interesting to compare the behavior of highly crystalline materials, for example, fibers in this regard.

References

1. G. Holden and R. Milkovich, Belgian Patent 627, 652 (July 29, 1963); U.S. Patent 3,265,765 (August 9, 1965, filed January 29, 1962); G. Holden, E.T. Bishop, and N.R. Legge, *Proc. Int. Rubber Conf.* 1967, Maclaren and Sons, London (1968), pp. 287–309; *J. Polym. Sci. Pt. C 26*, 37 (1969); N.R. Legge, *Chemtech 13*, 630 (1983)
2. F.S. Bates and G.H. Fredrickson, *Annu. Rev. Phys. Chem. 41*, 525 (1990)
3. R.P. Quirk, D.J. Kinning and L.J. Fetters, In *Comprehensive Polymer Science, Vol. 7, Specialty Polymers and Polymer Processing.* S.L. Aggarwal (Ed.) (1989) Pergamon Press, Oxford, U.K., p. 1
4. M. Morton, J.E. McGrath, and P.C. Juliano, *J. Polym. Sci. Pt. C 26*, 99 (1969)
5. M. Morton and L.J. Fetters, *Rubber Chem. Technol. 48*, 359 (1975)
6. L.J. Fetters, *Macromolecular Syntheses, Collective Vol. 1* (1977) John Wiley & Sons, New York, p. 463
7. M. Morton, *Block Polymers.* S.L. Aggarwal (Ed.) (1970) Plenum Press, New York, p. 1
8. M. Morton, *Anionic Polymerization: Principles and Practice* (1983) Academic Press, New York
9. R.P. Quirk and J.-J. Ma, *Polym. Int. 24*, 197 (1991)
10. R.P. Quirk and B. Lee, *Polym. Int. 27*, 359 (1992)
11. H.L. Hsieh and O.F. McKinney, *Polym. Lett. 4*, 843 (1966)
12. M. Morton and F.R. Ells, *J. Polym. Sci. 61*, 25 (1962)
13. R. Ohlinger and F. Bandermann, *Makromol. Chem. 181*, 1935 (1980)
14. D.J. Worsfold, *J. Polym. Sci. A1 5*, 2783 (1967)
15. M. Morton, *Encyclopedia of Polymer Science and Technology, Vol. 15* (1971) John Wiley & Sons, New York, p. 508
16. R.P. Quirk and T.-L. Huang, In *New Monomers and Polymers.* B.M. Culbertson and C.U. Pittman, Jr., (Eds.) (1984) Plenum Press, New York, p. 329
17. H.L. Hsieh, *Rubber Chem. Technol. 49*, 1305 (1976)
18. M. Morton, R.F. Kammereck and L.J. Fetters, *Br. Polym. J. 3*, 120 (1971); *Macromolecules 4*, 11 (1971)
19. R.N. Young, R.P. Quirk, and L.J. Fetters, *Adv. Polym. Sci. 56*, 1 (1984)
20. F. Bandermann, H.-D Speikamp, and L. Weigel, *Makromol. Chem. 186*, 2017 (1985)
21. L.J. Fetters and M. Morton, *Macromolecules 2*, 453 (1969)
22. T.E. Long, A.D. Broske, D.J. Bradley, and J.E. McGrath, *J. Polym. Sci. Polym. Chem. 27*, 4001 (1989)
23. L.-K. Bi and L.J. Fetters, *Macromolecules 8*, 90 (1975); *ibid. 9*, 732 (1976)
24. N. Hadjichristidis and L.J. Fetters, *Macromolecules 13*, 191 (1980)
25. D.B. Alward, D.J. Kinning, E.L. Thomas, and L.J. Fetters, *Macromolecules 19*, 215 (1986)

26. R.P. Quirk and F. Ignatz-Hoover, In *Recent Advances in Anionic Polymerization*. T.E. Hogen-Esch and J. Smid (Eds.) (1987) Elsevier, New York, p. 393

27. R.P. Quirk, B. Lee, and L.E. Schock, *Makromol. Chem. Macromol. Symp. 53*, 201 (1992)

28. R.P. Quirk and B. Lee, *Polym. Prep. (Am. Chem. Soc. Div. Polym. Chem.) 32(3)*, 607 (1991); B. Lee, Ph.D. Dissertation, University of Akron (1991)

29. R.P. Quirk, T. Yoo, and B. Lee, *J. Macromol. Sci. Pure Appl. Chem. A31*, 911 (1994)

30. J.F. Beecher, L. Marker, R.D. Bradford, and S.L. Aggarwal, *J. Polym. Sci. Pt. C 26*, 117 (1969)

31. G.E. Molau (Ed.), *Colloidal and Morphological Behavior of Block and Graft and Copolymers* (1971) Plenum Press, New York

32. B.R.M. Gallot, *Adv. Polym. Sci. 29*, 85 (1978)

33. I. Goodman (Ed.), *Developments in Block Copolymers, Vol. 1* (1982) Elsevier Applied Science, London, U.K.

34. M.J. Folkes (Ed.), *Processing, Structure and Properties of Block Copolymers* (1985) Elsevier Applied Science, New York

35. E.L. Thomas, D.B. Alward, D.J. Kinning, D.C. Martin, D.L. Handlin, Jr., and L.J. Fetters, *Macromolecules 19*, 2197 (1986)

36. L.J. Fetters, *J. Res. Natl. Bur. Stand. 69*, A 33 (1965)

37. D.J. Meier, *J. Polym. Sci. Pt. C 26*, 81 (1969)

38. S.H. Goh, Ph. D. Thesis, University of Akron (1971)

39. K.W. Scott, *J. Polym. Sci. 58*, 517 (1962)

40. R.P. Quirk, M.T. Sarkis, and D.J. Meier, In *Advances in Elastomers and Rubber Elasticity*. J. Lal and J.E. Mark (Eds.) (1986) Plenum Press, New York, p. 143

41. E.A. Grulke, In *Polymer Handbook*, 3rd edit. J. Brandrup and E.H. Immergut (Eds.) (1989) John Wiley & Sons, New York, p. 519

42. R.F. Fedors, *J. Polym. Sci. Pt. C 26*, 189 (1969)

43. H. Morawetz, *Macromolecules in Solution* (1961) Wiley–Interscience, New York, p. 41

44. J.C. Halpin, *Rubber Chem. Technol. 38*, 1007 (1965)

45. M. Morton, In *Multicomponent Polymer Systems, Advances in Chemistry Series No. 99*; N.A.J. Platzer (Ed.) (1971) American Chemical Society, Washington, D.C., p. 490

46. M. Morton, *Rubber Chem. Technol. 56*, 1096 (1983)

47. J.C. Saam, D.J. Gordon, and S. Lindsey, *Macromolecules 3*, 1 (1970)

48. M. Morton, Y. Kesten, and L.J. Fetters, *Appl. Polym. Symp. No. 26*, 113 (1975)

49. P.A. Small, *J. Appl. Chem. 3*, 71 (1953)

50. R.F. Fedors, *Polym. Eng. Sci. 14(2)*, 147 (1974)

51. M.F. Tse, Ph. D. Dissertation, University of Akron (1979)

52. J.D. Ferry, *Viscoelastic Properties of Polymers*, 2nd edit. (1970) John Wiley & Sons, New York, p. 406

53. M. Morton, N.K. Aggarwal, and M. Cizmecioglu, In *Copolymers, Polyblends and Composites*. D. Klempner and K.C. Frisch (Eds.) (1983) Plenum Press, New York; M. Morton, R.J. Murphy, and T.C. Cheng, In *Copolymers, Polymers and Composites, Advances in Chemistry Series No. 142*. N.A.J. Platzer (Ed.) (1975) American Chemical Society, Washington, D.C., p. 409

54. E.G. Shafrin and W.A. Zisman, In *Contact Angle, Wettability and Adhesion, Advances in Chemistry Series No. 43*. F.M. Fowkes (Ed.) (1964) American Chemical Society, Washington, D.C., p. 145

55. L.H. Lee, *J. Polym. Sci. Pt. A2 5*, 1103 (1967)

56. W.A. Zisman, Ref. 55, p. 4

57. L.H. Lee, *J. Appl. Polym. Sci. 12*, 719 (1968)

58. A.H. Ellison and W.A. Zisman, *J. Phys. Chem. 58*, 503 (1954)

59. A. Ghijsels and J. Raadsen, *Pure Appl. Chem. 52*, 1361 (1980)

60. M. Morton, N-C. Lee, and E.R. Terill, In *Elastomers and Rubber Elasticity, ACS Symposium Series No. 193*. J.E. Mark and J. Lal (Eds.) (1982) American Chemical Society, Washington, D.C., p. 101

61. J.C. Falk and R.J. Schott, *Macromolecules 4*, 152 (1971)

62. N-C. Lee, Ph. D. Dissertation, University of Akron (1982)

5   Polyolefin-Based Thermoplastic Elastomers

E.N. Kresge

5.1 Introduction   102

5.2 Random Block Polymers.   103
    5.2.1 Ethylene–Propylene Copolymers   103
    5.2.2 Ethylene–Higher α-Olefin Copolymers.   105
    5.2.3 Propylene–Higher α-Olefin Copolymers   106
    5.2.4 Random Stereoblock Polypropylene   107
        5.2.4.1 Heterogeneous Catalysts for Stereoblock Polypropylene.   107
        5.2.4.2 Homogeneous Catalysts for Stereoblock Polypropylene   109
        5.2.4.3 Crystallization of Stereoblock Polypropylene   111

5.3 Block Copolymers   112
    5.3.1 Ziegler–Natta Block Copolymers   113
    5.3.2 Hydrogenated Diene Block Copolymers.   115

5.4 Graft Copolymers   117
    5.4.1 Polyolefins Grafted with Crystalline Chains.   118

5.5 Polyolefin Blend Thermoplastic Elastomers.   119
    5.5.1 Morphology of Polyolefin Blend Thermoplastic Elastomers   120
    5.5.2 Mechanical Properties of Polyolefin Blends.   121
    5.5.3 Blends of Polyolefins with Other Polymers   123
    5.5.4 Applications for Polyolefin Blend Thermoplastic Elastomers   124

Acknowledgments   125

References   125

5.1   Introduction

Polyolefins are very widely used, both as elastomers and rigid thermoplastics. Their attributes of chemical inertness, low density, and low cost offer major advantages over many other polymers. The olefin monomers are basic chemicals and are readily available. Olefin monomer polymerization has been a considerable scientific and technological effort in the areas of catalysis, reactor engineering, and structure–property relationships. Compounding of polyolefins has been developed for ethylene–propylene–diene terpolymer (EPDM), ethylene–propylene random copolymer (EPM), butyl rubber, polyethylene, and isotactic polypropylene (iPP). Moreover, there were early insights into polymeric structures that could give elastic properties without chemical crosslinking [1]. All of these factors provided the early impetus for polyolefin-based thermoplastic elastomers (TPEs).

There are several distinct types of polyolefin-based TPEs. These include:

- Random block copolymers, for example, ethylene α-olefin copolymers;
- block copolymers, for example, hydrogenated butadiene–isoprene–butadiene block copolymer;
- stereoblock polymers, for example, stereoblock polypropylene;
- graft copolymers, for example, polyisobutylene–g–polystyrene and EPDM–g–pivalo-lactone; and
- blends, for example, blends of EPM with iPP and dynamically vulcanized blends of EPDM with a crystalline polyolefin.

All of these TPEs exhibit rubberlike characteristics and are melt processable in typical thermoplastic processing equipment. At present the polyolefin blends have gained the most widespread commercial importance and are marketed worldwide. Random block olefin copolymers that are based on metallocene polymerization catalysis have been introduced recently.

Most of the above polyolefin TPEs depend on crystallization of polymer chains to produce an elastomeric structure. In the random block copolymers, which are structurally similar to thermoplastic polyurethane (TPU) random block copolymers, ethylene sequences long enough to crystallize at use temperature act as physical crosslinks for the amorphous elastic chain segments. These block copolymers are similar to polystyrene–*block*-polybuta-diene–*block*-polystyrene (S–B–S) block copolymers in that the hard blocks associate into domains. However, these domains are crystalline whereas those formed by the polystyrene blocks are glassy. In stereoblock copolymers changes in intrachain tacticity provide crystal-line and amorphous sequences.

Graft copolymers have been produced in many varieties. Typically, a polyolefin rubbery backbone is grafted with polymer chains that will form a crystalline or a glassy phase resulting in a physically crosslinked network. In most block and graft polymers the elasticity is accomplished by having rubbery chains connected by reversible, physical crosslinks. The combination of physical crosslinks and trapped entanglements among the rubbery chains results in an elastic network somewhat like a chemically crosslinked rubber and the retroactive force is entropic in nature [2].

There are two distinct types of polyolefin blends that are TPEs: co-continuous phase blends and dynamically vulcanized blends. In the co-continuous phase blends both the elastomeric phase and the crystalline polyolefin phase are continuous and the longer time scale elastic properties depend mostly on the crystalline polyolefin phase. Both phases flow during processing. In the dynamically vulcanized blends the elastomeric phase is crosslinked and is discontinuous. The semicrystalline polyolefin phase in these blends is continuous and surrounds the crosslinked elastomeric phase. Elastic properties are the result of both the semicrystalline continuous polyolefin phase and the chemically crosslinked elastomer phase. The deformation and recovery in the semicrystalline phase are enthalpic in nature. During processing, flow takes place only in the polyolefin phase. Dynamically vulcanized blends are the subject of a separate chapter (Chapter 7).

5.2   Random Block Polymers

Random copolymers containing both crystallizable chain segments and amorphous chain segments are TPEs above the T_g of the amorphous phase. TPUs were the first TPEs to employ this type of chain structure. At temperatures above T_m, the melting point of the crystalline phase, they flow and can be fabricated. Random block copolymers based on polyethylene as the crystalline sequence have been used successfully as TPEs for some time. Thus ethylene vinylacetate copolymers produced by high-pressure free radical polymerization and containing 10 to 36 wt% vinylacetate exhibit rubberlike properties. Random block ethylene–propylene copolymers based on olefins containing crystalline ethylene sequences were introduced commercially in the 1960s [3].

Using the proper catalysts, crystalline isotactic propylene segments can also be synthesized by copolymerization with higher α-olefins to produce a TPE. The crystallinity in polypropylene can also be modified by introduction of atactic sequences in the polymer chains. These random stereoblock polymers have TPE properties.

5.2.1   Ethylene–Propylene Copolymers

The ethylene–propylene copolymers are synthesized with soluble Ziegler–Natta catalysts, for example, VOCl$_3$ with diethylaluminum chloride as the co-catalyst. The ethylene crystallinity depends on composition, interchain compositional distribution, and sequence distribution in the polymer. The stress–strain properties of an ethylene–propylene copolymer containing 16% crystallinity are shown in Fig. 5.1. The polymer undergoes yielding and drawing and further crystallization under high degrees of strain [4]. The yielding of the crystalline phase leads to permanent set on the first extension; however, the recovery properties resemble those of a crosslinked elastomer if the stress–strain behavior is confined to the strained region. This suggests that an interconnected crystalline phase is broken up into small dispersed domains during extension.

EPM has a T_g of ca. $-50\,^\circ$C. The amount of crystallinity, T_m and T_g all change with composition. Polymers with narrow intermolecular compositional distributions exhibit narrower melting behavior.

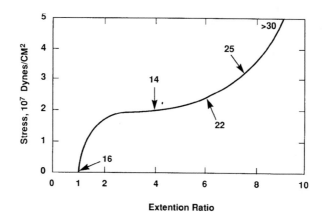

Figure 5.1   Ethylene–propylene copolymer. Dependence of degree of crystallinity on extension ratio. Stress–strain curve at $\sim 1\ \text{s}^{-1}$. (After Ref. [4])

The dependence of crystalline fraction and glass transition temperature on composition is shown in Fig. 5.2. The T_g of EPM is not detectable by Differential Scanning Calorimetry (DSC) above about 80 mol% ethylene. The fraction of crystallinity increases with ethylene content. There is considerable variation with composition owing to changes in sequence distribution that is controlled by the reactivity ratio of the catalyst used in the synthesis. In addition, some catalysts can also produce a broad intermolecular compositional distribution owing to species of different copolymerization activity.

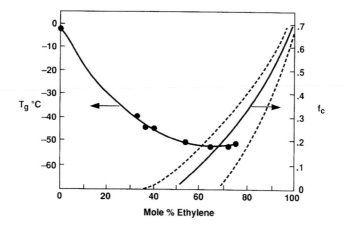

Figure 5.2   Glass transition temperature and crystallinity of ethylene–propylene copolymer. T_g is by DSC. Crystalline fraction, f_c, is by density and X-ray scattering. The f_c data show a range because of compositional and sequence distributions in the copolymer. T_g is undetectable by DSC over about 80% ethylene. (Data from Ref. [3])

5.2.2   Ethylene–Higher α-Olefin Copolymers

Copolymers of ethylene with higher α-olefins have been synthesized using Ziegler–Natta catalysts. Reactor, polymer recovery, and polymer handling limitations have made it difficult to produce elastomeric variations of these materials and this approach has been used mostly for production of linear low-density polyethylene. Higher contents of 1-butene and 1-hexene give semicrystalline, low-modulus polymers that are similar to ethylene–propylene copolymers in that the ethylene crystallinity provides the mechanism for thermoplastic processing. The polymers have been called *plastomers* because they have properties that can range from low-density polyethylenelike materials to TPEs with many of the characteristics of plasticized PVC.

Polymers of ethylene and higher α-olefins that are based on metallocene polymerization technology have recently been introduced. Polymers in the plastomer region have densities of 0.88 to about 0.90 g/cc and M_w of ca. 2×10^4 to ca. 2×10^5. The physical properties of a grade designed for flexible tubing applications are listed in Table 5.1. The major attributes of this TPE are clarity, kink resistance, and low level of extractables.

The commercial ethylene–α-olefin copolymers produced using metallocene catalysts are characterized by narrow molecular weight and compositional distributions. This results from the catalyst having a single type of polymerization site. The low T_g of the polymer ($\sim 55\,^\circ$C) and relatively high T_m result in rapid crystallization rates so that final properties are rapidly achieved and thermoplastic processing is easily carried out. Because of their easy processing, good physical properties, and the chemical inertness of polyolefins, plastomers of this type are expected to provide very useful TPEs with a range of rubberlike properties.

A random copolymer of ethylene and an aromatic α-olefin has been synthesized by hydrogenation of an emulsion styrene–butadiene copolymer (SBR) [5]. The free-radical emulsion polymerization produces a polymer containing styrene, *cis*-1,4-butadiene, *trans*-1,4-butadiene, and vinyl-1,2-butadiene units which upon hydrogenation yields a copolymer of ethylene, styrene, and 1-butene. Hydrogenation is carried out in the emulsion using a diimide-based process [6]. The polymer is a TPE that depends on ethylene crystallinity to provide a reversible crosslinking mechanism. The hydrogenated latex can also be used for casting films or dipping.

Table 5.1 Physical Properties of an Ethylene–α-Olefin Copolymer[a]

Property	Typical value	Method
Density	0.885 g/cc	ASTM D-1505
Melt index	3.8 dg/min	ASTM D-12.38
Internal haze	10% at 50 mils	ASTM D-1003
Hardness	83 A	ASTM D-2240
	35 D	
Tensile strength	20 MPa	ASTM D-638
Elongation at break	> 800%	ASTM D-638
T_m	68 °C	Peak melting point by DSC

[a] EXACT 4024, Exxon Chemical Co. Product data sheet

Table 5.2 Properties of Hydrogenated SBR[a]

	A	B
Percent	86.2	97.7
Percent crystallinity	6.0	12.3
Tensile strength (MPa)	N	12.4
Elongation at break (%)	> 1000	840
Modulus (MPa) at		
100%	1.5	2.6
200%	1.8	2.9
500%	2.2	3.7
Percent set[b]	20.0	72.5

[a] 15.7 wt% styrene, 12.9 wt% 1,2-vinyl butadiene in starting SBR
[b] Extended to 300% elongation, maintained for 10 min, then allowed to relax for 5 min before measurement

DSC data indicate a small increase in T_g on hydrogenation (for SBR-1502, from $-57\,^\circ$C to $-41\,^\circ$C). The T_m is broad, indicating a variation in methylene sequence lengths, with the highest melting at about 75 $^\circ$C.

The physical properties are listed in Table 5.2 for two copolymers. The *trans*-1,4-butadiene has been shown to hydrogenate faster than the vinyl side groups under these conditions, so the polymer with less saturation shows less polyethylene crystallinity. This is consistent with the stress–strain characteristics. The shape of the initial stress–strain curve is very similar to that shown in Fig. 5.1 for ethylene–propylene copolymers. The set data suggest that there is nonrecoverable plastic deformation of the crystalline regions.

Direct polymerization routes to make random copolymers of ethylene and aromatic α-olefins (e.g., styrene) would be expected to give TPEs with similar physical properties.

5.2.3   Propylene–Higher α-Olefin Copolymers

Like random ethylene–α-olefin copolymers, copolymers of propylene and butene with higher α-olefins have T_gs low enough to be useful elastomers. In contrast to similar copolymers based on ethylene, the polymerization catalysts employed must provide for stereo control to synthesize sequences that will crystallize [7].

γ-TiCl$_3$ with diethylaluminum chloride co-catalyst or TiCl$_4$ on MgCl$_2$ with a trialkyl aluminum co-catalyst produce isotactic propylene sequences. Copolymers of propylene with 20 to 60 wt% 1-hexene or 1-decene exhibit the properties of a TPE. The polymers have a crystalline melting of 130 $^\circ$C to 150 $^\circ$C and a T_g of -25° to $-30\,^\circ$C. Stress–strain properties are shown in Fig. 5.3 for a 1-propylene–1-hexene copolymer (50 wt%). As with the ethylene copolymers, there is yielding at low elongation followed by strain hardening. The polymers also show high permanent set on elongation, apparently due to yielding of the crystalline phase. As discussed later in this chapter in the section on blends, the properties of

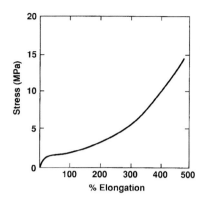

Figure 5.3   Stress–strain curve of a propylene–hexene copolymer (50% H). (After Ref. [7])

polypropylene–1-hexene copolymers can be improved by blending with isotactic polypropylene homopolymer.

5.2.4   Random Stereoblock Polypropylene

TPEs that contain random blocks of crystalline and amorphous chain units can be synthesized by introduction of nonstereospecific units in the chain. This is in contrast to the polymers discussed previously in this section, where copolymerization rather than stereospecificity is used to control the random crystalline block lengths. Since atactic polypropylene is noncrystalline and has a T_g of around $0\,°C$ it can serve as random rubbery blocks in isotactic polypropylene.

In very early work Natta [8] showed that isotactic polypropylene contained a small fraction of polymer that was low in crystallinity and was rubbery in nature. Isotactic polypropylene can be partially epimerized to yield random isotactic and atactic units in the chain [9].

Direct polymerization of random stereoblock polypropylene has recently been studied by two groups of researchers using either homogeneous [10] or heterogeneous [11] catalysts.

5.2.4.1   Heterogeneous Catalysts for Stereoblock Polypropylene

The work on heterogeneous catalysts was stimulated by the discovery that certain organo-transition metal catalysts supported on metal oxide produced elastic stereoblock polypropylene. Tetra(neophyl)zirconium or bis(arene)titanium complexes supported on partially hydroxylated Al_2O3 (0.4 to 0.8×10^{-3} mol OH/g) give catalysts with long lifetimes and high catalyst efficiencies [12]. The catalyst can either be preformed or can be generated in situ in the polymerization reactor by separate introduction of Al_2O_3 and the transition metal complex. The structures of the catalyst has been studied and the following reaction on the

Al$_2$O$_3$ surface gives the active polymerization sites:

Ca. 100 m²/g

Polymerizations are carried out either in liquid propylene or in hydrocarbon diluent with maximum productivity in the 60 °C to 80 °C temperature range. Molecular weights are high and can be controlled by addition of hydrogen. Interestingly, hydrogen is more effective at controlling molecular weight if it is added after the transition metal is reacted with the Al$_2$O$_3$ but before monomer is introduced. When hydrogen is reacted with tetra(neophyl)zirconium or aluminum, an additional molecule of *tert*-butyl benzene is liberated, resulting in one alkyl group on the zirconium. It is not clear if the hydrogen gives more active sites or higher productivity for the existing ones. Catalyst efficiencies of 6×10^5 g of polymer per gram of transition metal and inherent viscosities of 5 to 20 dL/g are obtained.

Many heterogeneous polymerization catalysts have a number of sites of different activity and chain tacticity control. This usually results in both broad molecular weight distributions that are multimodal and also in broad intramolecular tacticity distributions during copolymerization [13]. Solvent fractionation, which separates polymer on the basis of both crystalline melting point (isotactic content) as well as molecular weight, has been carried out on stereoblock polypropylene prepared with the terta(neophyl)zirconium/Al$_2$O$_3$ catalyst system (TN2/Al$_2$O$_3$). The data for fractionation with ether and heptane are shown in Table 5.3. The molecular weight distribution of each fraction is shown in Fig. 5.4. Isotactic polypropylene prepared with a typical catalyst is shown for comparison.

All fractions contained crystallinity, with the ether-soluble fraction having the lowest melting point and percent crystallinity as measured by heat of melting, ΔH_f. The heptane-insoluble fraction was the most crystalline and had the highest molecular weight. The fractions were still broad in molecular weight distribution and appeared multimodal.

^{13}C-NMR analysis of the fractions reveals that the crystallinity and the melting point correlate with the stereoregularity of the chains. The number of monomer reversals decreases (isotactic block length increases) in going from the ether-soluble to the less soluble fractions. These data indicate the polymers are heterogeneous and stereoblock with chains consisting of alternating isotactic and atactic sequences. The sequences are approximated by second-order Markov polymerization statistics, which were used to calculate block length distributions. The high molecular weight fractions of low stereoregularity have isotactic blocks that co-crystallize with the more stereoregular components to form an elastic network below T_m.

The stress–strain properties of an unfractionated polymer are similar to those of many random block copolymers. The tension set is due to a plastic deformation mechanism for some element of the morphology, for example, a continuous crystalline network. Figure 5.5 shows the stress–strain properties of the individual fractions.

Table 5.3 Fractionation of Stereoblock Polypropylene and Isotactic Polypropylene

Catalyst	Fraction	(%)	Inherent viscosity (dL/g)	T_m (°C)	Heat of fusion (J/g)	$M_n \times 10^3$	M_w/M_n
TNZ/Al$_2$O$_3$	Total	100	10.2	147	13.1	75	8.0
	Ether-soluble	28.2	3.8	52	1.6	81	7.5
	Heptane-soluble	17.9	4.8	130	27.7	39	24
	Heptane-insoluble	50.0	15.3	148	49.0	455	5.8
TNZ/Al$_2$O$_3$, H$_2$	Total	100	4.1	152	5.0	34	11
	Ether-soluble	49.0	2.5	53	1.8	22	12
	Heptane-soluble	33.0	3.3	137	29.2	22	22
	Heptane-insoluble	24.0	8.0	150	68.7	108	3.2
TiCl$_4$/MgCl$_2$/AlEt$_3$	Total	100	2.1	160	62.5	22	5.6
	Ether-soluble	21.0	0.82	—	—	19	3.5
	Heptane-soluble	19.3	0.78	140	68.0	28	3.7
	Heptane-insoluble	53.0	3.5	166	102.0	160	3.0

TNZ, tetra (neophyl) zirconium

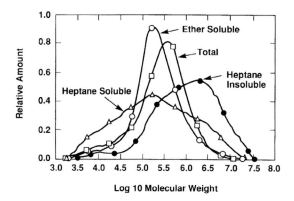

Figure 5.4   Stereoblock polypropylene molecular weight distribution. (After Ref. [7])

5.2.4.2   Homogeneous Catalysts for Stereoblock Polypropylene

Homogeneous metallocene catalysts can be used for stereochemical control in propylene polymerization [14] and one of their main attributes is that they can have a single, well-defined site for reaction. Catalysts of this type are capable of producing polymers with narrow molecular weight distributions and narrow tacticity distributions. The polymers often exhibit a more uniform distribution of crystalline blocks.

Chien and co-workers at the University of Massachusetts [10] polymerized propylene with the metallocene catalyst *rac*-[anti-ethylidene (1-η^5-tetramethylcyclopentadienyl)-(η^5-indenyl] dichlorotitanium activated with methylaluminoxane. TPEs were produced with this and similar metallocene catalysts.

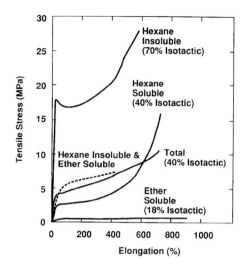

Figure 5.5   Stress–strain relationships in stereoblock polypropylene fractions. (After Ref. [7])

Polymerizations carried out at 25 °C and 50 °C resulted in polymers that were completely soluble in a single fractionation solvent. Lower temperature polymerization gave broader stereoregular distributions. Number average molecular weights of 1×10^5 were obtained. M_w/M_n was about 2.0, reflecting a most probable molecular weight distribution. The crystalline melting point was around 70 °C. The stress–strain properties of the polypropylene made at 25 °C are shown in Fig. 5.6.

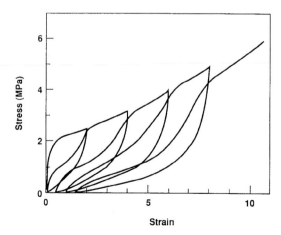

Figure 5.6   Stress–strain properties of stereoblock polypropylene made with homogeneous catalyst. (After Ref. [10])

5.2.4.3   Crystallization of Stereoblock Polypropylene

The small ΔT between T_g and T_m for stereoblock polypropylene results in slow crystallization rates. For the narrow MWD polymers produced with soluble catalysts ΔT is about 70°C. Dynamic mechanical property data, obtained by heating the polymer above T_m and immediately lowering the temperature to 40°C, showed that the moduli G' and G'' increased owing to crystallization but did not reach plateau values even after 2 h. DSC results for isothermal crystallization up to 960 min are shown in Fig. 5.7. As indicated by the dynamic mechanic properties, the crystalline fraction increases slowly with crystallization time. ΔH_f data obtained by DSC show a sigmoidal increase with time, suggesting a nucleation period followed by slow crystal growth due to low chain mobility.

The more heterogeneous stereoblock polypropylene produced by heterogeneous catalysts has similar crystallization characteristics. Fig. 5.8 gives DSC data for the individual fractions, the total polymers, and also for isotactic polypropylene. Since these polymers are in all likelihood a single phase in the melt, co-crystallization among the fractions would be expected.

The morphology of stereoblock polypropylene has been studied in some detail [12, 15, 16]. X-ray diffraction data on the narrow-MWD polymer show that the stereoregular chain segments crystallize as a mixture of α- and γ-phases. Stress–strain properties and dynamic mechanical data are consistent with the morphology proposed for TPUs, that is, a crystalline hard phase with the amorphous atactic regions forming an elastomeric second phase.

The morphology depends on co-crystallization of different size stereoregular blocks and must be very dependent on the degree of undercooling. Because of the appreciable length of the atactic sequences, under rapid crystallization rates they would be expected to be attached to isotactic sequences that are in different crystals.

The stereoblock polypropylenes have some practical limitations as TPEs. They exhibit a mechanical T_g at ca. 0°C and have a high tan δ at room temperature. The storage modulus starts to drop rapidly around 50°C to 70°C due to melting of the crystalline phase. Moreover, the slow crystallization rate limits the speed of thermoplastic processing.

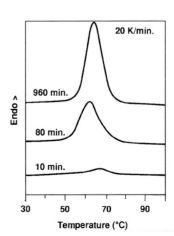

Figure 5.7   DSC analysis of homogeneous catalyst synthesized stereoblock polypropylene. (After Ref. [10])

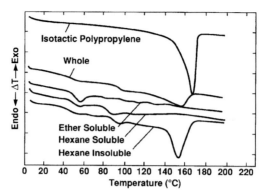

Figure 5.8   DSC analysis of heterogeneous catalyst synthesized stereoblock polypropylene. (After Ref. [7])

5.3   Block Copolymers

The idea of A–B–A block copolymers with crystalline end blocks and an elastomeric center block has been around for over 40 years. The success of polystyrene–*block*-polyisoprene–*block*-polystyrene (S–I–S) and S–B–S TPEs showed that triblock copolymers could have wide applications. Furthermore, much work was being conducted on Ziegler–Natta catalysis that could potentially be used to synthesize potential block copolymers based on polyolefins.

The basic requirements for synthesis of A–B–A olefin block polymers via a Ziegler–Natta process have been outlined by Boor [13] as follows:

1. All polymerization centers must become active simultaneously.
2. All centers must remain active and also not transfer.
3. All centers must be equal in polymerization rate.
4. The centers must be equally accessible (no diffusion control).
5. The centers must be stereoregulating to obtain crystalline polyolefin blocks (except for polyethylene).

While there are numerous patents and many articles describing the synthesis of olefin block copolymers, most fail to characterize the chain structure in detail. Boor stated in his review that "The reported block copolymers from propylene and ethylene are most likely mixtures of largely isotactic PP, some nearly pure PE, and either random block, or tapered copolymers, depending on the exact method of synthesis." Much of the early work was conducted with heterogeneous Ziegler–Natta catalysts which essentially have several active sites and do not initiate homogeneously. In addition, fractionation of the hydrocarbon polymers is difficult and depends largely on the crystalline fraction in the chain. To add to the confusion, high-impact isotactic polypropylene made in series reactors (with ethylene monomer added to the second reactor) has been called a "block copolymer," even though the proportion of A–B block chains formed is quite small.

Block copolymers can be readily synthesized by the indirect route of hydrogenating block copolymers containing 1,4-butadiene and isoprene to yield polyethylene and ethylene–

propylene copolymer type chains. These have been helpful in elucidating structure–property relationships (see Chapter 4).

"Living" carbocationic catalysts have been employed to make block olefin copolymers with elastic polyisobutylene segments [17]. This process results in a polymer with saturated central rubbery blocks and various glassy end blocks, such as polystyrene and polyindene. (See Chapter 13.)

5.3.1   Ziegler–Natta Block Copolymers

One of the early attempts to follow through on Tobolsky's idea of a TPE based on polyolefins was carried out by Kontos and co-workers [18]. Polymers were synthesized with Ziegler–Natta catalysts (e.g., $TiCl_4$ with lithium aluminum alkyls) with alternating feed composition. It was believed that the halftimes of the "living" macromolecules were on the order of several days and the block copolymers based on ethylene and propylene could be obtained in this manner. The stress–strain properties of these polymers showed rubberlike characteristics. However, from what has been learned more recently about the polymerization kinetics and heterogeneous nature of this catalyst system, the polymers are primarily blends of homo- and copolymers of ethylene and isotactic polypropylene.

Doi, Ueki, and Keii [19–21] described a living polymerization for syndiotactic polypropylene consisting of vanadium (acetylacetonate) and Et_2AlCl. The polymerizations were carried out at $-78\,°C$ to give a polymer with $M_w/M_n = 1.1$. Sequential monomer addition (at $-78\,°C$ in anisole) gave a sPP–EP–sPP block copolymer. The M_w/M_n of the block copolymers remained quite narrow in the range of 1.1 to 1.3. Evens [22] found that in the absence of anisole substantial amounts of homopolymer were produced along with block copolymers.

This route to A–B–A olefin block copolymers has limited utility owing to low activity, for example, in 5 hours, 1 g of catalyst produced 50 g of polymer. The catalyst failed to produce living polyethylene chains at $-78\,°C$. Further, above $-65\,°C$, none of the polymer chains were found to be living and broad molecular weight distributions were obtained [23].

Busico et al. [24] demonstrated that at low pressures (0.1 Bar) a mixture of isotactic polypropylene-*block*-ethylene–propylene and homopolymers could be obtained using $TiCl_3/Et_2AlCl$. It is not likely that this will turn out to be an attractive synthetic route.

Lock [25] reported that dimethyl bis(methylcyclopentadienyl) titanium (IV) and $TiCl_3$ gave polymerization results consistent with the existence of long-lived polymer chains. At $45\,°C$ polymer yield and molecular weight increased directly with polymerization time to give a constant number of moles of chains during polymerization times of up to 200 min. The number of polymer chains was also constant after removal of monomer and replacement with argon and then restarting of the polymerization. The catalyst produced polymers with M_w/M_n of 4 to 5 and this value was constant throughout the polymerization, suggesting multiple polymerization sites rather than significant transfer, delayed activation, or termination reactions.

The results for iPP–EP–iPP block copolymer synthesis indicate that the polymer contained mainly propylene crystallinity and small amounts of crystalline polyethylene. X-ray data showed 19% polypropylene, 3% polyethylene and 78% amorphous ethylene–

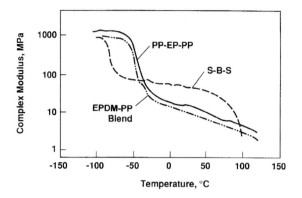

Figure 5.9   A–B–A block copolymer complex modulus vs. temperature. (After Ref. [25])

propylene segments. Fractionation of a polymer containing 14 wt% propylene and 86 wt% ethylene–propylene indicated that 66% of the ethylene–propylene blocks were connected to isotactic polypropylene blocks. It was not possible to distinguish between di- and triblocks.

In Fig. 5.9 the dynamic mechanical properties of an iPP–EP–iPP block copolymer are compared with those of an S–B–S and a dynamically vulcanized EPDM–polypropylene blend.

Optimum physical properties were found with 12 to 28 wt% end blocks with molecular weights of 30 to 90 × 10^3 g/mol and center blocks containing 55 wt% ethylene with molecular weights of 300 to 400 × 10^3 g/mol. Stress–strain properties are shown in Fig. 5.10. Table 5.4 compares properties of polymers with different end block concentration. Polymer rheology was not investigated but the molecular weights suggest rather high viscosities.

Wang and Huang [26] also investigated "PP–EP–PP." These polymers were synthesized using δ-TiCl$_3$-Et$_2$AlCl as the catalyst system. Since this catalyst system is heterogeneous and

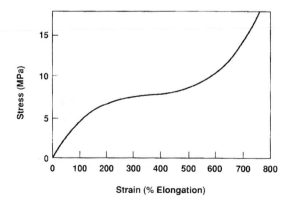

Figure 5.10   PP–EP–PP block copolymer stress–strain properties. (After Ref. [25])

Table 5.4 Properties of Ethylene–Propylene Block Copolymers

	Wt% polypropylene		
Properties at 25 °C	12	21	29
Shore A hardness	70	80	90
100% modulus (MPa)	2.1	4.0	6.5
Tensile strength (MPa)	13	14	21
Elongation at break (%)	800	800	700
Permanent set after 10 min recovery from 100% extension (%)	12	15	20

gives broad MWD and broad copolymer compositional distributions, the products are likely a mixture of homopolymers and copolymers with some undefined block copolymer content. Stress–strain properties were compared with blends of EPM and iPP. Had the comparisons been made at equal crystallinity (rather than equal propylene contents), the properties of the "PP–EP–PP" would be quite similar to those of such blends.

As discussed previously, it is difficult to directly synthesize well-defined block copolymers of the A–B–A type with Ziegler–Natta catalysts. It should also be noted that only one chain can be produced from a catalyst site. Thus, process economics may demand a mechanism that uses polymerization sites multiple times.

5.3.2   Hydrogenated Diene Block Copolymers

The well-characterized A–B–A olefin block copolymers are block copolymers polymerized by an anionic mechanism [27–31]. The end crystalline blocks consist of hydrogenated high 1,4-polybutadiene and the rubbery center blocks are either hydrogenated high, 1,4-poly-isoprene H(B–I–B) or hydrogenated 45% 1,2-polyolefin polybutadiene with about 45% 1,2 content H(B–B–B). These are excellent model systems.

The hydrogenations have been carried out with H_2–n-butyllithium–cobalt octoate catalyst [27], a Ziegler type catalyst prepared from Et_2Al and cobalt (II) octoate [28] or p-toluenesulfonylhydrazide. Hydrogenations are nearly complete and chain scission is minimal. One limitation of the synthetic method is that the polyethylene end blocks contain significant amounts of ethylene branches. This branching limits the T_m to less than 107 °C, giving the end blocks the crystallinity and properties similar to those of low-density polyethylene.

H(B–I–B), H(I–B–I), and H(B–I) are all semicrystalline, with the extent of crystallinity being independent of the architecture of the block arrangement but linearly dependent on butadiene content [29, 30], as shown in Fig. 5.11. DSC results indicate a slight increase in crystallinity and T_m of H(B–B–B) after tensile elongation to failure.

The stress–strain properties of H(B–I–B) are compared with H(B) and H(I) homopolymer in Fig. 5.12. All polymers have Mn of $\sim 200 \times 10^3$. The tensile properties of the H(B–I–B) polymers resemble those of S–B–S block copolymers at low "hard block" content. However, at high concentrations of styrene ($\sim 80\%$) S–B–S is a brittle material with short

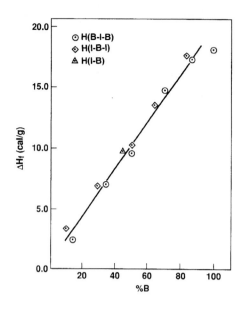

Figure 5.11 The linear dependence of ΔH_f on butadiene content in various block copolymers. (After Ref. [29])

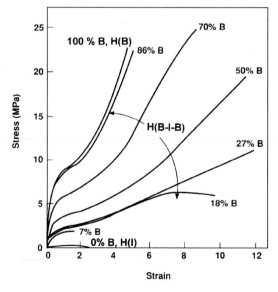

Figure 5.12 Comparison of the stress–strain properties of press-quenched films of H(B–I–B) to those from homopolymers H(B) and H(I). The composition of each polymer is denoted by the butadiene content next to the curve. (After Ref. [29])

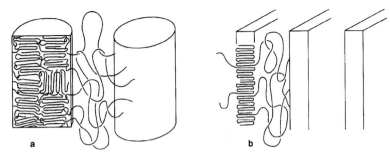

Figure 5.13   Molecular models for (a) the melt-crystallized form and (b) the solution-crystallized form of H(B–I–B) triblock polymer. (After Ref. [28])

elongation while H(B–I–B) with 80% B undergoes a ductile plastic deformation. Interestingly, the stress–strain properties for the inverted H(I–B–I) are similar to H(B–I–B) above 50% butadiene content. This is most likely due to the morphology of the block copolymer in that the crystalline phase is continuous in both polymers (although the morphology was not investigated). At lower butadiene content tensile properties fall off rapidly for the H(I–B–I) block polymer. Apparently, the H(B) is not a continuous phase at 29% and, of course, the H(I) chain ends are not captured into a physically crosslinked network.

The morphology of H(B–I–B) can be altered by the method of crystallization [28]. The triblock copolymers that are crystallized from the melt involve intermolecular crystallization. The polymers appear to be below their order–disorder transition near the melting point and are thus phase separated. This type of crystallization imparts good elastomeric stress–strain properties to the block copolymer. The material as shown in Fig. 5.12 behaves as a physically crosslinked elastomer, with a filler effect due to the ductile H(B) phase. In contrast, H(B–I–B) crystallized from a benzene solution is in a homogeneous medium. This promotes intramolecular crystallization of the H(B) phase and reduces physical crosslinking of the chains. The solution-crystallized polymer exhibits a high level of creep. The contrasting models for the solution and melt-crystallized H(B–I–B) are shown in Fig. 5.13.

5.4   Graft Copolymers

The two major polyolefin-based elastomers, EPDM and butyl rubber, are soluble and are inert to most types of reactions that lead to grafting, with the exception of those occurring on the olefin sites introduced by copolymerization of a diene. The olefin site or other sites produced by copolymerization or modification allow for many types of grafting reactions via chain initiation, chain termination, chain transfer, or chain coupling reactions. Consequently, there is a huge literature, mainly as patents, on the graft polymerization of olefin-based elastomers. Grafting insoluble side chains onto elastomeric backbones produces thermoplastic elastomers if the side chains can form a separate phase with the proper T_g or T_m. Grafting of crystalline polyolefins to elastomeric chains has been accomplished. Work carried out before 1970 is generally limited in the characterization of the graft copolymer molecular structure and morphology [32]. Graft copolymers are largely used as impact modifiers but have not become commercially significant TPEs.

5.4.1   Polyolefins Grafted with Crystalline Chains

In the late 1970s [33, 34] grafts of polypivalolactone (PPVL) on EPDM were reported and more recently data have been published on PPVL-grafted butyl rubber [35].

The graft copolymerization of PPVL is initiated by carboxylate anions randomly located on the chain. Solvent extraction revealed 2 to 10 wt% ungrafted PPVL. EPDMs with 7 to 42 wt% grafted PPVL and graft molecular weights of 350 to 2800 M_n were synthesized. The PPVL grafts formed crystalline domains 10 to ca. 100 nm in size, depending on molecular weight. T_ms of 120°C to 200°C were found. The stress–strain properties of a series of graft copolymers with increasing PPVL M_n and prepared from EPDM with a constant carboxylate anion content are shown in Fig. 5.14. At graft molecular weights corresponding to degrees of polymerization of greater than 3, good network properties were observed and compression set values were low.

The melt viscosity of EPDM–g–PPVL is considerably higher than that of the ungrafted backbone. This would be expected from the large differences in polarity of the graft and the backbone, since the polymer is probably phase seperated even above the T_m of PPVL chains. The activation energy of flow is remarkably high (110 to 170 kJ/mol vs. 2.5 kJ/mol for EPDM), reflecting the phase-separated nature of the melt and the comblike structure.

EPDM has been grafted with iPP using polymer–polymer coupling reactions and Ziegler–Natta graft copolymerization [36]. The Z–N grafts were made by polymerizing EPDM with vinyl side groups and then copolymerizing these side groups with polypropylene using a stereospecific catalyst. Most of the EPDM and nearly all of the iPP were found in graft copolymers. The materials were shown to be effective compatibilizers for mixtures of polypropylene with ethylene–propylene copolymers.

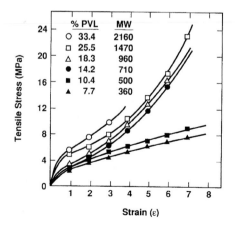

Figure 5.14   Effect of PPVL M_n and concentration of the stress–strain properties of EPDM–g–PPVL. Graft site concentration is held constant. (After Ref. [7])

5.5   Polyolefin Blend Thermoplastic Elastomers

Polyolefin blend TPEs, based primarily on EPM and iPP, have been used extensively for a number of years and are an important family of engineering materials [37]. The development of these TPEs follows that of S–B–S block copolymers and TPUs which were instrumental in showing the utility of thermoplastic processing methods for materials with rubberlike properties.

As indicated in the section on block copolymers, Ziegler–Natta type polymerization catalysts do not easily lend themselves to A–B–A block polymer synthesis, but it was found that blends of the proper morphology exhibit elastomeric properties and surprisingly good recovery after extension. When compounded and plasticized they afford a wide range of useful properties, excellent processing, and can be produced at moderate cost. Blending and compounding have long been an important aspect of rubber science and technology in thermosetting systems and TPEs based on the blending technology of poly(vinyl chloride) and acrylonitrile–butadiene copolymer date from the 1940s.

Hartman [38] carried out experiments in the late 1960s on blends of polyisobutylene with linear and branched polyethylene. The blending was reportedly carried out to provide grafting of butyl rubber onto the polyethylene and the properties reflected the blend ratios and the type of polyethylene used. These materials were introduced as ET Polymers by Allied Chemical Company on a developmental basis. Also, in the 1960s Kontos [18] using what is now recognized as blends of isotactic polypropylene and ethylene–propylene copolymers, achieved "plastic-rubber" properties.

Polypropylene and several other crystalline polymers were also found to be transform-able into elastic materials by changing their crystal structure [39–41]. With isotactic polypropylene it is possible to obtain 97% recovery from 100% extension by applying high stress during crystallization of fibers. The fibers were shown to have a nonrubberlike stress–temperature response and their elastic nature was due to their morphology. Electron microscopy showed close-packed lamellae with normals mainly parallel to the fiber axis. On extension these tilted and split apart, creating voids. This nonentropic elastic nature of isotactic polypropylene may play a role in blend technology.

In 1973 Fisher [42] made rubbery blends based on isotactic polypropylene and EPM. These were characterized by an ethylene–propylene phase that was either highly long-chain-branched or was partially cured. EPM and EPDM are more elastic at use temperature if the ethylene sequences are long enough to crystallize. As indicated in Section 5.2.1, where these polymers are strained, more crystallinity develops and recovery from extension is not high. The concept of using blends of high ethylene content EPM or EPDM with polyethylene and polypropylene to give TPEs was introduced in early blend work [43].

Plasticizer oils have been extensively used in thermoset rubber technology for many years. Oils have been used in crystalline polyolefins because they are not soluble in the crystalline regions. Gessler and Kresge [44] found that there were considerable processing advantages to incorporating oil in blends of EPM and isotactic polypropylene. In the melt, the oil partitions between the phases, greatly lowering the viscosity. This gives very good injection molding and extrusion characteristics. On crystallization the oil appears to be adsorbed by the amorphous rubber phase and is not rejected to the material's surface. The mechanical properties of these blends are similar to those without oil at the same volume fraction of polypropylene.

Other published blend work on polyolefin blends during the 1970s includes mill-mixed blends of butyl rubber with polypropylene over the entire compositional range. Physical properties were determined but no morphological data were presented [45]. Blends of EPM and isotactic polypropylene containing carbon black were described by Straub [46]. These materials are injection moldable and are electrostatically paintable. Blends of EPM and isotactic polypropylene were studied by Danesi and Porter [47] to establish relationships between morphology and physical properties as well as to determine the mechanisms of morphology development. Bicontinuous morphology, as well as a dispersed phase of the the minor phase in the major phase, were observed by optical microscopy. Jevanoff and co-workers [48] showed that a polypropylene "skin" can develop on extrusion of EPM–polypropylene blends. In this case it seems that the lower viscosity polypropylene phase migrated to the high shear rate region at the wall, while the internal regions of the extrudates retained a morphology where both phases were continuous.

Blends of semicrystalline EPDM with polyethylene were examined by Lindsay [49]. Here the polyethylene appeared to nucleate the crystallization of the EPDM phase, which has a lower melting point owing to shorter methylene segments. Tensile strengths were higher for the blends than for either polymer alone.

5.5.1   Morphology of Polyolefin Blend Thermoplastic Elastomers

To provide a TPE from nonthermoplastic elastomer materials by blending, it is critical to control the morphology of the system. This has been accomplished by choice of the mixing method, mixing conditions, rheological properties of the blend components, controlling the surface energy (polymer choice and/or polymeric compatibilizers), and chemical reactions during mixing.

In the simplest blends with polyolefins, such as high molecular weight EPM and isotactic polypropylene, intensive mixing results in two continuous phases as shown in Fig. 5.15. By adjusting the viscosity ratios, both phases can be kept continuous over a considerable range of volume fractions in this blend (e.g., 80/20 to 20/80).

At the typical copolymer monomer ratios and molecular weights of EPM copolymers used in blends, the elastomers are insoluble in polypropylene when held quiescently in the melt. This has been inferred from studies of the glass transition temperature of EPM–atactic polypropylene blends by DSC. These blends have a broad transition rather than the narrow change in heat capacity (C_p) found in soluble polymer systems [50]. Scattering studies [51] on deuterium-labeled EPM blended with isotactic polypropylene indicate a two-phase melt. Shear could affect solubility during mixing of simple blends, but the nodular nature of the phases and phase size is more consistent with a shear dispersion mechanism than a spinodal decomposition or crystallization on cooling from a thermodynamically soluble system. The polypropylene is nonspherulitic in nature with a well-defined X-ray diffraction pattern of monoclinic polypropylene. Blends of EPM and isotactic polypropylene have also been prepared by dissolving a 50 : 50 mixture in hot xylene and pouring the solution into an excess

Figure 5.15   Electron scanning micrograph of a biblend of EPM and polypropylene (70 : 30) prepared by intensive mixing after extraction of the EPM phase with heptane. Sample was fractured under liquid nitrogen. Bar is 1 μm

of 0 °C methanol [52]. It was postulated that when using this method the two phase co-continuous "network" structure obtained was the result of spinodal decomposition and a coarsening process.

For blends of high molecular weight polyisobutylene-based polymers with polypropylene, the solubility in the melt will be less than for blends of EPM with polypropylene, so it is likely that the morphology of these blends results from mixing a two-phase system.

For simple blends, the polypropylene phase is continuous and exhibits elastic properties owing to the open fiberlike microstructure. The EPM phase will be somewhat more elastic if it is semicrystalline or if it is highly branched. If this phase is continuous, it will also provide some elastic response, particularly at short time scales.

Several hypotheses for the formation of co-continuous phases have been formulated. Avgeropoulos et al. [53] showed that for EPDM–polybutadiene blends the volume fraction and torque ratio of the components determined the phase morphology. Paul and Barlow [54] found that phase inversion occurred when $\phi_1/\phi_2 = \eta_1/\eta_2$, where ϕ_1 and ϕ_2 are the volume fractions of polymer 1 and 2 and η_1 and η_2 are the viscosities. Lyngaae-Jorgensen and Utracki [55] found that the critical volume fraction for a continuous phase, $\phi_{cr} = 0.156$, is predicted by percolation theory for monodisperse spherical domains. This is in agreement with experiments on EPDM and isotactic polypropylene.

5.5.2   Mechanical Properties of Polyolefin Blends

The primary mechanical property that distinguishes elastomers from other materials and the characteristic that dictates their use in many applications is a low-modulus, high-elongation stress–strain curve. The distinction of thermoplastic as rubbery and nonrubbery is quite

Table 5.5 Properties of EPDM–Polyolefin Blends

Blend[a]							
EPDM[b], parts	80	70	60	80	60	80	60
Polypropylene[c], parts	20	30	40	—	—	—	—
Low-density polyethylene[d], parts	—	—	—	20	40	—	—
High-density polyethylene[e], parts	—	—	—	—	—	20	40
Physical properties							
Tensile strength (MPa)	8.3	10.5	13.9	5.8	8.0	8.5	10.2
Elongation at break (%)	220	150	80	290	190	210	130
Elongation set at break (%)	28	30	30	35	30	25	33

[a] Banbury mixer, about 7 min, maximum temperature about 200 °C
[b] Amorphous, high molecular weight ethylene–propylene–dicyclopentadiene (~ 5 wt%) terpolymer
[c] $\rho = 0.903$ g/cm^3, melt index $= 4.0$ g/10 min at 230 °C
[d] $\rho = 0.919$ g/cm^3, melt index $= 2.0$ g/10 min at 190 °C
[e] $\rho = 0.956$ g/cm^3, melt index $= 0.3$ g/10 min at 190 °C

arbitrary, but in general, materials are considered rubbery if they can be extended over 100% without failure and return to nearly their original dimensions in a short period of time. Thermoplastic polyolefin blends are produced with a spectrum of stress–strain properties via specific backbone polymer selection, morphology control, and compounding with filler and plasticizers.

The stress–strain properties of unfilled blends of a highly elastic amorphous EPDM [42] and various polyolefin resins are shown in Table 5.5. Blends containing high amounts of EPDM are quite rubbery in nature and have surprisingly low sets at break (20% to 35%). In contrast, when pure high molecular weight polyolefin resins are extended, they undergo a yield at low elongation followed by a typical drawing mechanism, and there is very little recovery after drawing. Table 5.6 lists the mechanical properties of blends similar to those above except a semicrystalline EPDM was used. This gave a significant improvement in properties, especially elongation to break. Not only is crystallinity in the unstrained state important, but also the increase in crystallinity during deformation, no doubt, has a large effect on the stress–strain properties of the blend.

The stress–strain curves for an unfilled blend of amorphous ethylene–propylene random copolymer and isotactic polypropylene are shown in Fig. 5.16.

Table 5.6 Properties of Semicrystalline EPDM–Polyolefin Blends

Blend[a]				
EPDM, parts	80	80	80	80
EPDM crystallinity (wt%)	12.9	2.7	12.9	2.7
High-density polyethylene[b], parts	20	20	—	—
Low-density polyethylene[c], parts	—	—	20	20
Physical properties				
Tensile strength (MPa)	15.0	5.4	14.5	7.6
Elongation at break (%)	730	940	720	880

[a] Mill mixed at 150 °C
[b] $\rho = 0.95$ g/cm^3
[c] $\rho = 0.92$ g/cm^3

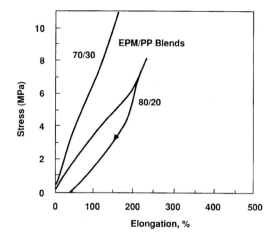

Figure 5.16   Stress–strain properties of blends of ethylene–propylene random copolymer and isotactic polypropylene. Return from 200% elongation is shown

5.5.3   Blends of Polyolefins with Other Polymers

TPEs produced from blends of isotactic polypropylene with other polymers have been extensively studied. Isotactic polypropylene was blended with propylene–1-hexene random copolymer to give a TPE. The blends had a single T_g (single tan δ peak) as well as a single melting point. The stress–strain properties are shown in Fig. 5.17 and the higher T_m induced by the isotactic polypropylene gives significantly improved mechanical properties at elevated temperature. Random stereoblock polypropylene also responds well in blends with isotactic polypropylene. The two polymers appear to co-crystallize and show a single DSC melting point. Stress–strain properties of the blends depend on composition, with 15 to 25 wt% isotactic polypropylene giving good elastic characteristics at 25 °C. Low-temperature properties are not attractive, however, with the T_g of ca. 0 °C.

   TPEs are also produced from blends of isotactic polypropylene with EVA [56]. The polymers are immiscible and form co-continuous phases from 50 to 70 wt% EVA. Dynamic mechanical spectroscopy of both uncrosslinked and dynamically crosslinked blends displayed two T_gs corresponding to EVA and isotactic polypropylene.

   Blends of EPDM and poly(butylene terephthalate) (PBT) have been studied as TPEs [57], both with and without dynamic vulcanization. Because of the high surface energy between EPDM and PBT it was necessary to use a compatibilizer to achieve elastomeric properties. This was accomplished by grafting the EPDM in an extruder with glycidyl methacrylate (GMA) using a peroxide initiator. Tensile properties for the blends are listed in Table 5.7.

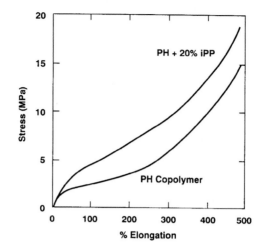

Figure 5.17   Stress–strain properties of propylene–hexene copolymer and a blend with isotactic polypropylene. (After Ref. [7])

Table 5.7   EPDM Blends with Poly(Butylene Terephthalate)

Blend	Tensile strength (MPa)	Percent elongation	Tensile set
1 : 1 EPDM–PBT	15.2	55	—
1 : 1 EPDM–PBT plus EPDM–g–GMA[a]	20.8	285	40
1 : 1 EPDM–PBT plus EPDM–g–GMA (Dynamically cured)[b]	22.8	285	15

[a] EPDM grafted with 3 wt% glycidyl methacrylate
[b] Crosslinked during extrusion with 0.5% to 2,5-dimethyl-2,5-di (*tert*-butylperoxy)-3-hexyne

The properties of the blends depended on compatibilization, and, as observed for EPDM–isotactic polypropylene blends, dynamic vulcanization enhanced tensile set, apparently by making the rubber phase more elastic.

5.5.4   Applications for Polyolefin Blend Thermoplastic Elastomers

Blends of isotactic polypropylene and ethylene–propylene copolymers that are designed for specific uses are fully compounded with fillers, reinforcing agents, antioxidants, colorants, plasticizers, etc. and are called TPOs (for thermoplastic polyolefins). There are three major market areas: automotive, wire and cable, and mechanical goods.

Automotive parts are the largest single market for TPOs. Excellent weatherability, low density, processing flexibility, and relatively low cost make them very common exterior and interior automotive parts, for example, air dams, bumper covers, fender extensions, grills, rub strips, conduit, grommats, and interior trim.

Table 5.8 TPO Suppliers and Products

Supplier	Trade name	Manufacturing location[a]
Advanced Elastomer Systems	Vistaflex	N, E, J
Ferro Corp	Ferroflex	N, E
Republic Plastics	ETA and RTA	N
RPI/Dexter	Deflex	N
A. Schulman	Polytrope	N
Teknor Apex	Telcar	N
DSM	Kelburou	E
British Vita	Vitacom TPO	E
Chemische Werke Huls	Vestolen	E

[a] N, North American; E, Western Europe; J, Japan

TPOs have replaced poly(vinyl chloride) and vulcanized elastomers in many electrical applications, such as flexible cords, booster cables, appliance wire, and low-voltage jacketing. TPO compounds are available that offer low smoke generation and flame resistance. Excellent electrical properties, water resistance, and ozone resistance are the main attributes in these applications.

TPOs are sold as pelletized compounds developed to meet specific applications and processing requirements. They can be processed on most common thermoplastic equipment, for example, injection molding, extrusion, injection blow-molding, vacuum forming, and as blow film.

General recommendations on TPO usage and processing have been reviewed [58, 59]. Specific recommendations should be obtained from suppliers of TPOs.

Currently, there are a number of suppliers that develop and manufacture TPOs based on blends of ethylene–propylene rubber and isotactic polypropylene. Table 5.8 lists the trade names, suppliers, and manufacturing locations of TPOs.

Acknowledgments

The author thanks his fellow researchers at Exxon Chemical Company for helpful discussions, particularly the late F.P. Baldwin, G. VerStrate, C. Cozewith, and S. Datta. D.N. Schulz and D. Lohse of Exxon Research and Engineering Company have given insightful comments. The helpful input of S. Abdou-Sabet, D.R. Hazelton, R.C. Puydak, and C.P. Rader of Advanced Elastomer Systems is acknowledged.

References

1. A.V. Tobolksy, *Rubber World 138*, 857 (1959)
2. E.N. Kresge, *Rubber World 208(2)*, 31 (1993)
3. F.P. Baldwin and G. VerStrate, *Rubber Chem. Technol. 45(3)*, 790 (1972)
4. G. VerStrate and Z.W. Wilchinsky, *J. Polym. Sci. A2 9*, 127 (1971)
5. D.K. Parker, R.F. Roberts, and H.W. Schiesol, Presented at ACS Rubber Div. Orlando, FL, October 26, 1993
6. Ibid, *Rubber Chem. Technol. 65*, 245 (1992)

7. C.-K. Shih and A.C.L. Su, *Thermoplastic Elastomers*. N.R. Legge, G. Holden, and H.E. Schroeder (Eds.) (1987) Hanser, Munich
8. G. Natta, *J. Polym. Sci. 34*, 531 (1959)
9. U.W. Suter and P. Neuenschwander, *Macromolecules 14*, 523 (1987)
10. G.H. Llinar, S.H. Dong, D.T. Mallin, M.D. Rausch, Y.G. Lin, H.H. Winter, and J.-C.W. Chien, *Macromolecules 25*, 1242 (1992)
11. J.W. Collette, C.W. Tullock, R.N. MacDonald, W.H. Buck, A.C.L. Su, J.R. Harrell, R. Mulhaupt, and B.C. Anderson, *Macromolecules 22*, 3851 (1989)
12. S.D. Ittel, *J. Macromol Sci.-Chem. A27(9-11)*, 1133 (1990)
13. J. Boor Jr., *Ziegler–Natta Catalysts and Polymerizations* (1979) Academic Press, New York
14. J.W.C. Chein, *Makromol. Chem. Macrmol. Symp. 63*, 209 (1992)
15. J.W. Collette, D.W. Ovenall, W.H. Buck, and R.C. Ferguson, *Macromolecules 22*, 3858 (1989)
16. Y.G. Lin, D.T. Mallin, J.C.W. Chien, and H.H. Winter, *Macromolecules 24*, 850 (1991)
17. J.P. Kennedy, S. Midha, and U. Tgrenogal, *Macromolecules 26*, 429 (1993)
18. E.G. Kontos, E.K. Easterbrook, and R.D. Gilbert, *J. Poly. Sci. 61*, 69 (1962)
19. Y. Doi, S. Ueki, and T. Keii, *Macromolecules 12*, 814 (1979)
20. Y. Doi, S. Ueki, and T. Keii, *Macromolecules 180*, 1359 (1979)
21. Y. Doi, and S. Ueki, *Macromol. Chem. Rapid Commun. 3*, 225 (1982)
22. G.G. Evens, 1981 MMI International Symposium on Transition Meta; Catalyzed Polymerization: Unsolved Problems, Midland, August 1981, Part A, p. 245
23. Y. Doi and T. Keii, *Adv. Poly. Sci. 73/74*, 201 (1986)
24. V. Busico, P. Corradini, P. Fontana, and V. Savino, *Makromol. Chem. Rapid Commun.* (a) *5*, 737 (1984), (b) *6*, 743 (1985)
25. G.A. Lock, *Thermoplastic Elastomers Based on Block Copolymers of Ethylene and Propylene, Advances in Polyolefins*. R.B. Seymour and T. Cheng (Eds.) (1985) Plenum Press, New York
26. L. Wang and B. Huang, *J. Poly. Sci. Pt. B, Physics 29*, 1447 (1991)
27. M. Morton, N.-C. Las, and E.R. Terrill, *Elastomeric Polydiene ABA Triblock Copolymers with Crystalline End Blocks in Elastomers and Rubber Elasticity, ACS Symposium Series No. 9330*. J.E. Mark and J. Lal (Eds.) (1980) American Chemical Society, Washington, D.C.
28. R. Sèguèla and J. Prud'homme, *Polymer 30*, 1447 (1989)
29. Y. Mohajer, G.L. Wilkes, J.C. Wang, and J.E. McGrath, *Effects of Variation of Composition and Block Sequence on Properties of Copolymers Containing Semicrystalline Block(s), Elastomer and Rubber Elasticity, ACS Symposium Series NO. 9330*. J.E. Mark and J. Lal (Eds.) (1982) American Chemical Society, Washington, D.C.
30. Y. Mohajer, G.L. Wilkes, J.W. Wang, and J.E. McGrath, *Polymer 23*, 1523 (1982)
31. J.C. Falk and R.J. Schlott, *Macromolecules 4*, 152 (1971)
32. R.J. Ceresa (Ed.) (1973) *Block and Graft Copolymerization*, John Wiley & Sons, New York
33. S.A. Sundet, R.C. Thamm, J.M. Meyer, W.H. Buck, S.W. Caywood, P.M. Sambra Manian, and B.C. Anderson, *Macromolecules 9*, 371 (1976)
34. R.C. Thamm and W.H. Buck, *J. Polymer. Sci. Polym. Chem. Ed. 16*, 539 (1978)
35. J.F. Harria, Jr. and W.H. Sharkey, *Macromolecules 19(12)*, 2903 (1986)
36. D.J. Lohse, S. Datta, and E.N. Kresge, *Macromolecules 24*, 561 (1991)
37. E.N. Kresge, *Rubb. Chem. Technol. 64(3)*, 469 (1991)
38. F.P. Hartman, C.L. Eddy, and G.P. Koo, *SPE J. 6*, 62 (1970)
39. R.G. Guynn and H. Brody, *J. Macromol Sci. Polym. Phys. 5*, 721 (1971)
40. B. Cayrol and J. Petermann, *J. Polym. Sci. Polym. Phys. Ed. 12*, 2169 (1974)
41. A. Moet., I. Palley, and E. Baer, *J. Appl. Phys. 51*, 5175 (1980)
42. (to Uniroyal Inc.), W.K. Fisher, U.S. Patent 3,835,201 (September 10, 1974)
43. M. Batiuk, R.M. Harman, and J.C. Healy, (to B.F. Goodrich Co.), U.S. Patent 3,919,358 (November 11, 1975)
44. A.M. Gessler and E.N. Kresge, (to Exxon Research and Eng. Co.), U.S. Patent 4,132,698 (January 2, 1979)
45. R.D. Deanin, R.O. Normandin, and C.P. Kannankeril, *Amer. Chem. Soc. Org. Coating Plast. Preprints 35*, 259 (1975)
46. R.M. Straub, (to E.I. DuPont), U.S. Patent 3,963,647 (June 15, 1976)

47. S. Danesi and R.S. Porter, *Polymer 19*, 448 (1978)
48. A. Jevanoff, E.N. Kresge, and L.L. Ban, Presentation at Polymer Blends Conference Plasticon 81, University of Warwick (1981)
49. G.A. Lindsay, C.J. Carman, and R.W. Smith, *Adv. Chem. Ser. 176*, 367 (1979)
50. E.N. Krege, In *Polymer Blends, Vol. 2*. D.R. Paul and S. Newman (Eds.) (1978) Academic Press, New York
51. D.J. Lohse, *Annu. Tech. Conf.-Soc. Plast. Eng. 43rd, 31*, 301 (1985)
52. N. Inabo, T. Yamada, S. Suzuki, and T. Hashimoto, *Macromolecules 21*, 407 (1988)
53. G.N. Averupoulos, F.C. Wissert, P.H. Biddison, and G.G.A. Bohm, *Rubber Chem. Technol. 49*, 93 (1976)
54. D.R. Paul and J.W. Barlow, *Adv. in chem. Series 211*, 3 (1986)
55. J. Lyngaae-Jorgenson and L.A. Utracki, *Makromolek. Chem.—Macromolecular Symposia 48–9*, 189 (1991)
56. S. Thomas and A. George, *Eur. Polym J. 28*, 11, 1451 (1992)
57. J.R. Campbell, F.F. Khouri, S.Y. Hobbs, T.J. Shea, and A.J. Moffet, *Polymer Preprints*, August 846 (1993)
58. D.J. Synnott, D.F. Sheridan, and E.G. Kontos, *EPDM–Polypropylene Blends in Thermoplastic Elastomers form Rubber Plastic Blends*. S.K De. and A.K. Bhowmick (Eds.) (1990) Ellis Horwood, New York
59. C.D. Shedd, In *Handbook of Thermoplastic Elastomers* 2nd edit. B.M. Walker and C. Rader (Eds.) (1988) Van Nostrand Reinhold, New York

6 Thermoplastic Elastomers Based on Halogen-Containing Polyolefins

G.H. Hofmann

6.1 Introduction 130

6.2 Melt Processable Rubber. 130
    6.2.1 Chemistry. 130
    6.2.2 Mechanical Properties. 131
    6.2.3 Chemical Resistance 133
    6.2.4 Melt-Processable Rubber Grades 134
        6.2.4.1 Extrusion Grades. 134
        6.2.4.2 Injection Molding Grades. 134
    6.2.5 Weather and Flame Resistance 134
    6.2.6 Electrical Properties 136
    6.2.7 Processing. 138
        6.2.7.1 Injection Molding 138
        6.2.7.2 Extrusion 138
        6.2.7.3 Calendering 140
        6.2.7.4 Extrusion Blow Molding 141
        6.2.7.5 Assembly 142
    6.2.8 Blends with Other Polymers 142
    6.2.9 Applications 142

6.3 Poly(vinyl chloride)–Nitrile Rubber Blends. 143
    6.3.1 Chemistry. 143
    6.3.2 Melt Compounding and Processing. 144
    6.3.3 Mechanical Properties. 145
    6.3.4 Chemical Resistance 147
    6.3.5 Applications 147

6.4 Poly(vinyl chloride)–Copolyester Elastomer Blends. 148
    6.4.1 Chemistry. 148
    6.4.2 Mechanical and Chemical Properties. 148
    6.4.3 Weather Resistance 148
    6.4.4 Melt Compounding 149
    6.4.5 Processing. 149
    6.4.6 Applications 150

6.5 Poly(vinyl chloride)–Polyurethane Elastomer Blends 150
    6.5.1 Chemistry. 150
    6.5.2 Melt Compounding and Processing. 150
    6.5.3 Weather Resistance 151
    6.5.4 Mechanical and Chemical Properties. 151
    6.5.5 Applications 151

Acknowledgments 151

References 152

6.1   Introduction

Thermoset elastomers, as a family, contain important members that have halogen atoms attached to the polymer backbone (e.g., polychloroprene, chlorinated polyethylene, and fluoroelastomers). Similarly, the family of thermoplastic elastomers (TPEs) contain such members. Chlorine is the halogen of most commercial significance found to date in both families of elastomers. This chapter is devoted exclusively to chlorine-containing TPEs since the newly emerging fluorine-containing TPEs are still in a relatively embryonic stage of development.

The vast literature on poly(vinyl chloride) (PVC) describes many plasticizers used to flexibilize PVC to give it some rubberlike characteristics while maintaining thermoplasticity. In PVC, the use of low molecular weight polymeric liquid plasticizers, as well as high molecular weight solid polymeric plasticizers, has been the subject of many previous reviews and articles [1–5]. This chapter includes only discussions of the blends of PVC with crosslinked or elastomeric polymers that give the properties of true TPEs. Reference to blends containing monomeric or nonelastomeric plasticizers is for comparison purposes only. The most important TPE based on PVC has been in use since the early 1970s and contains acrylonitrile–butadiene elastomer (NBR). For the most part, PVC–NBR products were proprietary compounds prepared for use primarily "in-house" by flexible PVC fabricators.

Alcryn® (DuPont registered trademark) melt-processable rubber (MPR) is a relative newcomer to the family of chlorine-containing TPEs. It was introduced to the plastics market in the late 1980s as a fully compounded pelletized product for direct fabrication purposes. It offered a number of property advantages over the existing TPEs based on PVC blended with NBR, copolyester (COP), and thermoplastic urethane (TPU). MPR pellets marketed to PVC fabricators, as well as to members of the general thermoplastics industry, allowed them to fabricate rubber goods directly on general-purpose thermoplastic equipment such as injection molding machines, extruders, calendars, etc.

6.2   Melt Processable Rubber

6.2.1   Chemistry

This family of TPEs is noted for its excellent rubberlike properties. Though thermoplastic, MPR parts perform, look, and feel like thermoset rubber. MPR is superior to many conventional midperformance vulcanized elastomers in its resistance to heat, oil, chemicals, and outdoor exposure. It maintains its key properties through many cycles of reprocessing; this, of course, is impossible with vulcanized thermoset rubbers.

These materials are described as "alloys of proprietary ethylene interpolymers and chlorinated polyolefins in which the ethylene polymer component has been partially cross-linked in situ" [6]. They are comprised of a blend of molecularly miscible polymers having a single T_g as shown in Fig. 6.1. Reinforcing agents such as carbon black or clay, plasticizers, and stabilizers have been added to give a functionally useful balance of properties over a wide range of end-use temperatures. These single phase, primarily amorphous polymer systems, are soft, flexible, have excellent recovery, and stress–strain curves essentially identical to those of typical thermoset elastomers. It is because of these inherent factors of rubbery character that they are defined as melt processable "rubber" as differentiated from the term "thermoplastic" elastomer used for two-phase systems such as thermoplastic vulcanizates (TPVs). In general, a TPV (e.g., Santoprene®) has a continuous matrix phase of thermo-plastic and a discontinuous elastomeric phase, resulting in a significantly different stress–strain curve.

6.2.2   Mechanical Properties

A rubber is soft, flexible, and elastic. Stress–strain curves are useful to measure flexibility and elasticity when comparing TPEs with rubber. The slope of the curve is a measure of the flexibility or stiffness of the material. Figure 6.2 compares the stress–strain curves of 70 Shore A TPV, MPR, vulcanized NBR, and vulcanized chloroprene (CR) rubber compounds. Note

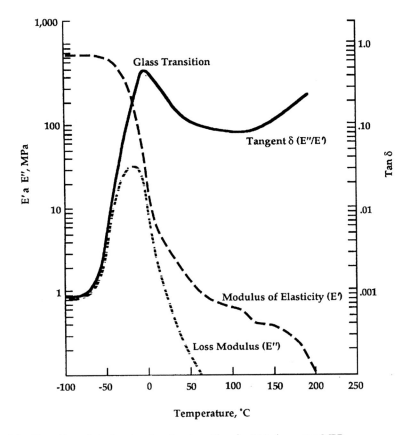

Figure 6.1   Rheovibron data showing the glass transition for 70A durometer MPR

6.2.4   Melt-Processable Rubber Grades

6.2.4.1   Extrusion Grades

In 1985, the MPR 1000 and 3000 series of products were first commercialized. The 3000 series materials are clay reinforced and available in a natural or off-white color. They are readily colorable using conventional color concentrates used for flexible PVC and are available in three hardnesses; Shore A 55, 65, and 75. The 1000 series are also available in three hardnesses. They are black grades, being reinforced with carbon black, and have somewhat enhanced physical properties over the 3000 series. Properties of all these grades are similar to those of general-purpose rubbers and are given in Table 6.2.

These grades produce relatively viscous melts as shown in Fig. 6.3. These grades are ideal for calendering, extrusion blow molding, and extrusion processing. They have been used, typically, for such extruded goods as hose, tubing, profiles, seals, gaskets, etc. A major market for these grades is coextrusions with PVC. MPR, being totally compatible with PVC, forms (when coextruded with PVC) a flexible rubber component strongly bonded to the rigid or semirigid PVC profile.

Injection molding of the 1000 and 3000 grades is not generally recommended. The high melt viscosity of these grades tends to limit them for use in simple parts with large cross-sectional areas. Molded-in stresses can be an additional consequence leading to distorted or warped parts. Compression molding can be performed on these grades, but prior fluxing of the pellets and sheeting out on a roll mill is recommended as a feedstock for the mold.

6.2.4.2   Injection Molding Grades

In 1989, additional grades designated as the 2000 series were introduced to the marketplace. These grades were designed to have enhanced flow (lower melt viscosity) for injection molding, and a viscosity–shear rate curve for a typical grade may also be found in Fig. 6.3. Six grades, three natural colored and three black, having Shore A hardnesses of 60, 70, and 80, can be readily injection molded into complex parts using equipment designed for flexible PVC molding compounds. Table 6.3 shows that most physical properties of the 2000 series products are similar to those of the 3000 series. An exception is the low-temperature properties which are superior to both the other series.

6.2.5   Weather and Flame Resistance

The weathering and ozone resistance of MPR are superior to those of other TPEs (styrenic, TPV, etc.) and also to those of many thermoset rubbers, including NBR, CR, and even ethylene-propylene-diene terpolymer (EPDM). Enhanced weather protection of the light-color grades can be achieved with suitable benzotriazole and hindered amine light stabilizers (HALSs), resulting in outstanding long-term property and appearance retention. Additional weathering stabilizers are unnecessary for the black grades of MPR. Because of this outstanding weatherability, MPR is finding wide usage in exterior construction applications such as roofing, glazing seals, and vent seals. Exterior components made of MPR are also

Table 6.2 Properties of Extrusion Grades of MPR

Properties	ASTM Test	Units	1000 Series			3000 Series		
			1060 BK	1070 BK	1080 BK	3055 NC	3065 NC	3075 NC
Mechanical								
Hardness, durometer A	D 2240	Shore A	60	70	78	56	65	75
Specific gravity	D 471	g/cm^3	1.19	1.23	1.25	1.18	1.26	1.35
Tensile properties								
Tensile strength	D 412	MPa	9.6	12.4	13.1	8.2	8.9	9.8
Elongation at break	D 412	%	300	270	210	440	400	360
100% Modulus	D 412	MPa	3.8	5.3	7.9	2.8	4.1	5.9
Torsional modulus	D 1043							
At 24 °C		MPa	1.9	2.2	2.8	1.3	2.1	3.4
At −20 °C		MPa	7.5	14.3	19.9	17.2	45.5	127.5
Tear strength								
Graves (Die C), at 24 °C	D 624	kN/m	26.3	28.0	24.5	28.9	35.9	49.0
Permanent set (tension)	D 412	%	8	10	8	6	9	11
Compression set, Method B	D 395							
After 22 h at 24 °C		%	15	15	15	17	17	23
After 22 h at 100 °C		%	55	55	55	65	69	67
Heat aging resistance								
Tensile properties after 7 days oven aging at 125 °C	D 573							
Tensile strength	D 412	MPa	10.6	13.1	14.0	8.7	8.9	10.5
Elongation at break	D 412	%	325	235	190	450	370	350
100% Modulus	D 412	MPa	3.8	5.3	9.4	2.4	4.5	6.6
Hardness, durometer A	D 2240	Shore A	67	70	77	58	65	74
Low-temperature properties								
Brittleness temperature	D 746	°C	−51	−53	−44	−54	−45	−30
Clash–Berg stiffness temperature (69 MPa)	D 1043	°C	−38	−34	−30	−28	−23	−17
Tabor abrasion, CS-17 wheel, 1000 g load	D 1044	mg/1000 cycles	7	7	5	<1	<1	<1
Chemical								
Fluid resistance—volume change								
After 7 days in water at 100 °C	D 471	%	12	8	10	15	14	13
After 7 days in ASTM oil No. 1 at 100 °C	D 471	%	−10	−9	−8	−12	−9	−6
After 7 days in ASTM oil No. 3 at 100 °C	D 471	%	27	25	23	25	30	29
After 7 days in ASTM Ref. Fuel B at 24 °C	D 471	%	30	30	29	30	32	36

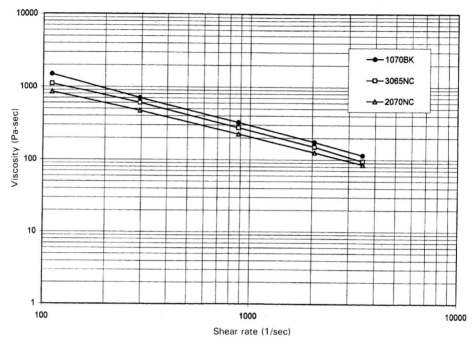

Figure 6.3   Effect of shear on viscosity at 190°C for three MPRs of similar hardness (65 to 70 Shore A)

being adopted for automobiles, trucks, agricultural equipment, marine equipment, and sporting goods.

It is relatively easy to make flame-retardant MPRs. The addition of antimony trioxide produces, during burning, the well-known synergistic reaction with the chlorine contained in the MPR, giving flame retardant properties. Flame-retardant MPR has utility in such applications as kitchen appliances, electrical boxes, and hospital and marine equipment.

6.2.6   Electrical Properties

MPRs are not targeted for electrical insulation applications. They are suitable for some low-voltage applications, but serve best as a durable protective jacketing material over electrical insulation. A major benefit can be obtained from MPRs in applications that take advantage of their surface conductivity. Such applications are in areas where static charge can lead to dust accumulation which, in turn, can lead to damage to electronic components, explosions in dust areas, where conveyor belting is used, tracking of dirt into "clean rooms," and paper sticking to drive rolls in duplicating machines.

Table 6.3 Properties of Injection Molding Grades of MPR

Properties	ASTM Test	Units	2000 Series					
			2060 NC	2070 NC	2080 NC	2060 BK	2070 BK	2080 BK
Mechanical								
Hardness, durometer A	D 2240	Shore A	60	70	78	60	70	80
Specific gravity	D 471	g/cm^3	1.12	1.20	1.26	1.10	1.14	1.17
Tensile properties								
Tensile strength	D 412	MPa	8.0	8.3	9.9	7.5	8.5	11.4
Elongation at break	D 412	%	430	375	340	400	300	285
100% Modulus	D 412	MPa	3.1	3.9	5.5	2.9	4.3	6.4
Torsional modulus	D 1043							
At 24°C		MPa	2.3	2.2	3.0	2.2	2.3	3.2
At −20°C	—	MPa	4.8	8.5	14.3	5.9	10.2	27.5
Tear strength								
Graves (Die C), at 24°C	D 624	kN/m	27.1	28.9	33.3	26.3	26.3	35.0
Permanent set (tension)	D 412	%	8	9	11	9	9	10
Compression set, Method B	D 395							
After 22 h at 24°C		%	12	14	17	12	13	14
After 22 h at 100°C		%	60	65	63	62	66	65
Heat aging resistance								
Tensile properties after 7 days oven aging at 125°C								
Tensile strength	D 573	MPa	6.5	5.5	5.5	7.6	8.4	11.0
Elongation at break	D 412	%	340	220	135	390	280	235
100% Modulus	D 412	MPa	2.7	3.5	4.4	2.7	4.4	5.2
Hardness, durometer A	D 2240	Shore A	60	65	71	63	70	76
Low-temperature properties								
Brittleness temperature	D 746	°C	−85	−85	−76	−87	−79	−86
Clash–Berg stiffness temperature (69 MPa)	D 1043	°C	−42	−40	−32	−40	−40	−17
Tabor abrasion, CS-17 wheel, 1000 g load	D 1044	mg/1000 cycles	5	9	10	5	5	3
Chemical								
Fluid resistance—volume change								
After 7 days in water at 100°C	D 471	%	8	7	8	8	6	5
After 7 days in ASTM oil No. 1 at 100°C	D 471	%	−21	−16	−14	−19	−17	−8
After 7 days in ASTM oil No. 3 at 100°C	D 471	%	18	21	25	20	22	30
After 7 days in ASTM Ref. Fuel B at 24°C	D 471	%	17	22	29	25	25	32

6.2.7   Processing

6.2.7.1   Injection Molding

MPR is essentially amorphous, having no detectable crystallinity. Because of its gelled network, significant flow can be induced only by application of a shearing force in combination with elevated temperature. Pseudoplastic flow, or shear thinning, is the mechanism by which shapes are fabricated. The viscosity vs. shear rate plots of Fig. 6.3 clearly show this shear thinning effect at a typical processing temperature for both injection molding and extrusion grades.

The rheology of the molding grades makes them ideally suited to the high shear conditions employed during injection molding. The combination of band heat and controlled shear, imparted by the rotating screw, will generate a properly fluxed uniform melt. The high shear produced in the gate and runner system during the injection part of the molding cycle facilitates filling the mold. Since MPR is completely amorphous, it does not exhibit any discontinuous volume change due to crystallization on cooling. MPR is susceptible to overpacking that can give warpage and shrinkage of the molded part. Thus it is essential that pressure be rapidly reduced as soon as the mold cavity is filled. MPR can be processed on a wide range of thermoplastic reciprocating screw injection molding machines and, under special circumstances, on modified rubber screw-ram injection presses. Rubber presses need to be modified to produce higher barrel temperatures and pressures than normally needed for rubber processing.

Proper screw selection is important for injection molding MPR. General purpose, gradual transition screws with compression ratios between 2.5 and 3.5 and an *L/D* of 20 : 1 are recommended. Screws equipped with full-flow ring check valves or smear tips may be used. Screws with a short compression zone (two flights) and long metering zones (six flights) with very shallow flights should be avoided since they tend to overheat the melt at high screw rpm. Flow passages throughout the machine must be carefully streamlined to eliminate melt stagnation and subsequent degradation. The use of corrosion-resistant materials will maximize equipment life. "Hastelloy" C-276 is recommended for screws and check rings; "Xaloy" 306 is recommended for barrel liners; and either hardened, type 420 stainless steel or electroless nickel plated steel are recommended for molds. Suggested injection molding conditions and settings are listed in Table 6.4. Molded parts of MPR rapidly develop strength in the mold, and parts can be demolded easily from hot molds. Large knockout pins are recommended because of the rubberlike nature of MPR. Uniform mold cavity temperature control is essential for good dimensional control of molded parts. Typically, mold shrinkage is slight [7].

6.2.7.2   Extrusion

The viscosities of all the MPR grades are relatively high (≥ 1000 Pa·s) at the low shear rates commonly found in extrusion. The lower viscosity molding grades, which also have excellent extrusion processability, show a greater viscosity response to temperature change than do the extrusion grades. In general, within a series, the softer grades have lower viscosities than the harder grades.

**Table 6.4 Reciprocating Screw Injection Molding Machine
Conditions**

	Temperature (°C)
Barrel	
Rear	170–180
Front	170–180
Nozzle	170–180
Mold	20–50
Molding stages	
Injection speed (cm³/s)	20
Injection pressure (MPa)	4–9
Injection time (first stage/boost) (s)	0.5–10
Second stage pressure (MPa)	2–6
Second stage time (s)	3–10
Cooling time (s)	2–20
Screw speed (rpm)	50–100
Back pressure (MPa)	0.2–0.6
Shot size	Control to fill mold

Rubber extruders, designed for low-viscosity gums, are generally inadequate for MPR extrusion. Best results are most often obtained with typical plastics extruders used for such common thermoplastics as plasticized nylons, PVC, or polyolefins. Length-to-diameter ratios of at least 20 : 1, and preferably 24 : 1, provide best quality. Extruder diameters up to 152 mm (6 in.) have been successfully used for MPR. Efficient cooling, to remove the heat generated by shearing of the material, is essential to obtain high productivity and to minimize degradation. Screw shear should be moderate and constant to obtain a good homogeneous melt. For most MPR extrusions, a simple three-zone screw, having a transition (compression) zone of at least one third of the screw length is recommended. Short compression zones should be avoided as they result in high localized shear and material overheating. Compression ratios of between 2.5 : 1 and 3.5 : 1 should be used, with the higher ratio recommended for the lowest viscosity grades and the lower ratio for the high-viscosity grades. When especially good mixing is required, as when color concentrates are incorporated, the use of a screw with a "barrier" section and/or a mixing element may be advantageous. A screw tip that is rounded or otherwise streamlined is necessary to avoid stagnation of polymer and subsequent degradation. Likewise, the breaker plate, adapter, and head designs should be streamlined. A screen pack may not be needed. If one is used it should be coarser or equal to 100 mesh to avoid excessive back pressure.

Since the grade of MPR, extruder size, and screw speed all affect the output and die swell, it is desirable to prototype each die design under the final conditions of use. The entrance to the die should be tapered at an angle of between 30 and 60 degrees to increase the flow velocity on the wall of the die and so optimize profile definition. A streamlined, well-balanced die requires shorter land lengths in the thin sections to prevent preferential flow through the large sections of the profile. The ratio of the land length to the thickness should be around 5 : 1. A minimum radius of 0.3 mm is recommended at sharp edges. The die swell of MPR can be 10% or greater, depending on die design. Increased die swell is given by the following guidelines:

- Increasing softness,
- decreasing thickness,
- increasing output, and
- decreasing land lengths.

The combination of heat and shear necessary to produce a uniform, properly fluxed melt depends on a number of factors, including the grade of MPR, the length/diameter ratio of the machine, and the compression ratio of the screw. Table 6.5 summarizes some general extruder configurations along with some typical temperature guidelines [8].

6.2.7.3   Calendering

MPRs are easily processed on calendering equipment which is capable of reaching the necessary temperature range to flux them. Unsupported and supported sheets, which display excellent heat, oil, and weather resistance, can be produced in a single pass without blisters or defects. MPRs can also be calendered to produce supported sheet using various substrates. Embossing is easily accomplished by using an embossing roll while the stock is still warm. In most circumstances, MPRs have sufficient external lubrication to achieve easy release of the sheet from the calender rolls. Only under very extreme conditions (very thin gages, high calendering speed, and high roll temperature) may it become necessary to add external lubricant to avoid roll sticking. Polymeric lubricants such as oxidized polyethylene waxes have shown very good efficiency without negative effects on performance or appearance of the calendered sheet. They should be incorporated during the pellet fluxing process.

Three- or four-roll calenders normally used for processing rubber or plastic can be used for processing MPR. They must be capable of maintaining roll temperatures in the range of 160 °C to 185 °C. Batch or continuous mixers for feedstock preparation must be capable of delivering fluxed stock at a temperature between 165 °C and 185 °C. Mills and/or strainers used for storage (buffers), or to deliver material to the calender, must also keep material at these temperatures. Top calender rolls should all be set to the same temperature in the range of 160 °C to 185 °C. The temperature of the lower rolls should be high enough to maintain tracking but low enough to prevent sticking (as low as 140 °C). Uneven speeds (about 1.05 : 1) between adjacent rolls will prevent blisters and roll speed may be adjusted to get desired production rates. A small uniform bank must be maintained at the calender nip. A small, pencil thin bank between the second and third rolls is sometimes useful to prevent blisters or blemishes in the sheet. The feed to the calender rolls must be hot and well fluxed and should be metered to provide a small, uniform rolling bank in the calender nip. Hot stock

Table 6.5 Extrusion Grades (1000 and 3000 Series) Temperature Profile

Length/diameter	Compression ratio	Temperature profile	Feed[a]	Transition meter[a]	Adapter/die[a]	Melt[b]
Long	High	Increasing	150 °C	160–170 °C	165 °C	180
		Flat	165 °C	165 °C	165 °C	180
Short	Low	Reverse	175 °C	170–160 °C	165 °C	180

[a] Temperature should be increased 10 °C to 15 °C when extruding the 2000 series
[b] Temperature should be increased 5 °C when extruding the 2000 series

can be fed as metered strip directly from an extruder or can be fed as a pig from an internal mixer or mill. Feed stock temperature must be maintained in the range of 160 °C to 185 °C [9].

6.2.7.4   *Extrusion Blow Molding*

As in any fabrication process for MPRs, systems that minimize holdup and possible degradation work best. Continuous extrusion, accumulator, and reciprocating screw systems all can be used for MPRs with proper residence time and temperature control. Extruders suitable for use with common thermoplastics (particularly those intended for PVC) are usually suitable for MPRs. Length-to-diameter ratios (*L/D*) of between 15 : 1 and 30 : 1 are usually satisfactory, with 20 : 1 or greater being preferred. Longer barrels provide a more uniform, well-fluxed melt. Shear is also necessary with MPRs to produce a uniform, well-fluxed melt. To provide the required shear, a section of the screw should have several flights that are no more than 0.100 in. (2.5 mm) deep. Any device that increases mixing and shear such as pins, torpedoes, dams, or barriers will improve plastication and broaden the processing window. The compression ratio should be between 2.0 and 3.0. Only moderate working is necessary; high shear screws could cause overheating and polymer degradation. A breaker plate of streamlined design supporting two 100-mesh (or coarser) screens is generally used for extrusion blow molding of MPRs. To maintain melt quality and uniformity, the accumulator should have as small a volume as is reasonable, smooth continuous contours, and good temperature control. MPRs are molded routinely at a blow ratio of 2 : 1 and have been formed satisfactorily at 3 : 1. It is good practice to size the parison die as close to final part dimensions as possible. Thus, parison diameter and wall thickness depend on the desired diameter and wall thickness of the part. The die should be streamlined to avoid polymer holdup. Generally, all operating variables must be optimized empirically for each machine and for each part.

Regardless of which extrusion blow molding process is utilized, a combination of barrel heat and shear is necessary to convert pellets of MPR into a properly fluxed, uniform melt. Melt temperature measured at the parison die or at the exit to the accumulator head should be between 160 °C and 185 °C. In general, barrel temperatures are set in a "reverse profile" (with the feed end set to a higher temperature than the head end) when extruding MPR. Additional heat to achieve the desired melt temperature is generated by shear. Screw speed is usually adjusted to achieve the output desired. Typical screw speeds range from 20 to 60 rpm. Increasing screw speed, with a suitable screw, will increase melt temperature due to shear heating. In this case barrel temperature set points may have to be lowered to maintain the desired melt temperature. This effect of shear is most noticeable with very high shear screws. With marginally low shear screws, increasing screw speed may actually decrease melt temperature because of reduced residence time within the barrel. If this occurs, barrel temperature settings may have to be increased to maintain the desired melt temperature. Parison drop time depends on part size, extruder size, and screw speed and is typically 1 to 2 s. Moderate pressures and blow rates generally yield well-defined parts while avoiding blowouts. Blow pressure typically ranges between 200 and 700 kPa gauge (30 to 100 psig). Chilled molds are recommended when extrusion blow molding MPRs. Excellent parts have been obtained with mold temperatures of 5 °C to 60 °C. Blow-molded parts do not show anisotropic (orientation) effects. Parts produced from MPRs develop strength rapidly in the absence of shear. They generally do not stick to the mold and have enough hot strength to demold easily in very short cycles—typically, 2 to 3 s [10].

6.2.7.5   Assembly

MPRs can be joined by mechanical fasteners, adhesives, or welding. Mechanical fasteners are quick but can be expensive and may not provide leaktight joints or the strength required for the particular application. MPRs have been satisfactorily adhered to various substrates including thermoset rubbers, metals, plastics, textiles, leather, and wood. Adhesive bonding provides good properties and sound joints but may not be desirable because of handling difficulties, material cost, environmental considerations, processing speed, or other reasons. Welding is often preferred because it is simple, fast, reliable, and capable of making joints with excellent strength. MPRs have been satisfactorily welded using both mechanical energy (ultrasonic), external heating (hot plate and gas), and electrical energy (radiofrequency power and electromagnetic induction).

6.2.8   Blends with Other Polymers

MPR is compatible in all proportions with PVC, certain types of copolyester (COP), and thermoplastic urethanes (TPUs). A study [11] of mixtures of these resins with MPR showed that (1) TPU improved toughness, abrasion, and oil resistance; (2) COP improved compression set, tear resistance, and low-temperature properties; and (3) PVC lowers cost and improves tear resistance. Mixtures of these components provide property combinations not available in any single component of the mixture. A practical example of this concept is the property enhancement of flexible PVC by the addition of MPR and COP. Improved resistance to plasticizer extraction by perchloroethylene dry cleaning solvent is one advantage. Another is resistance to stiffening by animal and vegetable fats, often found in food processing environments. These improvements are achieved without sacrificing the good resistance to detergents which is also necessary in such environments. A key mechanical property, flexural resistance, is improved by factors of up to five.

Various acrylic process aids, used alone or in combination with each other, give substantial increases, of as high as 100-fold, to the melt flow of high-viscosity MPRs. Some of these process aids do not significantly change the key physical properties while others increase hardness and stiffness as well as melt flow [12].

6.2.9   Applications

Several applications for MPRs in industry, agriculture, transportation, and construction have already been mentioned in previous sections of this chapter. Consumers often equate a glossy, shiny surface and a slippery, "cold" touch with plastics. In applications where a low-gloss, soft, "warm," high-friction feel is desirable, a true rubber is needed. MPRs meet this requirement while most TPEs do not. Thus automakers use MPRs for interior trim, knobs, grips, steering wheels, etc. Appliance makers use bright colored, "soft touch," flexible parts for kitchen appliances. The electronic industry values soft-touch grips and push buttons for handheld devices such as telephones, cameras, razors, etc. Home garden and commercial agricultural equipment, hose covers, seals, insulations, and even toys all require true rubberlike products to satisfy the consumer. Consumers prefer power tools that look like metal and rubber rather than plastic and vinyl. However, at this time no grades of MPR have FDA food contact approval.

Another major application area is profile coextrusions of rigid PVC with MPRs. These are used in architectural structures such as window seals and automotive profiles used for window lace and outer belt strip moldings. A prime example of an application capitalizing on essentially every beneficial characteristic of MPRs is their usage in oil bag constructions. These are large, flexible containers into which oil from a damaged tanker can be transferred. When not in use, the oil bag is extremely compact; but when it is launched into the sea, it can hold up to 20,000 cubic meters of oil. The bag's flexibility, plus its light weight, permits ease of handling, operation, and storage. The ease of producing this construction of strong fabric coated with a brightly colored thermoplastic MPR was a major consideration in selecting an MPR for this application. Another significant function of the bag is its ability to not only withstand oil and sea water but also to be weather and ozone resistant.

6.3   Poly(vinyl chloride)–Nitrile Rubber Blends

6.3.1   Chemistry

Before 1970, PVC was added to vulcanizates of elastomeric copolymers of acrylonitrile and 1,3-butadiene (NBR or nitrile rubber) to improve ozone and solvent resistance. Later, nitrile rubbers, already available to the rubber industry in bale form, were made available in powder form to the PVC industry. In 1983, the first powdered NBR developed specifically for the PVC industry became available. Since that time a number of others, including food grades and improved processing grades, have been commercialized.

Thermoplastic blends are created when the predominant polymer is PVC. As well as NBR, they also contain PVC plasticizers such as dioctyl phthalate (DOP), fillers, stabilizers, etc. A typical formulation is:

PVC	100	Parts
NBR	33	Parts
DOP	75	Parts
$CaCO_3$	20	Parts
Ba–Zn stabilizer	2	Parts
Expoxidized soya oil	2.5	Parts

These blends bridge the gap between conventional liquid plasticized PVC and conventional cured elastomers. Properly formulated PVC–NBR blends have rubberlike look and feel. They are flexible at low temperatures and have good tear strength. They show low compression set, good abrasion resistance, and minimum swelling or extraction when immersed in oils or solvents.

Important formulation variables, besides the ratio of PVC to NBR, are the acrylonitrile (ACN) content of the NBR, the Mooney viscosity of the NBR, the molecular weight of the PVC, the stabilizers and liquid plasticizers used, and the amount of reinforcing fillers added. In general, NBR elastomers containing 30% to 40% ACN are needed to obtain single-phase homogeneous blends, with optimum ACN content of about 33%. These are true plasticizers that provide the desired rubberlike properties. Reports of a single glass transition temperature (T_g), as measured by differential thermal analysis [13], intermediate between the T_gs of the two polymers, and the absence of two phases by electron photomicroscopy [14], supports the single-phase concept. Other examples have been reported of compositions with two glass transitions [15] and micro-domains rich in one or the other of the components [16]. This indicates that the morphological behavior of these blends is complex. The NBR may be either noncrosslinked or crosslinked to varying degrees. Noncrosslinked NBR gives the lowest viscosity blends, requiring the least energy to process. This can provide a wider processing margin before thermal degradation of the PVC occurs. The lower viscosity makes these blends most suitable for injection molding formulations. The crosslinked versions provide the least polymer memory (die swell), making them suitable for formulations that are calendered or extruded. The greater crosslink density also results in a greater resistance to compression set.

6.3.2   Melt Compounding and Processing

Because of the special handling and mixing equipment used in the PVC industry, free-flowing NBR powders are preferred. These powders have an average particle size of about 0.5 mm (0.02 in.) and contain about 10% of a partitioning agent consisting of either PVC, calcium carbonate, or silica [17]. They are designed to be free flowing, to be air conveyable, to resist separation from PVC blends, and to resist agglomeration during long-term storage. The NBR powder can be added to the PVC dry blending cycle after the PVC has absorbed all of the liquid plasticizers (this is known as the "dry point"). The temperature of the PVC blend to which the NBR is added should be no higher than about 40 °C to avoid rubber agglomeration [18]. The PVC–NBR powder blends can be processed in conventional thermoplastic mixing equipment, that is, low- and high-intensity mixers, single- and twin-screw extruders, continuous mixers, and kneaders. These blends can be either melt compounded into pellets before further processing or can be directly fabricated into finished shapes. Conventional PVC melt processing equipment is recommended for fabrication.

In recent years, fully compounded blends of PVC–NBR provided in pellet form have become available to rubber fabricators. They have been marketed as alternatives to flexible vinyl, mid-performance elastomers and to "in-house" PVC–NBR compound development. The rheology curve for a typical commercially available PVC–NBR compound (70 Shore A) is shown in Fig. 6.4 [19]. It shows lower viscosity than either the MPRs in Fig. 6.3 or typical "in-house" compounded PVC–NBR blends.

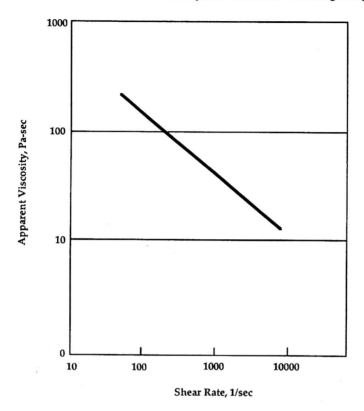

Figure 6.4   Commercial PVC–nitrile rubber compound (Sarlink 1170), capillary rheology at 180°C

6.3.3   Mechanical Properties

The physical properties of these compounds are typical for those found in mid-performance vulcanized rubbers. The stress–strain curves for three compounds [20] are shown in Fig. 6.5. Their behavior resembles that of a TPV rather than an MPR or a conventional vulcanized rubber (see Fig. 6.2). The stiffness of the compound increases slightly with increasing crosslinking and viscosity, and the elongation decreases somewhat. Other properties are shown in Table 6.6. Resistance to both low- and high-temperature compression set appears to be somewhat improved by use of highly crosslinked rubber. As for the remaining properties shown in this table, rubber selection does not appear to greatly affect hardness, tear resistance, or brittle point. Some drop-off in abrasion resistance, however, is evident at the highest crosslinking level.

In another study [14], using a higher level of nitrile rubber, a lower level of liquid plasticizer, and no filler, greater differences in physical properties were evident when comparing formulations containing crosslinked vs. uncrosslinked rubber. The formulation containing the crosslinked NBR showed significantly superior tensile strength, compression set resistance, abrasion resistance, and flexural properties. The molecular weight of the PVC also has a strong effect on the compound's physical properties; as it was increased, the tensile,

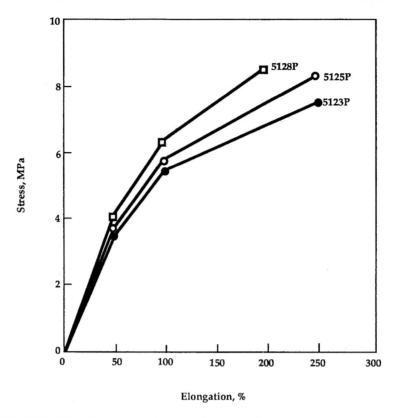

Figure 6.5   PVC–nitrile rubber compounds, stress–strain properties

Table 6.6 Flexible PVC Compounds Based on Powdered Nitrile Rubbers

	5123 P	5125 P	5128 P
NBR rubber properties			
AN content	33	33	33
Mooney viscosity (ML-4)	30	55	80
Precrosslinking	None	Medium	High
Shore A hardness			
Instantaneous	68	68	69
15-s Delay	64	64	63
Taber abrasion resistance, H-18 wheel, 1000 g load, 2000 rev g loss	0.657	0.644	0.713
Tear strength, ASTM D-624 die C (kN/m)	38.5	42.0	40.3
Brittlepoint, ASTM D-746 (°C)	−36	−37	−36
Compression set, ASTM D-395 B, (22 h, 25% compression)			
23 °C (%)	23	20	19
100 °C (%)	71	70	65

modulus, and hardness values of the blends also increased. Compression set also improves with increasing PVC molecular weight. Very soft compositions, having Shore A hardnesses below 50, can be produced by increasing the liquid plasticizer level. Liquid plasticizers are most often used in PVC–NBR compounds to improve low-temperature flexibility. The adipate and triglycol ester type plasticizers are more effective than phthalate ester types in maintaining both low-temperature flexibility and fuel B resistance of the compounds. The adipate, however, due to its volatility, gives inferior oven aging.

Physical properties of a typical PVC–NBR compound and two typical thermoset rubber compounds were compared at temperatures up to 121 °C [14]. At room temperature, the tensile strength of the PVC–NBR compound was comparable to that of the thermoset elastomers but was significantly less at higher temperatures. This behavior can be attributed to the crosslinked NBR in combination with the strong hydrogen bonding between the PVC and the rubber. These effects essentially add up to "match" the strength of the more highly crosslinked thermoset rubber. As the temperature increases, however, the hydrogen bonding decreases in the PVC–NBR blend, leaving only the effect of the crosslinked NBR. This contributes much less strength than in the case of the more highly crosslinked thermoset elastomers. Thus thermoset rubbers are required when high service temperature integrity is needed but at the sacrifice of recyclability of scrap material.

6.3.4   Chemical Resistance

Fluid resistance tests show fair resistance to Reference Fuel B with a slight reduction in volume swell as crosslinking is increased. These compounds appear to harden excessively in ASTM No. 2 oil and also show large reductions in volume numbers. Obviously, the DOP plasticizer is being extracted [20]. Reformulation with less extractable polymeric plasticizer should remedy this deficiency, but at a somewhat higher ingredients cost.

Because of the susceptibility of the NBR double bonds to oxidation, antioxidants are an essential ingredient in compounds requiring long-term endurance. Formulations without an antioxidant become embrittled and show loss of elongation and hardening when oven aged at 113 °C for 1 week. Retardation of the oxidation in the similar formulations containing an antioxidant show less change in elongation but hardening still occurs. The hardening is attributed primarily to volatility loss of the liquid plasticizer. Use of a nonvolatile polymeric plasticizer minimizes the increase in hardness.

6.3.5   Applications

The highly polar PVC–NBR compounds, when properly formulated, show excellent resistance to oil and chemicals, and excellent abrasion resistance and assembly capability either through adhesive bonding or by thermal, radiofrequency, or ultrasonic welding techniques. These compounds are used in applications requiring endurance under service conditions unacceptable for conventional plasticized PVC. Applications that have used PVC–NBR compounds, listed according to the fabrication technique, include seals and gaskets, hose and tubing and cable jacket (made by extrusion), boots and soling, grommets and plugs (made by injection molding) and sheeting, pads and flooring (made by calendering), and bellows and sleeves (made by extrusion blow molding). Products exposed to outdoor weathering require an adequate protection additive package which can include pigmentation, antioxidants, and ultraviolet light absorbers.

6.4   Poly(vinyl chloride)–Copolyester Elastomer Blends

6.4.1   Chemistry

Random block copolymers that consist of crystallizable tetramethylene terephthalate (4GT) hard segments and amorphous elastomeric polytetramethylene ether glycol (PTMEG–T) soft segments are described in Chapter 8. Blends of these copolyesters (COPs) with plasticized PVC show a high degree of compatibility, resulting in materials that combine the TPE properties of the COP with the excellent processing characteristics of plasticized PVC [21]. Blends are generally based on low-melting COPs which typically contain about 33% 4GT and melt below 180 °C.

6.4.2   Mechanical and Chemical Properties

Addition of these COPs to a plasticized PVC compound improves such properties as the low-temperature flexibility and impact resistance, abrasion and tear resistance, and resistance to oil and fuels.

As little as a 25% loading of COP produces considerable improvement in these properties, but also increases the hardness and room temperature torsional modulus [22].

The results shown above describe the effects of adding COP to a plasticized PVC. This, in effect, maintains the plasticizer/PVC ratio constant. If instead blends are prepared using a constant ratio of plasticizer to total resin (PVC plus COP), other relationships are observed. In this case, as the COP level increases both hardness and modulus decrease. At higher COP levels, the blends are stiffer at low temperatures and less resistant to oil and fuel extraction because the DOP/PVC ratio is increased. However, the critical property of heat distortion resistance is enhanced by addition of COP.

A typical cable jacket formulation, containing equal levels of PVC and COP, has significant property advantages over a conventionally plasticized PVC formulation. The blend has superior low-temperature flexibility, heat aging resistance, abrasion resistance, and greater elongation at break both before and after oil immersion. It also had good electrical properties such as dielectric strength and volume resistivity and good resistance to water, chemicals, and cut growth.

6.4.3   Weather Resistance

Protection from ultraviolet radiation is necessary for these compounds. A combination of 0.2 phr benzotriazole UV absorbers, 0.1 phr hindered amine light stabilizer, 0.2 phr hindered phenol antioxidant, and 0.03 phr rutile TiO_2 gives excellent protection against outdoor exposure. Of course, if colorability is not important, carbon black (at 2.5 phr) also affords excellent ultraviolet screening.

6.4.4   Melt Compounding

The plasticized PVC portion of the blend is prepared from suspension grade PVC by conventional industry methods, either in a high-intensity mixer or a heated ribbon blender. Ester type plasticizers such as DOP are most commonly used. Plasticizer levels between 10 and 50 parts per 100 of polymer (php) are recommended for the PVC–COP blends. This is less than the 1 : 1 level often used in plasticized PVC. Plasticizer levels below 10 php may give poor impact resistance and processability. Polymeric ester plasticizers or trimellitate plasticizer should offer improved heat and oil resistance over DOP. Ground limestone and whiting are the most common fillers for these systems. China Clay is also used for electrical insulation compounds. Filler levels 10 to 20 php are typical, although up to 50 php have been used. It is important to use neutral pH grade (i.e., pH 7 to 8) filler to avoid degradation of the COP. Although stearic acid and other fatty acids are used for both internal and external lubrication of plasticized PVC compounds, they are not recommended for use in these COP–PVC blends, since they also promote degradation of the COP. Instead, fatty acid amides (used at the 1% level) are recommended.

The COP can be mixed into the plasticized PVC powder blend before melt compounding, or the PVC and COP can be metered independently to the melt compounding device. Melt compounding can be carried out in conventional compounding machines including Banbury batch mixers, kneaders, and single- or twin-screw extruders with sufficient mixing sections. Maximum melt temperature should be less than 190 °C. The product is dried to less than 0.10% moisture, and packaged in moisture-barrier packaging.

6.4.5   Processing

PVC–COP blends can be processed on conventional injection molding and extrusion equipment. Gradual transition screws, such as those used for polyethylene, are preferred to avoid overworking the material. Compression ratios of 3.0 : 1 to 3.5 : 1 are recommended, as well as length/diameter ratios of at least 18 : 1 for molding and 24 : 1 for extrusion. Relatively high metering zone depths, from 2.5 to 3.0 mm for a 6.35 cm (2.5 in.) screw, are also recommended.

Since COPs are subject to hydrolysis at processing temperatures, it is absolutely essential that the PVC–COP compound be dry for processing. PVC is inherently thermal unstable and so heat history during processing should be kept to a minimum. Temperatures at the feed zone in the region of 130 °C to 155 °C are suggested with a gradual rise in temperature toward the nozzle to about 150 °C to 170 °C. The preferred melt temperature range for a 50 : 50 PVC–COP blend is 160 °C to 170 °C. Increased melt temperature, particularly for injection molding, may be required at higher levels of copolyester or if the plasticizer level is lower than 30 php. In any event, maximum melt temperature should never exceed 190 °C. As with all PVC-containing materials, corrosion-resistant materials of construction are recommended so as to maximize equipment life.

6.4.6   Applications

PVC–COP blends offer a lower cost alternative to thermoplastic urethanes in protective jackets for hose and tubing. At the same time they offer an extended operating temperature range over 100% PVC sheathing. In automotive primary wiring, PVC–COP blends can withstand higher under-the-hood temperatures than 100% PVC systems. For sport shoe soling, PVC–COP blends cost less than thermoplastic urethanes but have equivalent low-temperature properties and abrasion resistance. They also do not suffer from the cracking problems associated with 100% PVC soling that is caused by plasticizer migration. PVC–COP blends are also used as protective jackets in retractile cords, for telephones, and domestic electrical appliances because of the better memory characteristics compared to plasticized PVC. They do not lose their retractile characteristics because of tension set or temperature. Although material costs are higher than for plasticized PVC, PVC–COP blends offer a lower cost finished product. Because of the better memory characteristics of the blend, cord covers can be reduced in thickness. This also enables the constructor to use fewer strands of larger diameter copper conductors in place of the multiple strands for thinner conductors, which offers additional savings. For record player turntable mats, PVC–COP blends again offer a lower cost alternative to TPU while overcoming the plasticizer migration problems of plasticized PVC. In fire hose jackets, these blends also offer a low-cost alternative to TPU while maintaining performance characteristics.

6.5   Poly(vinyl chloride)–Polyurethane Elastomer Blends

6.5.1   Chemistry

Polyurethane thermoplastic elastomers (TPUs), described in detail in Chapter 2, are block copolymer combining sequences of low glass transition, amorphous or low melting "soft" block with a rigid "hard" segment that has a crystalline melting point above room temperature. There are several choices for both the "soft" and "hard" segments. Most of these TPU types are very compatible with PVC. The blends exhibit only one major glass transition whose position on the temperature scale is raised with increasing levels of PVC [23].

6.5.2   Melt Compounding and Processing

Because of the heat sensitivity of PVC, only the softest TPUs (those with hardness values of Shore 80A and below) can be safely melt compounded with it. The Shore 80A grades of TPU require about a 200 °C processing temperature, while the softer grades allow a somewhat lower temperature. The melt compounding and subsequent processing recommendations for

PVC–TPU blends are essentially the same as given in the previous section for PVC–COP blends. PVC–TPU blends are similarly subject to hydrolysis; so it is essential that before processing the material be dried to less than 0.03% moisture content to maintain optimum chemical and mechanical properties.

6.5.3   Weather Resistance

For black applications, up to 5% carbon black is recommended for protection of PVC–TPU blends against UV. For natural color, benzotriazole type UV absorber is recommended at up to 2% loading. Antioxidants such as hindered phenolic and organosulfur types can also be incorporated to extend the material's outdoor life.

6.5.4   Mechanical and Chemical Properties

PVC–TPU blends combine the wear resistance and toughness of the TPU with the stiffness and high modulus of the PVC. A broad range of hardnesses can be formulated by adjusting the plasticizer level in the PVC and using various hardness grades of TPU. A PVC–TPU blend containing 30% TPU is equivalent to a commercial plasticized PVC compound in all respects and displays better abrasion resistance and low-temperature flexibility. The addition of 15% TPU doubles the Ross flex resistance at $-29\,°C$ [24].

The oil resistance of PVC–TPU blends is also improved. Immersion in ASTM No. 3 oil for 7 days at room temperature has minimal effect on volume swell and causes no decrease in tear strength. Increasing the TPU content improves the flexural performance. A compound containing 30% TPU is markedly better than a commercial plasticized PVC product of the same hardness with respect both to flex life and also to cut growth resistance after oil immersion. This trend continues as the TPU content is further increased. There appears to be an optimum TPU content of 40% for oil resistance and an optimum TPU content of 50% for the other properties [20].

6.5.5   Applications

Blends of PVC–TPU have been successfully used for shoe sole and shoe heel applications. They can be used in applications where the essential requirements of wear resistance, oil resistance, good flexibility, good compression set, and reasonable cost are required. They have generally better performance than plasticized PVC but cost less than pure TPU.

Acknowledgments

The author wishes to thank John F. Hagman, Ronald E. Myrick, and Robert E. Stuart for contributing much of the technical data in this chapter. A special thanks goes to Margaret R. Maney and to Susan I.W. Hansen, who helped prepare the final manuscript.

References

1. L.M. Robeson and J.E. McGrath, *Polym. Eng. Sci. 17*(5), 300 (1977)
2. C.F. Hammer, In *Polymer Blends, Vol. 2.* D.R. Paul and S. Newman (Eds.) (1978) Academic Press, New York, p. 219
3. G.H. Hofmann, In *Polymer Blends and Mixtures, NATO ASI Series No. 89.* D.J. Walsh, J.S. Higgins, and A. Maconnacie (Eds.) (1985) Martinus Nijhoff, Dordrecht, p. 117
4. G.H. Hofmann, R.J. Statz, R.B. Case, In *Proceedings of the SPE 51st ANTEC, Vol. XXXIX,* 2938 (1993)
5. R.E. Asay, M.D. Hein, and D.L. Wharry, *J. Vinyl Tech. 15*(2), 76 (1993)
6. J.G. Wallace, In *Handbook of Thermoplastic Elastomers.* B.M. Walker and C.P. Rader (Eds.) (1988) Van Nostrand Reinhold, New York, p. 143
7. Alcryn® Injection Molding Guide, No. H-08774
8. Alcryn® Extrusion Guide, No. H-33446
9. Alcryn® Calendering Guide, No. H-30622
10. Alcryn® Extrusion Blow Molding Guide, E-95214
11. R.E. Myrick, In *Proceedings of the SPE 52nd ANTEC, Vol. XL* (1994)
12. G.H. Hofmann, In *Proceedings of the SPE 47th ANTEC, Vol. XXXV,* 1752 (1989)
13. V.R. Landi, *Appl. Polym. Symp. 25,* 223 (1974)
14. M.K. Stockdale, *J. Vinyl Tech. 12*(4), 235 (1990)
15. Y.G. Oganesove, V.S. Osipchik, K.G. Mindiyarov, V.G. Rayevskii, and S.S. Voyutskii, *Polym. Sci. USSR* (Eng. Trans.) *11,* 1012 (1969)
16. M. Matsuo, C. Nozaki, and Y. Jyo, *Polym. Eng. Sci. 9,* 197 (1969)
17. P.W. Milner and G.R. Duval, In *Thermoplastic Elastomers 3* (1991) Rapra Technol. Ltd, Sudsbury, U.K., p. 7
18. G.R. Duval and P.W. Milner, In *PVC 87,* Brighton, U.K., April 28–30, 1987
19. Product Bulletin, Sarlink® Thermoplastic Elastomers, DSM Thermoplastic Elastomers, Inc.
20. B. Kliever and R.D. DeMarco, *Rubber Plast. News* February 15, 25 (1993)
21. R.W. Crawford and W.K. Witsiepe, U.S. Patent 3,718,715 (February 27, 1973)
22. M. Brown, *Rubb. Indus.* 102, June (1975)
23. D.J. Hourston and I.D. Hughes, *J. Appl. Polym. Sci. 26*(10), 3467 (1981)
24. H.W. Bonk, A.A. Sardanopli, H. Ulrich, and A.A.R. Sayigh, *J. Elastoplastics 3,* 157 (1971)

7 Thermoplastic Elastomers Based on Dynamically Vulcanized Elastomer–Thermoplastic Blends

A.Y. Coran and R.P. Patel

7.1 Introduction 154

7.2 Preparation of Elastomer–Plastic Blends by Dynamic Vulcanization 156

7.3 Properties of Blends Prepared by Dynamic Vulcanization 157
    7.3.1 Polyolefin-Based Thermoplastic Vulcanizates 157
        7.3.1.1 EPDM–Polyolefin Thermoplastic Vulcanizates 157
        7.3.1.2 Diene Rubber–Polyolefin-Based Thermoplastic Vulcanizates 162
        7.3.1.3 Butyl Rubber–Polypropylene-Based Thermoplastic Vulcanizates 162
    7.3.2 NBR–Nylon Thermoplastic Elastomer Compositions 166
        7.3.2.1 Effect of Curatives 167
        7.3.2.2 Effect of Elastomer Characteristics 167
        7.3.2.3 Effects of NBR–Nylon Proportions 169
        7.3.2.4 Effect of Plasticizers 169
        7.3.2.5 Effect of Filler 170
        7.3.2.6 Overall Assessment of NBR–Nylon Thermoplastic Elastomeric
            Compositions 170
        7.3.2.7 Other Thermoplastic Vulcanizates 171
    7.3.4 Characteristics of Elastomers and Plastics for Correlations with Blend Properties . . 173
        7.3.4.1 Dynamic Shear Modulus 174
        7.3.4.2 Tensile Strength of the Hard-Phase Material 174
        7.3.4.3 Crystallinity 174
        7.3.4.4 Critical Surface Tension for Wetting 175
        7.3.4.5 Critical Entanglement Spacing 176
    7.3.5 The Correlation Between Blend Properties and the Characterization of the
    Blend Components 177
    7.3.6 Technological Compatibilization of NBR–Polyolefin Blends by Elastomer–Plastic
    Graft Formation 178
        7.3.6.1 Resistance to Hot Oil and Brittle Point 182
        7.3.6.2 Overall Assessment of Compatibilized NBR–Polypropylene Thermoplastic
            Vulcanizates 182
    7.3.7 Blends of Thermoplastic Vulcanizates Based on Dissimilar Plastics 183
        7.3.7.1 Description of the Dissimilar-Plastics Compositions 183
        7.3.7.2 Blends of Thermoplastic Vulcanizates 183

7.4 Technological Applications 185
    7.4.1 Processing—Fabrication Technology 185
    7.4.2 End-Use Applications 187
        7.4.2.1 Emerging Applications 188

References 189

7.1   Introduction

Elastomer–thermoplastic blends have become technologically useful as thermoplastic elastomers in recent years [1–3]. They have many of the properties of elastomers, but are processable as thermoplastics [4]. They do not need to be vulcanized during fabrication into end-use parts. Thus, they offer a substantial economic advantage with respect to the fabrication of finished parts.

For many end uses, the ideal elastomer–plastic blend comprises finely divided elastomer particles dispersed in a relatively small amount of plastic. The elastomer particles should be crosslinked to promote elasticity (the ability of the blend composition to retract forcibly from a large deformation). The favorable morphology should remain during the fabrication of the material into parts, and in use. Because of these requirements for the ideal case, the usual methods for preparing elastomer–plastic blends by melt mixing, solution blending, or latex mixing [5] are not sufficient.

The best way to produce thermoplastic elastomeric compositions comprising vulcanized elastomer particles dispersed in melt-processable plastic matrices is by the method called dynamic vulcanization. It is the process of vulcanizing an elastomer during its melt mixing with a molten plastic [6–9]. This chapter describes the dynamic vulcanization process and the products that can be produced thereby. The scope of applicability of dynamic vulcanization is discussed. End-use applications of some of the products that can be prepared by dynamic vulcanization are also described.

Dynamic vulcanization is a route to new thermoplastic elastomers that have many properties as good or even, in some cases, better than those of elastomeric block copolymers. Yet, the new materials are prepared from blends of existing polymers. Thus, new, improved thermoplastic elastomers can be prepared from "old" polymers. Entirely new processes and materials can be avoided. Moreover, high new-product "entrance fees" (caused by such barriers as environmental concerns, capital costs, and the necessity for high-volume polymerization units and competitive divisors) can also be avoided.

This technology has led to a significant number of new thermoplastic elastomeric products commercialized during the last half of the 1980s [10]. It is important to note that the commercialization of the dynamic vulcanization technology was aided by the discovery [11] of preferred compositions based on Lewis-acid-catalyzed methylol–phenolic dynamic vulcanization systems for the new thermoplastic elastomers.

There is much commercial interest in dynamic vulcanization since the introduction of proprietary products (e.g., Santoprene® thermoplastic elastomer) prepared by the dynamic vulcanization of blends of olefin rubber with polyolefin resin. If the elastomer particles of such a blend are small enough and if they are fully vulcanized, then the properties of the blend are greatly improved. Examples of the improvements are as follows:

- Reduced permanent set,
- improved ultimate mechanical properties,
- improved fatigue resistance,
- greater resistance to attack by fluids, for example, hot oils,
- improved high-temperature utility,
- greater stability of phase morphology in the melt,

- greater melt strength, and
- more reliable thermoplastic fabricability.

In short, dynamic vulcanization can provide compositions that are very elastomeric in their performance characteristics. These thermoplastic vulcanizate compositions can be readily fabricated into finished parts in thermoplastic processing equipment.

Because of the surprisingly beneficial effects of complete dynamic vulcanization (in contrast to prior and *partial* dynamic vulcanization [7], the work described in this chapter opens a broad new field of thermoplastic–elastomer research.

As stated above, in some respects, the new materials can out-perform block copolymer-type thermoplastic elastomers. This is because the particulate elastomeric domains are comprised of fully vulcanized elastomer. As a result, the thermoplastic vulcanizates can perform well with respect to hot-oil resistance, compression set (at elevated temperatures), and high-temperature utility. A factor contributing to the greater high-temperature utility is that during fabrication of a finished part, many of the vulcanized rubber particles physically interact with one another to form "network" of vulcanized elastomer. When scrap is reground and reworked in the melt, the clusters of touching, loosely bound together particles disintegrates and melt processability is restored.

Thermoplastic vulcanizates (TPVs) have been prepared from a great number of plastics and elastomers; however, only a limited number of elastomer–plastic combinations give technologically useful blends, even after dynamic vulcanization. The results of a study of 99 elastomer–plastic combinations (based on 11 elastomers and 9 plastics) are also reviewed in this chapter. The goal of this work was to define the practical scope of compositions that can be prepared by dynamic vulcanization. This was accomplished by an analysis that related mechanical properties of the dynamically vulcanized blends to characteristics of their elastomeric and plastic components. The conclusion of that work was that the best elastomer–plastic thermoplastic vulcanizates are those in which (1) the surface energies of the plastic and elastomer are matched, (2) the entanglement molecular length of the elastomer is low, and (3) the plastic is at least 15% crystalline.

Were it not for the large difference between the surface energy of acrylonitrile–butadiene copolymer (NBR) and that of polypropylene, a blend of dynamically vulcanized NBR with isotactic polypropylene should be a good choice for an oil-resistant thermoplastic elastomer. In early work with these materials only marginally good compositions were obtained. Now it has been found that NBR–polypropylene-based thermoplastic elastomers, which are resistant to hot oil and have excellent strength-related properties, can be prepared by the dynamic vulcanization of technologically compatibilized NBR–polypropylene blends. Technological compatibilization of the blend (before it is dynamically vulcanized) is brought about by the presence of a small amount (about 1%) of compatibilizing agent in the blend. The compatibilizing agent is a block copolymer containing segments similar to each of the polymers that are to be compatibilized. It acts as a macromolecular surfactant and its presence, during mixing, permits the formation of very small droplets of the elastomer which later, during dynamic vulcanization, become very small particles of vulcanized NBR. Research results related to the dynamic vulcanization of technologically compatibilized elastomer–plastic blends, which can lead to a wide variety of new product development opportunities, are considered in this chapter. In addition, it is even possible to use a mixture of two thermodynamically incompatible plastics with two thermodynamically incompatible rubbers if the plastics are compatibilized by the action of block copolymers and if each of the rubbers can give a useful composition with one of the plastics in the compatibilized blend.

7.2   Preparation of Elastomer–Plastic Blends by Dynamic Vulcanization

Polymer blends, in general, have been prepared commercially by melt mixing, solution blending, or latex mixing. Elastomer–plastic blends of the type discussed here, containing rather large amounts of elastomer, have generally been prepared by melt-mixing techniques. Melt mixing avoids problems of contamination, solvent or water removal, etc. In general, Banbury mixers, mixing extruders, and the newer twin-screw mixers are suitable for melt mixing elastomer with plastics. However, for the purposes of this discussion, emphasis will be on laboratory melt-mixing techniques.

The procedures given below are based on the use of either a small Brabender mixer or a Haake Rheomix. In each case the mixer was fitted with cam-type rotors and optimum batch sizes were between 55 and 75 g. In a process he named dynamic vulcanization, Gessler [6] prepared "semirigid" elastomer-plastic compositions containing minor proportions of partially vulcanized elastomer. Fischer [7] used a dynamic vulcanization process to prepare compositions containing varying amounts of partially vulcanized ethylene–propylene–diene terpolymer (EPDM) elastomer. Large proportions of elastomer were generally used and soft compositions could be obtained. An organic peroxide was used to crosslink the elastomer in the presence of polypropylene, which was greatly damaged by the action of the peroxide.

Later, it was found that very strong, elastomeric compositions of EPDM and polypropylene can be prepared by dynamic vulcanization, provided that peroxide curatives are avoided [8, 9]. If enough plastic phase is present in the molten state, then the compositions are processable as thermoplastics. Plasticizers or extender oils and fillers can be used to expand the volume of the elastomer phase. In the molten state, a suitable plasticizer can expand the volume of the plastic or "hard" phase. If the hard phase material is a crystalline material such as polypropylene, then, on cooling, its crystallization can force the plasticizer out of the hard phase into the elastomer phase. Thus, the plasticizer may act as both a processing aid at melt temperatures and also a softener at the lower temperatures of use.

The dynamic vulcanization process has been applied to many elastomer–plastic combinations. It can be described as follows. Elastomer and plastic are first melt mixed, usually in an internal mixer. After a well-mixed blend has been formed, vulcanizing agents (curatives, crosslinkers) are added. Vulcanization then occurs while mixing continues. The more rapid the rate of vulcanization, the more intense the mixing must be to ensure good fabricability of the blend composition. It is convenient to follow the progress of vulcanization by monitoring mixing torque or mixing energy requirement during the process. After the mixing torque or energy curve goes through a maximum, mixing can be continued somewhat longer to improve the fabricability of the blend. After discharge from the mixer, the blend can be chopped, extruded, pelletized, injection molded, etc.

7.3   Properties of Blends Prepared by Dynamic Vulcanization

Dynamic vulcanization, as described herein, is the process of vulcanizing elastomer during its intimate melt mixing with a nonvulcanizing thermoplastic polymer. Small elastomer droplets are vulcanized to give a particulate vulcanized elastomer phase of stable domain morphology during melt processing and subsequently.

As stated earlier, the effect of the dynamic vulcanization of elastomer–plastic blends is to produce compositions that have improvements in permanent set, ultimate mechanical properties, fatigue resistance, hot oil resistance, high-temperature utility, melt strength, and thermoplastic fabricability. Permanent set of these compositions can be improved by only slight or partial vulcanization of the elastomer. Such compositions can be produced either by the partial vulcanization of the elastomer before its mixture with plastic or by dynamic vulcanization [7] (during mixing with plastic). However, the other improvements can be obtained only (at least in the case of EPDM–polyolefin compositions) by dynamic vulcanization in which the elastomer is technologically fully vulcanized. The term "fully vulcanized" refers to a state of cure such that the crosslink density is at least 7×10^{-5} mol per milliliter of elastomer (determined by swelling) or that the elastomer is less than about 3% extractable by cyclohexane at $23\,°C$ [8].

7.3.1   Polyolefin-based Thermoplastic Vulcanizates

7.3.1.1   EPDM–Polyolefin Thermoplastic Vulcanizates

The dynamic vulcanization of blends of EPDM elastomer with polypropylene and with polyethylene has been described [9]. Mechanical properties, hardness, tension set values, and other parameters associated with unfilled compositions are given in Table 7.1. The general recipe of the compositions was as follows:

EPDM (Epsyn® 70A)	100
Polyolefin resin	X
Zinc oxide	5
Stearic acid	1
Sulfur	Y
TMTD (tetramethylthiuram disulfide)	$Y/2$
MBTS (2-benzothiazolyl disulfide)	$Y/4$

where X, the number of parts by weight of polyolefin resin, and Y, the amount of sulfur, were varied.

Not all of the compositions of Table 7.1 were prepared by dynamic vulcanization, however. In the first four compositions, for comparison purposes, the elastomer was first press cured and then ground, by tight roll-milling, to various particle sizes. The ground elastomer particles were then mixed with molten polypropylene. Compositions 5 to 20 of Table 7.1 were obtained by dynamic vulcanization.

Particle sizes were determined by optical microscopy. However, the particles of the

Table 7.1 Properties of Unfilled Thermoplastic Compositions

Composition number	Resin type[a]/ parts per 100 parts of rubber (phr)	Sulfur (phr)	Method of preparation[c]	Crosslink density ("moles" ×10^5/ml)	Rubber particle size, (μm) d_n	d_w	Hardness (Shore D)	Young's modulus (MPa)	Stress at 100% strain (MPa)	Tensile strength (MPa)	Elongation at break (%)	Tension set (%)
1	PP/66.7	2.00	S	16.4	72	750	43	97	8.4	8.6	165	—
2	PP/66.7	2.00	S	16.4	39	290	41	102	8.4	9.8	215	22
3	PP/66.7	2.00	S	16.4	17	96	41	105	8.4	13.9	380	22
4	PP/66.7	2.00	S	16.4	5.4	30	42	103	8.4	19.1	480	20
5	PP/66.7	2.00	D	16.4	~1–2		42	58	8.0	24.3	530	16
6	PP/66.7	1.00	D	12.3	—		40	60	7.2	18.2	490	17
7	PP/66.7	0.50	D	7.8	—		39	61	6.3	15.0	500	19
8	PP/66.7	0.25	D	5.4	—		40	56	6.7	15.8	510	19
9	PP/66.7	0.13	D	1.0	—		35	57	6.0	9.1	407	27
10	PP/66.7	0.00	—	0.0	—		22	72	4.8	4.9	190	66
11	PP/33.3	1.00	D	12.3	—		29	13	3.9	12.8	490	7
12	PP/42.9	2.00	D	16.4	—		34	22	5.6	17.9	470	9
13	PP/53.8	2.00	D	16.4	—		36	32	7.6	25.1	460	12
14	PP/81.8	2.00	D	16.4	—		43	82	8.5	24.6	550	19
15	PP/122	2.00	D	16.4	—		48	162	11.3	27.5	560	31
16	PP/233	5.00	D	14.5	—		59	435	13.6	28.8	580	46
17	None[b]/0	2.00	S	16.4	—		11	2.3	1.5	2.0	150	1
18	PP[b]/∞	0.00	—	—	—		71	854	19.2	28.5	530	—
19	PE/66.7	2.00	D	16.4	—		35	51	7.2	14.8	440	18
20	PE/66.7	0.00	—	0.0	—		21	46	4.1	3.5	240	24

[a] PP, Polypropylene, Profax® 6723; PE, polyethylene; Marlex® EHM 6006

[b] Compositions 17 and 18 are control compositions purely of cured rubber and polypropylene, respectively

[c] S, static (conventional) vulcanization; D, dynamic vulcanization

Figure 7.1   Scanning electron microphotograph of OsO₄ stained sample of a 73 Shore A hardness commercial grade of completely vulcanized EPDM–polypropylene-derived thermoplastic vulcanizate

dynamic vulcanizates were so small that their diameters could only be determined to be in the range of 1 to 2 μm. (More recently, electron microscopy has revealed that the commercial grades of EPDM–polypropylene-derived thermoplastic elastomer contain elastomer particles in the 1- to 2-μm range. An example is given in Fig. 7.1). For the measurement of the crosslink density of the elastomer, samples of the elastomer alone were press cured under conditions selected to simulate the conditions of dynamic vulcanization. Crosslink densities of the press-cured samples were then determined on the basis of solvent-swelling measurements by using the Flory–Rehner equation [12]. It may be argued that the presence of polyolefin might change the crosslink density, especially in the case of the dynamically cured compositions. However, we have determined that the polyolefin is essentially completely extractable from the cured elastomer by boiling decalin or xylene. This would indicate that there is little or no interaction of the thermoplastic resin with the sulfur-based curing system. Measurements of mechanical properties and hardness were by usual means.

The effect of the elastomer particle size on the tensile strength and ultimate elongation is shown in Fig. 7.2. This is a composite stress–strain curve constructed from the data associated with compositions 1 to 5 of Table 7.1. Each X denotes a stress and strain at rupture or failure. From Fig. 7.2, ultimate strength, ultimate elongation, and energy to break (area under curve from origin to an appropriate X) are apparent. The average elastomer particle size associated with each X is noted. The ultimate properties are an inverse function of elastomer particle diameter. The best compositions were prepared by dynamic vulcanization. There appears to be no other means of producing rubber particles of such small size.

Major effects of changes in crosslink density are shown in Fig. 7.3. Only a small amount of crosslink formation is required for a large improvement in tension set. Tensile strength improves rather continuously as the crosslink density of the elastomer phase increases, but the compositions remain fabricable as thermoplastics even at high elastomer crosslink densities. However, only small changes in the stiffness of the compositions occur with great changes in the extent of cure.

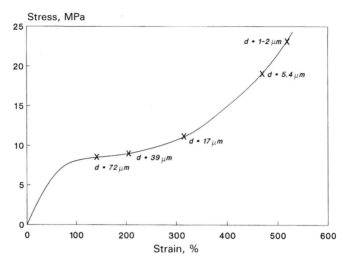

Figure 7.2   Effect of vulcanized rubber particle size on mechanical properties (x denotes failure)

As the proportion of polypropylene to elastomer increases, the compositions become less elastomeric and more like the plastic (Fig. 7.4). Modulus, hardness, tension set, and strength increase. The shape of the plot of strength against the proportion of polypropylene is interesting. Strengths are low until at least 30 parts of polypropylene per 100 parts of elastomer are used. Then as the amount is further increased, strength rapidly increases until the polypropylene content is about 50 phr. Further addition of polypropylene increases the strength only slightly.

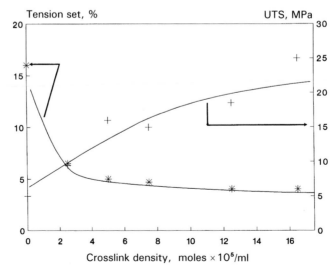

Figure 7.3   Effect of crosslink density on tensile strength and tension set

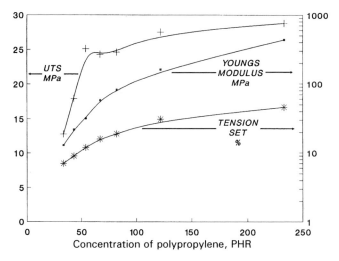

Figure 7.4   Effect of polypropylene content on EPDM–polypropylene thermoplastic vulcanizate properties

Dynamic modulus in both vulcanized and unvulcanized states is shown as a function of temperature in Fig. 7.5. The composition is that of No. 5 of Table 7.1. The dynamic shear moduli were determined by means of a torsion pendulum [13]. An effect of dynamic vulcanization is to prevent a complete loss of elasticity and strength at the melting point of the resin, polypropylene. Provided that there is enough crosslinked rubber in the composition, it continues to exhibit sufficient strength for the torsion pendulum modulus measurements even after the plastic phase melting point has been surpassed.

Another effect of dynamic vulcanization, indicated by the results of Fig. 7.5, is a smaller decrease in modulus as the temperature is increased. The unique performance of the thermoplastic vulcanizates, even above the plastic melting points, suggests a variety of high-temperature applications. The data given for compositions 19 and 20 of Table 7.1

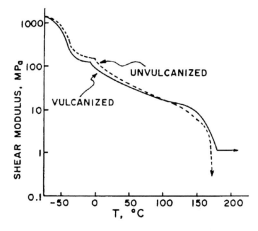

Figure 7.5   Effect of temperature on the stiffness of EPDM–polypropylene blend compositions

indicate that EPDM–polyethylene blends are also greatly improved by dynamic vulcaniza-tion. However, on the basis of comparisons between compositions based on polyethylene with those based on polypropylene, one concludes that the best compositions are prepared from polypropylene.

The morphology of an EPDM–polypropylene dynamic vulcanizate is that of a dispersed phase morphology. This morphology has been observed to be independent of the elastomer/thermoplastic ratio or the molecular weights of the constituent polymers. In contrast, a variety of morphologies can be obtained for simple blends depending on polymer ratios, molecular weights of the polymers, or the mixing conditions [14]. These conclusions are demonstrated by the photomicrographs of Fig. 7.6a,b.

The data of Table 7.2 demonstrate the effects of black loading and oil extension. The compositions are variations of No. 15 of Table 7.1. The effect of the filler is to strengthen the composition somewhat and to give some stiffening (with respect to hardness and stress at 100% strain but not with respect to the "zero strain" or Young's modulus). Both oil extension and carbon black or other filler loading can give compositions of lower cost but excellent quality. From data such as those in Table 7.2, and from measurements of hot oil resistance, fatigue life, etc., it has been concluded that these oil-extended, filled compositions will perform similarly to conventional elastomer vulcanizates.

As noted previously, the commercialization of thermoplastic elastomers prepared by dynamic vulcanization was greatly aided by the discovery [11] that compositions based on Lewis acid-catalyzed methylol–phenolic resin type cure systems are surprisingly superior to those prepared from the accelerated sulfur system mentioned previously. This is especially true with respect to the fabricability of the compositions based on the phenolic curing systems.

7.3.1.2   Diene Rubber–Polyolefin-Based Thermoplastic Vulcanizates

Thermoplastic elastomers based on blends of polyolefins with diene rubbers such as butyl rubber, natural rubber, NBR, styrene-butadiene copolymer (SBR), etc. have been described by Coran and Patel [15–17]. These compositions have fairly good initial tensile properties, and their thermal stability is somewhat better than that of thermoset diene rubbers. Much of the commercial development work has centered around TPVs based upon NR–polypropylene blends. Campbell et al. [18] have prepared compositions of partially vulcanized natural rubber. Recently, Payne et al. [19] have reported improved mechanical properties of fully vulcanized naturall rubber–polypropylene-based TPVs of different hardnesses. They found that unlike thermoset natural rubber, these TPVs have very good resistance to cracking induced by ozone (Table 7.3). The low-temperature brittle point increases as the natural rubber content decreases. These compositions have fairly good retention of tensile properties in hot air at 100°C for up to 1 month (Table 7.4).

7.3.1.3   Butyl Rubber–Polypropylene-Based Thermoplastic Vulcanizates

Butyl rubber and halogenated butyl rubbers have very low air and moisture permeability. This makes them well suited for use in tire, sporting goods, and medical applications. This also provides an opportunity to develop TPVs of low air and moisture permeability.

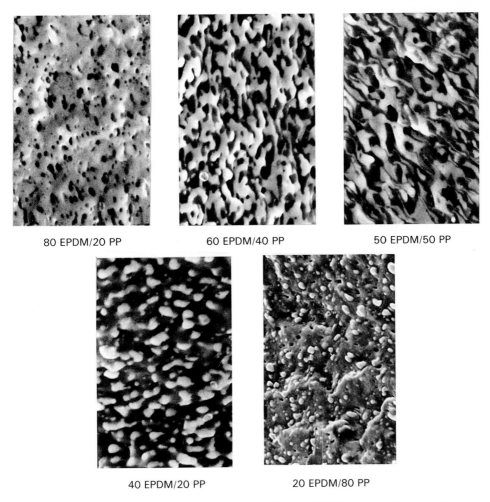

80 EPDM/20 PP	60 EPDM/40 PP	50 EPDM/50 PP
40 EPDM/20 PP	20 EPDM/80 PP	

Figure 7.6a   SEM photomicrographs of uncured blends of various EPDM–polypropylene proportions

In the early 1960s, Gessler [6] first reported the preparation of toughened (semihard to hard, not rubbery) polypropylene compositions by dynamic vulcanization. These compositions were prepared by melt blending a mixture of chlorobutyl rubber, polypropylene, and vulcanizing agent (zinc oxide). The work was later expanded by Coran and Patel [20], who produced soft, rubbery thermoplastic compositions with crosslinked particulate rubber phases (Fig. 7.7). Since the crosslinked rubber of such a composition is in the dispersed particulate form, one would not expect typical butyl rubber impermeability from these TPV compositions. However, by careful selection of compounding ingredients, it was possible to prepare butyl–polypropylene or halobutyl–polypropylene-based TPVs of very low permeability (Table 7.5). The air and water permeabilities of butyl-based dynamic vulcanizates are compared to those of conventional thermosets and of EPDM–polypropylene-based TPVs in Fig. 7.8.

4.93μ

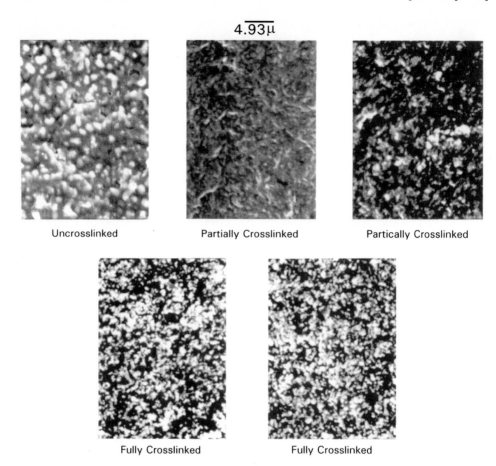

| Uncrosslinked | Partially Crosslinked | Partically Crosslinked |

| Fully Crosslinked | Fully Crosslinked |

Figure 7.6b   SEM photomicrographs of 40:60 EPDM–polypropylene TPVs prepared with different curative levels

Recently, Puydak and Hazelton [21] prepared TPVs from rubber–rubber–plastic triblends. In most of these compositions the best choice for the plastic phase was polypropylene. One of the elastomers is crosslinked under dynamic vulcanization conditions during molten-state mixing with the second elastomer and the plastic. Transmission electron micrographs of such compositions are shown in Fig. 7.9. These TPVs offer some advantages over two-component TPVs (Table 7.6).

Table 7.2 Effect of Carbon Black and Extender Oil[a]

Carbon black (phr)	Extender oil (phr)	Tensile strength (MPa)	Stress at 100% strain (MPa)	Young's modulus (MPa)	Elongation at break (%)	Shore hardness	Tension set (%)
0.0	0.0	27.5	11.3	162	560	48D	31
80.0	0.0	31.0	14.3	120	410	51D	30
0.0	80.0	15.2	6.4	47	550	29D	19
80.0	80.0	23.0	7.2	23	530	33D	16
80.0	160.0	15.2	4.8	11.5	490	74A	13

[a] Carbon black is N 327 and oil is Sunpar 2280.

Table 7.3 Mechanical Properties of Different Natural Rubber-Polypropylene-Based TPVs

Property		ASTM#			
Hardness, Shore	D2240	60A	70A	90A	50D
Tensile strength (MPa)	D412	5.0	7.6	11.4	20.8
Stress at 100% strain (MPa)	D412	2.1	3.7	6.5	10.5
Ultimate elongation (%)	D412	300	380	400	620
Tension set (%)	D412	10	16	35	50
Tear strength (KN/m)	D624	22	29	65	98
Compression set (22 h)	D395				
at 23 °C (%)	Method B	24	26	32	45
at 100 °C (%)		30	32	38	63
Brittle point (°C)	D746	−50	−50	−45	−35
Ozone resistance at 40 °C (100 ppm of O_3[a])	D518	10	10	10	10
Specific gravity	D297	1.04	1.04	1.02	0.99

Table 7.4 Effect of Hot Air Aging on Percent Retention of Tensile Properties of NR-Polypropylene-Based TPVs

TPE hardness	Property	Aging time (days at 100 °C)			
		1	7	15	30
60A	Tensile strength	99	91	80	40
	Stress at 100% strain	104	65	80	68
	Ultimate elongation	98	110	126	85
70A	Tensile strength	100	87	76	43
	Stress at 100% strain	100	90	86	80
	Ultimate elongation	98	110	113	56
90A	Tensile strength	103	91	86	66
	Stress at 100% strain	107	103	104	99
	Ultimate elongation	92	93	93	60
50D	Tensile strength	101	95	80	66
	Stress at 100% strain	108	109	102	103
	Ultimate elongation	93	93	91	70

Table 7.5 Air Permeability of Thermoset Butyl Rubber vs. TPVs

Elastomer	Relative air permeability
Butyl–PP TPE	1.45
EPDM–PP TPE	4.44
Thermoset butyl (inner liner)	1.00

ASTM D1434 at 35 °C; sample thickness: 0.76 mm

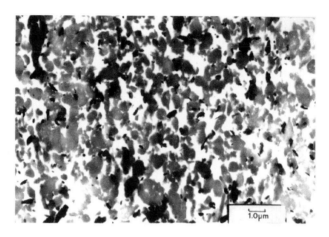

Figure 7.7   Morphology of butyl rubber–polypropylene blend-based TPV

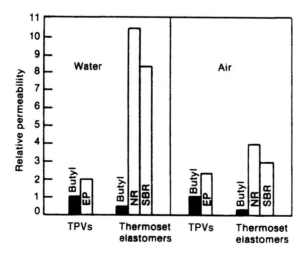

Figure 7.8   Comparison of permeability of thermoset rubbers with that of Butyl Rubber–polypropylene and EPDM–polypropylene blend-based TPVs

7.3.2   NBR–Nylon Thermoplastic Elastomer Compositions

As in the case of the EPDM–polypropylene compositions, NBR–nylon compositions were prepared by melt mixing the polymers and other components in a (Brabender) laboratory size mixer [22]. The temperatures for mixing and molding varied with the melting point of the nylon used in each composition.

Nitrile elastomers can be grouped into two categories: those that are self-curing (i.e., cure at elevated mixing temperatures in the absence of curative), and those that are resistant to self-curing. (Self-curing at high temperatures is a result of thermal-oxidative instability.) To determine whether or not an elastomer is self-curing, a sample can be mixed at 225 °C. Self-curing elastomers generally gel and crumble (scorch) within 1 to 8 min, whereas non-self-curing samples can be mixed for 20 min without crumbling.

7.3.2.1   Effect of Curatives

The effect of curatives such as *m*-phenylenebismaleimide is complicated by the fact that some nitrile elastomers tend to self-cure at the temperature of mixing. This is particularly true when high-melting nylons are used with self-curing NBR. In such cases, the effect of adding curative on the physical properties of the NBR–polyamide vulcanizate is minimized, since the properties of the composition are improved by the crosslinking of the elastomer just from mixing. A much greater curative effect is observed with the non-self-curing elastomer; however, the best properties are obtained with the self-curing type of NBR.

It was found that the addition of a dimethylol phenolic compound also substantially improves the properties of NBR–nylon blends. The curative *m*-phenylenebismaleimide induces considerable gel formation in the elastomer phase and the improved product properties have been associated with such gelation. However, it is surprising that, in the case of the dimethylol phenolic curative, high-strength blends are obtained even when the gel content of the NBR is as low as 50%. A portion of the dimethylol phenolic compound is believed to react with the nylon to give chain extension or a small amount of crosslinking. This could increase the viscosity of the molten nylon to where it is more like that of the elastomer, and thus mixing, homogenization, and elastomer particle size reduction are greatly improved. In addition, some of the crosslinking may be between the molecules of the nylon and those of the elastomer, giving nylon–NBR graft molecules. This should improve homogenization and interfacial adhesion.

Table 7.6 Synergism on Combining Elastomers in DVAs

TPE Property	Neoprene (CR)	Butyl (IIR)	CR + IIR (1 : 1)
Tensile strength (MPa)	4.7	7.1	7.5
Ultimate elongation (%)	170	220	250
Hardness (Shore A)	58	60	66
ASTM No. 3 Oil Swell (70 h at 100 °C, %)	21	75	44
Tear strength (kN/m)	12.3	19.3	24.5
Compression set (22 h at 100 °C, %)	66	47	52

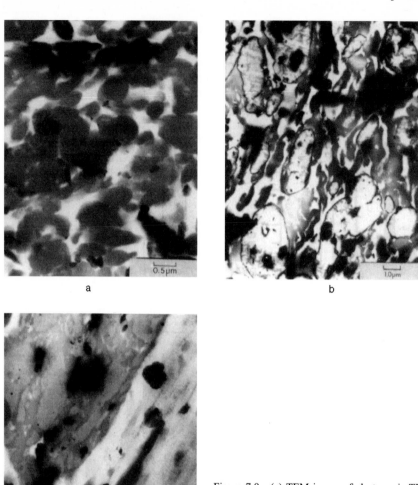

Figure 7.9   (a) TEM image of elastomeric TPV showing micron sized cured rubber particles in a continuous PP matrix; (b) TEM image of a TPV containing two elastomers, neoprene (large, irregular particles outlined in black) and butyl (small dark gray particles); (c) TEM image of a TPV with a crosslinked phase (lighter gray) surrounded by an uncured phase (darker gray) in a continuous polymer phase (white)

The effect of the curatives on tension set is widely variable. This is suggested by the data of Table 7.7, where the various curatives give a wide range of set values. This is in contrast to what was observed with EPDM–polypropylene compositions, where curatives invariably reduced set values. The reason for this variation of the effect of curative on tension set is not understood, but it could relate to the extent to which the curatives promote molecular linkages between the nylon and elastomer, rather than curing the elastomer.

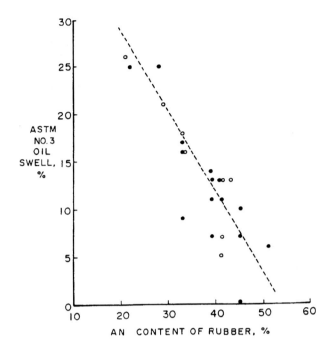

Figure 7.10   Effect of acrylonitrile (AN) content on hot (150°C) oil (ASTM No. 3) swelling. Self-curing rubber ○; non-self-curing rubber ●

Table 7.7 Cured NBR–Nylon Blends with Different Types of Curatives[a]

Curative type	Tensile strength (MPa)	Stress at 100% strain (MPa)	Elongation at break (%)	Tension set (%)	Shore D hardness	True stress at break (MPa)
None (control)	3.1	2.5	290	72	17	12.3
Accelerated sulfur[b]	8.3	7.4	160	15	35	21.7
Activated bismaleimide[c]	8.5	3.7	310	51	28	34.9
Peroxide[d]	7.9	6.1	220	31	32	25.3

[a] Blends comprise 40 parts of nylon 6/6–6/6–10 terpolymer (mp 160 °C) and 60 parts Chemigum® N365 non-self-curing NBR 39% AN)

[b] Accelerated sulfur system contains 5 parts of ZnO, 0.5 parts of stearic acid, 2 parts of tetramethylthiuram disulfide, 1 part of morpholinothiobenzothiazole, and 0.2 parts of sulfur per 100 parts of rubber

[c] Activated bismaleamide is 3 parts of m-phenylenebismaleamide and 0.75 part of 2,2-bisbenzothiazolyl disulfide per 100 parts of rubber

[d] Peroxide is 0.5 part of 2,5-dimethyl-2,5-bis(tert-butylperoxy)hexane (90% active), Lupersol® L-101

7.3.2.2   Effect of Elastomer Characteristics

Elastomers of differing viscosities, differing nitrile contents, and differing tendencies toward self-curing were studied. There appears to be no simple relationship between the strength of a composition and the characteristics of its elastomer phase. However, as mentioned previously, self-curing elastomers tend to give the best strengths, and the effect of curative addition is greatest for the non-self-curing elastomers. The effect of acrylonitrile content on hot oil resistance is similar to what is observed in the usual NBR vulcanizates, that is, high acrylonitrile content gives low oil swelling. This is true for both the self-curing and non-self-curing types of NBR. Oil swell data for all of the compositions are plotted in Fig. 7.10.

7.3.2.3   Effects of NBR–Nylon Proportions

These effects are similar to those obtained for the EPDM–polyolefin compositions. Increases in the amount of elastomer in the compositions reduce stiffness and strength, but improve resistance to permanent set. Also, extensibility can be increased somewhat. If more than 50% of the composition is elastomer, compositions having tension set values less than 50% are obtained. However, excessive amounts of elastomer can result in poor fabricability.

7.3.2.4   Effect of Plasticizers

Plasticizers can be added to compositions of NBR and nylon. The effect is to soften the compositions and improve fabricability. The melting point of the nylon phase and its crystallinity can either increase or decrease, although the expected effect is for both to be decreased [23, 24]. However, another expected effect is for plasticizer to decrease the viscosity of the nylon phase and this may promote crystallization from the melt and enable formation of more perfect crystals. In some cases these two effects tend to cancel. Plasticizer addition can either increase or decrease ultimate elongation. Tensile strength generally decreases with the incorporation of plasticizer. Effects of plasticization are indicated by the data in Table 7.8. Note that plasticizers that are more compatible with the nylon phase tend to give compositions of better mechanical integrity.

7.3.2.5   Effect of Filler

Small amounts of clay have little effect on hardness, stiffness, or strength, although extensibility is reduced. Young's modulus actually decreases. The effects are similar to those reported for carbon black-filled compositions containing EPDM and polypropylene. Again, it is thought that the filler goes into the elastomer phase and it has the effect of both stiffening this phase and increasing its volume. These effects are opposite and largely cancel. Another effect of filler is to severely reduce the thermoplasticity and therefore to reduce the expected fabricability. To obtain the full benefit of filler, plasticizers can be used to regain both thermoplasticity and extensibility.

7.3.2.6   Overall Assessment of NBR–Nylon Thermoplastic Elastomeric Compositions

The assessment is complicated by the large number of variations that are possible. There are many types of nylon (polyamide) resins with a wide range of melting points, polarities, etc. Also, the variations in the types of NBR are great (nitrile content, viscosity, susceptibility to self-curing, etc.). In addition, the effects of the different curing systems are widely varying. Nevertheless, conclusions can be drawn.

Table 7.8 The Effect of Plasticizers on the Properties of Nylon-NBR Compositions[a]

Plasticizers (compatible phase[b])	Change in tensile strength (%)	Change in elonga- tion at break (%)	Change in hardness (D units)	Change in melting pt (°C) Peak	Final	Change in crystal- linity (%)
Methyl phthalyl ethyl glycolate (R)	−25	+15	−7	+7	+1	−18
Butyl phthalyl butyl glycolate (R)	−27	−31	−10	−3	−7	−4
C_7-C_9 trialkyl trimelitate (R)	−28	−35	−9	−2	−6	0
Dioctyl phthalate (R)	−26	−38	−5	+1	−4	0
Dibutyl sebacate (R)	−34	−23	−10	+1	−6	−11
N-Ethyl-o- and p-toluenesulfonamide (N)	+6	+12	−7	−7	−12	−12
2-Propyl-4,4′-bisphenol (N)	0	+27	−7	−8	−12	−22
Nonylphenol (N)	0	+31	−7	−3	−6	−5

[a] Recipe: 50 parts nylon 6/6–6/6–10 terpolymer copolymer (Zytel® 63), 50 parts of NBR (Hycar® 1092-80), 10 parts of plasticizer, 2.5 parts of ZnO, 1.0 part of TMTD, and 0.50 part of bisbenzothiazolyl disulfide. Properties without plasticizer: UTS = 11.6 MPa, UE = 260%, Shore D hardness = 45, melting point = 152 °C (peak) or 168 °C (final), heat of fusion (proportional to crystallinity) = 11.6 J/g of nylon
[b] The symbols N or R indicate primary compatibility with nylon or rubber, respectively
[c] Melting points and relative crystallinity were determined by differential calorimetry (Perkin–Elmer Differential Scanning Calorimeter DSC-1B, 10 °C/min., 20-mg sample)

A variety of nylons of differing melting points and NBRs of differing acrylonitrile contents can be used in broad ranges of proportions. The compositions can be further altered by the incorporation of fillers and plasticizers. Thus many types of NBR–nylon based elastomeric materials, fabricable as thermoplastics and exhibiting fairly good strength and excellent hot oil resistance, can be produced in a wide range of hardnesses.

7.3.2.7   Other Thermoplastic Vulcanizates

A large number of elastomer–plastic combinations have been used in the preparation of thermoplastic vulcanizates by dynamic vulcanization. In one study, such compositions were compared in a systematic way [25]. Compositions were prepared as before, by vulcanization of blends during melt mixing. Various curative systems were used, and these were selected, for each composition, following optimizational experimentation. Types of curative systems that were used are as follows:

- Dimethylol phenolic (P),
- bismaleimide (M),
- bismaleimide–MBTS (M–M),
- bismaleimide–peroxide (M–O),
- organic peroxide (O),
- organic peroxide–coagent (O–C),
- accelerated sulfur (S), and
- soap-sulfur or –sulfur donor (SO–S).

In addition, thermal-oxidative stabilizers were used when appropriate. All of the compositions contained elastomer and plastic in a weight ratio of 60 : 40. The 60 : 40 ratio was chosen

Table 7.9 Curatives in Rubber–Pastic Blends[a]

Rubber–Plastic	PP	PE	PS	ABS	SAN	PMMA	PBT	PA	PC
IIR	P	P	P	S	S	P	P	P	P
EPDM	S	S	M–O	S	M–O	P	M–O	M–O	M–O
PTPR	P	S	P	P	P	S	M	M	P
NR	P	M–M	M–M	M–M	M–M	M–M	M–M	M–M	M–M
BR	M	M	M	M	M	M	M	M	M
SBR	P	M–M	M	M–M	M–M	M	M	M	M
EVA	O	M–O	M–O	M–O	M–O	M–O	O	O	O
ACM	SO–S	SO–S	SO–S	SO–S	SO–S	SO–S	SO–S	SO–S	SO–S
CPE	M–O	O	O	O	O	O	O	O	O
CR	S	S	M–M	M–M	M–M	S	M	M–M	—
NBR	M–O	M–M	O	O	O	M	—	M	M–M

[a] Each composition is identified corresponding to a rubber–plastic (row–column) combination. There are 11 rubbers (rows) and 9 plastics (columns), which give $9 \times 11 = 99$ combinations. Abbreviations for plastics are defined in Table 7.13. Abbreviations for rubbers are standard (ASTM D 1418); in addition, PTPR is poly(*trans*-pentenamer) (Bayer), EVA is ethylene-vinyl acetate copolymer rubber, and CPE is chlorinated polyethylene rubber. Curative symbols are identified in the text

for screening [9] of elastomer–plastic pairs because, when good compositions were obtained at this ratio, they were soft enough and elastic enough (tension set less than 50%) to be considered elastomeric. The compositions studied are given in Table 7.9, in which the types of curing systems are identified by the parenthetic symbols given in the above list of curing systems.

Ultimate tensile strength, ultimate elongation, and tension set values of the various compositions are given in Tables 7.10, 7.11, and 7.12. Those values of tensile strength that are given in parentheses for CPE–PBT, CR–PBT, and CR–PA are in doubt since the elastomers (CPE and CR) are insufficiently stable to withstand processing at the high melt temperatures for PBT (polyester) and PA (high-melting nylon). The value for EVA–PBT is in doubt because of the instability of the peroxide curative, which was probably used up before its complete mixture with the molten blend. For similar reasons, other tensile strength values may be low; however, for the purposes of this work only the parenthetic values were removed from consideration in attempts to correlate the measured properties of the elastomer–plastic compositions with characteristics of the elastomer and plastic components. Of course, the stability of a component in the presence of the others is in itself a characteristic expected to correlate with blend properties. In fact, one might say that an elastomer is technologically incompatible with a particular plastic if the plastic must be processed at temperatures higher than the temperature at which the elastomer becomes unstable.

In the case of the ultimate elongation values given in Table 7.11, the parenthetic values are in doubt for the reasons stated previously, and again such values were not used in correlations between blend properties and characteristics of the components. Many of the tension set values are missing from Table 7.12 because the measurement is impossible with compositions that cannot be stretched to an elongation of at least 100%. Other values are missing because the work was done early in the program before tension set was routinely measured. The tension set value of 17% obtained for ACM–PC (acrylate elastomer-polycarbonate resin) appears excessively low (high elastic recovery). Young's modulus for

Table 7.10 Ultimate Tensile Strength σ_b of 60/40 Rubber–Plastic Thermoplastic Vulcanizates

Rubber–plastic	PP	PE	PS	ABS	SAN	PMMA	PBT	PA	PC
IIR	21.6	14.9	0.9	1.7	4.3	5.4	1.4	4.0	1.3
EPDM	24.3	16.4	7.9	3.2	5.6	6.0	12.2	7.7	15.7
PTPR	22.7	12.1	6.9	11.0	13.4	4.7	12.1	10.8	2.5
NR	26.4	18.2	6.2	5.8	8.4	1.8	10.9	5.7	6.7
BR	20.8	19.3	11.6	9.9	8.3	3.5	12.8	16.3	2.1
SBR	21.7	17.1	15.8	10.8	8.1	5.7	21.7	14.6	7.3
EVA	17.8	18.9	12.7	9.6	12.9	9.3	(3.4)	10.9	9.6
ACM	4.0	4.2	11.4	9.4	7.7	6.2	14.6	16.1	5.2
CPE	12.3	10.5	14.0	13.7	17.9	17.0	(13.0)	17.3	20.8
CR	13.0	13.8	15.5	12.8	12.5	8.9	(13.5)	(3.2)	14.7
NBR	17.0	17.6	7.7	13.6	25.8	10.8	19.3	21.5	18.2

[a] Values are in MPa; see footnote to Table 7.9

this composition is also very low 275 psi (1.9 MPa). It appears that elastomer is the only continuous phase, yet the composition is moldable as a thermoplastic. This behavior could be explained if the elastomer did not cure in the presence of molten polycarbonate resin, or if the molten polycarbonate decomposed in the presence of the acrylate polymer, possibly by transesterification. This would be another type of technological incompatibility.

7.3.4   Characteristics of Elastomers and Plastics for Correlations with Blend Properties

Properties of elastomer–plastic blends have been correlated with various properties and characteristics of elastomers and plastics described below. Values are given in Table 7.13.

Table 7.11 Elongation at Break ε_B of 60/40 Rubber–Plastic Thermoplastic Vulcanizates[a]

Rubber–Plastic	PP	PE	PS	ABS	SAN	PMMA	PBT	PA	PC
IIR	380	312	3	18	7	6	156	34	161
EPDM	530	612	69	18	5	6	102	30	66
PTPR	210	280	35	15	10	10	47	60	5
NR	390	360	85	56	14	58	62	42	21
BR	258	229	73	64	12	5	52	121	5
SBR	428	240	89	70	12	15	102	201	19
EVA	319	349	166	102	109	59	(126)	160	81
ACM	18	20	20	144	18	21	135	163	140
CPE	314	221	140	197	151	146	(159)	160	135
CR	141	390	67	96	7	5	65	(6)	91
NBR	204	190	20	164	196	56	350	320	130

[a] Values are in %. See footnote to Table 7.9

Table 7.12 Tension Set ε_S of 60/40 Rubber–Plastic Thermoplastic Vulcanizates[a]

Rubber–Plastic	PP	PE	PS	ABS	SAN	PMMA	PBT	PA	PC
IIR	23	28	—	—	—	—	—	—	—
EPDM	16	—	—	—	—	—	—	—	—
PTPR	20	27	—	—	—	—	—	—	—
NR	24	—	—	—	—	—	—	—	—
BR	27	—	—	—	—	—	—	—	—
SBR	30	—	—	—	—	—	—	—	—
EVA	36	36	70	—	—	—	—	26	—
ACM	—	—	—	—	—	—	41	56	17
CPE	55	58	—	65	91	82	40	59	85
CR	33	37	—	—	—	—	—	—	—
NBR	31	—	—	—	55	—	25	44	—

[a] Values are in %. See footnote to Table 7.9

7.3.4.1   Dynamic Shear Modulus

This property (G') was taken as a measure of stiffness. It was measured by means of a torsion pendulum using specimens whose dimensions were selected to give test frequencies between 0.5 and 2 Hz. It was selected rather than Young's modulus because of convenience. When considering such widely varying materials as elastomers and hard plastics, it is difficult to find a convenient test condition (rate of loading) appropriate for both elastomers and plastics. The shear moduli of the hard and soft phases (along with elastomer–plastic proportions) were correlated with shear moduli of blends.

7.3.4.2   Tensile Strength of the Hard-Phase Material

This property was considered because it represents a limit for the strength of the elastomer–plastic blend. Yield stress rather than stress at break was used as tensile strength for crystalline materials since breaking occurs only after necking and drawing. (Generally, the elastomeric blends do not exhibit drawing-necking behavior.) The values in Table 7.13 were determined in the same way as for the elastomer–plastic blends, by using molded samples that had been equilibrated against laboratory air. For nylon many of the literature values relate to dried samples and are therefore somewhat higher than the values shown here.

7.3.4.3   Crystallinity

The weight fractions of crystallinity (W_c) of many of the plastics are also given in Table 7.13. The values in the tables are approximations based on the densities of the materials. The reasons for considering crystallinity were empirical; however, interesting correlations between hard-phase crystallinity and certain blend properties have been obtained.

Table 7.13 Approximate Polymer Characteristics

	σ_B (MPa)	G^a (MPa)	$\gamma_c{}^b$ (mN/m)	$N_c{}^c$ (number of chain atoms)	$W_c{}^d$ (weight fraction)
Polypropylene (PP)	30.0	520	28	—	0.63
Polyethylene (PE)	31.7	760	29	—	0.7
Polystyrene (PS)	42	1170	33	—	0.00
ABS	58	926	38	—	0.00
SAN	58	1130	38	—	0.00
Polymethylmethacrylate (PMMA)	61.8	—	39	—	0.00
Polybutylene terephthalate (PBT)	53.3	909	39	—	0.31
Nylon-6,9 (PA)	46	510	39	—	0.25
Polycarbonate (PC)	66.5	860	42	—	0.00
IIR	—	0.46	27	570	0.00
EPDM	—	0.97	28	460	0.00
Poly-*trans*-pentenamer (PTPR)	—	—	31	417	0.00
IR (NR)	—	0.32	31	454	0.00
BR	—	0.17	32	416	0.00
SBR	—	0.52	33	460	0.00
Ethylene-vinylacetate elastomer (EVA)	—	0.93	34	342	0.00
ACM	—	—	37	778	0.00
Chlorinated polyethylene	—	—	37	356	0.00
CR	—	—	38	350	0.00
NBR	—	0.99	39	290	0.00

[a] G' was determined by torsion pendulum at about 1 Hz. Rubbers were not vulcanized
[b] $\gamma_c{}^b$ is critical surface tension for wetting
[c] N_c is critical molecular length for entanglement
[d] W_c is crystallinity
[e] ABS was considered as SAN containing BR particles: thus σ_B for ABS was considered to be the same as for SAN. The somewhat increased elastomer concentration (over 60 wt%) should have only a small effect on the ultimate properties

7.3.4.4 *Critical Surface Tension for Wetting*

This parameter (γ_C) was originally used as an estimate of polymer surface energy by Zisman [26]. It was estimated by determining contact angles of various liquids against a given polymer surface. The contact angles were plotted as functions of the surface tension of the test liquids and the surface tension of liquid corresponding to an extrapolated contact angle of zero was taken as the critical surface tension for wetting (or spreading). At one time, it was believed that γ_C was approximately the surface tension γ_S of solid polymer. At any rate, the difference between the critical surface tension for wetting (for the elastomer and the plastic) should be a rough estimate of the interfacial tension between the elastomer and plastic during melt mixing. Interfacial tension is a factor that determines, at least in part, the droplet size of one liquid dispersed in another [27]. Lower surface tensions give smaller droplets (which should result in smaller particles of one polymer dispersed in the other after mixing and cooling).

The interfacial tension between two immiscible monomeric liquids is approximated by the difference between the two surface tensions. Unfortunately this is not the case for polymers. However, there is a hypothetical surface tension γ_X, which is characteristic of each polymer listed in Wu's review of interfacial tension between molten polymers [28]. If the value of γ_X for one polymer is subtracted from that of the other, the interfacial tension is estimated fairly reliably. The hypothetical values γ_X correlate well with γ_C. It is also interesting that the critical surface tension γ_C for wetting correlates with solubility parameter and that differences between solubility parameters $(\sigma_1 - \sigma_2)$ of the polymers of a two-phase system correlate with interfacial tension. Indeed, Helfand and Sapse [29] have given a theoretical basis for this.

From all of this we conclude that $\Delta\gamma_C$, the difference between critical surface tensions for wetting of each of two polymers, may be at least a qualitative estimate of the interfacial tension γ_{12}. The lower the difference $\Delta\gamma_C$ (which we sometimes call the surface energy mismatch), then, the smaller should be the particles of one molten polymer dispersed in the other. Also a low surface energy mismatch should give better wetting, better interfacial adhesion, and increased diffusion of the polymers across the interface.

Some of the values of γ_C listed in Table 7.13 were taken from the literature [30]. Those values not available were estimated on the basis of the correlation between solubility parameter σ and γ_C.

7.3.4.5   Critical Entanglement Spacing

The critical entanglement spacing N_c is defined as the number of polymer chain atoms that corresponds to a molecular weight sufficiently large for entanglements to occur between molecules of undiluted polymer. It has been measured as the molecular weight where the slope of a plot of log viscosity vs. log molecular weight changes from 1.0 to 3.4, the change being associated with intermolecular entanglements.

The reason for considering entanglement spacing as a parameter to correlate with blend properties was empirical. It was found that dynamically vulcanized elastomers that have low values of N_c gave the higher quality blends with plastics [25]. Although this was an empirical observation, one might speculate why such elastomers give the best blends (with respect to ultimate properties).

It has been observed that, when polymers are blended together, fibrous structures appear which then break up into polymer droplets [31, 32]. We believe it likely that polymers whose molecules are more entangled might be drawn into finer "fibers," during the early phase of mixing, to give emulsions of polymer droplets of smaller size. Of course, after dynamic vulcanization, these droplets would become very small vulcanized rubber particles.

Another explanation could be that a tendency for entanglement might promote adhesion, if some of the entanglement occurred across the interface between the molecules of the different polymers. After vulcanization and after cooling of the composition such entanglement-derived locked-in loops should improve interfacial adhesion. A schematic visualization of this is shown in Fig. 7.11.

Values of N_c obtained under the same conditions, for all of the elastomers, are not available in the literature. However, it is possible to calculate values of N_c from the chemical structure of the elastomer molecules by using a modified method of Aharoni [33]. The calculated values appear in Table 7.13.

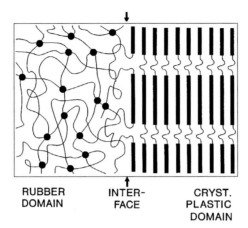

RUBBER   INTER-   CRYST.
DOMAIN   FACE   PLASTIC
                                 DOMAIN

Figure 7.11   Schematic diagram of adhesion promoted by interfacial entanglements of loops between crystalline domains

7.3.5   The Correlation Between Blend Properties and the Characteristics of Blend Components

The stress at break σ_B was correlated with the component characteristics as a relative ultimate tensile strength σ_B/σ_H, where σ_H is the strength of the hard phase (plastic) material, believed to be a limiting factor.

The effects on relative tensile strength σ_B/σ_H, ultimate elongation ε_B, and tension set ε_B are plotted according to regression equations in Fig. 7.12. In each case a property is plotted as a function of one of the three characterizing parameters ($\Delta\gamma_{SH}$, N_c, or W_c) with the other two variables held constant, each at a desirable level.

If we accept Fig. 7.12 as an overall view of the effects, certain conclusions can be drawn: (1) An increase in the crystallinity of the plastic material component improves both mechanical strength and elastic recovery. (2) Elastomers of lower N_c give compositions of greater mechanical integrity. (3) Compositions in which the surface energies of the elastomer and plastic phases are closely matched are strong and extensible. As stated previously, matching of surface energies should give lower interfacial tensions, resulting in smaller elastomer particles that act as smaller stress concentrator-flaws, and so improve both the high strength and extensibility.

Thus, based on a few characteristics of the pure elastomers and plastics, elastomer–plastic combinations can be selected, with a good probability of success, to give thermoplastic vulcanizates (by dynamic vulcanization) of good mechanical integrity and elastic recovery. The best compositions are prepared when the surface energies of the elastomer and plastic material are matched, when the entanglement molecular length of the elastomer is low, and when the plastic material is crystalline. It is necessary that neither the plastic nor the elastomer decomposes in the presence of the other at temperatures required for melt mixing.

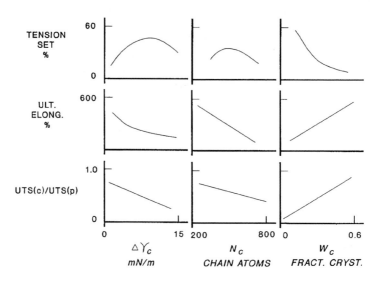

Figure 7.12   Effect of pure-component characteristics on the properties of thermoplastic vulcanizate compositions. The ratio σ_B/σ_H is the relative tensile strength. The value ε_B is the ultimate elongation (in %), and ε_S is the tension set (in %). The value $\Delta\gamma_{SH}$ is the difference between the critical surface tension for wetting γ_C (in mN/m) of the rubber and that of the plastic; N_c is the critical molecular length (chain atoms of rubber molecules) for entanglement; W_c is the weight fraction of crystallinity of the hard-phase material

Also, a curing system is needed, appropriate for the elastomer under the conditions of melt mixing.

7.3.6   Technological Compatibilization of NBR–Polyolefin Blends by Elastomer–Plastic Graft Formation

Were it not for their gross mutual incompatibility (in the thermodynamic sense), a combination of a polyolefin resin with NBR might be a good choice of materials from which to prepare oil-resistant thermoplastic elastomer compositions by dynamic vulcanization. Early work with these materials demonstrated only marginal success in obtaining good mechanical properties for such compositions [34]. This was probably due to the large surface energy difference between the two types of polymers. Mutual wetting between the polymers appeared incomplete; relatively large particles of cured elastomer dispersed in polyolefin resin formed during mixing and dynamic vulcanization. An approach to technological compatibilization, in addition to dynamic vulcanization, was thus sought [35, 36].

It is now generally accepted that a block copolymer can compatibilize mixtures of the "parent" homopolymers. The block copolymers act as macromolecular surfactants to promote and stabilize the emulsion of the molten homopolymers [37–39]. Figure 7.13 is an idealized visualization of this.

It has been found that a dimethylol phenolic compound (such as "phenolic resin" curative SP 1045) can be used to technologically compatibilize a mixture of polyolefin and NBR. This compatibilization could be the result of the formation of a block copolymer of the type visualized in Fig. 7.13. Such a compatibilizing block copolymer could be formed by the following reaction scheme:

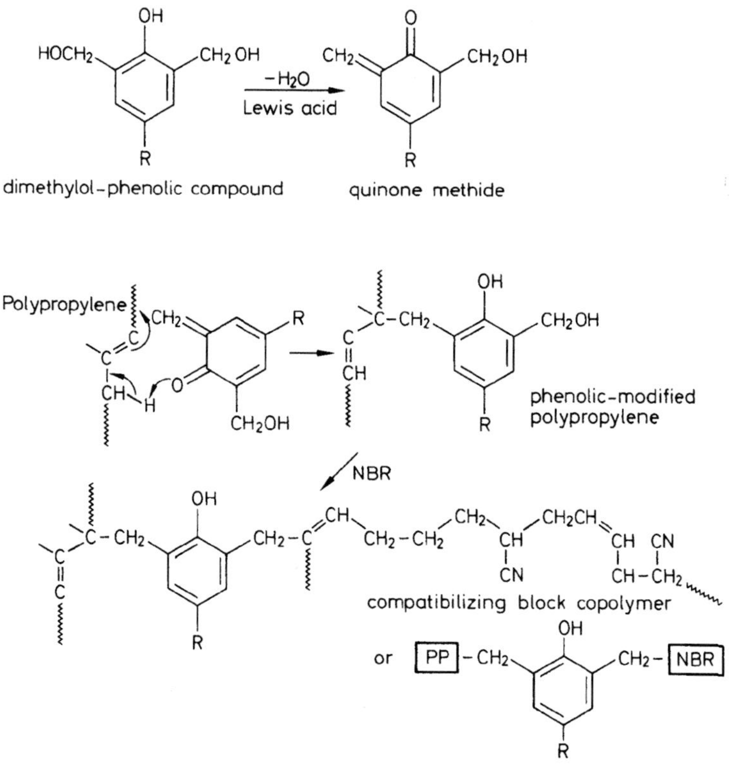

dimethylol–phenolic compound   quinone methide

The scheme requires the presence of olefinic unsaturation in the polypropylene molecules. Indeed, to satisfy demands of strict polymerization stoichiometry, there is, on the average, one double bond per polypropylene molecule unless hydrogen is used to control molecular weight. The scheme is similar to that proposed long ago for the phenolic-resin curing of diene elastomers [40–42].

In practice, the polyolefin resin is treated with about 1 to 4 parts of a phenolic curative (e.g., SP 1045) per 100 parts of polyolefin resin (e.g., polypropylene) in the presence of 0.1 to 0.5 parts of a Lewis acid (e.g., $SnCl_2$) at a temperature of about 180°C to 190°C. The phenolic-modified polyolefin is then melt mixed with NBR for a sufficient time for compatibilization to occur with the formation of a blend of improved homogenization. Then with continuing mixing, curative for the elastomer is added. (This can be additional dimethylol phenolic resin curative.) If the NBR contains a small amount (ca. 5%) of an amine-terminated liquid NBR, ATBN (e.g., Hycar® 1300X16, B.F. Goodrich Co.), then the properties of the compatibilized blend are even better.

The formation of polymer–polymer grafts can be accomplished by a number of other chemical means, in addition to the above use of dimethylol phenolic derivatives. In some cases the results are even better. Such a case is the use of maleic-modified polypropylene to form the block-polymeric compatibilizing agent by reaction with the amine-terminated liquid NBR. In this case, polypropylene is modified by the action of either maleic acid or maleic anhydride in the presence of decomposing organic peroxide [43, 44]. During the process, the molecular weight of the polypropylene becomes greatly reduced as the molecules thereof acquire pendant succinic anhydride groups:

Polypropylene + (maleic anhydride) —R*/peroxide→ maleic-modified polypropylene

or COOH / COOH

If part of the polypropylene in a NBR–polypropylene composition is maleic-modified, and if part of the NBR is amine terminated, then compatibilizing amounts of NBR–polypropylene block copolymers form in situ during melt mixing:

NHR—NBR → compatibilizing block copolymer

$$\left(\boxed{PP} \diagdown \begin{array}{c} COOH \\ CH_2-CONR-\boxed{NBR} \end{array} \right)$$

Only a small amount of compatibilizing block copolymer is needed to obtain a substantial improvement in the properties of a blend. The data of Table 7.14 relate to compositions in which 10% of the polypropylene was modified by the action of 5 parts of maleic anhydride in the presence of 0.87 parts of Lupersol®-101 peroxide per 100 parts of polypropylene at 180°C to 190°C. Varying amounts of the NBR were replaced by amine-terminated liquid NBR. Since the maleic-modified polypropylene was generally in stoichiometric excess, it can be assumed that essentially all of the amine-terminated elastomer was grafted to some of the polypropylene. After each compatibilized blend was prepared, it was dynamically vulcanized and subjected to the usual treatment for molding and testing.

The data of Table 7.14 indicate that improved blend properties are obtained when as little as 0.16% of the elastomer is grafted to polypropylene. Also, after about 2% to 3% of the elastomer is grafted to the polyolefin, additional graft formation gives almost no further improvement. It should be noted that the mechanical properties of the compatibilized dynamically vulcanized blends of NBR and polypropylene can be about as good as those of dynamically vulcanized EPDM and polypropylene.

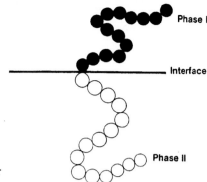

Phase I

Interface

Phase II

Figure 7.13   Idealized compatibilizing block copolymer molecule

Table 7.14 Properties of Compatibilized-Blend Dynamic Vulcanizates as a Function of the Amount of Rubber Grafted to Plastic

	1	2	3	4	5	6	7	8
Recipe[a]								
Polypropylene[b]	50	45	45	45	45	45	45	45
Maleic-modified polypropylene[c]	—	5	5	5	5	5	5	5
NBR[b]	50	50	49.22	46.88	43.75	37.5	25	—
NBR Masterbatch[d]	—	—	0.78	3.12	6.25	12.5	25	50
SP 1045[e]	3.75	3.75	3.75	3.75	3.75	3.75	3.75	3.75
SnCl$_2$·2H$_2$O	0.50	0.50	0.50	0.50	0.5	0.50	0.50	0.50
ATBN[f] as % of NBR (% rubber grafted to polypropylene[f])	0.00	0.00	0.16	0.62	1.25	2.5	5	10
Properties								
Tensile strength σ_B (MPa)	8.8	12.0	12.4	15.2	22.0	25.5	25.7	26.7
Stress at 100% strain (MPa)	—	12.0	12.1	12.0	12.3	12.3	12.5	12.9
Young's modulus E (MPa)	209	200	212	223	185	188	184	237
Elongation at break ε_B (%)	19	110	170	290	400	440	430	540
Tension set ε_s (%)	—	—	45	40	40	40	42	45
True stress at break σ_B* (MPa)	10	25	33	59	110	138	136	171
Breaking energy[g] (J/cm^3)	1.2	11	18.4	34.6	54.9	64.1	61.7	86.5
Improvement in breaking energy due to compatibilization (%)	—	—	67	215	399	483	461	686

[a] Parts by weight

[b] Polypropylene is Profax® 6723; NBR is Hycar® 1092-80 nitrile rubber; ATBN is Hycar® ATBN 1300x16 amine-terminated liquid nitrile rubber

[c] See text

[d] 90% by weight of Hycar® 1092-80, 10% by weight of Hycar® ATBN 1300x16 liquid nitrile rubber

[e] SP 1045 is a dimethylolphenolic vulcanizing agent

[f] A quantitative reaction is assumed between ATBN and maleic-modified polypropylene (which is assumed to be present in excess)

[g] Breaking energy values were obtained from the stress-strain curves

Table 7.15 Compatibilized NBR–Polypropylene Composition Blended with EPDM–Polypropylene Composition

	1	2	3
Wt % NBR–polypropylene composition[a]	100	50	0
Wt % EPDM–polypropylene composition[b]	0	50	100
Properties			
Tensile strength (MPa)	22.6	15.9	8.6
Stress at 100% strain (MPa)	11.2	7.9	4.4
Elongation at break (%)	585	510	415
Hardness (A scale)	93	87	68
Tension set (%)	48	23	10
True stress at break (MPa)	155	97	44.3
ASTM No. 3 oil volume swelling (%)[c]	22	32.5	62.5
Brittle point (°C)	−24	−47	< −60

[a] Recipe for the NBR–polypropylene composition is: Profax® 6723 polypropylene, 45 (parts by weight); maleic-modified polypropylene, 5; Hycar® 1092-80 NBR/Hycar® ATBN 1300x16 liquid NBR (90 : 10), 50; SP-1045 phenolic curative, 3.75; SnCl$_2$·2H$_2$O, 0.50; Naugard® 495 stabilizer (added after vulcanization), 1.0
[b] Commercially available Santoprene® 201-73 (Advanced Elastomers Systems)
[c] Hot-oil swelling, 70 h at 100 °C

7.3.6.1   Resistance to Hot Oil and Brittle Point

The hot oil resistance and brittle point of a compatibilized NBR–polypropylene thermoplastic vulcanizate prepared by dynamic vulcanization are shown in Table 7.15 (Stock 1). Though the hot oil resistance is excellent, low-temperature performance is somewhat lacking.

It has been found that the low-temperature brittle point can be reduced with a minimum sacrifice of hot oil resistance by blending the compatibilized NBR–polypropylene composition with a commercially available thermoplastic elastomer based on polypropylene and vulcanized EPDM. The two thermoplastic elastomer compositions are mutually compatible, since they are both based on a continuous phase of polypropylene. The results given in Table 7.15 indicate that the mechanical properties of the blend are about the average of the two components but surprisingly the hot oil resistance is better than the average.

7.3.6.2   Overall Assessment of Compatibilized NBR–Polypropylene Thermoplastic Vulcanizates

The mechanical properties of these compatibilized blends can approach those of dynamically vulcanized EPDM–polypropylene blends and, in addition, excellent hot oil resistance can be achieved. The NBR-based compositions may not have good enough resistance to low-temperature embrittlement for some applications, but compositions based on EPDM and polypropylene can be blended with the NBR–polypropylene composition to improve the brittle point without causing severe losses of other attributes.

7.3.7   Blends of Thermoplastic Vulcanizates Based on Dissimilar Plastics

In this section we describe thermoplastic elastomer compositions comprising technologically compatibilized blends of thermoplastic vulcanizates (TPVs) based on nylon and on polypropylene. Both the nylon- and the polypropylene-based thermoplastic vulcanizates were previously prepared by dynamic vulcanization. The compositions were technologically compatibilized by the presence of chemically modified polypropylene which presumably reacted with small portions of the nylon to form compatibilizing amounts of nylon–polypropylene graft-linked copolymer [35]. Components of the blends of thermoplastic vulcanizates are described below.

7.3.7.1   Description of the Dissimilar-Plastics Compositions

A 50 : 50 blend of EPDM and polypropylene was dynamically vulcanized by 5 parts (per 100 parts of combined polymers) of SP 1045 dimethylol phenolic curative in the presence of 1 part of $SnCl_2 \cdot 2H_2O$ at 180 °C to 190 °C. The composition was designated EPDM–polypropylene TPV.

A 65 : 35 blend of NBR and nylon 6/6–6 copolymer (melting point, 213 °C) was treated with 1.3 parts (per 100 parts of combined polymers) of SP 1045 phenolic curative during its mixing at 215 °C. (Flectol®·H antidegradant was used at a level of 2 parts per 100 parts of polymer.) The composition was designated NBR–NY TPV.

A 50 : 50 blend of a millable, vulcanizable polyurethane rubber (Adiprene®·C) and nylon 6/6–6/6–10 terpolymer (melting point 163 °C—Zytel®·63) was dynamically vulcanized at 180 °C by the action of 1.0 m-phenylenebismaleimide in the presence of 0.5 parts of L-101 organic peroxide per 100 parts of combined polymers. The product was designated PU–NY TPV.

A 50 : 50 blend of Hydrin® 400 rubber poly(epichlorohydrin), (ECH) and the same nylon 6/6–6/6–10 terpolymer was dynamically vulcanized at 170 °C to 180 °C by the action of 1.67 parts of zinc stearate, 1.0 part of bisbenzothiazole disulfide (MBTS), and 0.4 part of sulfur per 100 parts of combined polymers, in the presence of 1.0 part of Flectol®·H antidegradant. The product was designated ECH–NY TPV.

A functionalized polymer was prepared by melt mixing 100 parts of polypropylene with 5 parts of maleic acid in the presence of L-101 peroxide (added after the acid and polymer were well mixed) at 180 °C for about 3 min. The product was designated MA-mod. PP.

7.3.7.2   Blends of Thermoplastic Vulcanizates

Blends of the various thermoplastic vulcanizates (with and without the chemically modified, compatibilizing, functionalized polypropylene) were prepared by melt mixing the ingredients in a Brabender or Haake Rheomix internal mixer at about 10 °C above the melting point of the nylon used in each case.

Properties of molded sheets of the mixtures of thermoplastic vulcanizates are given in Table 7.16. These test results indicate that nylon-based TPVs containing a variety of different types of rubbers can be used in the compatibilized blends. A substantial improvement in properties is obtained in each case. (Note the comparison between the control stocks

Table 7.16 Compatibilization of Polypropylene–Nylon Compositions[a]

Recipe	1	2	3	4	5	6
Dynamically vulcanized EPDM–PP composition	50	50	50	50	50	50
Dynamically vulcanized NBR–nylon composition	50	50	—	—	—	—
Dynamically vulcanized polyurethane–nylon composition	—	—	50	50	—	—
Dynamically vulcanized epichlorohydrin rubber–nylon composition	—	—	—	—	50	50
Polypropylene	10	—	10	—	10	—
Maleic-modified polypropylene	—	10	—	10	—	10
Properties						
Tensile strength (MPa)	16.1	26.9	9.2	19.6	11.9	22.0
Stress at 100% strain (MPa)	14.0	16.5	—	11.8	9.3	12.0
Young's modulus (MPa)	132	147	173	157	206	199
Elongation at break (%)	170	310	73	270	230	340
Tension set (%)	40	39	—	52	40	50
True stress at break (MPa)	43	110	16	73	39	97

[a] Component blends are defined in the text

containing added unmodified polypropylene [No. 1, No. 3, and No. 5] and the experimental stocks containing the maleic-modified polypropylene [No. 2, No. 4, and No. 6].)

It is believed that the maleic acid-modified polypropylene molecules contain pendant succinic anhydride groups that react with amine groups in the nylon to form compatibilizing amounts of a graft-linked nylon–polypropylene block copolymer [44]. It should be noted that a number of other chemical modifications of polypropylene have also given results similar to those obtained with maleic acid-modified polypropylene [35].

From this work, one can conclude that compositions that have excellent mechanical properties can be prepared by melt mixing thermoplastic vulcanizates that have been previously prepared by dynamic vulcanization. Excellent mixed-TPV compositions can be obtained even when the rubbers and plastics are mutually grossly incompatible with respect to thermodynamic considerations. In such cases, however, it appears to be necessary that a compatibilizing agent be present in the mixture to promote the interaction between the thermoplastic materials.

The rubber associated with one of the thermoplastic components can differ greatly from the rubber associated with the other thermoplastic component. Thus compositions can be produced that have good mechanical properties and that can contain both differing thermoplastic resins and differing rubbers. As a result, the possible combinations of components for TPV compositions has been greatly expanded.

7.4   Technological Applications

The elastomer–plastic blends discussed here are intended for use as thermoplastic elastomers. These are materials that have many of the properties of conventional vulcanized (thermoset) elastomers, but are processable and can be fabricated into parts by the rapid techniques used for thermoplastic materials.

Thermoplastic processing is far more economically attractive than traditional multistep elastomer processing. In the case of conventional thermoset elastomer processing, the producer of elastomeric articles purchases elastomer, fillers, extender oils or plasticizers, curatives, antidegradants, etc.; these ingredients must then be mixed and uniformly dispersed; after a stock is mixed, it is then shaped by extrusion, calendering, etc.; the crudely shaped preform is then vulcanized in its final shape by heating it in a mold contained by a press and heated for vulcanization which can take a long period of time. Mold flash or overflow, as well as rejected parts, are not economically reprocessable.

On the other hand, a part produced from a thermoplastic elastomer is shaped or molded into its final shape in a single step. Mold flash and rejected parts can simply be ground and reused. Detailed comparisons have shown that thermoplastic processing is more economical than thermoset elastomer processing [4] and this provides the economic incentive for the development of thermoplastic elastomers.

The commercial compositions based on melt-mixed blends of completely dynamically vulcanized EPDM and polypropylene, as described herein, have many of the excellent properties of polyurethane and polyether–ester copolymer thermoplastic elastomers. In addition, the properties of these commercial grades of completely vulcanized EPDM–polypropylene thermoplastic vulcanizate materials compare favorably with those of conventional vulcanized specialty rubbers (CR, EPDM, and CSM). With respect to fatigue life, the commercial EPDM–polypropylene thermoplastic vulcanizates out-perform the conventional specialty elastomers. Comparisons between selected properties of three thermoplastic elastomers and those of a conventional elastomer are given in Table 7.17.

7.4.1   Processing—Fabrication Technology

The processing of a rubber–plastic blend composition into a finished part is a function of the melt rheology of the composition, the temperature, and shear rate. Processing is also a function of the strength of the molten material under the strain due to its processing, that is, a function of its resistance to melt fracture. Melt fracture, during extrusion, can give rise to very poor surface textures or even functionally useless parts.

The melt rheology of an elastomer–plastic blend composition is related to that of the plastic material in the idealization given in Fig. 7.14. At high shear rates, the viscosity–shear rate profiles are similar for both products. However, at very low shear rates, the viscosity of the blend can be very high. In the case of dynamically vulcanized elastomer–plastic melt-mixed blends containing high levels of elastomer, the viscosity can approach infinity when the shear rate approaches zero (presumably due to cured elastomer particle–particle interference). Under the conditions of melt extrusion, the molten material undergoes rapid flow in the die. Then, as the material passes out of the die, the rate of deformation drops to

Table 7.17 Properties of Various Types of Elastomer Compositions

Property	Partially vulcanized EPDM–PP blend[a]	Completely vulcanized EPDM–PP blend[b]	Neoprene Vulcanizate	Ester–ether copolymer thermoplastic elastomer[c]
Hardness (A Scale)	77	80	80	92
Tensile strength (MPa)	6.6	9.7	9.7	25.5
Ultimate elongation (%)	200	400	400	450
Volume swelling (% in ASTM No. 3 oil, 74 h at 100 °C)	—[d]	50	35	30
Compression set (ASTM Method B, 22 h at 100 °C, %)	70	39	35	33
Upper use temperature (°C)	100	125	110	125
Type of processing[e]	TP	TP	CV	TP

[a] TPR®-1700 thermoplastic elastomer (Uniroyal)
[b] Santoprene® thermoplastic elastomer (Advanced Elastomer Systems)
[c] Hytrel® thermoplastic elastomer (DuPont)
[d] Sample disintegrated
[e] TP, thermoplastic processing; CV, conventional vulcanization

zero, and, since the viscosity approaches infinity, little or no die swell is observed. The dimensions of extrusion profiles of such materials are thus easily controlled. This shows that the melt viscosity of an elastomer–plastic blend is highly shear rate sensitive.

The effect of temperature on the viscosity of an EPDM–polypropylene thermoplastic elastomer at high levels of elastomer is illustrated in Fig. 7.15. The viscosity of this type of blend is relatively temperature insensitive. (However, compositions based on other types of thermoplastic show more temperature sensitivity). Figures 7.14 and 7.15 indicate that, in processing certain elastomer plastic blends by flow techniques such as extrusion and injection molding, shear rates should be kept high enough to facilitate adequate flow.

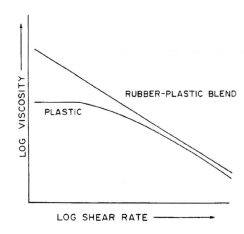

Figure 7.14   Relationship between viscosity and the shear rate for a plastic and its blend with rubber

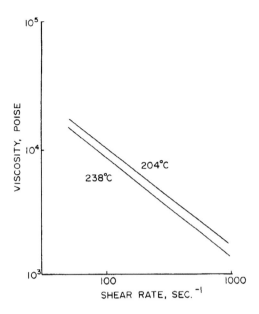

Figure 7.15   Effect of temperature on the EPDM–polypropylene viscosity-shear-rated relationship for a typical thermoplastic vulcanizate

The high melt viscosity of these products can be advantageous in processing [4]. It can provide high melt integrity or "green strength" and permit the retention of shapes of parts produced by extrusion or blow molding. The high melt viscosity and low die swell are also helpful in calendaring sheet and film products. For injection molded parts, fast injection rates (under high pressure) give lower viscosities owing to the high sensitivity of viscosity to shear rate. Thus, fast injection rates facilitate rapid and complete mold filling. When the mold is filled, the shear rate is reduced to zero. This increased viscosity, which can approach infinity, enables more rapid extraction of the part from the mold. The overall effect is a faster injection molding cycle. In addition, the moderate temperature sensitivity of the viscosity of such a composition gives a "broad temperature window" for processing. Typical injection molding and extrusion conditions for an EPDM–polypropylene blend are given by Tables 7.18 and 7.19.

In addition to the above (injection molding, calendering, extrusion, and blow molding), foaming, thermoforming, and compression molding of olefinic elastomer–plastic blends have been reviewed [1]. However, it should be noted that processing conditions vary widely with equipment mold designs, specific blend compositions, etc. The best conditions for the production of a given part must be confirmed by prototype runs in factory equipment.

7.4.2   End-Use Applications

New techniques of melt-mixing blends of elastomers and plastics, dynamically vulcanized, and employing compatibilizing agents have greatly expanded the number of useful

Table 7.18 Injection Molding Conditions for EPDM–Polypropylene-Based Thermoplastic Elastomeric Blends

Rear-zone barrel temperature (°C)	180–220
Center-zone barrel temperature (°C)	205–220
Nozzle temperature (°C)	205–220
Molt temperature (°C)	20–65
Injection pressure (MPa)	35–140
Hold pressure (MPa)	30–110
Back pressure (MPa)	0.7–3.5
Screw speed (rpm)	25–75
Injection speed	Moderate to fast
Injection time (s)	5–25
Hold time (s)	15–75
Total cycle time (s)	20–100

Table 7.19 Extrusion Conditions for EPDM–Polypropylene-Based Elastomeric Blends

Rear-zone barrel temperature (°C)	175–210
Center-zone barrel temperature (°C)	175–210
Front-zone temperature (°C)	190–220
Adapter temperature (°C)	200–225
Die temperature (°C)	205–225
Melt temperature (°C)	205–235
Screw speed (rpm)	10–150

thermoplastic elastomers. It is probable that this expansion will continue and accelerate. Potential and proven applications of these more recently developed thermoplastic elastomers are listed here.

For mechanical rubber goods applications: caster wheels, convoluted bellows, flexible diaphragms, gaskets, seals, extruded profiles, tubing, mounts, bumpers, housings, glazing seals, valves, shields, suction cups, torque couplings, vibration isolators, plugs, connectors, caps, rollers, oil-well injection lines, handles, and grips.

For under-the-hood automotive applications: air conditioning hose cover, fuel-line hose cover, vacuum tubing, vacuum connectors, body plugs, seals, bushings, grommets, electrical components, convoluted bellows, steering gear boots, emission tubing, protective sleeves, shock isolators, and air ducts.

For industrial hose applications: hydraulic (wire braid), agricultural spray, paint spray, plant air–water, industrial tubing, and mine hose.

For electrical applications: plugs, strain relief, wire and cable insulation and jacketing, bushings, enclosures, connectors, and terminal ends.

7.4.2.1   Emerging Applications

Because of the high level of product development activity sustained by the industry, a number of new thermoplastic elastomeric compositions are emerging. Efforts are directed toward the development of thermoplastic elastomeric blend compositions that are more resistant to hot oil, compositions that are more resistant to higher temperatures, and, eventually, compositions that are more resistant to both hot oil and higher temperatures. The combination of the two techniques discussed in this chapter have opened the door to many new product opportunities.

As stated throughout this chapter, dynamic vulcanization greatly improves the properties of elastomer–plastic blends. However, in addition to dynamic vulcanization, technological compatibilization by the incorporation of block-copolymer compatibilizers has greatly increased the number of elastomer–plastic combinations that can give reliably good mechanical properties. It is now not necessary that the elastomer and thermoplastic be of similar surface energy. Thus, combinations can be selected on the basis of the properties of the individual elastomer and thermoplastic components. For example, a high-melting crystalline plastic can, at least in principle, be combined with an elastomer that has a very low brittleness temperature to produce a thermoplastic elastomeric material with a very wide range of end-use temperatures. Alternately, the high-melting crystalline plastic can be combined with an oil-resistant elastomer of good thermal-oxidative stability to give a thermoplastic elastomeric material suited for applications in highly aggressive environments.

References

1. C.P. Rader, *Handbook of Thermoplastic Elastomers*. 2nd Ed. B.M. Walker and C.P. Rader (Eds.) (1988) Van Nostrand Reinhold, New York, pp. 85–140
2. E.N. Kresge, *Polymer Blends, Vol. 2*. D.R. Paul and S. Newman (Eds.) (1978) Academic Press, New York, p. 293
3. E.N. Kresge, *J. Appl. Polym. Sci. Appl. Polym. Symp. 39*, 37 (1984)
4. G.E. O'Connor and M.A. Fath, *Rubber World*, December (1981), January (1982)
5. B.D. Gesner, In *Encyclopedia of Polymer Science and Technol., Vol. 10*. H.F. Mark and N.G. Gaylord (Eds.) (1969) Wiley–Interscience, New York, p. 694
6. A.M. Gessler, U.S. Patent 3,037,954 (June 5, 1962)
7. W.K. Fischer, U.S. Patent 3,758,643 (September 11, 1973)
8. A.Y. Coran, B. Das, and R.P. Patel, U.S. Patent 4,130,535 (December 19, 1978)
9. A.Y. Coran and R.P. Patel, *Rubber Chem. Technol. 53*, 141 (1980)
10. S. Abdou-Sabet and R.P. Patel, *Rubber Chem. Technol. 64*, 769 (1991)
11. S. Abdou-Sabet and M.A. Fath, U.S. Patent 4,311,628 (1982)
12. P.J. Flory, *Principles of Polymer Chemistry* (1953) Cornell University Press, Ithaca, New York, p. 576
13. L.E. Nielson, *Rev. Sci. Instr. 22*, 690 (1951)
14. S. Abdou-Sabet and R.P. Patel, Paper presented at American Chemical Society, Rubber Division, Washington, D.C., October 9–12, 1990
15. A.Y. Coran, and R.P. Patel, U.S. Patent 4,104,210 (August 1, 1978)
16. A.Y. Coran, and R.P. Patel, U.S. Patent 4,183,876 (January 15, 1980)
17. A.Y. Coran, and R.P. Patel, U.S. Patent 4,271,049 (June 2, 1981)
18. D.S. Campbell, D.J. Elliott, and M.A. Wheelans, *NR Technol. 9*, 21 (1978)
19. M.P. Payne, D.S.T. Wang, R.P. Patel and M.M. Sasa, Paper No. 34, Presented at the Rubber Division, ACS Meeting, Washington, D.C., October 10–12, 1990
20. A.Y. Coran, and R.P. Patel, U.S. Patent 4,130,534 (December 19, 1978)
21. R.C. Puydak and D.R. Hazelton, *Plast. Eng. 44*, 37–39 (1988)
22. A.Y. Coran, and R.P. Patel, *Rubber Chem. Technol. 53*, 781 (1980)
23. H.L. Wagner and P.J. Flory, *J. Am. Chem. Soc. 74*, 195 (1952)
24. P.J. Flory, *Principles of Polymer Chemistry* (1953) Cornell University Press, Ithaca, New York, p. 568
25. A.Y. Coran, R.P. Patel, and D. Williams, *Rubber Chem. Technol. 55*, 116 (1982)

26. W.A. Zisman, *Adv. Chem. Ser. 43*, 1 (1964)
27. T. Mikami, R.G. Cox, and S.G. Mason, *Int. J. Multiphase Flow 2*, 112 (1975)
28. S. Wu, *Polymer Blends, Vol. 1*, D.R. Paul and S. Newman (Eds.) (1978) Academic Press, New York, p. 244
29. E. Helfand and A.M. Sapse, *J. Chem. Phys. 62*, 1327 (1975)
30. G.I. Crocker, *Rubber Chem. Technol. 42*, 30 (1969)
31. G.N. Avgeropoulos, F.C. Weissert, P.H. Biddison, and G.G.A. Boehm, *Rubber Chem. Technol. 49*, 93 (1976)
32. G.R. Hamed, *Rubber Chem. Technol. 55*, 151 (1982)
33. S.M. Aharoni, *J. Appl. Polym. Sci. 21*, 1323 (1977)
34. A.Y. Coran and R.P. Patel, U.S. Patent 4,104,210 (August 1, 1978)
35. A.Y. Coran and R.P. Patel, U.S. Patent 4,355,139 (October 19, 1982)
36. A.Y. Coran and R.P. Patel, *Rubber Chem. Technol. 56*, 1045 (1983)
37. D.R. Paul, *Polymer Blends, Vol. 2*, D.R. Paul and S. Newman (Eds.) (1978) Academic Press, New York, p. 35
38. O. Olabisi, L.M. Robeson and M.T. Shaw, *Polymer Miscibility* (1979) Academic Press, New York, p. 321
39. N.G. Gaylord, *Adv. Chem. Ser. 142*, 76 (1975)
40. S. Van der Meer, *Rev. Gen. Caoutch. Plast. 20*, 230 (1943)
41. C. Thelamon, *Rubber Chem. Technol. 36*, 268 (1963)
42. A. Giller, *Kant. Gummi. Kunstst. 19*, 188 (1966)
43. M. Veda Minoura, S. Mizunuma, and M. Oba, *J. Appl. Polym. Sci. 13*, 1625 (1969)
44. F. Ide and A. Hasegawa, *J. Appl. Polym. Sci. 18*, 963 (1974)

8   Thermoplastic Polyether Ester Elastomers

R.K. Adams, G.K. Hoeschele, and W.K. Witsiepe*

8.1 Introduction 192

8.2 Early Fiber Research 192

8.3 Synthetic Methods 195

8.4 Polymer Structure and Morphology 197
    8.4.1 The Crystalline Region 201
    8.4.2 The Amorphous Region 204
    8.4.3 Overall Morphology 205
    8.4.4 Orientation and Stress–Strain Behavior 210

8.5 Properties of Commercial Copolyether Ester Elastomers 211
    8.5.1 Mechanical Properties 211
    8.5.2 Melt Rheology 211
    8.5.3 Degradation and Stabilization of Polyether Ester Elastomers 212

8.6 Structural Variations 213
    8.6.1 Hard Segment Modifications 213
    8.6.2 Polyether Soft Segment Modifications 217

8.7 Polymer Blends with Polyether Ester Elastomers 219

8.8 Commercial Aspects 221
    8.8.1 Major Producers 221
    8.8.2 Typical Applications 221

References 222

*Deceased April 28, 1996

distribution of units. The results of this calculation are shown in Fig. 8.4. Fifty-two per cent of the total polymer or 90% of the 4GT is in blocks of three or more 4GT units. This is the block length now believed necessary for crystallization of poly(tetramethylene terephthalate) units in the copolymer matrix.

To understand the relationship of the various articles in the field, particularly those dealing with the structure of the polymers, it is important to know how the authors define the soft segment and hence the concentration of hard segment. This definition will affect the statistics of groups along the polymer chain as well as the relationship of the mole fraction to weight fraction of hard segment. Early articles such as those by Cella [34] and Buck [35] defined the soft segment as the terephthalate of the polyether involved, for example, PTMOT, $MW = 1130$ assuming an actual PTMO molecular weight of 1000. This convention is still usually followed in literature involving commercial applications. Perego [36] and Valance and Cooper [37] define the same soft segment as PTMO less one butanediol unit, $MW = 910$, with the leftover material, quite logically, counted as tetramethylene terephthalate and included as hard segment. Thus Cella would define a certain polymer as having 58 wt% hard segment while Valance and Cooper would consider the same polymer to contain 66.1 wt% hard segment (see Table 8.1). To confound matters further, Wegner [38] and Bandara and Droescher [39] define the soft segment as the entire polyol molecule but do not include the linking terephthalate segment, demanded by the chemistry, in calculation of either the soft segment weight fraction or the hard segment weight fraction.

Vallance and Cooper [37] determined the crystalline content of a series of 4GT–PTMOT polymers with increasing hard segment content by calorimetry. The results are given in Table 8.1.

In this table, A, B and C are the mole ratios of DMT, 4GT and PTMO(1000); u_2 is the overall tetramethylene terephthalate weight fraction (PTMO $= 910$); w_2 is the weight fraction of total sample that is crystalline; and T_h is the highest observed melting temperature. Samples were quenched in ice after compression molding. w_1 is the weight fraction of

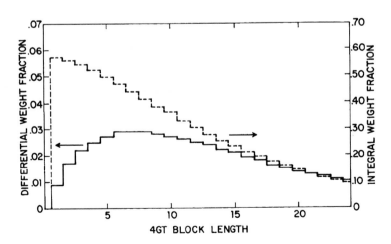

Figure 8.4   Calculated tetramethylene terephthalate block-length distribution of polymer containing 58 wt% tetramethylene terephthalate. (Reprinted with permission from Ref. [34], *J. Polym. Sci.* Copyright 1973, John Wiley & Sons)

Table 8.1 Crystallinity of 4GT–PTMO Copolymer

A/B/C	%4GT	μ_2	p	DP_n	w_2	w_1	$T_h(K)$	ρ_1
3.6/2.6/1	33.6	0.465	0.782	4.59	0.113	0.887	478	1.058
5.9/4.9/1	48.8	0.588	0.855	6.90	0.226	0.774	486	1.098
7.5/6.5/1	55.9	0.645	0.882	8.47	0.284	0.716	487	1.112
9.8/8.8/1	63.1	0.703	0.907	10.75	0.330	0.670	489	1.131
17.3/16.3/1	76.0	0.807	0.945	18.18	0.403	0.597	492	1.162
28.0/27.0/1	84.0	0.871	0.966	29.41	0.423	0.577	493	1.190
1.0/1.0/0	100	1.000	1.0	—	0.475	0.525	495	1.238

Adapted from Ref. [37] and published with permission from *Macromolecules*. Copyright 1984, American Chemical Society

amorphous material, ρ_1 is the density of the amorphous material, p is the mole fraction of tetramethylene terephthalate linkages in the copolymer, and DP_n is the number average degree of polymerization of the hard blocks calculated from $1/(1 - p)$.

The calculations for u_2 define the soft segment as PTMO (MW = 910) as discussed previously. Everything else is considered to be tetramethylene terephthalate, which has a weight fraction of u_2 and a mole fraction of p. If the soft segment is defined as PTMOT, then the number average degree of polymerization of the hard segment is 3.6 vs. the 4.6 given in Table 8.1. DP as calculated by Cella and Buck and Wegner and Droescher will normally be 1 less than that calculated by Cooper and Perego for the same polymer. Finally, Cooper may refer to polymers by the PTMOT convention while calculating hard segment mole fraction by the PTMO = 910 convention.

The amount of crystalline 4GT present in these samples was found to be relatively low, at best less than half of the 4GT present. It follows that much of the 4GT is in the amorphous phase. For example, the copolymer having a weight fraction of 0.703 total ester has an amorphous phase containing 59 wt% uncrystallized tetramethylene terephthalate sequences and only 41 wt% polyether. Thus there is not a clean break between regions of polyether and polyester and so the observed density of the amorphous phase increases with total hard segment content.

The melting points of the polymers follow the melting point depression theory proposed by Flory [3]. This is shown in Fig. 8.5, where $\ln p$ is plotted against reciprocal temperature, T_h, K. The point representing the shortest hard segment with a number average DP of 4.6 (PTMO = 910) is probably above the line because of a combination of end effects, and the low concentration of segments of crystallizable lengths.

The substitution of PTMO glycol having a higher molecular weight obviously increases the average block lengths of the hard segments and, with low concentrations of 4GT units, can improve certain polymer properties, such as melting point. Replacing a portion of the terephthalic acid with a second acid such as isophthalic acid or a portion of the tetramethylene glycol with a second low molecular weight glycol decreases the 4GT hard segment block length. This lowers the melting point and reduces the hardness of a polymer while the weight fraction of PTMO glycol is kept constant. The melting points of such mixed polymers are also consistent with Flory's melting point depression theory [40].

variously interpreted. The following highlights attempt to simplify a very complex subject in the available space.

Cella [34] and Buck [35] proposed a model of disoriented, interpenetrating, and co-continuous crystalline and amorphous domains. An important feature of this model was the existence of short tie molecules connecting thin crystalline lamellae. These tie molecules were thought to bear the load at initial stages of deformation. Many features of this model are still considered to be correct.

Seymour [58], on the other hand, proposed a model analogous to that of semicrystalline homopolymers on the basis of small-angle light scattering and polarized light microscopy studies. The polymers were found to exhibit a spherulitic supercrystalline structure that distorted into an ovoid configuration in response to stress. This was interpreted to mean that the crystalline structure as a whole was bearing the load. As a result of wet chemical studies, the lamellae were found to be chain folded with a thickness of seven hard segment units.

On the basis of both electron microscopy and polarized light microscopy studies with a staining procedure that better defined domain boundaries, Zhu, Wegner and Bandara [43, 46] found that the crystalline morphology was highly variable and depended on the rate of cooling and also on any thermal post treatment. Slow crystallization, with supercoolings less than $30\,^\circ$C, favored formation of a spherulitic structure while rapid quenching yielded a structure similar to that proposed by Cella. On annealing, the spherulites transformed themselves into a dendritic structure. Both types of structures were built up from lamellae thought to vary in thickness. The mechanical properties and T_g of the materials varied with the volume fraction of crystallization after annealing. Wegner et al. [47] postulated a nucleation process where the lamellar thickness was determined by the most frequent hard segment length and in which both shorter and longer sequences were rejected to the amorphous region.

Wegner, Droescher and others [48–53] prepared a series of hard segment oligomers and polymers with known hard segment length, and Schmidt and Droescher [54] correlated the reciprocal degree of polymerization and reciprocal melting point of these materials with the number average degree of polymerization of random polymers. This, of course, is the relationship proposed by Flory and Vrij [55] for melting point depression of oligomers. However, the values calculated from the plot for the heat of fusion and, to a lesser extent, the melting point of pure 4GT were not in agreement with the known properties of 4GT.

Briber and Thomas [44], largely on the basis of very detailed electron microscopy studies, proposed a different explanation. That is, fractionation of hard segment lengths was due to coupling of the sequence distribution with the stable crystal size at a given crystallization temperature. The lamellar thickness increased with the temperature of crystallization and thus shorter sequences were unable to participate in the crystallization/annealing process. Long sequences were able to make a second pass through the same or an adjacent crystal. They also concluded that the lateral dimensions of the crystals were determined by the availability of hard segment material of sufficient length to traverse the crystal axis normal to the lamellar surface. They also confirmed that the unit cell was substantially that reported earlier for 4GT homopolymer [54] and that the lamellar surface consisted of short hard segment sequences and polyether sequences with a limited amount of adjacent chain folding.

Droescher and Bandara [39] concluded, based on SAXS long period, that lamellar thickness did not differ much before or after annealing. However, the "specific internal surface" of the crystalline material and the chord lengths of both the crystalline and amorphous regions increased markedly on annealing. This was interpreted to mean that the increased crystalline volume after annealing was due to growth in the size of the existing lamellae rather than to the formation of additional small lamellae. The growth of the lamellae was concluded to be in the lateral dimensions rather than the chain direction. They proposed a fringed micelle model for the lamellae (see Fig. 8.9), which had a crystalline core of three or four hard segments and a boundary region of about 9 Å in thickness. Melting point was determined by the degree of perfection of the crystals rather than primarily by their thickness. It will be noted that this model does not suggest tight folding of polymer chains to adjacent unit cells.

Vallance and Cooper [37], on the basis of small- and wide-angle X-ray scattering, DSC, and dielectric spectroscopy studies, also found that the single crystallites were only 20 Å to 40 Å thick in the chain (001) direction of the crystal and 60 Å to 110 Å in the transverse directions (010 and 100). The crystallized mass fraction tended to increase with overall ester content but was never more than 50% of the overall polyester content. The distance separating individual crystallites in the (001) direction was 115 Å and it was assumed that soft segments and uncrystallized hard segments connected (001) faces of adjacent crystallites and held them in place with respect to each other. At overall ester contents equal or less than 70%, the oblate individual crystals coalesced to form extended, jogged, multicrystalline structures. Evidence was found for an amorphous transition zone of decreased electrical conductivity in the dielectric spectra of interfacial polarizations. Since segment mobility adjacent to the (001) face is lower, this is not surprising and is consistent with the 9 Å boundary region described by Droescher and Bandara [39]. Beyond that, there is also good agreement between many of the crystallite measurements described in the two articles even though the samples were often made using differing techniques.

Cooper and Miller [59–61] used small angle neutron scattering (SANS) to study the chain conformations of hard and soft segments as well as the whole chain in bulk polymer samples. The procedure makes use of partial deuterium labeling to provide a contrast mechanism for SANS experiments but does not alter the chemistry of the system. Based on close agreement between calculated and observed values for the radius of gyration (R_g), they

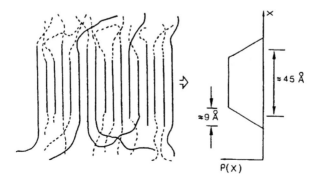

Figure 8.9   Schematic representation of the lamellar structure of bulk-annealed block copolyether ester. (Reprinted with permission from Ref. [39], *Colloid Polym. J.* Copyright 1983, Dr. Dietrich Steinkopff Verlag)

concluded that the polyester hard segments are chain folded to adjacent cells at room temperature with a repeat distance of three chain units. As the temperature increases above the T_g (about 50 °C), the hard segments undergo a melting recrystallization process whereby the high-energy, tightly folded, chain conformation is replaced by a lower energy, more fully extended, conformation. The morphological change resulting from this process is the thickening of crystalline lamellae. At still higher temperature, hard segment material is extracted from the amorphous phase with thickening and lateral growth of existing lamellae. This is in agreement with previous SAXS observations [35, 62], but should not be construed to mean that much of the very large increase in L (long period) after annealing near the melting point is due to thickening of crystalline lamellae.

At room temperature, the polyether soft segments were found to be extended, that is, taut, compared with the random coil model. The radius of gyration of the soft segment (R_g) was found to decrease with increasing temperature. This is caused by relaxation of taut, soft segment tie molecules, while chain-folded, hard segment crystals melted and recrystallized into a more extended chain conformation. The overall result is that the R_g of the entire chain increases with increasing temperature as a result of these combined effects [61]. Figure 8.10 is a model of polyether ester elastomers proposed by Cooper and Miller [59].

The overall morphology thus is that of a two-phase system consisting of a pure 4GT crystalline phase and an amorphous phase. The amorphous phase contains a substantially homogeneous mixture of polyether soft segments and 4GT hard segments rejected from the crystalline phase. The amorphous phase has a single T_g which can be correlated with its composition by the Gordon–Taylor relationship [56]. The crystalline phase has a unit cell equivalent to the α form of pure 4GT. A more or less continuous crystalline super-structure is formed of single crystal lamellae tied together by short sequences of hard segment. The individual lamellae are platelets with the polymer chain direction normal to the flat surface. In rapidly cooled specimens, the platelets are three or four hard segment units thick, (i.e., 35 Å

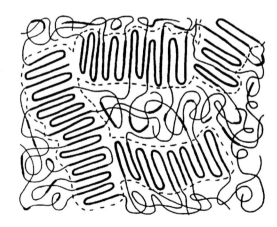

Figure 8.10   Model of the microstructure and chain conformation in the polyether–polyester block copolymer materials. (Reprinted with permission from Ref. [59], *Rubber Chem. Technol.* Copyright 1985, American Chemical Society)

to 45 Å), with a 9 Å thick transition zone to the amorphous region. The faces of the lamellae are tied to each other through the amorphous region with a combination of polyether and uncrystallized 4GT tie linkages. While many of the tie linkages are in a random coil, certain linkages are slightly extended and under stress. If force is applied to the system, the initial extension is entropic by extension of the tie molecules. With additional force, the unit cells go to the more extended β conformation, and finally, with further force, the crystalline structure breaks down completely, forming a system of banded fibers in the stress direction with the lamellae normal to the stress direction and the polymer chains parallel. Further extension is entropic until rupture occurs. On heating of the original material above its T_g, melting and recrystallization of the tightly folded chain hard segments occurs with the formation of more extended chain, thicker crystallites.

While there is general agreement on the morphology of 4GT–PTMOT cooled rapidly to ambient temperature, the morphology of material annealed at elevated temperature is more controversial. The "long period" spacing, L, determined by SAXS is a measure of the average distance between crystallites and as such is the sum of the thickness of the crystallites and the amorphous regions. During the annealing process at elevated temperature or on crystallization at elevated temperature, the value of L increases with temperature, along with the volume of crystalline material in the sample. This is especially true on annealing close to the polymer melting point. Early workers believed that this increase in L was largely due to thickening of the crystalline lamellae. Bandara and Droescher [39, 62] and Fakirov [63, 64], working with 4GT–PEOT, have confirmed the increase in L and the increased crystallinity but find by other means that the lamellar thickness is relatively unaffected by the process. Fakirov also showed that the increase in L is proportional to the square root of the molecular weight of the long-chain glycol and thus its unperturbed end-to-end distance. This suggests that the increase is in the amorphous phase rather than in the crystalline phase. It has also been suggested that the long period may be due to some phenomena other than distances between crystal faces, for example, in the case of the PEOT polymers, the presence of phase separation in the amorphous regions.

Taking advantage of the slow rate of crystallization of 4GI hard segments, Cooper and co-workers [42] were able to monitor the phase separation/crystallization process of 4GI–PTMOI from the supercooled liquid to the completion of crystallization over a range of compositions and temperatures. The initial supercooled melt is homogeneous with a single T_g that depends on the composition. Even though the high molecular weight hard and soft segment polymers are incompatible, it is apparent that short sequences are compatible at temperatures well below the melting point of the hard segment polymer. Phase separation of the hard segments then proceeds, with the formation of a crystalline phase and with a liquid phase depleted of hard segment. The composition of the amorphous phase is homogeneous but with regions possessing nonequilibrium mobility, as, for example, interfacial and/or tie molecules. Crystallization was detected by SAXS as the weight fraction of crystalline material approaches 0.05. The modulus of the copolymer increases from very low to high values in line with the weight fraction of crystalline material present (see Fig. 8.6). The presence of a second ester segment as in 4GT–4GI–PTMOT reduces the availability of crystallizable hard segment with a corresponding reduction in modulus. Melting point depression of the semicrystalline samples may be predicted by using Flory's random copolymer crystallization equation but the calculated heats of fusion are lower than expected [40, 42, 65, 66].

Due to the incompatibility of even short sequences of PEOT and 4GT, Fakirov *et al.* [67] found four different domains in 4GT–PEOT copolymers. These are a crystalline 4GT phase, an amorphous 4GT phase, an amorphous PEOT phase, and, depending on the composition and sample thermal history, a mixed amorphous phase. The amorphous phase of 2GT–PTMOT copolymers was also found to have a second glass transition temperature and hence also to have a second heterogeneous amorphous phase.

The lamellar-network type morphology proposed by Cella for 4GT–PTMOT has also been proposed for polymers 2GT–PTMOT, 4GN–PTMON (N = 2,6-naphthalene dicar-boxylic acid) and 4GT–PPOT (PPO = poly(propylene oxide) glycol) [35].

8.4.4   Orientation and Stress–Strain Behavior

Figure 8.11 is a typical stress–strain curve for a medium hardness 4GT–PTMOT copolyether ester at low strain rate. While the quantitative aspects of the curve depend on the hard segment content of the polymer, its shape reflects the morphology of these polymers. In the range of 30 to 90 wt% 4GT units, deformation at low elongation (about 10%) is largely reversible and results in part from reorientation of the continuous crystalline superstructure of the polymers and in part by transformation of the α unit cell structure to the β form. The tetramethylene chain is kinked and compact in the α form but fully extended in the β form, resulting in an 11.6% increase in the length of the unit cell. This transformation is thought to be fully reversible with release of stress [45]. The high initial or Young's modulus shown in region I of Fig. 8.11 is due to this pseudoelastic deformation. At greater elongation, in region II, a drawing process occurs in which the original crystalline superstructure is reorganized. The original disoriented lamellae become oriented perpendicular to the direction of stress with the

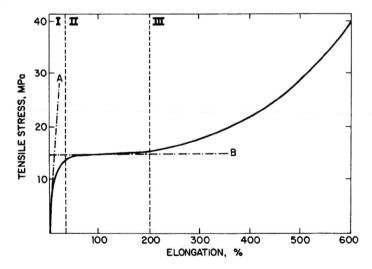

Figure 8.11   Stress–strain curve for polyester 58 wt% 4GT–PTMO(1000)T. (A) Slope = Young's modulus; (B) yield stress. (Reprinted with permission from Ref. [34], *J. Polym. Sci.* Copyright 1973, John Wiley & Sons)

polymer chains in the direction of stress. The process continues up to about 300% extension and results in irreversible disruption of the crystalline matrix. Samples strained to these intermediate levels exhibit considerable permanent set. At strains beyond about 300% only very limited increases in orientation occur and the stress is transmitted primarily by the rubbery amorphous phase until the sample breaks.

8.5   Properties of Commercial Copolyether Ester Elastomers

8.5.1   Mechanical Properties

Rather complete compilations of mechanical properties, chemical properties and processing information for commercial polyester elastomers have been published [68, 69].

Commercial thermoplastic polyester elastomers are available in Shore D hardnesses ranging from 30 to above 75. In terms of hardness, stiffness, and resilience, these elastomers bridge the gap between cured elastomers and rigid plastics. Their stress–strain properties in tension have been discussed previously. Their behavior toward compression and shear is similar, in that distortion from low strains is largely reversible, while higher strains result in permanent distortion. All grades have brittle temperatures below $-70\,^\circ$C. Surprisingly, a polymer containing 95 wt% 4GT units exhibits a brittle temperature below $-40\,^\circ$C, even though 4GT homopolymer has a brittle temperature of $0\,^\circ$C [13]. All grades exhibit good resistance to impact in the notched Izod test, with the softer grades not breaking even at $-40\,^\circ$C. The softer grades have outstanding resistance to flex cut growth.

The commercial grades of 4GT–PTMOT differ largely in the amount of polyether contained in the copolymers and the presence or absence of a second ester segment. The effect of 4GT concentration is shown in Table 8.5.

8.5.2   Melt Rheology

The apparent melt viscosity of polyester elastomers decreases only to a limited extent with increasing shear rate. Over the range of shear rate common for calendering, extrusion, and injection molding (10 to 7000 reciprocal seconds) the melt viscosity drops from 1000 to about 300 Pa-s. The viscosity of a typical polyurethane of similar hardness will drop from 7000 to 300 Pa-s over the same range. Because of their low melt viscosity at low shear rate, the polyester elastomers are readily used in low-shear procedures such as calendering, impregnation of porous substrates, melt casting, and rotational molding [71]. Viscosity can be controlled by adjusting the process temperature. Special viscosity grades are available from resin manufacturers.

Blow molding of standard grades of polyester elastomers suitable for extrusion or injection molding is difficult to carry out because of their low melt viscosity, low melt tension, and low die swell. Methods suggested to improve these properties to levels sufficient to facilitate blow molding have been discussed in Section 8.3. Of special interest, however, is the

Table 8.5 Physical Properties of 4GT–PTMO(1000[a])T Copolymers

PTMO in the copolymer, wt%[a]	0	10	25	35	50	60
Hardness Shore D	80	74	63	55	45	38
Density (mg m^{-3})	1.3	1.27	1.23	1.2	1.16	1.12
Glass transition temperature[b] (°C)	55	44	2	−23	−50	−60
Melting temperature (°C)	223	221	210	202	183	192
Heat of fusion (kJ kg^{-1})	54	60	41	38	29	25
T_B (%, ASTM 412)	200	360	540	570	700	650
Tensile strength (MPa, Din 53504)	52	45	38	32	21	17
Young's modulus (MPa)	2600	830	300	170	90	53
Loss tangent (G''/G')		0.011	0.076	0.062	0.035	0.03	0.03
Storage modulus G' (log MPa)		2.74	2.7	2.17	1.94	1.6	1.25

All data are unpublished results from the Akzo Fibers and Polymers Research Laboratories, Arnheim, The Netherlands. Reprinted with permission from *Polymer Handbook*. Copyright 1989, John Wiley & Sons
[a] $M_n = 1000$ except for the 60 wt% copolymer in which $M_n = 2000$
[b] Taken at peak tan δ

use of small amounts of multifunctional epoxides or polymeric epoxides. Suitable polymeric additives include copolymers of ethylene or ethyl acrylate with glycidyl methacrylate [72, 73]. These materials react with carboxyl chain ends on the copolyether ester to form higher viscosity, tougher materials. The reaction can be accelerated, if needed, with catalysts such as the sodium salts of phenols [29] or tertiary amines [28]. Viscosity may also be increased in blends with ionomers such as ethylene–carboxylic acid copolymers [31a].

8.5.3   Degradation and Stabilization of Polyether Ester Elastomers

In the absence of antioxidants, copolyether ester elastomers are rapidly degraded in air at elevated temperatures, with a reduction in polymer inherent viscosity. The reaction is free radical in nature, and probably involves attack at the carbon atoms on the polymer chain α to ether oxygen. Chain scission with the formation of formaldehyde [74] follows. The formaldehyde is oxidized to formic acid, and this in turn causes additional chain cleavage.

With proper stabilization [75, 76], the copolyether esters are relatively resistant to air oxidation. A complete stabilizer package will include a free radical trap, an agent to decompose hydroperoxides, and an agent to trap formaldehyde. The free radical trap will usually be a hindered phenolic antioxidant such as either Irganox 1010[®] or a secondary aromatic amine such as Agerite White[®]. The aromatic amines are superior radical scavengers but badly discolor the stock. Phosphite esters [77] or aliphatic thioethers are used as peroxide scavengers. Formaldehyde can be scavenged with amides [78], urethanes [79], or ureas [80] containing an ⟩N–H linkage. The best stabilizer systems will increase the time to embrittlement at 140 °C from just a few hours to about a month [75]. Light-colored heat-resistant stocks have also been proposed [81].

The polyester elastomers are also subject to oxidative degradation when exposed to ultraviolet light. The reaction is believed to involve formation and subsequent degradation of hydroperoxides and a mechanism has been proposed [82]. Yellowing was associated with photooxidation of polyester sequences [83].

In applications where a black stock is acceptable, the addition of carbon black at 0.5 to 3.0 wt% is very effective as a UV screen. For white or light-colored stocks, typical UV absorbers such as substituted benzotriazoles, either alone or in combination with phosphate esters, are useful [84], as are combinations of triazine antioxidants with benzophenone UV-light absorbers [85]. Derivatives of 2,2,6,6-tetramethyl-4-piperidinyl alcohol, or the corresponding N-methyl derivatives [76], have been shown to provide excellent protection against degradation by light in white or light-colored polyester elastomers; however, the presence of most of these derivatives is highly detrimental to heat aging even in the presence of increased amounts of antioxidants. Pentamethyl-4-piperidinyl esters of malonic acid with one or two 3,5-di-tert-butyl-4-hydroxybenzyl substituents do not reduce thermal stability significantly [86]. On the other hand, if photodegradation is desired for environmental reasons, the process can be accelerated by the addition of fluorescent whitening agents to the compound [87].

The 4GT–PTMOT copolyether esters are much more resistant to hydrolysis than aliphatic polyesters but in certain applications where exposure to water at elevated temperature is involved, further stabilization is required. Because of their polar nature, polyester elastomers are permeable to water and are subject to degradation by hydrolysis at elevated temperatures, even though the rate is much less than that for a polymer containing aliphatic acid ester groups such as a polyesterurethane. The addition of 1 or 2 wt% of a hindered aromatic polycarbodiimide increases the useful service life of polyester elastomers several fold in the presence of hot water or steam [30].

The melt stability of the polyester elastomers is quite adequate for normal processing. However, in situations where higher than normal processing temperatures and times may be needed, for example, filling a large mold by melt casting, the melt stability can be improved by incorporating a minor amount of a polyepoxide [29]. The polyepoxide reacts at least partially with carboxyl end-groups already present in the polyester as well as additional carboxyl groups formed by degradation of the melt.

8.6   Structural Variations

The previous discussion has focused largely on the system of polymers containing 4GT hard segments and PTMO(1000)T soft segments in varying amounts. The use of other hard and soft segments is considered in these sections.

8.6.1   Hard Segment Modifications

Data for some alternative aromatic acids are included in Table 8.6 [88].

Where rupture is involved, the polymer based on isophthalic acid has better physical properties than the terephthalate-based control, but the modulus is much lower. The isophthalate-based elastomer requires several hours to crystallize and is quite low melting. The melting point and crystallinity of 4GT and 4GI homopolymers vary with crystallization

Table 8.6 Physical Properties of 50 wt% 4G Dicarboxylate/50 wt% PTMO(1000) Dicarboxylate Copolyether Esters

Polymer	4GT	4GI	4GN	4GTP
Average hard segment block length	5.14	56.14	4.20	3.04
η_{inh} (l/g)	0.18	0.16	0.15	0.17
Stress at 100% (MPa)	11.7	7.2	10	2.8
Tensile strength (MPa)	48.4	58.6	51	60.3
Elongation at break (%)	755	720	660	390
Permanent set (%)	370	126	280	3
Split tear strength (kN/m)	48	54	117	47
Shore D hardness	48	39	49	31
Compression set B (22 h/70 °C), %	52	91	49	100+
Clash-Berg $T_{10,000}$ (°C)[a]	−33	−31	−14	2
Melting point (°C)	189	85/112	202	

N, 2,6-naphthalene dicarboxylate; TP, *m*-terphenyl-4,4″-dicarboxylate
[a] An index of low temperature stiffening. The temperature at which the bending modulus exceeds 10,000 psi, or 700 kg/cm^2
Source: Ref. [88]

conditions and have been a matter of controversy in the literature. Recent values indicate the extent of crystallinity of the two materials is similar but the melting point of 4GT is much higher (246 °C vs. 165 °C) and related to the lower heat of fusion of 4GI (140.5 J/g [89, 39] vs. 1.21 J/g [90] for 4GI) rather than to the entropies of fusion [90], which are similar.

The elastomer based on 2,6-naphthalene dicarboxylic acid has excellent properties in general but poorer low temperature stiffening characteristics. The tear strength of the material is more than double that of the terephthalate polymer. A full series of elastomers was also evaluated using 2G through 10G diols in combination with 2,6-naphthalene dicarboxylic acid and PTMO(1000). These polymers also exhibited surprisingly good properties [88]. One is reminded of the exceptional properties of polyurethanes based on symmetrical naphthalene diisocyanates. Copolyetheresters having a 4GN hard segment have been suggested for use as elastic fibers [91].

The elastomer prepared from *m*-terphenyl dicarboxylic acid is slow crystallizing and the values in Table 8.6 represent its properties shortly after molding. The polymer was transparent and did not exhibit a melting point when the data in the table were obtained. After the polymer was aged for 28 days it became opaque and showed two DSC endotherms at 63 °C and 158 °C. The 100% modulus and hardness of the polymer increased on crystallization, but tensile strength, elongation at break, and permanent set remained substantially unchanged. The very low permanent set of 3% for both samples is most surprising as is the apparent very rubbery shape of the stress–strain curve. The existence of a glassy hard segment phase was suggested to explain the unusual properties of the uncrystallized polymer [88]. Since permanent set is not increased after crystallization, it must be concluded that the crystallites in this polymer are not disrupted at high elongations and that permanent stress crystallization does not occur.

Copolyether esters having hard segments from *cis*- or *cis-trans*-1,4-cyclohexanedimethanol and *trans*-1,4-cyclohexyldicarboxylic acid were found to be suitable for blow molding owing to their high melt viscosity [92]. Polymers made using *cis-trans* diacid were too low melting to be useful. Elastic fibers have been made from the same dibasic acid but in combination with 1,4-cyclohexylene glycol. Both monomer units should have at least 40% *trans* isomer content [93].

Data for a series of polyester elastomers with variable short-chain glycol terephthalate hard segments are listed in Table 8.7 [88]. It has been shown that the modulus at 300% extension of copolyether ester elastomers increases uniformly with the melting point of the hard segment homopolymer [12]. For that reason it is not surprising that modulus values tend to show an odd–even behavior, with short-chain glycols having an even number of carbon atoms yielding a higher modulus than their neighbors. The anomalous polymer is 3GT. Its rupture properties are unexpectedly low but, on the plus side, its tension set at break is less than expected. Perhaps there is some characteristic of its crystal structure that prevents realignment under stress-induced crystallization. It has been shown that oriented fibers of 3GT homopolymer respond differently to stress than either 2GT or 4GT [94]. The rupture properties of a copolyether having trimethylene dibenzoate hard segments were quite good [95]. When combined with a soft segment containing copolymerized 3-methyltetrahydro-furan, the set at break was only 16% and it had higher unload power from 60% extension than either 2GBB, 4GBB, or 6GBB hard segments. The average mole ratio of hard segment to soft segment was close to 1.

Based solely on the data in Table 8.7, it would be difficult to choose between 2GT- and 4GT-based polyester elastomers at this or higher hardness [96]. Indeed, softer copolymers having a lower concentration of 2GT hard segment might look much more attractive than their 4GT counterparts—if fully crystallized. However, the rate of crystallization of the 2GT copolyether esters is much too slow. A polymer containing about 58 wt% 4GT whose equilibrium hardness is 55 Shore D will reach 53 hardness in about 6 min or less. A similar 2GT polymer will have an induction period at very low hardness of about 2 h and then take an additional 16 h before nearing equilibrium hardness. While the hardening rate of the 2GT-

Table 8.7 Properties of 50 wt% Alkylene Terephthalate–50 wt% PTMO(1000)T Copolymers as a Function of Short Chain Glycol Structure

Polymer	2GT	3GT	4GT	5GT	6GT	10GT
η_{lnh} (l/g)	0.13	0.17	0.18	0.16	0.15	0.13
Yield strength (MPa)				4.8		
Stress at 100% (MPa)	11.4	11.7	11.7	4.5	5.2	6.1
Tensile strength (MPa)	45.5	22.8	48.4	15.4	13.5	15.5
Elongation at break (%)	675	660	755	880	750	640
Permanent set (%)	275	195	370	210	370	370
Split tear strength (kN/m)	42	15	48	25	18	7.8
Shore D hardness	46	48	48	32	33	35
Compression set B (%)	52	48	52	90	86	81
Class-Berg $T_{10,000}$ (°C)	−38	−36	−33	−50	−53	−51
T_m (°C)	224	198	189	106	122	106
Hard segment homopolymer, T_m (°C)	280	225	236	140	160	130

based elastomer can be improved with nucleating agents and heated molds [97, 98, 99], it does not begin to approach the ease of processing that exists for 4GT-based elastomers.

Polymers based on 5GT, 6GT, and 10GT are relatively low melting and their physical properties are generally inferior to those observed for the polymers based on the lower glycols.

Elastic fibers containing 1,4-cyclohexanedimethyl terephthalate hard segments were found to have low tension set in combination with excellent stress–strain and stress decay properties. The hard segment homopolymer had a melting point of 318 °C for the *trans* isomer of the glycol and 256 °C for the *cis* isomer. Elastic properties were developed at low hard segment concentrations with examples containing as little as 11 wt% hard segment. This was in part due to use of a crystallization-resistant PTMO copolymer soft segment that permitted use of high molecular weight polyols. The processing temperature was high (260 °C), perhaps too high for general-purpose molding [6].

In the course of developing relatively soft polyester elastomers that have outstanding tear strength at high tearing rates, a series of polymers were made using mixtures of terephthalic acid with a minor amount of a second aliphatic or aromatic dicarboxylic acid [8, 12]. The compositions and properties of these polymers are presented in Table 8.8 along with two control polymers containing only 4GT units [12]. The table has been modified by including the calculated average 4GT block length assuming that the polymers are random. The presence of the second acid unit modifies the morphology of the copolymer to reduce both the polymer melt temperature and also the volume of crystalline 4GT present, while at the same time increasing the amount of uncrystallized hard segment in the amorphous phase. Interpolating between datum points, it has been found that the 4GT crystalline fraction in polymers containing 60 wt% aromatic ester is roughly halved if there are two aromatic ester units in the ratio 4:1 4GT–4GI [40]. It is likely that the added 4GT in the amorphous phase crystallizes under stress, thus enhancing tear strength.

Table 8.8 Physical Properties of 4GT–PTMO(1000)T Polymers Containing a Second Short-Chain Ester Group

4GT weight fraction	0.40	0.40	0.40	0.40	0.50	0.40
Second short chain ester, weight fraction	0.10	0.10	0.10	0.10		
Second short chain ester, structure	4G10	4GP	4GI	4GTP		
Average 4GT block length	2.18	2.03	2.03	2.56	5.14	3.42
η_{inh} (l/g)	0.16	0.17	0.18	0.19	0.16	0.16
Stress at 100% (MPa)	7.2	7.2	7.6	8.3	12.4	8.3
Stress at 300% (MPa)	8.6	9.1	9.8	11.9	15.2	11
Tensile strength (MPa)	30.9	46	51.3	50.7	51	17.2
Elongation at break (%)	860	760	780	520	780	730
Split tear strength (kN/m)	26.3	33.3	38.5	114	50.8	26.3
Clash-Berg $T_{10,000}$ (°C)	−53	−46	−44	−10	−26	−52
Shore D hardness	41	43	40	49	49	44
Compression set B (22 h/70°C, %)	54	45	70	80	54	53

4GP, tetramethylene phthalate

Returning to Table 8.5, it appears that Young's modulus and storage modulus are largely a function of the degree of crystallinity. The same is approximately true for the 100% modulus in Table 8.6. The behavior of the polymers with regard to properties involving rupture follows a different pattern which appears to depend on the amount of crystallizable material in the amorphous phase. It is likely that tensile and tear strengths are enhanced by stress crystallization of the hard segments in the amorphous phase.

Other advantages for the use of mixed-acid, short-chain glycol segments are improved solvent resistance and rupture properties that are not as highly dependent on polymer molecular weight. On the negative side, the mixed acid polymers are lower melting and slower crystallizing.

With the exception of solvent resistance, the use of a second diacid or diol provides only limited improvements in properties for polymers containing a total of 50 wt% or more 4GT units. Increased concentrations of 4GT units lead to increased concentrations of these units in both the crystalline phase and the amorphous phase.

Mixed short-chain glycols are less often used because of greater difficulty in controlling the evaporation of a short-chain glycol mixture during finishing. An exception is a group of polymers containing a combination of 4GT and CHDM-T hard segments. If the mole ratio of 4GT–*trans*-CHMD-T is less than 1, the melt temperature will be greater than that of 4GT alone. A reasonable compromise is a ratio of 0.6, where the very high melting point of CHDM-T is reduced and that of 4GT is increased. The products were reported to have improved thermal and physical properties compared to polymers based on 4GT alone but little data were given [100].

8.6.2   Polyether Soft Segment Modifications

The effects of changing the structure of the polyether segment in polyester elastomers are shown in Table 8.9 [101]. The table has been modified from the original by including the

Table 8.9 57 wt% 4GT–Polyether-T Copolymers: Properties as a Function of Polyether Structure

Polyether glycol	PTMO	PEO	PPO	PPO	EOPPO[a]
Glycol molecular weight	975	985	1005	1000	1150
Copolymer properties					
η_{inh} (l/g)	0.17	0.17	0.09	0.17	0.11
Stress at 100% (MPa)	13.4	13.2	11.7	14.5	11.7
Tensile strength (MPa)	47.2	42.6	27.2	42.1	32.4
Elongation at break (%)	660	705	640	600	675
Set at break (%)	365	190	355		350
Tear strength (kN/m)	62	31	18	56	23

After Ref. [10] *Rubber Chem. Tech.* Reprinted with permission. Copyright 1977, American Chemical Society

[a] PPO capped with 15 wt% ethylene oxide to improve its reactivity

properties of a second PPO(1000) glycol-based elastomer finished to higher molecular weight by solid-phase polymerization [13]. Secondary hydroxyl end groups on PPO-based glycols prevent reaching high molecular weight by the ordinary processes. This problem can be minimized by capping the PPO end groups with ethylene oxide (EOPPO). This provides a more reactive and more stable glycol [102]; even so this glycol is normally used in combination with a branching monomer to obtain acceptable copolymer molecular weight.

A second method for improving the reactivity and stability of PPO-based glycols involves amination of PPO-glycol to replace the terminal hydroxyl groups with primary amine groups. The resulting diamine is reacted with a trifunctional aromatic acid anhydride, such as trimellitic anhydride, to provide a polyether soft segment terminated with carboxylic acid end groups. The diacid can then be used in much the same way as PEO or PTMO to prepare copolyether esters [103].

Table 8.9 shows that elastomers based on PTMO(1000), PEO(1000), and PPO(1000) have similar properties when compared at similar inherent viscosities. However, the PTMO-based copolymer has the highest tensile and tear strengths. The EOPPO(1000) copolymer appears to be equivalent to the PPO(1000) copolymer with about the same inherent viscosity. The hydrolytic stability of elastomers based on PTMO and PPO is about equivalent and these polymers are much more resistant to hydrolysis than the PEO-based copolymer. The PEO copolymer has a water swell about 10 times that of the PTMO copolymer [101]. The water swell of 15% 2GT–PEOT(1540) copolymer approaches 300% but can be reduced in a controlled manner by copolymerization with 4GT–PTOMT [104]. The more polar PEO copolymers have lower oil swell than similar PTMO copolymers [101].

In harder copolymers containing 76 wt% 4GT, the PTMO(1000)-based elastomer is similar to elastomers based on PEO(1000), PPO(1000), and EOPPO(1000) except for a much higher tear strength. When the four glycols are compared in a soft elastomer based on a 40 : 10 mixture of terephthalic and isophthalic acid in a copolyether ester containing 50 wt% 4GT–4GI, the PTMO(1000)-based copolymer has much better tensile and tear strengths while the stress at 100% elongation is substantially the same for all the copolymers [101].

The properties of polyester elastomers are also affected by the molecular weight and molecular weight distribution of the polyether soft segments. The effect of increasing the molecular weight of the soft segment at constant polyester concentration is shown in Table

Table 8.10 57 wt% 4GT–PTMOT: Properties as a Function of PTMO Molecular Weight

PTMO molecular weight	600	813	975	1993	2899
Mole ratio 4GT–PTMOT	4.41	5.69	6.67	12.8	18.26
Polymer properties					
η_{inh} (l/g)	0.16	0.18	0.17	0.17	0.18
Stress at 100% (MPa)	14.9	13.1	13.4	14	15.8
Tensile strength at break (MPa)	42.1	51	47.2	53.1	29.8
Elongation at break (%)	620	670	660	645	440
Tear strength (kN/m)	91	60	62	52	40
Clash-Berg T_{1000} (°C)		−9	−7	19	

After Ref. [101], *Rubber Chem. Tech.* Reprinted with permission. Copyright 1977, American Chemical Society

8.10. Over the molecular weight range of 600 to 2899, the mole ratio, and hence the length of hard segments, increases from about 4 to 18. At an average length of 4 units, much of the polyester would be uncrystallizable and in the amorphous phase, while oligomers having 18 units would be about as high melting as homopolymer. The other factor coming into play is the crystallizability of PTMO glycol, which at room temperature begins to crystallize at 2000 molecular weight. These ester hard segments in the amorphous phase will stiffen the copolymer at 600 MW with the effect dropping as the amount of crystallizable hard segment increases. At higher PTMO glycol molecular weights (starting around 2000), partial crystallization of the polyether stiffens the polymer. At 2899 molecular weight, it is possible that partial phase separation occurred during polymerization, resulting in the reduction in tear strength, tensile strength, and elongation at break actually observed. The variation of properties of copolyether esters based on PEO glycol is similar to that observed with polymers based on PTMO but with soft segment crystallization occurring at somewhat higher molecular weights [101]. Crystallization of PEO soft segments was also investigated by calorimetric, dynamic mechanical, and spin-probe measurements. Depending on the type of hard segment, the soft segment melting point was about 40 °C for 6000 molecular weight and −12 °C for 1000 MW PEO [105].

Even in those cases where soft segment crystallization does not occur under static conditions, it may occur when the polymer is stressed. Polyether ester fibers based on PTMO tend to crystallize under stress, thereby preventing full recovery when the stress is released. The problem is heightened by the need to use high molecular weight glycols in these applications but can be alleviated by using a copolyether glycol. Thus elastic fibers having 15 wt% copolymerized 8-oxabicyclo(4 : 3 : 0) nonane in the PTMO reduced the tension set of 10 wt% CHMDT–PTMO(4400)T from about 90% to 30% when tested at 20 °C. The hard segment, in this case, is especially effective. The average sequence length is only 1.82 [6]. Fibers based on a PTMO soft segment with 3 to 20 wt% copolymerized 3-methyltetra-hydrofuran were found to have much improved tension set and unload power from 100% elongation compared with fibers based on homopolymer PTMO. Hard segments of 2GT, 2GN, 4GT, and 4GN were compared. PTMO molecular weights were as high as 5000 [106]. Generally similar results were reported using 3-methyloxetane as the comonomer [107].

PEO may also be made crystallization resistant by copolymerization with propylene oxide or by using bisphenol A as an initiator [108].

8.7   Polymer Blends with Polyether Ester Elastomers

The good melt stability and low melt viscosity of polyester elastomers facilitate their use in polymer blends, particularly when there is a need to improve resistance to impact at low temperature, to provide elastomeric character, or to compatiblize the mixture. A bewildering variety of blends of various sorts are the subject of much of the current copolyether ester patent literature. We will only make a superficial survey here.

Blends of the homopolymer 4GT with substantial amounts of soft or medium hardness polyester elastomers (4GT–PTMO(1000)T) are stiffer and have a higher yield strength at

processing overcomes a disadvantageous total materials cost. Compared with engineering plastics, the copolyether esters offer an unusual combination of strength, with flexibility over a useful range of strain and dynamic properties. In addition, they have a wide range of temperature serviceability, combining impact resistance at low temperature with resistance to creep at elevated temperatures. They also resist flex fatigue, with minimal heat buildup in flexing applications. They especially replace ridged plastics in applications where improved shock absorption, impact strength, flexing, sealing, spring characteristics, or silent mechanical operation is required.

The hose and tubing market was one of the first to capitalize on the heat and chemical resistance of the copolyether esters. The high strength of the resins permits lighter weight, thin section designs. Power drive belts of copolyether esters have sufficient strength to compete with fiber-reinforced rubber. These belts are preferred in applications subject to high shock loading at machine startup. With thermoplastics processing, new belt cross-sections are possible, for example, injection molded toothed belts for office equipment.

One heavy-duty application involves railroad draft gear requiring large shock-absorbing capacity. The copolyether ester offers high impact strength, low hysteresis, and excellent compressive properties. Impact strength is 10 times that of cast urethane in this application.

Because of their high water vapor transmission rate, higher hardness PEO-based polyether esters are used as waterproof, breathable films in the manufacture of outer garments, tents, and foot wear [132].

New technology applications involve jacketing for fiber optics cable where mechanical strength combined with environmental resistance are key advantages. Industry is adopting thin wire insulation, switches, and connectors based on copolyether ester elastomers [68]. Potential exists for medical applications where the exceptional blood–tissue compatibility of the copolyether esters would be important [133, 134]. Elastic fiber applications, the original goal of this work, have not come to commercial fruition, although they continue to be the subject of numerous patents and papers [135] and include specific uses such as heat-bondable sheath core fibers [136], antistatic monofilament sheath core hair brush bristles [137], and finally silklike elastic fabrics prepared from blends of copolyether ester fibers and polyester fibers [138].

References

1. D. Coleman, *J. Polym. Sci. 14*, 15 (1954). D. Coleman (to ICI), British Patent 682,866 (November 19, 1952)
2. O.B. Edgar and R. Hill, *J. Polym. Sci. 8*, 1 (1952)
3. P.T. Flory, *J. Chem. Phys. 17*, 233 (1949)
4. W.H. Charch and J.C. Shivers, *Text Res. J. 29*, 536 (1959)
5. J.C. Shivers (to DuPont), U.S. Patent 3023192 (February 27, 1962)
6. A. Bell, C.J. Kibler, and J.G. Smith (to Eastman Kodak), U.S. Patent 3243413 (March 29, 1966)
7. A.A. Nishimura and H. Komogata, *J. Macromol. Sci., Chem.* A1, 617 (1967)
8. W.K. Witsiepe (to DuPont), U.S. Patent 3651014 (March 21, 1972)
9. W.K. Witsiepe (to DuPont), U.S. Patent 3763109 (October 2, 1973)
10. W.K. Witsiepe (to DuPont), U.S. Patent 3755146 (October 16, 1973)
11. M. Brown and W.K. Witsiepe, *Rubber Age 104*, 35 (1972)

12. W.K. Witsiepe, *ACS Advances in Chemistry Series 129*, 39 (1973)
13. G.K. Hoeschele, *Chimia 28*, 544 (1974); G.K. Hoeschele and W.K. Witsiepe, *Angew. Makromol. Chem. 29/30*, 267 (1973)
14. J. Hsu and K.Y. Choi, *J. Appl. Polym. Sci. 33(2)*, 329–51 (1987)
15. L. Yurramendi, M.J. Barandiaran, and J.M. Asua, *Polymer 29*, 871 (1988); *J. Macromol. Sci., Chem. A24*, 1357 (1987)
16. T. Ito and H. Susuu (to Nippon Synthetic Chem. Ind.), Japanese Patent 91-255847 (September 6, 1991)
17. G.K. Hoeschele, R.H. McGirk, and R. Heath (to Du Pont), U.S. Patent 5120822 (June 9, 1992)
18. R.J. McCready (to General Electric), U.S. Patent 4544734 (October 1, 1990)
19. Y. Takanawo, I. Okino, and Y. Nakatani (to Kanagafuchi Kagaku Kogyo Kabushiki Kaisha), U.S. Patent 5331066 (July 19, 1994)
20. C. Tanaka and M. Hiratsuka, *Preprint (G3K05) of Polym. Symp., Soc. Polym. Sci. Japan* (October 1, 1982)
21. C.K. Shih and J.M. McKenna, IUPAC Meeting Amherst, MA, 1982
22. W. Meiyan, Z. Dong, C. Chuanfu, and Q. Chunqin, *Proc. of China—U.S. Bilateral Symp. on Polym. Chem. and Phys.*, October 5–10, 1979, Beijing Science Press, China 1981, distr. by Van Nostrand Reinhold Co
23. C. Tanaka, Y. Futura, and N. Naito (to Toray Industries), U.S. Patent 4251652 (February 17, 1981)
24. G.K. Hoeschele (to DuPont), U.S. Patent 3801547 (April 2, 1974)
25. S. Tamura, T. Matsuki, J. Kuwata, and H. Ishii (to Du Pont-Toray), Japanese Patent 02269118 (November 2, 1990)
26. T. Matsuki, J. Kuwata, and H.Takayama (to DuPont Toray), Japanese Patent 01095127 (April 13, 1989)
27. Z. Roslaniec, J. Slonecki, and H. Wojcikiewicz (to Politechnika Szczecinska), Polish Patent 158339 (August 31, 1992)
28. G.K. Hoeschele (to DuPont), U.S. Patent 3723568 (March 27, 1973)
29. K.T. Kim, D.W. Woo, M.S. Baek, and Y.C. Lee (to Cheil Synthetics), European Patent 577508 (June 30, 1993)
30. M. Brown, G.K. Hoeschele, and W.K. Witsiepe (to DuPont), U.S. Patent 3835098 (September 10, 1974)
31. T. Shime and H. Yamada (to Teijin Limited), U.S. Patent 3433770 (March 18, 1969)
31a. Chi-Kai Shih (to DuPont), U.S. Patent 4010222 (March 1, 1977)
32. A. Higashiyama, Y. Yamamoto, R. Chujo, and M. Wu., *Polym. J. (Tokyo) 24*, 1345 (1992)
33. K.H. Frensdorff, *Macromolecules 4*, 369 (1971)
34. R.J. Cella, *J. Polym. Sci. Symp. 42 (2)*, 727 (1973)
35. W.H. Buck, R.J. Cella, Jr., E.K. Gladding, and J.R. Wolfe, *J. Polym. Science Symp. 48*, 47 (1974)
36. G. Perego, M. Cesari, and R. Vitali, *J. Appl. Polym. Sci. 29*, 1157 (1984)
37. M.A. Vallance and S.L. Cooper, *Macromolecules 17*, 1208 (1984)
38. G. Wegner, in *Thermoplastic Elastomers—A Comprehensive Review*. N.R. Legge, G. Holden, and H.E. Schroeder (Eds.) (1986) Hanser, Munich
39. U. Bandara and M. Droescher, *Colloid and Polym. J. 261*, 26 (1983)
40. J.L. Castles, M.A. Vallance, J.M. McKenna, and S.L. Cooper, *J. Polym. Sci., Polym. Phys. Ed., 23(10)*, 2119 (1985)
41. Xu Zhonge, Yuang Ping, Zhong Jingguo, Jiang Erfang, Wu Meiyan, and L.J. Fetters, *J. Appl. Polym. Sci. 37*, 3195 (1989)
42. J. Castles Stevenson and S.L. Cooper, *Macromolecules 21*, 1309 (1988)
43. Li-Lan Zhu and G. Wegner, *Makromol. Chem. 182*, 3625 (1981)
44. R.M. Briber and E.L. Thomas, *Polymer 26, 8* (1985); *ibid, 27*, 66 (1986)
45. J. Desborough and I.H. Hall, *Polymer 18*, 825 (1977)
46. Li-Lan Zhu, G. Wegner, and U. Bandara, *Makromol. Chemie 182*, 3639 (1981)
47. G. Wegner, T. Fujii, W. Meyer, and G. Lieser, *Angew. Makromol. Chem. 74*, 295 (1978)
48. H.W. Hasslin, M. Droescher, and G. Wegner, *Makromol. Chem. 179*, 1373 (1978)
49. R. Bill, M. Droescher, and G. Wegner, *Makromol. Chem. 179*, 2993 (1978)
50. H.W. Hasslin, M. Droescher, and G. Wegner, *Makromol. Chem. 181*, 301 (1980)
51. H.W. Hasslin and M. Droescher, *Makromol. Chem. 181*, 2357 (1980)
52. R. Bill, M. Droescher, and G. Wegner, *Makromol. Chem. 182*, 1033 (1981)
53. F.G. Schmidt and M. Droescher, *Makromol. Chem. 184*, 2669 (1983)

54. M. Droescher, U. Bandara, and F. G. Schmidt, *Makromol. Chem., Suppl.* 6, 107 (1984)
55. P.J. Flory and A. Vrij, *J. Amer. Chem. Soc.* 85, 3548 (1963)
56. M. Gordon and G.S. Taylor, *J. Appl. Chem.* 2, 493 (1952)
57. A. Lilaonitkul, J. West, and S.L. Cooper, *J. Macromol. Sci. Phys.* B12 (4), 563 (1976)
58. R.W. Seymour, J.R. Overton, and L.S. Corley, *Macromolecules 8*, 331 (1975)
59. S.L. Cooper and J.A. Miller, *Rubber Chem. Technol.* 58, 899 (1985)
60. J.A. Miller, J.M. McKenna, G. Pruckmayr, J.E. Epperson, and S.L. Miller, *Macromolecules 18*, 1727 (1985)
61. S.L. Cooper, J.A. Miller, and J.G. Homan, *J. Appl. Cryst.* 21, 692 (1988)
62. U. Bandara and M. Droescher, *Angew. Makromol. Chem. 107*, 1 (1982)
63. S. Fakirov, A.A. Apostolov, and C. Fakirov, *Int. J. Polym. Mater. 18*, 51 (1992)
64. A.A. Apostolov and S. Fakirov, *J. Macromol. Sci., Phys. B31(3)*, 329–55 (1992)
65. R.A. Phillips, J.M. McKenna, and S.L. Cooper, *J. Polym. Sci. Pt B Polym. Phys. 32(5)*, 791-802 (1994)
66. R.A. Phillips and S.L. Cooper, *Polymer 35(19)*, 4146–55: (1994)
67. S. Fakirov, A.A. Apostolov, P. Boeseke, and H.G. Zachman, *J. Macromol. Sci. Phys. 29*, 379 (1990)
68. T.W. Sheridan, In *Handbook of Thermoplastic Elastomers*, 2nd edit. B.M. Walker and C.P. Rader (Eds.) (1988) Van Nostrand Reinhold, New York, pp. 181–223
69. W. Hoffman and R. Koch, *Kunststoffe 79*, 606 (1989)
70. K.B. Wagner, In *Polymer Handbook*, 3rd edit. (1989) J. Brandrup and E.H. Immergut (Eds.) John Wiley & Sons, New York, V-107
71. R. Jakeways, I.M. Ward, M.A. Wilding, I.H. Hall, and M.G. Pass, *J. Polym. Sci. Polym. Phys. Ed. 13*, 799 (1975)
72. Y. Nishiya, Y. Makabe, and Y. Yamamoto (to Toray Indiustries, Inc.), Japanese Patent 04275357 (September 30, 1992)
73. T. Shiaku, K. Hijikata, F. Kenji, and M. Suzuki (to Polyplastics Kk.), Japanese Patent 05112701 (May 7, 1993)
74. G.K. Hoeschele, *Angew. Makromol. Chem. 58/59*, 299 (1977)
75. J.J. Zeilstra, *Angew. Makromol. Chem. 137*, 83 (1985)
76. Jiang-qing Pan and Jie Zhang, *Polym. Degrad. Stab. 36(1)*, 65 (1992)
77. A. Chiolle and G.P. Maltoni (to ECP Enichem Polimeri), European Patent 510545 (October 10, 1992); A. Chiolle, G.P. Maltoni, and R. Stella (Ausimont SPA), U.S. Patent 5106892 (April 21, 1992)
78. G.K. Hoeschele (to DuPont), U.S. Patent 3896078 (July 22, 1975)
79. G.K. Hoeschele (to DuPont), U.S. Patent 3904706 (September 9, 1975)
80. G.K. Hoeschele (to DuPont), U.S. Patent 3856749 (December 24, 1974)
81. M.D. Golder and S.J. Gromelski, Jr. (to Hoescht Celanese), European Patent 0144175 (September 28, 1988)
82. M.H. Tabankia, J.L. Philippart, and J.L. Gardette, *Polym. Degrad. Stab. 12*, 349 (1985)
83. M.H. Tabankia and J.L. Gardette, *Polym. Degrad. Stab. 19*, 113 (1987)
84. Anon., *Res. Discl. 215*, 78 (1982)
85. M.D. Golder and B.M. Mulholland (to Hoechst Celanese), U.S. Patent 5032631, (July 16, 1991)
86. G.K. Hoeschele (to DuPont), U.S. Patent 4185003 (January 22, 1980)
87. I. Morita and H. Imanaka (to Toyo Boseki), Japanese Patent 06107922 (April 19, 1994)
88. J.R. Wolfe, Jr., *ACS Adv. Chem. Ser. 176*, 129 (1979)
89. K.H. Illers, *Colloid Polym. Sci. 258*, 117 (1980)
90. R.A. Phillips, J.M. McKenna, and S.L. Cooper, *J. Polym. Sci. Pt B Polym. Phys. 32(5)*, 791–802 (1994)
91. Anon. (to Teijin Ltd.), Japanese Patent 04240211 (August 27, 1992)
92. B. Davis, T.F. Gray, and H.R. Musser (to Eastman Kodak), U.S. Patent 4349469 (September 14, 1982)
93. G.D. Figuly (to DuPont), U.S. Patent 4985536 (January 15, 1991)
94. R. Jakeways, I. Ard, M.A. Wilding, I.H. Hall, and M.G. Pass, *J. Polym. Sci. Polym. Phys. Ed. 13, 799* (1975)
95. R.N. Greene (to DuPont), U.S. Patent 5128185 (July 7, 1992)
96. A.B. Ijzermans, F.J. Pluijm, F.J. Huntjens, and J.F. Repin, *Br. Polym. J.* 7, 211 (1975)
97. J.J. Zeilstra, *J. Appl. Polym. Sci. 31(7)*, 1977 (1986)
98. H. Herlinger, P. Hirt, and R. Kurz, *Chemiefasern/Textilind. 39(10)*, 1066 (1989)
99. Nan-I Liu (to General Electric), European Patent 318788 (June 7, 1989)

100. R.J. McCready and J.A. Tyrell (to General Electric), U.S. Patent 4714755 (December 22, 1987)
101. J.R. Wolfe, Jr., *Rubber Chem. & Technol. 50*, 688 (1977)
102. J.J. Zielstra and T. Brink (to AKZO NV), U.S. Patent 4687835 (August 18, 1987); G.K. Hoeschele (to DuPont) U.S. Patent 4205158 (May 27, 1980)
103. R.J. McCready (to General Electric), U.S. Patent 4552950 (November 12, 1985)
104. K. Takahashi, H. Nakasuji, and H. Okuda (to Polyplastics Co.), European Patent 197789 (October 15, 1986)
105. R.N. Greene (to DuPont), U.S. Patent 5116937 (May 26, 1992)
106. F. Lembicz and J. Slonecki, *Makromol. Chem.* 191, 1363 (1990)
107. G.R. Goodley, R.N. Greene, and C. King (to DuPont), U.S. Patent 4906729 (March 6, 1990)
108. Anon (to Daicel Chemical Industries), Japanese Patent 60026027 (February 8, 1985)
109. N.W. Hayman, J.R. Wright, R.I. Hancock, K. Jones, R.H. Still, S.R.K. Dauber, R. Peters, and T.H. Shaw (to Imperial Chem. Indust.), European Patent 315325 (May 10, 1989)
110. M. Brown and R.M. Prosser (to DuPont), U.S. Patent 3907962 (September 23, 1975)
111. K.P. Gallagher, X. Zhang, J.P. Runt, G. Huynh-ba, and J. S. Lin, *Macromolecules 26*, 588 (1993)
112. J. Runt, Lei Du, L.M. Martynowicz, D.M. Martynowicz, D.M. Brezny, and M. Mayo, *Macromolecules 22*, 3908 (1989)
113. K. Hirobe and A. Somemiya (to Kanegafuchi Kagaku Kogyo), U.S. Patent 4840984 (June 20, 1989)
114. W.F.H. Borman and Nan-i Lui (to General Electric), World Intellectual Property Organization Patent 9304124 (March 4, 1993)
115. K. Horiuchi, K. Yonetani, and S. Inoue (to Toray Ind.), Japanese Patent 61148261 (July 7, 1986)
116. R.J. Statz (to DuPont), European Patent 341731 (September 9, 1993)
117. S.H. Agarwal (to General Electric), U.S. Patent 4857604 (August 15, 1989)
118. H. Endo, K. Hashimoto, K. Igi, M. Ishii, T. Matsumoto, Y. Murata, O. Yoshifumi, T. Okamura, K. Tanaka, S. Taniguchi, and S. Shinichi (to Idemitsu Petrochemical) , U.S. Patent 4657973 (April 14, 1987)
119. M.R. McCormick, T.A. Morelli, W.J. Peascoe, S.F. Rasch, J.A. Tyrell, and M.T. Wong (to General Electric), U.S. Patent 4992506 (February 12, 1991)
120. Nan-I Liu (to General Electric), U.S. Patent 4814389 (March 21, 1989)
121. M. Hongo, H. Shigemitsu, N. Yamamoto, and A. Yanagase (to Mitsubishi Rayon), European Patent 308871 (March 29, 1989)
122. D.M. Blakely and R.W. Seymour (to Eastman Kodak), U.S. Patent 5118760 (June 2, 1992)
123. M. Kunitomi, H. Ishii, and Y. Yamamoto (to Toray Ind. Inc.), World Intellectual Property Organization Patent 8912660 (December 28, 1989)
124. W.P. Gergen (to Shell Oil Co., USA), U.S. Patent 4818798 (April 4, 1989)
125. K. Hasegawa, T. Teramoto, T. Nakajima, and T. Konomoto (to Japan Synthetic Rubber Co.), European Patent 506465 (September 30, 1992)
126. J.R. Wolfe, Jr. (to DuPont), European Patent 293821 (December 7, 1988)
127. S. Thomas, B.R. Gupta, and S.K. De, *J. Vinyl Technol. 9(2)*, 71 (1987)
128. R.W. Crawford and W.K. Witsiepe (to DuPont) U.S. Patent 3718715 (February 2, 1973)
129. D.J. Hourston and I.D. Hughes, *Rubber Conf. 77, 1*, 1 (1977)
130. S. Radhakrishnan and D.R. Saini, *J. Appl. Polym. Sci. 52*, 1577 (1994)
131. *Worldwide Rubber Statistics*, International Institute of Synthetic Rubber Producers, Houston (1994)
132. C.M.F. Vrovenraets and D.J. Sikkema (to Akzo), U.S. Patent 4493870 (January 15, 1985); G.J. Ostapchenko (to DuPont), U.S. Patent 4725481 (February 16, 1988)
133. S. Wang, *Polym J. (Tokyo) 21*, 179 (1989)
134. D. Bakker, C.A. Van Blitterswijk, S.C. Hesseling, W.T. Daems, and J.J. Grote, *J. Biomed. Mater. Res. 24*, 277 (1990)
135. G.C. Richeson and J.E. Spruell, *J. Appl. Polym. Sci. 41*, 845 (1990)
136. M. Yoshida, T. Yamaguchi, and T. Motohiro (to Teijin Ltd.) Japanese Patent 05156561 (June 22, 1993)
137. W.B. Bond (to DuPont) European Patent 160320 (November 6, 1985)
138. T. Hibino and Y. Okamoto (to Unitika) Japanese Patent 05311567 (November 22, 1993)

9 Thermoplastic Elastomers Based on Polyamides

R.G. Nelb and A.T. Chen

9.1 Introduction 228

9.2 Segmented Block Copolymers 228
    9.2.1 Structure. 228
    9.2.2 Morphology. 229

9.3 Polyamide Thermoplastic Elastomers 230
    9.3.1 Synthesis 230
        9.3.1.1 Polyesteramide, Polyetheresteramide, and Polycarbonate-esteramide
                 Synthesis. 231
        9.3.1.2 Polyether-*block*-amide Synthesis. 233
    9.3.2 Morphology. 236

9.4 Structure–Property Relationships 239
    9.4.1 Hard and Soft Segment Considerations 240

9.5 Physical Properties 241
    9.5.1 Tensile Properties 241
    9.5.2 High Temperature Tensile Properties 242
    9.5.3 Dry Heat Aging 243
    9.5.4 Humid Aging. 246
    9.5.5 Chemical and Solvent Resistance 246
    9.5.6 Tear Strength 247
    9.5.7 Abrasion Resistance 249
    9.5.8 Compression Set 249
    9.5.9 Flex Properties 250
    9.5.10 Adhesion. 251
    9.5.11 Weatherability. 251
    9.5.12 Electrical Properties 251

9.6 Processing Conditions 252

9.7 Applications 254

9.8 Summary 254

References 255

especially for the amorphous portion. At elevated temperatures and with a lower percentage of hard segments acting as physical crosslinks, the modulus and the tensile strength are reduced. For long-term aging, the thermal oxidative stability of the soft segment is also important.

To extend the upper temperature range of a segmented elastomer, it is necessary to introduce physical crosslinks that soften or melt at higher temperatures. Increasing the amount of hard segment can often accomplish this but it also leads to higher hardness and reduces the elastic properties. Alternatively, the extension of the use temperature can be accomplished by using a more crystalline hard segment with a higher melting point. This type of hard segment will be less soluble in the soft segment matrix and will help to maintain the integrity of the hard domains at high temperatures. For crystalline hard segments, the effectiveness of the physical crosslinks is more dependent upon the T_m than the T_g of the hard domains because the crystallinity suppresses the effect of the T_g. Consequently, the use of a high melting, crystalline hard segment was the approach taken in the design and development of the PEA, PEEA, and PCEA thermoplastic polyamide elastomers. The introduction of crystalline hard segments also benefits other properties such as solvent resistance.

9.3   Polyamide Thermoplastic Elastomers

9.3.1   Synthesis

The polyamide elastomers discussed in this chapter can be classified according to the composition of polyamide used for the hard segment and by the composition of the soft segment. The elastomers developed by Dow Chemical contain hard segments based on semiaromatic amides and soft segments based on aliphatic polyesters, aliphatic polyethers, or aliphatic polycarbonates. In this chapter, these three types of elastomers are designated polyesteramides (PEAs), polyetheresteramides (PEEAs), and polycarbonate-esteramides (PCEAs), respectively. The amide elastomers developed by Atochem contain aliphatic amide-based hard segments and polyether-based soft segments. They are called polyether-*block*-amides (PE-*b*-As) to distinguish them from the PEEA elastomers. (Note: Although this nomenclature typifies a diblock polymer, PE-*b*-As are segmented block copolymers.)

The elastomers discussed in this chapter also differ in the polymer forming reactions used to produce them. The PE-*b*-A elastomers are made by the esterification reaction between carboxylic acid-terminated aliphatic amide blocks and hydroxyl-terminated polyether diols. In contrast, the polymer forming reaction of the PEA, PEEA, and PCEA elastomers is the reaction between an aromatic isocyanate and aliphatic carboxylic acid to form the amide moiety. Carboxylic acid-terminated polyester or polyether diols form the soft segments. Thus, the amide hard segments are formed at the same time as the polymer. These two approaches are shown schematically in Fig. 9.1. Because the semiaromatic amide does *not* undergo the ester–amide interchange reaction either under the reaction conditions or during later processing, an ester-based soft segment can be employed. In the case of the aliphatic amide, however, the ester–amide interchange reaction would randomize the blocks and the polymer would lose the properties associated with segmented block polymers.

PE-b-A:

HO$_2$C-polyamide-CO$_2$H + HO-polyether-OH $\longrightarrow$

[polyamide-ester-polyether-ester]

PEA or PCEA:

OCN-Ar-NCO + HO$_2$C-R-CO$_2$H + HO$_2$C-polyester-CO$_2$H $\longrightarrow$

[(poly-Ar-amide)polyester]

PEEA:

OCN-Ar-NCO + HO$_2$C-R-CO$_2$H + HO$_2$C-ester-polyether-ester-CO$_2$H $\longrightarrow$

[(poly-Ar-amide)-ester-polyether-ester]

Figure 9.1   Schematic of polyamide elastomer polymerization

9.3.1.1   *Polyesteramide, Polyetheresteramide, and Polycarbonate-esteramide Synthesis*

The formation of the amide group is the reaction occurring during the polymerization of monomers into the PEAs, the PEEAs, and the PCEAs. Polyamides are typically synthesized by the condensation reactions of either diamines with dicarboxylic acids (through the amine salts) [Eq. (9.1)] or by the ring opening polymerization of cyclic lactams [Eq. (9.2)]. For example, nylon 6/6 and nylon 6/10, and nylon 6 are made via these routes, respectively. These polyamides are generally limited to aliphatic species because of the low reactivity of the amine salts of aromatic diamines and the limited availability of cyclic lactams. Polyamides, including aromatic polyamides, can also be produced by the reaction of diamines with diacid chlorides [Eq. (9.3)] (or occasionally esters). However, acid chlorides are expensive monomers and generate corrosive gas (hydrogen chloride).

$$n\mathrm{H_2N-R-NH_2} + n\mathrm{HO_2C-R'-CO_2H} \rightarrow \mathrm{H-[\!-NH-R-NHCO-R'-CO-]}_n\mathrm{-OH} + n\mathrm{H_2O}$$

$$(9.1)$$

$$n(\overset{\overset{\displaystyle -\!\!-\!\!-\!\!C=O}{\vert}}{\mathrm{CH_2)}_x\mathrm{-NH}} \rightarrow \mathrm{HO_2C-[\!-(CH_2)}_x\mathrm{NHCO-]}_n\mathrm{-(CH_2)}_x\mathrm{NH_2} \qquad (9.2)$$

$$n\mathrm{H_2N-R-NH_2} + n\mathrm{ClOC-R'-COCl} \rightarrow \mathrm{H-[\!-NH-R-NHCO-R'-CO-]}_n\mathrm{-OH} + n\mathrm{HCl}$$

$$(9.3)$$

$$\mathrm{R, R' = Ar \quad or \quad Aliphatic}$$

Amides can also be synthesized from isocyanates and carboxylic acids with the loss of carbon dioxide [Eq. (9.4)]. The use of an aromatic isocyanate in this reaction affords an amide identical to one that would result from an aromatic diamine used in the amine salt method discussed above, without the complication of low amine reactivity. The mechanism of this amide forming reaction is quite complex and has been reported in the literature [3–10].

$$R-NCO + R'-CO_2H \rightarrow R-NHCO-R' + CO_2 \tag{9.4}$$

This reaction been adapted to the synthesis of semiaromatic polyamides [10–15]. With the appropriate reaction conditions [10], this reaction can produce high molecular weight polyamides with no undesirable by-products. The formation of high molecular weight polyamides by this method is evidence of the efficiency of the reaction sequence, since greater than 99% conversion is required in condensation polymerizations to achieve high polymers [16]. The versatility of this reaction has allowed the synthesis of a wide range of polymers ranging from engineering thermoplastics [12, 13] to TPEs [14, 15].

PEA, PEEA, and PCEA elastomers are synthesized [14, 15, 17] by the condensation of the aromatic diisocyanate, MDI, with aliphatic dicarboxylic acids and a polyester or polyether prepolymer with a M_n of 500 to 5000 and terminated with aliphatic carboxylic acid. Two moles of carbon dioxide are lost for every mole of the diisocyanate consumed in the reaction (Fig. 9.2). The added dicarboxylic acid serves as the hard segment chain extender and forms the semiaromatic amide hard segment with the MDI. It is analogous to the 1,4-butanediol used to increase the length of the urethane block in polyurethane elastomers (see Chapter 2).

The dicarboxylic acids generally used as hard segment chain extenders are adipic (C-6) and/or azelaic (C-9) acids. This choice is based on economics and the melting point of the MDI-amides. MDI-amides derived from dicarboxylic acids with less than six carbon atoms have melting points at or above the decomposition temperature of the elastomer, while MDI-amides based on carboxylic acids with more than nine carbon atoms are too costly. Also, the melting points of the MDI-amides based on long-chain dicarboxylic acids are too low to offer any advantages in upper use temperatures. Aromatic dicarboxylic acids, although readily available, result in fully aromatic amide hard segments whose melting points are too high to be processed readily. Thus, hard segments based on MDI-adipamide ($T_m = 325\,°C$) and MDI-azelamide ($T_m = 285\,°C$) offer a good balance of cost and performance.

Polyester, polyether, or polycarbonate prepolymers terminated with aliphatic carboxylic acids form the soft segment block for PEA, PEEA, and PCEA, respectively. These prepolymers are obtained by the esterification reaction of commercially available dihydroxyl-terminated polyols with an excess of aliphatic carboxylic acid, or directly by the

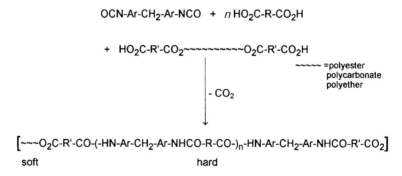

Figure 9.2 Typical PEA, PEEA, and PCEA polymer reaction scheme

reaction of a short-chain diol with an excess of a dicarboxylic acid. The excess of carboxylic acid ensures that all the hydroxyl groups have been reacted and the ratio of carboxylic acid to polyol controls the molecular weight of the prepolymer. This esterification is commonly carried out in the presence of a catalyst such as *p*-toluenesulfonic acid or an organo titanate.

The soft segments generally used in PEEA elastomers are based on commonly available polyoxyalkylene glycols such as polyoxyethylene glycol, polyoxypropylene glycol, and polyoxytetramethylene glycol. Polyester polyols such as hexamethylene adipate glycol or tetramethylene azelate glycol are also used to make PEA elastomers. A special subset of the PEA elastomers are the PCEAs. These use soft segments based on aliphatic polycarbonate glycols such as hexamethylene carbonate glycol to impart improved property balances.

The polymerization is usually carried out homogeneously in a polar solvent [one that is nonreactive with isocyanates at elevated temperatures (200 °C to 280 °C)] by the controlled addition of the diisocyanate to a solution of the other co-monomers. The polymer is recovered by precipitation or by solvent removal under vacuum. These elastomers can also be prepared by a one- or two-step reactive extrusion process [18, 19]. Alternatively, low molecular weight, carboxylic acid-terminated polymer can be made via the solution process and then converted to high molecular weight by adding the deficient amount of diisocyanate in a reactive extrusion process [20]. In all cases, provisions must be made for venting the carbon dioxide gas released as a result of the amide formation.

The PEA, PEEA, and PCEA elastomers produced by this method are transparent, with a pale yellow-brown color. The elastomers are soluble in polar amide solvents such as dimethylformamide (DMF), dimethylacetamide (DMAC), and *N*-methylpyrollidinone (NMP). The inherent viscosities of these elastomers typically range between 0.8 dl/g and 1.5 dl/g, measured as a 0.5 wt% solution in NMP.

The amide content of the elastomer and the crystallinity of the hard segment domains can be changed by varying the amounts and type of dicarboxylic acid chain extenders in the formulation and/or by changing the molecular weight of the polyester, polyether, or polycarbonate soft segments. The value of the hard segment T_m relative to the T_m of the analogous homopolyamide generally reflects the extent of crystallinity in the hard domains. By controlling the weight percent amide hard segment and the molecular weights of the hard and soft segments, these elastomers can be made to span the range of durometer hardness from Shore 80A to Shore 70D. The type of amide used and the type of soft segment can be chosen to fine tune the desired performance characteristics of these elastomers. Table 9.1 lists several different PEAs, PEEAs, and PCEAs that have been synthesized and characterized.

9.3.1.2   Polyether-block-amide Synthesis

The polymer forming reaction used to synthesize the PE-*b*-A elastomers is the formation of an ester moiety. There are two primary methods for the formation of aliphatic esters. One is the transesterification reaction between short-chain esters of carboxylic acids and hydroxyl functional compounds [Eq. (9.5)]. A second related method is the direct esterification of carboxylic acid- and hydroxyl-containing compounds [Eq. (9.6)]. Appropriate catalysts increase the reaction rates and reduce the severity of the reaction conditions. Both of these

Table 9.1 Glass Transition Temperatures and Melting Points of PEA, PEEA, and PCEA

	Hardness (Shore A)	Hard segment extender	Amide (%)	T_g (soft) (°C)	T_m (hard) (°C)
PEA-1	88A	Adipic	25	−40	270
PEA-2	94A	Azelaic	35	−28	230
PEA-3	94A	Adipic	33	−34	275
PEA-4	55D	Azelaic/Adipic	37	−33	236
PEA-5	60D	Azelaic/Adipic	39	−33	238
PEA-6	70D	Azelaic/Adipic	42	−34	240
PEEA-1	92A	Azelaic/Adipic	31	−50	251
PEEA-2	92A	Azelaic	31	−40	264
PEEA-3	90A	Adipic	31	−40	290
PCEA-1	88A	Azelaic	35	−40	230
PCEA-2	92A	Adipic	35	−38	252
PCEA-3	92A	Azelaic	35	−30	230

PEA, polyesteramide; PEEA, polyetheresteramide; PCEA, polycarbonate-esteramide

pathways are well established means to produce high molecular weight polymers from the analogous difunctional monomers.

$$RCO_2R' + R''-OH \rightarrow R-CO_2R'' + R'OH \tag{9.5}$$

$$RCO_2H + R'-OH \rightarrow R-CO_2R' + H_2O \tag{9.6}$$

The polymerization reaction to form the PE-*b*-As involves the esterification reaction of α,ω-dicarboxylic acid-terminated aliphatic amide blocks with polyoxyalkylene glycols. Thus, there is one ester moiety linking the amide and the ether blocks in PE-*b*-As. To prepare a pure PE-*b*-A amide elastomer, an amine-terminated polyether could be used in the place of the polyoxyalkylene glycol. However, the amine-terminated polyols have limited availability and are expensive to produce.

The polyamide hard segments of the PE-*b*-A elastomers are derived from dicarboxylic acid-terminated aliphatic polyamides that are prepared from cyclic lactams, amine salts of aliphatic amines and carboxylic acids, and α,ω-amino acids. Examples of the polyamides that are used as hard segments include nylon 6, nylon 11, nylon 12, nylon 4/6, nylon 6/6, nylon 6/10, nylon 6/11 and nylon 6/12 [21–23]. These blocks are prepared prior to the polymerization step by a melt condensation reaction under the same conditions typically used for the formation of high molecular weight polyamides. To ensure that the terminal groups of the amide blocks are carboxylic acid, additional dicarboxylic acid is added, Fig. 9.3. This dicarboxylic acid also serves as a chain stopper to control the molecular weight of the ensuing polyamide oligomer by changing the stoichiometric balance of acid and amine monomers. The amount of additional dicarboxylic acid necessary to obtain the desired molecular weight can easily be calculated using well known statistical expressions [24].

The polyoxyalkylene glycols used for the soft segment are generally commercially available materials, such as polyoxyethylene glycol, polyoxypropylene glycol, and polyoxytetramethylene glycol. Their molecular weights range from 200 to 3000.

The PE-*b*-A elastomers are prepared by the melt polycondensation of the dicarboxylic acid amide block and the polyoxyalkylene glycol at elevated temperature (200 °C to 300 °C)

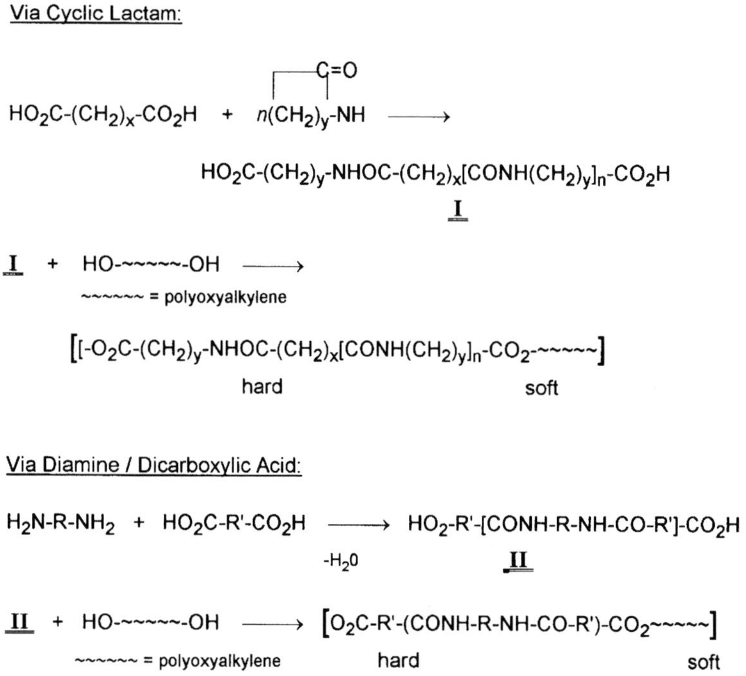

Via Cyclic Lactam:

$$HO_2C-(CH_2)_x-CO_2H \quad + \quad n(CH_2)_y-NH \xrightarrow{\hspace{1cm}}$$

(with cyclic lactam showing $C=O$ bridging)

$$HO_2C-(CH_2)_y-NHOC-(CH_2)_x[CONH(CH_2)_y]_n-CO_2H$$

<u>I</u>

$$\underline{I} \quad + \quad HO\text{-}\sim\sim\sim\sim\text{-}OH \xrightarrow{\hspace{1cm}}$$

$\sim\sim\sim\sim$ = polyoxyalkylene

$$\left[[-O_2C-(CH_2)_y-NHOC-(CH_2)_x[CONH(CH_2)_y]_n-CO_2-\sim\sim\sim\sim]\right]$$

hard   soft

Via Diamine / Dicarboxylic Acid:

$$H_2N-R-NH_2 \quad + \quad HO_2C-R'-CO_2H \xrightarrow{\hspace{1cm}} HO_2-R'-[CONH-R-NH-CO-R']-CO_2H$$

$-H_2O$ <u>II</u>

$$\underline{II} \quad + \quad HO\text{-}\sim\sim\sim\sim\text{-}OH \xrightarrow{\hspace{1cm}} [O_2C-R'-(CONH-R-NH-CO-R')-CO_2\sim\sim\sim\sim]$$

$\sim\sim\sim\sim$ = polyoxyalkylene   hard   soft

Figure 9.3   Typical PE-*b*-A polymer reaction scheme

and under high vacuum (<2 mm Hg) for 2 or more hours. The addition of metallic tetraalkoxide catalysts, such as tetrabutyltitanate, increases the rate of the esterification reaction and increases the molecular weight of the polymer produced [22, 25]. It has been reported that zirconium or hafnium tetraalkoxides are more efficient in this reaction [26]. These catalysts may be used in combination with alkaline or alkaline-earth alcoholate cocatalysts [27]. The process can be done either batchwise or continuously by using up to three agitated thin film reactors configured in series [28]. The thin film reactors offer the advantages of better heat transfer and continuous surface renewal to facilitate the removal of the water of condensation. High molecular weight polymer can be produced with much shorter residence times at the elevated polymerization temperatures which, in turn, improves the color of the polymer.

PE-*b*-A elastomers designed for molded products generally have amide segments with molecular weights in the range of 800 to 5000. The ether soft segments have a molecular weight range of 400 to 3000 and make up from 5% to 50% (by weight) of the elastomer. Softer, more elastomeric materials can be obtained by using amide and ether segment molecular weights of 500 to 2000 and 1000 to 3000, respectively. In this case the soft segments typically comprise 60% to 80% (by weight) of the elastomer [29].

The PE-*b*-A elastomers range from nearly transparent to opaque white materials, depending upon the amide content of the sample. Typical inherent viscosities of 1.5 dl/g are obtained as measured at a concentration of 0.5 g/dl in *m*-cresol at 25 °C. The variety of aliphatic amides and their molecular weight that can be used as hard segments allows the

The average molecular weight of the segments also affects the observed values of the T_g and T_m.

The effect of phase separation on the polymer hardness is also exemplified in Table 9.1. PEEA-2, which has the same hard segment content and hard segment length as PEA-2, is several Shore hardness points softer because of its greater phase separation.

In the PE-*b*-A elastomers, the hard and soft phases are not very compatible and, in the softer grades of the PE-*b*-A elastomers, a distinct soft segment T_m is evident in the DSC shown in Fig. 9.5. This soft segment T_m is not evident in higher hardness formulations, because of forced compatibility and the smaller proportion of soft segment present.

Infrared dichroism is a technique that can indicate the orientation of the hard and soft segments when under stress by measuring the infrared absorption of functional groups characteristic of the hard or soft segments [31]. When analyzed by this technique (see Fig. 9.6), PEA-2 shows behavior typical of a segmented elastomer with crystalline hard segments [32]. The soft segments, monitored through the C−H stretch of the methylene group at 2930 and 2860 cm^{-1}, align parallel to the applied stress, resulting in a positive orientation function at all strain levels. Because the hard segments are crystalline, the orientation function observed is quite different. At low strain levels, the crystallites align perpendicular to the applied stress because of the parallel alignment of the lamellae. This gives rise to a negative orientation function, monitored using the N−H stretch at 3360 cm^{-1}. As the stress levels increase, the crystallites are disrupted and the hard segments are able to align parallel to the stress and the orientation function becomes positive. This is in contrast to the less crystalline polyurethanes which show positive orientation functions at all strain levels [32].

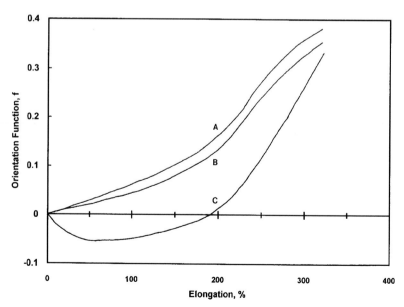

Figure 9.6   Infrared dichroism of PEA-2. (A) CH$_2$ 2860 cm^{-1}; (B) CH$_2$ 2930 cm^{-1}; (C) NH 3310 cm^{-1}

The stress-induced crystallization of the soft segment can be observed visually by a whitening of the sample when it is strained above 300%. In Fig. 9.7, a DSC thermogram of a sample, PEA-2, frozen while strained, indicates a T_m for the soft segment at 40 °C. On heating the frozen sample above 40 °C or relaxing the sample at room temperature, the soft segment readily reverts back to the amorphous state.

Limited experiments with pulsed proton nuclear magnetic resonance (NMR) relaxation time measurements, as well as data from small-angle light scattering experiments, also result in data supportive of a segmented type of polymer.

9.4   Structure–Property Relationships

The physical properties exhibited by thermoplastic, segmented block copolymers, such as PEA, PEEA, PCEA, and PE-*b*-A, are dependent on such parameters as the chemical composition of the hard and soft segments and also their respective lengths and weight ratios. In the following section, the effects that various chemical and compositional variables have on the performance of the elastomer are summarized. Specific examples are discussed in later sections but it will become apparent here that complex interactions exist between many

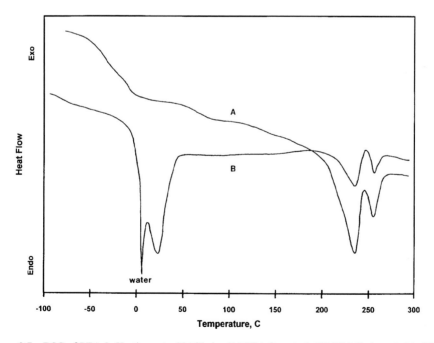

Figure 9.7   DSC of PEA-2. Heating rate: 20 °C/min. (A) PEA-2 control; (B) PEA-2 elongated to 370% at 50 cm/min, then cooled under strain to −78 °C before scanning. The water peak at 0 °C is an artifact of the freezing process

of these parameters and that these influence the ultimate performance of these elastomer systems.

9.4.1   Hard and Soft Segment Considerations

The chemical composition of the hard and soft segments and the hard segment content have the greatest effect on the performance of segmented thermoplastic elastomers. The hard segment content is defined as the weight percent of hard segments in the total elastomer formulation. Table 9.3 indicates which of these three parameters are most important in determining the performance properties of segmented TPEs. (Note that these are only generalizations and exceptions may occur.)

In these amide elastomers, the effects of the hard segment factors are generally associated with the level of crystallinity of the amide hard segments and the melting point of these crystallites. The more crystalline amide segments will have higher melting points at shorter block lengths (lower block molecular weight) and this directly affects the ceiling use temperature of the elastomer. Highly crystalline hard segments also tend to be more completely phase separated, resulting in softer materials. The level of crystallinity also affects the chemical resistance, since more crystalline materials are less soluble and less affected by solvents.

The chemical composition of the soft segment, which is often the larger component by weight in these elastomer systems, affects the thermal oxidative stability. For example, the polyether-based elastomers are generally more susceptible to oxidative chain scission than the ester-based. In contrast, many of the elastomers with polyether-based soft segments (with the exception of the short-chain polyethers such as polyoxyethylene glycol) are more hydrophobic than the esters and are less susceptible to hydrolysis. Even within the carboxylate ester series, hydrolytic stability decreases as the percent of ester groups increases. The molecular weight of the segment can affect the degree of compatibility (short chains cause forced compatibility), which in turn affects the degree of phase separation, resulting in the hardness changes mentioned previously [33, 34]. The chemical structure is the main influence on the T_g of the soft segment. The degree of phase mixing affects the degrees of freedom of the soft segment chains, which also affects the T_g. Thus, several factors define the lower limits or the "leathery" region of the elastomer's performance range.

Table 9.3 Correlations of Thermoplastic Elastomer Parameters and Physical Properties

Property	Hard segment composition	Soft segment composition	Amide content
Hardness	√		√
Degree of phase separation	√	√	√
T_m	√		
Tensile properties	√		√
Thermal oxidative stability		√	√
Chemical resistance	√	√	√
Hydrolytic stability		√	√
Low-temperature properties		√	√

The ratio of hard to soft segments, or the percentage amide content, affects many of the elastomers' properties as the polymer slowly changes from a soft polyether or ester to a hard polyamide. This change in hardness of the polymer is observed by handling the elastomer samples.

9.5   Physical Properties

It should be apparent from this summary that the physical properties observed in an elastomer are the result of complex interactions of many factors. In this section, many of the physical properties of the PEA, PEEA, PCEA, and PE-*b*-A TPEs are presented and explained by one or more of the molecular parameters mentioned in the previous section.

9.5.1   Tensile Properties

The tensile properties of several formulations of PEA, PEEA, PCEA, and PE-*b*-A are tabulated in Tables 9.2 and 9.4. A typical stress–strain curve for PEA is shown in Fig. 9.8. The initial moduli of these polyamide elastomers, which depend on the amide content, are much higher than for many other TPEs in the same hardness range. This is a result of the higher load bearing ability of the crystalline portion of the amide segment domains. The deformation of the hard segment in this low-strain region is mostly reversible. At higher strain levels, however, disruption and reorganization of the crystalline lamellae occurs and the stress–strain curve levels out as chain slippage relieves some of the stress [35]. This change is irreversible and results in the observed tensile set.

Table 9.4 Tensile Properties of PEA, PEEA, and PCEA (ASTM-412)

Sample	Modulus (MPa)			Tensile strength (MPa)	Elongation (%)	Tensile set (%)
	50%	100%	300%			
PEA-1	8.5	10.8	18.2	28.1	470	—
PEA-1A[a]	8.7	12.0	21.0	31.0	495	—
PEA-2	11.3	13.2	19.2	26.2	470	50
PEA-2A[b]	12.6	16.4	28.8	31.0	370	40
PEA-4	14.7	18.9	—	36.0	295	34
PEA-5	19.0	22.7	33.1	33.1	300	100
PEA-6	27.0	30.1	—	42.2	265	92
PEEA-2	9.2	11.5	18.4	18.6	300	16
PEEA-2[b]	9.6	12.6	16.7	18.8	410	50
PCEA-2	10.2	13.5	18.8	20.5	390	—
PCEA-3	15.8	16.5	—	21.9	270	—

[a] Sample annealed at 175 °C for 4 h
[b] Sample annealed at 200 °C for 3 h

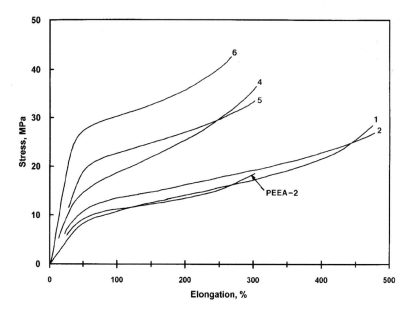

Figure 9.8   Typical room temperature stress–strain curves (ASTM D-412) of PEA-(1–6) and PEEA-2 elastomers, as indicated. Compositions as shown in Table 9.1

Finally, the load is borne by the soft segment chains which become oriented and crystallize as shown by the DSC studies. This reinforces the elastomer and causes an increase in the stress until catastrophic failure occurs at break [36, 37]. As an indication of the integrity of the crystalline amide hard segment domains on a microscopic scale, PEA does not show a tensile yield even at a hardness as high as Shore 70D, and therefore it has lower permanent set than some of the other crystalline TPEs. The tensile set values after 100% strain for PEA-2 are 11% after 1 min of relaxation and only 7% after a 10-min relaxation period. A harder material, PEA-4, exhibits only slightly higher tensile set values of 14% and 12% under the same conditions. The hard segment domains of the PE-b-A are more easily distorted, resulting in higher values of permanent tensile set at high levels of strain (>20%). Their behavior is similar to that observed with the copolyester elastomers.

The tensile properties of PEA and PEEA show an improvement after being annealed above the T_g of the amide hard segment. In addition to the relief of molded-in stresses, annealing also promotes the reorganization of the short-range ordered amide segments in the hard segment domains into larger and more perfect crystallites. As a result of this change, the area of the higher melting endotherm shown in the DSC in Fig. 9.4 increases at the expense of the lower melting endotherm. The net result is often higher modulus values, higher tensile strengths, and better elongations, as indicated in Table 9.4.

9.5.2   High Temperature Tensile Properties

The purpose behind the development of the PEA, PEEA, and PCEA elastomers was to synthesize a TPE suitable for use at elevated temperatures. The success of this endeavor can

Table 9.5 Tensile Properties of PEA at Elevated Temperatures (ASTM-412, D-3196)

Sample	Test temp. (°C)	Modulus (MPa) 50%	100%	300%	Tensile strength (MPa)	Elongation (%)
PEA-1	RT	8.7	12.0	21.0	31.0	495
	150	6.5	8.9	13.3	14.8	340
PEA-2	RT	11.3	13.2	19.2	26.2	470
	100	7.4	8.3	9.7	14.6	480
	150	5.5	5.9	6.3	7.7	320
PEA-4	RT	14.7	18.9	—	36.0	295
	100	5.8	8.1	15.8	20.6	390
	150	3.7	5.1	9.1	9.6	310
PEA-5	RT	22.2	22.7	33.1	33.1	300
	100	8.4	8.4	12.4	20.0	500
	150	3.2	3.2	3.7	5.0	480
PEA-6	RT	30.1	30.1	—	42.2	265
	100	9.0	9.0	14.3	26.9	490
	150	6.1	6.1	9.4	16.2	540

All samples have been annealed for 3 h at 200 °C except PEA-1, which was annealed for 4 h at 175 °C

be determined by performing ring tensile tests in an environmental chamber (ASTM D-412, D-3196) at several test temperatures. Table 9.5 lists the tensile properties measured for several PEA formulations and Fig. 9.9 depicts typical stress–strain curves obtained at 150 °C. These data also indicate that the high-temperature performance of a segmented elastomer is closely dependent upon both the hard segment crystallinity and also upon its melting point. It is also noteworthy that these materials retain useful tensile properties under conditions where most other TPEs could not even be tested. The properties retained by the softer formulations are particularly significant since satisfactory high temperature performance is most difficult to achieve in a soft TPE.

The relatively low hard segment melting points (T_m) of the PE-b-A elastomers limit their upper use temperature to about 80 °C. The actual temperature depends on the amide content and the requirements of the actual application.

Since many application areas for TPEs do not require high elongations, the initial moduli are often the primary design criteria. The retention of the modulus at several test temperatures is illustrated for PEA-2 in Fig. 9.10 for 10% and 20% strain.

9.5.3   Dry Heat Aging

The PEA, PEEA, and PCEA elastomers are also very resistant to long-term dry heat aging, even at 150 °C in the absence of any added heat stabilizers. When compared to unaged control samples in Table 9.6, some of the room temperature tensile properties actually improved after aging for 5 days at 150 °C. This is due to an annealing effect rather than to oxidative crosslinking, since the samples are still soluble after aging. Similar studies have been carried

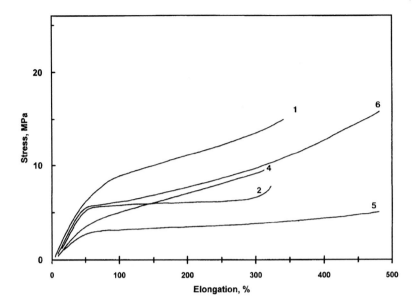

Figure 9.9   Typical stress–strain curves for PEA-1,-2,-4,-5,-6 elastomers at 150 °C, as indicated

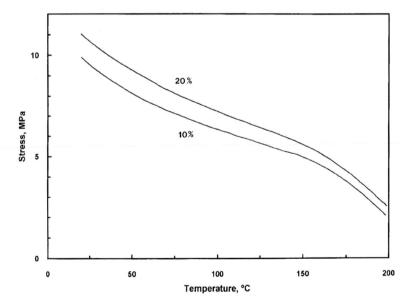

Figure 9.10   Tensile stress of PEA-2 as a function of temperature at 10% and 20% strain levels, as indicated

Table 9.6 Dry Heat Aging of PEA and PEEA

Tensile properties after 120 h at 150 °C (percent retention)

Sample	Modulus (MPa)			Tensile strength (MPa)	Elongation (%)
	50%	100%	300%		
PEA-1	5.8 (68)	8.9 (74)	18.0 (100)	26.3 (94)	430 (91)
PEA-2	12.9 (114)	16.0 (121)	26.2 (136)	29.7 (113)	390 (83)
PEEA-2	10.8 (117)	14.6 (127)	—	18.0 (97)	260 (87)

Tensile properties after 120 h at 175 °C (percent retention)

Sample	Modulus (MPa)			Tensile strength (MPa)	Elongation (%)
	50%	100%	300%		
PEA-1	3.6 (42)	5.8 (54)	16.7 (92)	19.1 (68)	320 (68)
PEA-2	12.0 (106)	15.9 (120)	—	26.9 (103)	300 (64)
PCEA-3	16.9 (107)	18.6 (113)	—	23.0 (105)	250 (93)

Samples were annealed at 200 °C for 3 h. No added heat stabilizers

out for the ester-based PEA elastomers at 175 °C with only small losses in properties. The thermal aging performance is also related to the amount of amide hard segment present, declining as the percent amide decreases. With the addition of stabilizers [38], PEA-2 exhibits improved heat resistance. This becomes evident when the samples are tested at elevated temperatures after dry heat aging, as demonstrated in Table 9.7.

At 150 °C, there is little difference between the thermal aging performance of the ether- and ester-based elastomers. However, under the more severe conditions of 175 °C, the ester- and carbonate-based elastomers show better retention of tensile properties compared to the ether-based equivalents. This is due to the inherent oxidative instability of the ether linkage. The PE-*b*-A elastomers follow the same trend. Unstabilized nylon 6-based elastomers (hardness = 55D) become very fragile and friable after only 1 day at 150 °C but with the

Table 9.7 Dry Heat Aging of Stabilized PEA-2; Tensile Properties after 120 h at 150 °C

Sample	Test temp. °C	Modulus (MPa)			Tensile strength (MPa)	Elongation (%)
		50%	100%	300%		
Neat	RT	13.7	16.6	—	23.6	275
	100	9.0	11.1	—	13.5	210
	150	7.9	—	—	8.8	80
Stabilized	RT	13.4	16.1	27.4	32.0	350
	100	10.0	12.0	15.7	15.8	315
	150	8.7	10.2	—	10.8	120

Samples were aged, then tested at the indicated temperatures

addition of amine-based antioxidants and cuprous and potassium iodide heat stabilizers, the elastomers retain about 80% of their original tensile strength after 15 days at 150 °C. [39].

9.5.4   Humid Aging

Table 9.8 lists the retention of properties for PEA, PEEA, and PCEA after 3 to 30 days at 85 °C and 100% relative humidity. The data show that the chemical composition of the soft segment is very important in determining the performance of an elastomer after humid aging. Most ester-based TPEs are susceptible to hydrolysis, which reduces their properties by lowering the polymer molecular weight, and PEA is no exception. However, the presence of a sacrificial stabilizer such as a polymeric carbodiimide-based additive does improve the humid aging characteristics of these polymers. The PEA formulations become less susceptible as the hardness increases since the amount of ester present is reduced. There is no indication of any hydrolysis of the semiaromatic amide hard segments under these conditions.

Elastomers that have ether-based soft segments, such as PEEA, are much less sensitive to the presence of water since the ether linkage does not hydrolyze. The PEEA polymers show excellent retention of their original tensile properties in humid aging tests, in contrast to the PEA elastomers. The carbonate-ester-based PCEA elastomers exhibited even better performance because of the greater hydrolysis resistance of a carbonate ester.

The PE-*b*-A elastomers, which also have ether-based soft segments, have good resistance to hydrolysis during water immersion at 100 °C, as shown in Table 9.9. However, even the performance of these elastomers can be greatly improved by reducing the carboxylic acid end group content of the polymer [40]. This is accomplished by adding a minor amount of a monocarboxylic polyamide oligomer during the polymerization reaction. This component acts as a chain terminator and reduces the amount of the carboxylic acid normally found on chain ends. Under the test conditions, this terminal carboxylic acid can catalyze the hydrolysis of both the ester linkages and the aliphatic amide hard segments.

9.5.5   Chemical and Solvent Resistance

The nature of the hard segment has a large effect on the chemical and solvent resistance of a segmented block elastomer. Since the soft segments commonly used are soluble in common organic solvents, they will swell in these solvents and lose their integrity. Thus, it is the hard segment domains that maintain the integrity of the swollen polymer. The semicrystalline amide hard segments used in all these polyamide elastomers have low solubility in many solvents. Thus, PEA-2 has excellent resistance in oils, fuels, and grease. PEA-2 also fares well in phosphate-based hydraulic fluids; however, chlorinated solvents have a much greater adverse effect. As the proportion of the hard segment increases in the harder formulations, the chemical resistance also improves. This effect can be seen in the results of seven day immersion tests shown in Table 9.10.

Table 9.8 Humid Aging of PEA, PEEA, and PCEA; Tensile Properties after Exposure to 100% Relative Humidity at 85 °C

Sample	Time (days)	Modulus (MPa)			Tensile strength (MPa)	Elongation (%)	Change	
		50%	100%	300%			% Vol	% Wt
PEA-1	Ctrl	8.5	11.1	18.3	28.2	495		
	3	8.1	10.3	15.0	19.0	500		
	7	7.2	9.0	11.2	10.8	280		
PEA-2	Ctrl	13.1	16.1	25.0	27.2	345		
	3	10.4	13.5	17.8	20.0	425	0.32	0.21
	7	—	—	—	8.7	50	0.31	0.23
PEA-2[a]	Ctrl	14.1	20.4	30.0	30.8	305		
	3	11.1	15.6	24.9	27.6	380	0.98	0.26
	7	12.1	13.9	—	14.8	190	0.21	0.08
	10	12.0	15.3	—	18.0	245	0.39	0.28
PEA-4	Ctrl	18.6	22.2	—	30.0	219		
	3	16.7	20.6	—	24.3	210		
PEA-5	Ctrl	28.1	29.1	—	31.4	110		
	3	27.5	27.6	—	27.6	100		
PEEA-1	Ctrl	7.9	11.4	14.9	15.4	310		
	3	8.3	11.0	—	14.5	240	−0.41	−0.45
	7	7.9	11.1	—	13.2	190	−0.72	−0.71
PEEA-2	Ctrl	8.7	12.8	—	17.3	277		
	3	8.8	12.1	—	15.2	220	−0.24	−0.06
	7	9.0	12.2	—	15.7	270	−0.17	−0.22
PEEA-3	Ctrl	9.2	11.7	15.6	15.6	300		
	3	9.4	11.6	15.5	15.5	300	−0.24	
	7	8.2	11.1	—	13.0	200	−0.51	
PCEA-2	Ctrl	10.2	13.5	18.8	—	280		
	3	10.1	13.1	18.8	21.5	320	−0.05	
	7	9.3	12.3	18.8	20.6	360	−0.02	
	10	10.0	13.2	19.2	20.6	370	+0.70	
	14	10.6	14.3	18.2	19.9	360	+0.28	
PCEA-3	Ctrl	14.0	16.0	—	21.8	250		
	10	13.9	16.0	—	18.1	200	−0.45	
	20	14.6	16.6	—	18.7	220	+0.07	
	30	13.2	14.9	—	18.5	210	+0.47	

[a] PEA-2 stabilized with 0.75 wt% Staboxol P

9.5.6   Tear Strength

PEA-2 has exceptionally good tear strength (ASTM D-624, Die C) with values of 151 and 51 kN/m (860 and 290 pli) at room temperature and 150 °C, respectively. Since the Die C tear strength generally increases with tensile strength, the harder grades such as PEA-4 have even higher values of 169 and 58 kN/m (960 and 330 pli) at the same test temperatures. This same trend is observed for the PE-*b*-A elastomers in Table 9.11. PEA-2 also has a value of 31.5 kN/m (180 pli) for split tear strength (ASTM D-470).

Table 9.12 Abrasion Resistance of PEA and PE-*b*-A Elastomers, Taber Abrasion (ASTM D-1044)

	Weight loss (mg/1000 revs.)		
Sample	CS-17 wheel	H-18 wheel	H-22 wheel
PEA-2	4	89	60
PE-*b*-A-1	46	94	
PE-*b*-A-2	25	81	
PE-*b*-A-3	14	70	
PE-*b*-A-4	11	65	
PE-*b*-A-5	12	46	
PE-*b*-A-6	—	—	

PE-*b*-A data from Atochem technical literature
PE-*b*-A samples injection molded, aged 14 days at 23 °C, 50% relative humidity

Over the course of the test, reorganization in the hard segment domains occurs leading to compression set values somewhat higher than found with crosslinked rubber materials. Factors that favor well-organized crystal structures in the hard domains generally help to reduce the compression set. Results are summarized in Table 9.13. It has also been demonstrated that the compression set performance can be improved by the addition of small amounts of additives based on carbodiimides [41].

9.5.9   Flex Properties

The T_g of the soft segments relative to the test temperature is an important parameter in the flex fatigue of a TPE since the polymer becomes "leathery" as the T_g is approached. Formulations with greater phase separation usually have lower T_g's and often have improved performance in flex fatigue tests. This effect is illustrated in Table 9.14, where the results for PEA-1 (T_g −40 °C) and PEA-2 (T_g −28 °C) are compared in the Ross flex cut growth test (ASTM D-1052).

Table 9.13 Compression Set of PEA and PE-*b*-A Elastomers (ASTM D-395 A, B)

	Compression set (%)		
Sample	Method A	Method B	Temp (°C)
PEA-1	—	36	RT
PEA-1	—	79	100
PEA-2	2	40	RT
PE-*b*-A-1	62	—	70
PE-*b*-A-2	54	—	70
PE-*b*-A-3	21	—	70
PE-*b*-A-4	10	—	70
PE-*b*-A-5	6	—	70
PE-*b*-A-6	6	—	70

Table 9.14 Ross Flex for PEA (ASTM D-1052)

		Test temperature	
Sample		$-20\,°C$	$-35\,°C$
PEA-1	Cycles[a]	1,050,000	1,011,000
	% Cut growth	0	0
PEA-2	Cycles	26,800	12,300
	% Cut growth	800	1,000[b]

[a] Test stopped at one million cycles
[b] Sample failed

9.5.10   Adhesion

The adhesive characteristics of PEA-4 have been measured in a lap shear test using wire brushed, *unprimed* aluminum sheets. The test pieces were prepared by compression molding a 0.18-mm film of the PEA between the aluminum sheets. Lap shear strengths ranging from 7.6 to 9.2 MPa (1100 to 1300 psi) were obtained at room temperature.

9.5.11   Weatherability

In addition to outdoor exposure to ultraviolet (UV) radiation, elastomers used in indoor applications are also exposed to significant levels of UV radiation through the increased use of fluorescent lighting. PEA-2, with no additional stabilization, has excellent resistance to UV radiation under moisture condensing conditions (ASTM G-53), as shown in Table 9.15. Since the finished color of PEA is yellow-brown, there is no discernible discoloration after 2500 h of exposure. Several stabilized grades of the PE-*b*-A elastomers have improved service life and less color shift as a result of exposure to UV radiation.

9.5.12   Electrical Properties

The electrical properties of PEA-2 are summarized in Table 9.16. These values generally qualify these elastomers as insulating materials for low-voltage applications.

Table 9.15 Weatherability; QUV Exposure of PEA-2 (ASTM G-53)

	Modulus (MPa)			Tensile strength	Elongation	Set	Shore
Time (h)	50%	100%	300%	(MPa)	(%)	(%)	hardness
0	14.3	16.1	25.6	30.5	340	40	93
40	14.3	16.6	—	23.7	280	30	94
100	14.3	16.3	23.6	25.0	300	40	94
300	14.8	16.8	24.1	24.5	300	40	95
500	14.2	16.5	24.2	25.0	320	40	93
1000	15.0	16.5	23.7	25.0	350	50	93
2500	14.8	16.4	22.7	24.1	350	50	93

Table 9.16 Electrical Properties of PEA-2 at 22 °C and 50% Relative Humidity

Dielectric constant (ASTM D-150)	Frequency (Hz)
10.26	60
9.30	10^3
5.67	10^6
Dissipation factor (ASTM D-150)	
0.092	60
0.066	10^3
0.0100	10^6
Surface resistivity (ASTM D-257)	3.09×10^{12} ohms
Volume resistivity (ASTM D-257)	8.13×10^{10} ohm-cm

9.6   Processing Conditions

The polyamide elastomers discussed in this chapter have been melt processed on injection molding, blow molding, and extrusion equipment, which includes profile, wire coating, and film extrusions. Attributes of the amide elastomers that facilitate melt processing include a processing window larger than that of thermoplastic polyurethanes and also good melt strength. Unlike polyurethanes, however, no bond dissociation and recombination occurs during melt processing. Typical starting conditions for the extrusion and injection molding of PEA-2 and PE-*b*-A elastomers are listed in Tables 9.17 and 9.18. Since the processing temperature depends on the hard segment T_m (Tables 9.1 and 9.2), the zone temperatures may need to be adjusted accordingly for different polymers. The apparent melt viscosity for PEA-2 as a function of shear rate at several temperatures is shown in Fig. 9.11.

Mold shrinkage has been measured for PEA-2 using a 125 mm by 125 mm by 1.6 mm plaque (5 in by 5 in × 1/16 in). Shrinkage parallel to the flow is 1.5% and perpendicular shrinkage is 1.0%. If the mold surface is kept scrupulously clean, there is no sticking and no need for a mold release.

To obtain the optimal properties, these polyamide elastomers must be dried before to processing. As in any amide- or ester-based polymer, absorbed moisture can lead to hydrolytic chain scission at the elevated processing temperatures and subsequent loss of molecular weight. Thus, it is essential that the resin have a moisture content of less than 0.02% before processing. This level can be attained for the PEA and PEEA elastomers by drying for 4 to 6 h at 100 °C to 110 °C in a dehumidifying hopper dryer (dew point −30 °C to −40 °C). The PE-*b*-A elastomers with hardnesses greater than 40D should be dried for 4 h at 80 °C in a dehumidifying drier, while those less than 40D should be dried for 6 h at 70 °C. It is also recommended that the feed hopper be purged with dry nitrogen to maintain the low moisture level.

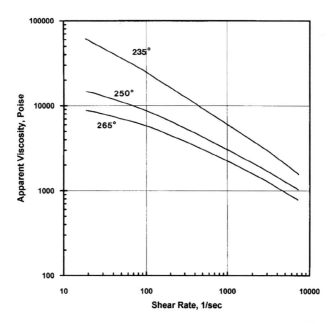

Figure 9.11   Apparent melt viscosity of PEA-2 as a function of shear rate at indicated temperatures

Table 9.17 Typical Processing Conditions for PEA-2

Injection molding			Extrusion	
Temperature			Temperature	
Zone	°C		Zone	°C
Rear	230–240		Feed	230–240
Middle	240–250		Transition	240–250
Front	240–250		Metering	245–255
Nozzle	250–260		Die	245–255
Mold	75–85			
Melt	245–255		Melt	250–255
Screw speed	RPM		Screw speed	RPM
	80–100			50
Pressure	MPa			
Injection	8.3			
Hold	3.4			
Back	0.7			
Cycle time	s			
Injection	2			
Hold	5–8			

Table 9.18 Typical PE-*b*-A Injection Molding and Extrusion Temperatures

	Injection molding		Extrusion melt (°C)
	Melt (°C)	Mold (°C)	
PE-*b*-A-1	180–220	20–40	170–210
PE-*b*-A-2	200–240	20–40	190–220
PE-*b*-A-3	200–240	20–40	210–230
PE-*b*-A-4	240–280	20–40	210–230
PE-*b*-A-5	240–280	20–40	210–230

Data from Atochem technical literature

9.7   Applications

The polyamide-based elastomers are one of the newest additions to the class of TPEs and their full application potential is still being discovered. Because of the high service temperatures and the good thermal aging and chemical resistance characteristics, the PEA TPEs are expected to fill the gap between the thermoplastic polyurethanes and the silicone-based polymers. This balance of properties allow these materials to be considered for automotive under-the-hood type applications. Another potential application is for high-temperature insulation in the wire and cable market.

The wide performance range of the PE-*b*-As allows these materials to be used in a variety of markets. Athletic footwear, hose and tubing, powder coating, and automotive are but a few of the markets in which PE-*b*-A elastomers are in use. The addition of fillers and other modifiers extends their usefulness even further [42].

When polyoxyalkylene glycols are used as the soft segments in polyamide elastomers, the hydrophilicity of the resulting elastomer can be controlled by adjusting the amount of ethylene oxide in the polyoxyalkylene glycols. Consequently, many specialized applications, such as the antistatic packaging or humidity-sensitive sensors, can be achieved by using tailor-made polyamide elastomers that have an appropriate amount of polyoxyethylene or ethylene oxide capped polyoxypropylene glycols as the soft segments [43–45].

Polyamide elastomers have very good compatibility with many other polymers. This character, in combination with the excellent processability and thermal stability, makes polyamide elastomers one of the best materials for preparing polymer alloys with desirable properties and many polymer blends using polyamide elastomers have been reported [46–55].

9.8   Summary

The effectiveness of the physical crosslinks formed by the hard segment domains in segmented copolymers is illustrated by the high upper use temperature. The goal of increasing the temperature limit of TPEs has been met by the incorporation of high melting, semicrystalline polyamides as the hard segments. This has been accomplished by the reaction of aromatic diisocyanates with dicarboxylic acids and derivatives to provide semiaromatic polyamides. The result is a series of polyesteramide, polycarbonate-esteramides, and polyetheresteramides that have excellent tensile properties at 150 °C continuous service temperature, and that can withstand even higher temperature excursions. In addition, these polymers also have good thermal aging and fuel, oil, and grease resistance.

Using more conventional chemistry, the PE-*b*-A elastomers can be made from a variety of monomers, resulting in a virtually limitless opportunity to optimize the properties for a particular application. These materials also take advantage of the stability of the amide moiety that leads to excellent chemical resistance, tensile properties, and thermal aging.

Heating above the melting point of the amide hard segments melts the physical crosslinks, allowing the hard segments to flow. Thus, these new elastomers are truly thermoplastic and they can be melt processed readily by conventional means into a wide variety of useful end products.

References

1. G.M. Estes, S.L. Cooper, and A.V. Tobolsky, *J. Macromol. Sci. Rev. Macromol Chem. C4(2)*, 313 (1970)
2. T.K. Kwei, *J. Appl. Polym. Sci. 27*, 2891 (1982) and references cited therein
3. W. Dieckmann and F. Breest, *Ber. 39*, 3052 (1906)
4. H. Staudinger, *Helv. Chem. Acta 5*, 87 (1922)
5. W. D'Olieslager and I. DeAquirre, *Bull. Soc. Chim. Fr. 1*, 179 (1967)
6. P. Babusiaux, R. Longeray, and J. Dreux, *Julius Liebigs Ann. Chem. 3*, 487 (1976)
7. M.F. Sorokin, S.M. Marukhina, and V.N. Stokozenko, *Tr. Mosk. Khim. Tekhnol. Inst. 86*, 25 (1975)
8. M.F. Sorokin, S.M. Marukhina, and V.N. Stokozenko, *Deposited Doc. 1975, VINITI 1908–75*, p. 33
9. M.F. Sorokin, S.M. Marukhina, J.V. Galkina, and V.N. Stokozenko, *Tr. Mosk. Khim. Tekhnol. Inst. 86*, 27 (1975)
10. J.T. Chapin, B.K. Onder, and W.J. Farrissey, Jr., *Polymer Preprints 21 (2)*, 130 (1980)
11. K.B. Onder and C.P. Smith, U.S. Patent 4,156,065 (May 22, 1979)
12. K.B. Onder, P.S. Andrews, W.J. Farrissey, and J.N. Tilley, *Polymer Preprints 21 (2)*, 132 (1980)
13. K.B. Onder, W.J. Farrissey, Jr., J.T. Chapin, and P.S. Andrews, SPE 39th Annual Technical Conference (ANTEC), Boston, May 4–7, 1981, p. 883
14. A.T. Chen, W.J. Farrissey, and R.G. Nelb, II, U.S. Patent 4,129,715 (December 12, 1978)
15. R.G. Nelb, II, A.T. Chen, W.J. Farrissey, Jr., and K.B. Onder, SPE 39th Annual Technical Conference (ANTEC), Boston, May 4–7, 1981, p. 421
16. P.J. Flory, *Principles of Polymer Chemistry*, Cornell University Press, 1953, p. 93
17. A.T. Chen, American Chemical Society, Rubber Division National Meeting, October 17–20, 1989
18. R.G. Nelb, II and R.W. Oertel, III, U.S. Patent 4,420,603 (December 13, 1983)
19. H.W. Bonk, R.G. Nelb, II and R.W. Oertel, III, U.S. Patent 4,420,602 (December 13, 1983)
20. R.G. Nelb, II, K. Onder, K.W. Rausch, and J.A. Vanderlip, U.S. Patent 4,672,094 (June 9, 1987)
21. G.E. Deleens, P. Foy, and E. Maréchal, *Eur. Pol. J. 13*, 337 (1977)
22. P. Foy, C. Junghlut, and G.E. Deleens, U.S. Patent 4,230,838 (October 28, 1980)
23. P.F. Van Hutten, E. Walch, A.H. Veeken, and R.J. Gaymans, *Polymer 31*, 524 (1990)
24. L.C. Case, *J. Polym. Sci. 29*, 469 (1958)
25. G. Deleens, P. Foy, and E. Maréchal, *Eur. Pol. J. 13*, 343 (1977)
26. G. Deleens, J. Ferlampin, and M. Gonnet, U.S. Patent 4,252,920 (February 24, 1981)
27. P. Foy, C. Jungblut, and G. Deleens, U.S. Patent 4,332,920 (June 1, 1982)

10A.1   Introduction

The discovery of Surlyn® ionomer resins in 1961 is one of the continuing series of technical advances that has characterized DuPont's involvement with ethylene polymers for well over 40 years. Following the discovery of free-radical polyethylene by ICI in England during the 1930s, both DuPont and Union Carbide were licensed to produce this polymer during World War II. Small-scale equipment for high-pressure experimentation became available and a widely ranging exploratory program on the copolymerization of ethylene with other monomers was carried out by members of DuPont's Central Research Department. It was established that potentially valuable copolymers could be obtained from such comonomers as vinyl acetate [1], vinyl chloride [2], vinylidene chloride [3], vinylidene fluoride [4], carbon monoxide [5], sulfur dioxide [6], and many others [7]. It was recognized that most of these comonomers had the effect of increasing the elasticity of the ethylene polymers.

In the early 1960s, research on specialized polymers, based mainly on ethylene, was initiated. Ethylene–vinyl acetate polymers were found to have excellent potential as wax modifiers and ingredients in hot-melt adhesives [8]. J.B. Armitage found that high-quality copolymers of ethylene with methacrylic acid could be obtained by careful control of polymerization conditions [9]. These polymers exhibited excellent adhesion to aluminum foil. Copolymerization work was also in progress with carbon monoxide, sulfur dioxide, acrylamide, and acrylic esters. Emphasis was placed on the use of comonomers to enhance adhesion, compatibility with other polymers, and controlled crosslinking.

10A.2   Ionomer Discovery

Beginning in late 1960, the author's research activities included a search for new chemical methods to crosslink ethylene polymers. Electron-beam radiation and peroxide treatment were already well known, but there were economic and safety problems, so polar functional groups were considered. First experiments were with ethylene–vinyl acetate and a few promising reactions were found, but the crosslinking reagents were difficult to handle under commercial conditions. Attention was then focussed on ethylene–methacrylic acid, which had been prepared on a commercial scale in a well-stirred reactor under the conditions specified in the Armitage patent [9].

The use of epichlorohydrin as a difunctional crosslinking reagent was first tried. If successful, the crosslink would be flexible, and would contain a hydroxyl group. As a preliminary step, the ethylene–methacrylic acid copolymer was converted to its sodium salt by reaction with a stoichiometric quantity of sodium methoxide. Addition of methanolic sodium methoxide to a stirred solution of the polymer in xylene at about 90 °C resulted in immediate gelation. The polymeric product was recovered from the gel by macerating it with a large excess of acetone in a Waring blender. The filtered precipitate was vacuum dried and molded for physical testing. The molded sample was clear, in contrast to the translucent starting material. Hand flexing indicated that the sodium salt was not only stiffer than the precursor, but also far more resilient, much more like a cured elastomer than a limp polyolefin. Infrared spectroscopy gave a scan consistent with conversion of the free acid to the salt form.

The melt index of the sodium salt was very low (below 0.1 g/10 min) [10] but a smooth, clear extrudate was obtained, and there was no indication of the fractured appearance typical of covalently crosslinked polyethylene. There was much speculation with respect to the type of crosslinking, and decarboxylation of the sodium salt was suggested. This question was resolved when a sample was heated in xylene. It swelled, but did not dissolve. When hydrochloric acid was added, the polymer dissolved. After isolation and drying, its infrared spectrum and melt index were the same as those of the starting material.

10A.3   Development of Ionomer Technology and Applications

Since the ionically crosslinked polymer was so different from conventional polyethylene, it was further studied to establish its properties and possible end-uses.

Three copolymers, containing 5%, 10%, and 18% by weight of methacrylic acid, had been produced on a commercial scale for end-use development, and thus there was no shortage of material for our laboratory work. Experiments on a heated roll mill showed that neutralization could be effected very rapidly in the melt, with no necessity for workup or drying. The acid copolymers were banded on the mill at 125 °C and the metallic reagent added as an oxide, hydroxide, or methylate, either dry or as a concentrated solution. As neutralization proceeded, the melt usually became so elastic that characteristically loud snapping noises were emitted. Homogeneity was assured by normal cutting and folding procedures.

The effects of varying the degree of neutralization on physical properties were investigated by the author and C.A. Carrere. The stress–strain curves of compression-molded samples were quite different from that of branched polyethylene, with stress increasing past the yield point (Fig. 10A.1). They combined features of elastomers with those of semicrystalline plastics. Stiffness increased with neutralization to a plateau at about 40% (Fig. 10A.2). However, tensile strength continued to increase at higher levels of neutralization (Fig. 10A.3).

Ionic crosslinking was also extended to other cations at this time. A value of about 77% neutralization was selected for most cations, since the tensile strength usually reached a plateau at this point. Soluble hydroxides of Group 1 were used successfully. Within Group II, magnesium and strontium hydroxides were effective reagents, but zinc presented a problem initially. When zinc oxide was used in a milling experiment, some neutralization occurred smoothly, but unreacted zinc oxide was still present after extended periods of milling, as indicated by a white, opaque melt. D.L. Funck suggested the use of acetates as reagents, the acetic acid being removed by volatilization, and this approach was immediately successful. Acetates of zinc, lead, copper, barium, cobalt, and nickel all gave clear melts and quantitative ionic crosslinking was achieved. Following up on this approach, in the case of zinc it was soon found that addition of a few drops of acetic acid to an opaque melt of acid copolymer–zinc oxide on the mill immediately resulted in a clear, elastic product [11].

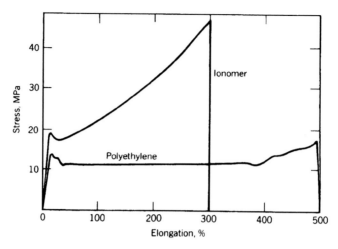

Figure 10A.1   Stress–strain curves of ionomer and conventional polyethylene, density 0.920. Test speed is 5 cm/min

That all these cations gave similar solid-state and melt properties was unexpected. Most of us anticipated that the divalent cations would bind the chains together to give intractable products, but this was not the case, and a similar pattern of melt flow vs. neutralization was found (Fig. 10A.4).

Early experiments on water absorption were very simple in nature, involving boiling molded slabs for 1 h. As shown in Table 10A.1, the nature of the cation had a large effect.

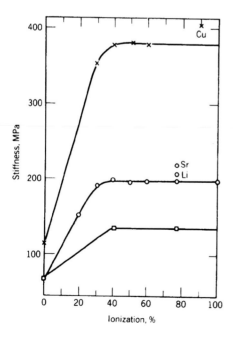

Figure 10A.2   Stiffness (Tinius Olsen) vs. degree of ionization (monobasic acid). The cation is sodium unless otherwise indicated. ■ 1.7 mol% COOH; ● 3.5 mol% COOH; ✕ 5.9 mol% COOH

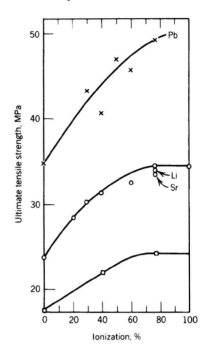

Figure 10A.3   Ultimate tensile strength vs. degree of ionization (monobasic acid). The cation is sodium unless otherwise
indicated. □ 1.7 mol% COOH; ○ 3.5 mol% COOH; ✗
5.9 mol% COOH

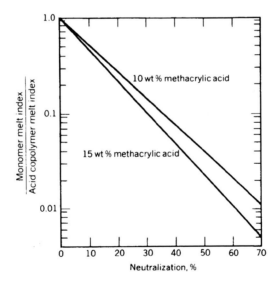

Figure 10A.4   Effect of percent neutralization on melt index of ionomers

262 R.W. Rees [Refs. on p. 270]

Table 10A.1 Effects of Various Cations on Water Adsorption

Metal ion	Weight gain (%) after 1 h boil
None	0.13
Sodium	2.25
Potassium	2.35
Lithium	0.5
Magnesium	1.5
Zinc	0.22
Strontium	0.16
Lead	0.13

Although the full implication of water absorption on processing conditions, packaging requirements, and performance would require much more detailed work, it was tentatively decided that products based on sodium and zinc, representing a large difference in water absorption, would be emphasized in future work. Hindsight reveals that this was a sound decision, since both sodium and zinc ionomers have been in the Surlyn® product line continuously since 1965.

Although small-scale tests in the melt indexer showed clearly that the ionically cross-linked copolymers were thermoplastic, their behavior in conventional plastics industry process equipment remained questionable. The roll-mill was used to prepare a few kilos of sodium salt, and this was injection molded, using a simple 1-oz plunger machine that injected into V-block molds, capable of being changed between shots. It was, therefore, possible to mold a variety of different articles, including test specimens, small gears, clothespins, and chain links (Example X, Ref. [11]). A few problems were identified, including a tendency toward mold sticking and excessively high melt viscosity. Also, the degree of clarity inherent in the polymer was not obtained in some moldings. It was observed that as the melt cooled it became hazy at about 70 °C, and then cleared again as cooling continued. In thick sections, this haze could be "frozen-in"; thus, when optimum clarity was desired, it was preferable to remove moldings at about 60 °C and complete cooling by immersion in an ice-water bath.

A 1-in. diameter single-screw extruder was used to prepare the first sample of ionomer blown film, with the melt temperature set at 225 °C to keep the melt viscosity within reasonable limits. A stable bubble was obtained and a quantity of very thin, 12 μm (0.5 mil) film was collected. The impact resistance was found to be exceptionally high, comparable to that of biaxially oriented polyethylene terephthalate (Example VIII, Ref. [11]). In addition, heat-sealing trials gave strong seals over a wide range of conditions. Small packages were made and subjected to drop tests during the following months. Damage resistance was observed to be excellent.

Larger samples were subsequently converted into film by a variety of methods. Optical quality was well short of commercial standards, but the material was adequate to demonstrate that it would perform extremely well in skin packaging. The high melt strength of the polymer allowed it to be drawn down tightly around sharp objects without puncturing. Some spectacular exhibits were made by skin-packaging fish hooks on paper-board backing, the film being drawn down so snugly around the hooks that it was invisible. Thus a key end-use for the Surlyn® product was recognized within the first year.

0.6 mm (15 mil) sheets were extruded through a flat die and given a smooth surface by passage between the chrome-plated rolls of a "three-roll finisher." Frozen-food trays were molded, using a conventional vacuum forming device. It was observed that drawability was excellent, because of the high melt strength, and satisfactory parts were obtainable under a wide range of conditions.

In addition to the film and sheet areas, attention was directed to molded articles that could use the somewhat elastomeric characteristics of the new polymers. Work was in progress on a high-performance plastic golf ball and it appeared that the impact toughness of the ionic copolymers might qualify them for this demanding application. Formulation work by the author, T.G. Smith, and Mario Pagano gave products with the desired density and resilience. This early work, in 1961–2, resulted in a one-piece ball which performed well on the golf course, but the "click" was much too loud and clearly unacceptable. Emphasis was then placed on softer compositions, containing other monomers to lower crystallinity. Later in the 1960s, the use of Surlyn® ionomers as golf ball covers was developed successfully in cooperation with Ram Corp.

In view of the clarity and toughness of the ionically linked copolymers, they were evaluated as safety-glass interlayers, replacing poly(vinyl butyral). Early testing showed very good break-height performance relative to the commercial butyral of 1961, but adhesion was below acceptable limits and internal haze did not meet the stringent standards set by automotive customers. Further work showed that good break-height and adhesion could be combined in a single product [12], but the problem of achieving satisfactory optical characteristics would require a lengthy development program.

Another potential automotive market application was in tire chains. In dry road tests, ionic copolymer chains outlasted their steel equivalents, and ran much quieter. Winter-road tests revealed shortcomings in the chain-link design, so this development program was useful only in terms of demonstrating ability to withstand unusual punishment.

Further efforts were directed to scaleup. V.C. Wolff extended the roll-mill techniques to large-scale equipment capable of producing 20 kg (44 lb) in a single run. A.A. McLeod developed a diffusion method that worked well for sodium salts. Pellets of the ethylene–methacrylic acid copolymers were submerged in aqueous or methanolic solutions of sodium hydroxide at 50 °C to 100 °C, and then melt homogenized before use. An experiment by the author on neutralization in an extruder indicated that the ionic copolymer could be obtained by this route, but efficient means were needed to remove volatiles from the melt. Further work by V.C. Wolff and R.L. Saxton on the extrusion–reaction concept resulted in a process that could be used to prepare a variety of ionic copolymer types in quantities up to 50 kg (110 lb) (Examples 27 to 34 of Ref. [13]).

Since direct synthesis is normally preferable to post-synthesis reactions from the economic standpoint, considerable effort in our studies was devoted to ethylene copolymer-izations with salts of unsaturated acids. To obtain a partially neutralized copolymer, most experiments involved ethylene, methacrylic acid, and sodium methacrylate. High-pressure, free-radical polymerizations were carried out with vigorous agitation under conditions specified by the author and V.C. Wolff. High molecular weight polymers were obtained, but their properties were quite unlike those of the post-synthesis neutralized materials, and

Table 10A.4 Ionomers Derived from Terpolymer Precursors

| Composition by Weight | | | | | | | |
Ethylene	Vinyl acetate	Methacrylic acid	Cation	Melt index (g/10 min)	Modulus (MPa)	Tensile strength (MPa)	Wt. elongation (%)
70	20	10	—	9.0	13.8	10.3	530
70	20	10	Na^+	1.2	56	43.4	410
70	20	10	Mg^{2+}	0.04	37.44	45	280
65	25	10	—	12.5	12.9	9.6	610
65	25	10	Na^+	1.7	18.5	20.9	410
65	25	10	Mg^{2+}	0.007	37.4	37.1	300

To convert MPa to psi, multiply by 145

Many of the terpolymer ionomers combined melt processability, low modulus, and high tensile strength, so they were recognized as promising thermoplastic elastomers. Further research by I.C. Kogon, K.F. King, and others confirmed the high strength and toughness but also showed that resistance to compression set was poor, even at comparatively low temperatures. The ionic mobility resulted in reforming of the ionic crosslinks in the compressed samples, so that when the load was released, very little recovery occurred.

A typical experiment by I.C. Kogon consisted of milling a stoichiometric amount of magnesium oxide into a 65 : 25 : 10 ethylene–vinyl acetate–methacrylic acid terpolymer on a cold (25 °C) rubber mill, followed by compression molding at 140 °C. An attractive, strong slab was obtained, having a Yerzley resilience value of 67%, but compression set exceeded 100% at 70 °C. Addition of SRF carbon black effected only a slight decrease in compression set. A white film composition, neutralized with sodium ions, and containing titanium oxide and talc, was calendered into attractive film having excellent resistance to dirt pickup and ozone. Development effort was focussed on end-uses requiring the toughness and flexibility of cured elastomers without the necessity for good compression set resistance. A strong tendency to creep under load has been found to be characteristic of several processable ionomeric ethylene copolymer elastomers and has also been reported in ionomeric butadiene copolymers [16].

In a study by K.F. King of chlorinated ionic copolymers, interesting solubility behavior was found. The chlorinated ethylene–methacrylic acid copolymers, containing about 30% chlorine, could be dissolved in xylene and converted to the potassium salt by adding methanolic potassium hydroxide, without phase separation. However, once isolated as films by evaporation of the solvent, the ionic polymers could not be redissolved. This behavior was potentially advantageous for use in paints and other surface coatings.

The concept of using ionic crosslinking to enhance compatibility was explored by the author during 1961–2. Acid-containing polymers of methyl methacrylate, styrene, acrylonitrile, and other commercially important monomers were prepared by free-radical copolymerization. When these were blended with carboxylated ethylene polymers, tensile properties were generally poor because of gross incompatibility of the polymeric phases. Introduction of metal ions, by reaction with zinc acetate, for example, increased strength by a factor of 2 to 5 [17]. In the case of the poly(methyl methacrylate–ethylene) copolymer blends, partial neutralization yielded a clear, visually homogeneous material. This was not true in the case of poly(styrene–ethylene) copolymer blends, but worthwhile improvements in strength were observed.

Extension of ionic crosslinking to a broad spectrum of polymer types was also investigated. Data on poly(styrene–methacrylic acid) and analogous copolymers were reported in U.S. Patent 3,322,734. The elongation at break of an acidic styrene copolymer increased from 6% to 10% as a result of partial neutralization with sodium as the counterion. Likewise, in the case of methyl methacrylate copolymers, neutralization increased elongation from 3.5% to 6.0%. Other mechanical properties were only slightly changed. In further work on acrylic systems by K.L. Howe, it was concluded that the effects of neutralization on physical properties was less dramatic in rigid, glassy polymers than in flexible, semicrystalline, or elastomeric materials.

10A.5   Diamine Ionomers

A second family of ionically crosslinked polyolefins was under investigation concurrently with the Surlyn® work. These interesting, reversibly crosslinked polymers were diamine salts of ethylene–methacrylic acid [18]. To prepare them, a diamine was added to a well-stirred xylene solution of the polymers at about 130 °C. No gelation or viscosity increase occurred, and the product was recovered by methanol precipitation. On molding into a slab, the polymer was found to be clear and resilient, much like the corresponding sodium salt. The stiffness and yield point had increased substantially, while ultimate tensile strength and melt viscosity were unchanged. These effects are summarized in Table 10A.5.

It was soon found that the diamine salts could be prepared by injecting an amine, such as hexamethylenediamine, into the melt of the acid copolymer precursor, while it was being

Table 10A.5 Properties of Amine Ionomers Derived from Ethylene–Methacrylic Acid

Percent acid by wt.	Diamine	Weight % diamine added	Melt index	Modulus (Mpa)	Ult. tensile strength (MPa)	Ult. elongation
10	—	—	5.8	69	23.5	550
10	Hexamethylenediamine	10	5.8	262	23	390
10	Hexamethylenediamine	15	5.8	289	24.2	380
10	Decamethylenediamine	10	5.0	216	22.9	340
10	bis(p-aminocyclohexyl) methane	10	4.7	291	23.9	380
18	—	—	6.3	110	34.5	600
18	Hexamethylenediamine	10	6.3	345	33.2	480
18	Hexamethylenediamine	18	6.3	448	32	480
18	Diethylenetriamine	18	1.7	287	31	390

To convert MPa to psi, multiply by 145

pumped through the mixing section of a single-screw extruder [19]. Thus, substantial quantities of the diamine ionomer could be made, provided that the maximum temperature was kept below 190 °C. As freshly prepared, the polymer was transparent and water-white. However, after extended storage, a brown coloration developed because of oxidation of the diamine. Demonstration of good processability by conventional melt techniques was rapidly accomplished. The problem of high melt viscosity encountered with the metal cations was absent. On the other hand, the desirable melt strength of the metal ionomers was also lacking in the amine salts. Weathering studies showed that the polymers containing hexamethylene-diamine were much better than conventional ethylene polymers in resisting physical degradation by UV radiation, probably because the amine acted as a sacrificial antioxidant. These findings led to a development program aimed at flexible glazing for automotive and recreational uses. Some progress was made in identifying stabilizers for the system, but color development proved to be a persistent problem.

The relationship between diamine chain length, acid content, and physical properties was explored, and it was concluded that short diamines were effective only at high acid levels, while longer molecules such as decamethylenediamine could be used in all acid-containing polymers. This suggested that intramolecular bonding was important. In the area of amine strength, it was concluded that dissociation constants above 10^{-8} were needed. Diamines containing ring structures, such as bis(p-aminocyclohexyl) methane, gave excellent proper-ties. Heating the amine crosslinked polymers under vacuum split out water, resulting in covalent, nonreversible crosslinks.

Complex ionomers containing metal cations in addition to diamines were prepared, and evaluated as film resins. It was observed that clarity was excellent, but no other outstanding properties were identified, so these polymers were not included in the DuPont commercial offerings. Many years later, compositions obtained by adding diamines to commercial sodium ionomers were patented by Advanced Glass Systems, Inc. for use in safety glass interlayers [20]. Laminates are now produced commercially for uses requiring outstanding resistance to penetration. E. Hirasawa and co-workers at DuPont-Mitsui Polychemical have published extensively on ionomers containing zinc ions in combination with diamines [21–23].

10A.6   Structural Studies on Ionomers

Following the early exploration of ionomer properties, interest developed during 1962–3 in obtaining a better understanding of their structures. The author and R.P. Schatz conducted light microscopy experiments that showed that the crystallites seen in a 10% methacrylic acid copolymer disappeared completely after neutralization. Transmission electron micrographs, obtained by H.A. Davis, confirmed the remarkable morphological change that accompanied neutralization. X-ray diffraction studies by F.C. Wilson uncovered many interesting phenomena. He found an "ionomer peak" corresponding to a spacing of about 25 Å. Surprisingly, the actual level of crystallinity was not greatly affected by neutralization, while the visible crystallites disappeared. These results, taken in combination with physical property data, suggested that ionic clustering was a key feature of ionomer fine structure. The first brief publication on ionomers by the author introduced the concept of clustering

[14]. Other early studies on the physical chemistry of ionomers included infrared spectroscopy by the author, Differential Thermal Analysis (DTA) melting point measurements by D.L. Brebner, and torsion pendulum studies by E.T. Pieski.

Following publication of basic information on ionomer structure in 1965 [24], interest within DuPont was sufficiently high that a more detailed study was undertaken by R. Longworth, working in collaboration with F.C. Wilson, H.A. Davis, and others. In a symposium during the 1968 San Francisco ACS meeting, the first of a lengthy series of debates on ionomer structure took place [25].

10A.7   Ionomers in Blends and Alloys

The unusual low temperature toughness and resilience of the Surlyn® ionomers has spurred interest in their potential as toughening agents and polymeric modifiers. An extensive study of tougheners for nylon and polyester thermoplastics by B.N. Epstein and co-workers during the 1970s showed that certain ionomer types were highly effective in producing high-impact nylon resins [26].

More recently, R.P. Saltman and co-workers have developed new types of high-performance, elastomeric thermoplastics. These are partially grafted materials, obtained by blending (1) high melting thermoplastics such as nylon 66, (2) flexible plastics or rubbers, and (3) a polymeric grafting agent. The extent of grafting is critical in optimizing products in which ease of processing is combined with outstanding heat aging stability. Two U.S. patents describe partially grafted thermoplastic compositions in which the flexible components are ionomeric [27, 28]. Excellent results have been reported with nylon, polyester, or polypropylene as the hard components in these interesting alloys.

10A.8   Product Development

During the period of 1962–4, product development work by V.C. Long, D. DeVoe, R.L. Saxton, B. Borgerson, and others resulted in a Surlyn® product line of sodium and zinc ionomers that formed the basis of a successful commercial venture. The early problem of excessively high melt viscosity was solved by lowering the molecular weight of the free acid precursors. It was found that a wide variation in the degree of neutralization was needed for diverse applications, so inevitably a rather extensive line of ionomer products evolved.

Acknowledgments

Although the development of Surlyn® and related products was never the subject of a vast industrial task force, important contributions were made by many individuals in addition to those mentioned above. Special recognition is due to D.J. Vaughan who was the Research

Supervisor directly involved with ionomer activities during the first critical months, and R.H. Kinsey who was closely involved, both in Research and Marketing, for over 20 years and became DuPont's leading expert on Surlyn® products.

References

1. M.J. Roedel (to E. I. du Pont de Nemours & Co.), U.S. Patent 2,377,753 (June 5, 1945)
2. M.M. Brubaker, J.R. Roland, and M.D. Peterson (to E. E. de Pont de Nemours & Co.), U.S. Patent 2,497,291 (February 14, 1950)
3. W.E. Hanford and J.R. Roland (to E. I. du Pont de Nemours & Co.), U.S. Patent 2,397,260 (March 26, 1946)
4. T.A. Ford (to E. I. du Pont de Nemours & Co.), U.S. Patent 2,468,954 (April 26, 1949)
5. D.D. Coffman, P.S. Pinkney, F.T. Wall, W.H. Wood, and H.S. Young, *J. Am. Chem. Soc.* 74, 3391 (1952)
6. M.M. Brubaker and J. Harman (to E. I. du Pont de Nemours & Co.), U.S. Patent 2,241,900 (April 26, 1938)
7. E.T. Pieski, In *Polythene*, A. Renfrew and P. Morgan (Eds.) (1960) Wiley-Interscience, New York
8. A. Oken (to E. I. du Pont de Nemours & Co.), U.S. Patent 3,189,573 (June 15, 1965)
9. J.B. Armitage (to E. I. du Pont de Nemours & Co.), U.S. Patent 4,351,931 (September 28, 1982)
10. ASTM D1238-79 condition E
11. R.W. Rees (to E. I. du Pont de Nemours & Co.), U.S. Patent 3,264,272 (August 2, 1966)
12. R.W. Rees (to E. I. du Pont de Nemours & Co.), U.S. Patent 3,344,014 (September 26, 1967)
13. R.W. Rees (to E. I. du Pont de Nemours & Co.), U.S. Patent 3,404,134 (October 1, 1968)
14. R.W. Rees, *Mod. Plastics 42*, 209 (1964)
15. R.M. Busche and D.L. Funck (to E. I. du Pont de Nemours & Co.), U.S. Patent 3,272,771 (September 13, 1966)
16. A.V. Tobolsky, P.F. Lyons, and N. Hata, *Macromolecules 1*, 515 (1968)
17. R.W. Rees (to E. I. du Pont de Nemours & Co.), U.S. Patent 3,437,718 (April 8, 1969)
18. R.W. Rees, *Polym. Prepr. Am. Chem. Soc. Div. Polym. Chem. 14*, 796 (1973)
19. R.W. Rees (to E. I. du Pont de Nemours & Co.), U.S. Patent 3,471,460 (October 7, 1969)
20. W.N. Smith (to Advanced Glass Systems), U.S. Patent 4,732,944 (March 22, 1988)
21. E. Hirasawa, Y. Yamamoto, K. Tadano, and S. Yano, *Macromolecules 22*, 2776 (1989)
22. S. Yano, H. Yamamoto, K. Tadano, Y. Yamamoto, and E. Hirasawa, *Polymer 28*, 1965 (1987)
23. K. Tadano, E. Hirasawa, H. Yamamoto, and S. Yano, *Macromolecules 22*, 226 (1989)
24. R.W. Rees and D.J. Vaughan, *Polym. Prepr. Am. Chem. Soc. Div. Polym. Chem. 6*, 287 (1965)
25. *Polym. Prepr. Am. Chem. Soc. Div. Polym. Chem. 9*, 515, 583 (1968)
26. B.N. Epstein (to DuPont), U.S. Patent 4,174,358 (November 13, 1979)
27. R.T. Saltman (to E. I. Dupont de Nemours and Co.), U.S. Patent 4,871,810 (October 3, 1989)
28. R.T. Saltman (to E. I. Dupont de Nemours and Co.), U.S. Patent 5,091,478 (February 25, 1992)

10B Research on Ionomeric Systems

W.J. MacKnight and R.D. Lundberg

10B.1 Introduction. 272

10B.2 Theory. 272

10B.3 Morphological Experiments 274
    10B.3.1 Scattering Studies. 274
        10B.3.1.1 X-Ray Scattering. 274
        10B.3.1.2 Neutron Scattering. 277
    10B.3.2 Electron Microscopy 278
    10B.3.3 Summary of Morphological Information 279

10B.4 Recent Developments: Synthesis 280
    10B.4.1 Halato-Telechelic Ionomers 280
        10B.4.1.1 Telechelic Polyisobutylene Sulfonate Ionomers 281
        10B.4.1.2 Halato-Telechelic Carboxylate Ionomers 281
    10B.4.2 Sulfonated Polypentenamers 282
    10B.4.3 Copolymerization of Sulfonate Monomers 282
    10B.4.4 Block Copolymer Ionomers. 284
    10B.4.5 Polyurethane Ionomers. 284

10B.5 Recent Developments: Properties 285
    10B.5.1 The Glass Transition Temperature T_g 285
    10B.5.2 Mechanical Properties 286

10B.6 Preferential Plasticization. 288

10B.7 Ionic Interactions in Polymer Blends. 288

10B.8 Applications of Ionomeric Elastomers 291
    10B.8.1 Thermoplastic Elastomers 291
    10B.8.2 Adhesives 292
    10B.8.3 Miscellaneous Applications 292

10B.9 Conclusions. 292

10B.10 Future Developments 293

References. 293

10B.1   Introduction

In 1946 McAlevy [1] discovered a family of elastomers based on the chlorosulfonation and chlorination of polyethylene and with a substantial degree of ionic crosslinking. These elastomers were introduced commercially in the early 1950s by DuPont. The materials, suitably cured with various metal oxides, gave rise to ionic, or a combination of ionic and covalent crosslinks, depending on the system used, and were commercially available under the trade name Hypalon®.

Brown, in 1954, presented an article [2] on carboxylic elastomers such as poly(butadiene–acrylonitrile–methacrylic acid) vulcanized by metal salts. A terpolymer of this type containing about 0.01 equivalent of carboxyl per 100 rubber, vulcanized with 0.2 equivalents of zinc oxide per 100 rubber, showed a tensile strength of 62 MPa (9,000 psi) and elongation at break of 550%. There were some adverse properties of these metal oxide vulcanizates related to high compression set and rapid stress relaxation. The action of metal salts in crosslinking butadiene–methacrylic acid copolymers was discussed also by Brown [3, 4] in 1957 and 1963.

In Chapter 10A R.W. Rees of DuPont describes the early research in the 1960s leading to Surlyn. It is also notable that an article by Rees in 1964 [5] contains the designation of these polymers as "ionomers," a term coined by DuPont, and also importantly contains the first reference to clustering in ionomers. Two articles presented by Rees and Vaughan [6] in 1965 discussed the effect of ionic bonding on polymer structure and physical properties.

More recently, new families of ionic elastomers have emerged that possess a wide variety of properties leading to different applications. An overview of available ionic elastomers or flexible plastics is summarized in Table 10B.1.

We have previously reviewed [7, 8] the manufacture/synthesis, properties, and applications of a number of ionic elastomers, including many of those listed in Table 10B.1. In this chapter the material given in these earlier reviews will be updated and amplified and new polymers will be considered.

10B.2   Theory

The first successful theoretical attempt to deduce the spatial arrangement of salt groups in ionomers was that of Eisenberg [24a, 24b]. In that work it was assumed that the fundamental structural entity is the contact ion pair. On the basis of steric considerations, it was then shown that only a small number of ion pairs (the "multiplet") can associate without the presence of intervening hydrocarbon and that there is a tendency for multiplets to associate further into "clusters" that contain a considerable quantity of hydrocarbon material. This association is favored by electrostatic interactions between multiplets and opposed by forces arising from the elastic nature of the backbone chains. In the original formulation of the theory, Eisenberg assumed that the chains on average would undergo no dimensional changes as a result of the clustering phenomenon.

Table 10B.1 Commercial and Experimental Ionomers

Polymer system	Trade name	Manufacturer	Applications
Commercial			
Ethylene–methacrylic acid copolymer	Surlyn®	DuPont	Modified thermoplastic
Ethylene–acrylic acid copolymer	Iotek®	Exxon	Modified thermoplastic
Butadiene-acrylic acid copolymer	Hycar®	Goodrich	High green-strength elastomer
Chlorosulfonated polyethylene	Hypalon®	DuPont	Specialty elastomer, covalently crosslinked
Perfluorosulfonate ionomers	Nafion®	DuPont	Multiple membranes
Telechelic polybutadiene	Hycar®	Goodrich	Specialty elastomer uses
Sulfonated ethylene–propylene–diene terpolymer	—	Exxon	Drilling mud additive, elastomer coating applications
Sulfonated polystyrene	—	Exxon	Drilling mud additive
Experimental		*References*	*Comment*
Sulfonated butyl elastomer		[9]	High green-strength elastomer
Sulfonated polypentenamer		[10]	Model ionomer system
Telechelic polyisobutylene sulfonated ionomers		[11]	Model ionomer system
Alkyl methacrylate–sulfonate copolymers		[12]	High green-strength elastomer
Isoprene–butadiene sulfonate copolymers		[13]	Pressure-sensitive adhesives
Acid-amine ionomers		[14–16]	Polymer blends
Metal sulfonate ionomers/aminated complexes		[17, 18]	Polymer blends
Metal carboxylate ionomers/aminated complexes		[19]	Polymer blends
Zwitterionic polysiloxane		[20]	Polymer blends
Sulfonated block copolymer ionomers		[21]	Improved thermoplastic elastomer
Carboxylate block copolymer ionomers		[22]	Model block ionomers
Polyurethane ionomers		[23]	Model ionomer

Forsman [25] later removed this restriction and showed that the chain dimensions must actually increase as a result of association, a result confirmed by experiment [26]. There can be little doubt that the properties of ionomers can be interpreted on the basis of the existence of multiplets and clusters even though the precise structures of these units may remain obscure. Naturally, the ratio of the concentration of salt groups present as multiplets to those present as clusters differs with different backbones, acid types, and neutralizing species.

More recently Eisenberg, Hird and Moore [24b] have proposed a new morphological model for ionomers, here referred to as the EHM model. The model is based on the existence of multiplets, which reduce the mobility of the polymer chains in their vicinity. The thickness of the restricted mobility layer surrounding each multiplet is postulated to be of the order of the persistence length [26] of the polymer. Such isolated multiplets act as large crosslinks, thus increasing the glass transition temperature of the material. As the ion content is increased, the regions of restricted mobility surrounding each multiplet overlap to form larger contiguous regions of restricted mobility. When these regions become sufficiently large, they exhibit phase-separated behavior and are termed clusters. In contrast to previous theories, this model does not propose a condensation of multiplets to form a cluster, but they are close together by virtue of the proximity of the ion pairs in the polymer chains. Eisenberg et al. [24b] suggest their approach accounts for the "ionic" peak detected in X-ray scattering data and the separate T_g often seen in ionomers above a critical ion concentration.

10B.3   Morphological Experiments

10B.3.1   Scattering Studies

10B.3.1.1   X-Ray Scattering

The X-ray scattering results have been of central importance in the interpretation of the structure of ionomers. Figure 10B.1 compares the X-ray scattering observed for low-density polyethylene, an ethylene–methacrylic acid copolymer, and its sodium salt over a range of Bragg angles from $2\theta \cong 2°$ to $2\theta = 40°$. The presence of polyethylenelike crystallinity in all three samples is readily apparent from the 110 and 200 peaks arising from the orthorhombic polyethylene unit cell. The acid copolymer and the ionomer exhibit less crystallinity than the parent polyethylene but are quite similar to each other. The ionomer contains a new feature, however, consisting of a peak centered at approximately $2\theta = 4°$. This peak, which will be referred to as the ionic peak, appears to be a common feature of all ionomers, regardless of the nature of the backbone and also of the presence or absence of backbone crystallinity. The ionic peak, in addition, possesses the following characteristics:

1. The ionic peak occurs in all ionomers regardless of the nature of the cation, being present with lithium as well as heavy metals, divalent or trivalent cations, quaternary ammonium ions, etc.
2. Both the magnitude and the location of the ionic peak depend on the nature of the cation. Thus at a given ionic concentration the peak occurs at lower angles for cesium cations than for lithium cations. In addition, the magnitude of the ionic peak is several thousandfold greater for cesium than for lithium.
3. The ionic peak is relatively insensitive to temperature. Thus it was found that, for the ionomer depicted in Fig. 10B.1, the ionic peak persisted to at least 300 °C.
4. The ionic peak is destroyed or moved to lower angles when the ionomer is saturated with water. The scattering profile in the vicinity of the ionic peak in the water-saturated ionomer is different from that of the parent acid copolymer.

It is well known that the interpretation of any scattering data is model dependent. The procedure is to assume a reasonable model, fit the experimentally observed data, and deduce model parameters from the best fit.

Several models for the distribution of salt groups in ionomers have been proposed based mainly on analysis of the ionic peak. Those appearing up to 1979 are summarized in [28]. They consist mainly of two approaches: (1) that the peak arises from structure within a scattering entity; and (2) that the peak arises from interparticle interference effects.

As a representative of the first approach may be cited the "shell-core" model [29] originally proposed in 1974 and later elaborated [30, 31]. In essence the "shell-core" model,

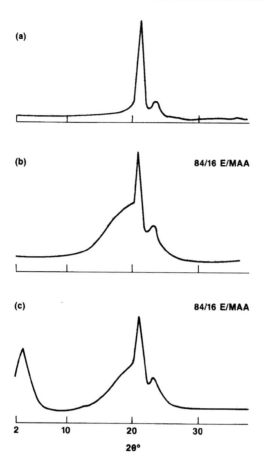

Figure 10B.1   X-Ray diffraction scans of (a) low density polyethylene, (b) ethylene-methacrylic acid copolymer (5.8 mol% acid), (c) the 100% neutralized sodium salt of (b). From R. Longworth and D.J. Vaughan, *Nature, 218*, 85 (1968)

depicted in Fig. 10B.2, postulates that in the dry state a cluster of ~ 1 nm (10 Å) in radius is shielded from surrounding matrix ions not incorporated into clusters by a shell of hydrocarbon chains. The surrounding matrix ions which cannot approach the cluster more closely than the outside of the hydrocarbon shell will be attracted to the cluster by electrostatic forces. This mechanism establishes a preferred distance between the cluster and the matrix ions. This distance is assumed to be of the order of 2 nm (20 Å) and accounts for the spacing of the ionic peak.

Figure 10B.3 schematically illustrates the interparticle interference model, where it is assumed that the peak arises from a preferred interparticle distance and the "shell-core" model and its variants such as the "lamellar shell-core model."

Recently, Yarusso and Cooper [32] have proposed a new interpretation of the ionic peak which rests on the liquidlike scattering from hard spheres described originally by Fournet [33]. Yarusso and Cooper studied sulfonated polystyrene ionomers and concluded that the

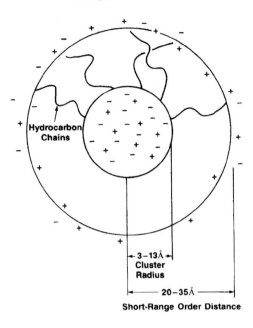

Figure 10B.2   "Shell-core" model for clusters. (Reprinted with permission from Ref. [29])

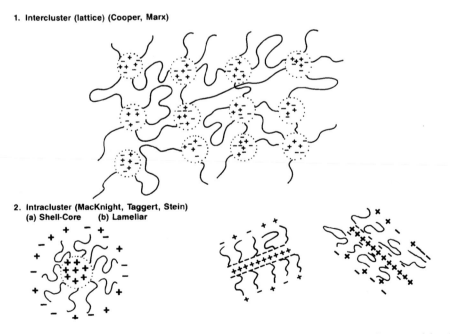

Figure 10B.3   Origin of SAX peak: (1) intercluster (lattice) (Cooper, Marx); (2) (a) shell core model and 2 (b) lamellar shell core model

Fournet model was quantitatively capable of modeling the ionic peak. They found that, in the zinc neutralized polystyrene sulfonates, about half of the ionic groups are aggregated in well-ordered domains ("clusters") with the remainder dispersed in the matrix ("multiplets"). The clusters are about 2.0 nm (20 Å) in diameter and clusters approach each other no more closely than 3.4 nm (34 Å) center to center. It is clear that this model, although based on quite different physical principles, yields structural parameters very similar to those obtained from the "shell-core" model and consistent with what may be termed the "standard model" for ionomer structure of multiplets and clusters.

The EHM model also envisages the ionic peak to arise from interparticle scattering [24b]. In this case interference effects between multiplets are responsible for the peak. The multiplets assume preferred distances from one another because of the regions of restricted mobility that are postulated to exist around them. Once again, the structural parameters derived from the EHM model are similar to those derived from either the Yarusso–Cooper model or the shell-core model.

10B.3.1.2 Neutron Scattering

Small-angle neutron scattering (SANS) has assumed great importance in the investigation of polymer morphology. One of its most impressive accomplishments is the measurement of single chain dimensions in bulk. This is generally achieved by selectively labeling a small fraction of the polymer chains by replacing hydrogen with deuterium, to take advantage of the much higher coherent neutron scattering cross-section of the deuteron compared to the proton.

Several SANS studies of ionomers have appeared on both deuterium-labeled and unlabeled systems [10, 17–20]. The earlier work [34] showed that an ionic peak, similar to that observed by X-rays and discussed above, could be discerned in some cases, especially when the sample was "decorated" by the incorporation of D_2O. It was also tentatively concluded [35] that the radius of gyration (R_g) of the individual chains is not altered when the acid is converted to the salt in the case of polystyrene–methacrylic acid copolymers. Subsequent SANS experiments were performed on sulfonated polystyrene ionomers over the range of 0% to 8.5% sulfonation [26]. The samples were prepared by first mixing small amounts (up to 3%) of anionically polymerized (narrow molecular weight distribution) perdeuteropolystyrene with polystyrene. The sulfonation reaction was accomplished subsequently by treating the blend with concentrated sulfuric acid and acetic anhydride in dichloroethane. This procedure ensured that the deuterated and protonated polymers would contain exactly the same concentration and distribution of sulfonate groups, a result difficult or impossible to achieve by copolymerization. The neutron scattering data were treated by conventional means to obtain R_gs and apparent molecular weights. These results are collected in Table 10B.2. It is apparent from Table 10B.2 that aggregation of ionic groups is accompanied by considerable chain expansion. As already noted, this is consistent with the theory of Forsman [25].

The molecular weights measured by neutron scattering listed in Table 10B.2 are generally too high, even allowing for the fact that they are weight average rather than number average values. This is unlikely to be the result of segregation of deuterium-tagged chains, since there are no trends in the molecular weight data for any of the derivatives and it has been well established that poly(perdeuterostyrene) is molecularly dispersed in poly-

Table 10B.2 Radius of Gyration, R_g, and Molecular Weight of Polystyrene Sodium Sulfonate Ionomers Measured by SANS[a]

Mol% sodium sulfonate	R_g (nM)	$\overline{M}_w \times 10^4$
0	8.6 ± 0.2	1.5
1.9	10.5 ± 0.5	2.3
4.2	11.1 ± 0.5	2.3
8.5	12.3 ± 0.9	2.7

[a] For the starting polystyrene, $M_n = 90,000$ with $M_w/M_n = 1.05$
For polyperdeuterostyrene $M_n = 100,000$ with $M_w/M_n = 1.06$. (From Ref. [26])

styrene of the same molecular weight. It is possible that inaccuracies in the molecular weights are caused by errors in the reference measurements or sample concentration, together with difficulties in calculating the contrast factor K for the ionomer samples. In any case, the determination of R_g is independent of the absolute intensity calibration required in obtaining the molecular weight.

In a separate investigation, a series of polypentenamer sulfonate ionomers was studied [36]. In this case, contrast was achieved by adding measured amounts of D_2O to the samples. Figure 10B.4 shows the results for the 17 mol% polypentenamer cesium sulfonate. For the dry film there is no evidence of a scattering maximum. However, as small amounts of D_2O are added the SANS peak becomes detectable. The Bragg spacing of the small-angle X-ray ionic peak observed for the dry 17% cesium derivative is essentially the same as the SANS peak at low D_2O concentrations. Above a D_2O/SO_3 ratio of about 6, the SANS ionic peak moves markedly to lower angles. The results are consistent with a phase-separated model where absorbed water is incorporated into the ionic clusters, remaining separated from the matrix even at saturation.

Recently, Register and Cooper [27] have reexamined the dependence of R_g on salt group content in carboxylated telechelic ionomers. They find that R_g does not depend on salt group content whether or not the salt groups are present as clusters or multiplets. If this result is substantiated, it would appear that the Forsman theory will have to be revised. It should be noted that if the EHM model is correct, chain dimensions should increase with increasing salt group content.

10B.3.2 Electron Microscopy

Reference [28] reviews a number of electron microscopy studies of ionomer morphology in the period up to 1979. None of these studies made a convincing case for the direct imaging of ionic clusters. This is because of the small size of the clusters (<5 nm (50 Å) based on scattering data) and difficulties encountered in sample preparation. The entire problem was reexamined in 1980 [37]. In this study ionomers based on ethylene–methacrylic acid copolymers, sulfonated polypentenamer, sulfonated polystyrene and sulfonated poly(ethylene–propylene–diene monomer) (EPDM rubber) were examined. The transfer theory of imaging was used to interpret the results. Solvent casting was found to produce no useful

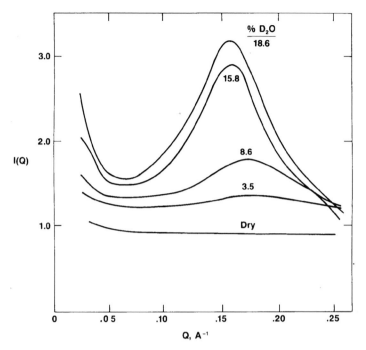

Figure 10B.4   Neutron scattered intensity vs. scattering vector, Q, for the 17% cesium ionomer of sulfonated polypentenamer. Numbers above each curve indicate weight percent D_2O. (Reprinted with permission from Ref. [36])

information about ionic clusters. Microtomed sections also showed no distinct domain structure, even in ionomers neutralized with cesium. However, microtomed sections of sulfonated EPDM appeared to contain 300 nm (3000 Å) phase-separated regions. Osmium tetroxide staining of these EPDM sections showed domains averaging less than 3 nm (30 Å) in size and primarily inside those regions. Unfortunately, the section thickness prevented an accurate determination of the size distribution or the detailed shape of these domains and hence the selection of the most appropriate model of domain structure.

10B.3.3   Summary of Morphological Information

There is a considerable body of experimental and theoretical evidence that salt groups in ionomers exist in two different environments, termed multiplets and clusters. The multiplets are considered to consist of small numbers of ion dipoles (perhaps up to 6 or 8) associated together to form higher multipoles–quadrupoles, hexapoles, octapoles, etc. These multiplets are dispersed in the hydrocarbon matrix and are not phase separated from it. Thus in addition to acting as ionic crosslinks, they affect such properties of the matrix as the glass transition temperature, water sensitivity, etc. The clusters are considered to be small (<5 nm or 50 Å) microphase separated regions rich in ion pairs but also containing considerable quantities of hydrocarbon. They possess at least some of the properties of a separate phase, including relaxation behavior associated with a glass transition temperature and have a minimal effect on the properties of the hydrocarbon matrix (they may have some reinforcing effect). The proportion of salt groups that resides in either of the two environments in a particular ionomer is determined by the nature of the backbone, the total concentration of salt groups, and their chemical nature. The details of the local structure of the clusters is not known and neither is the mechanism by which the clusters interact with low molecular weight polar impurities such as water. Up until the present time, attempts to image the clusters directly by electron microscopy have been unsuccessful.

The EHM model is also successful in interpreting the morphological results.

10B.4   Recent Developments: Synthesis

Ionomers are typically prepared by copolymerization of a functionalized monomer with an olefinic unsaturated monomer or by direct functionalization of a preformed polymer. Typically, carboxyl-containing ionomers are obtained by direct copolymerization of acrylic or methacrylic acid with ethylene, styrene, and similar comonomers by free-radical copolymerization. The resulting copolymer is generally available as the free acid, which can be neutralized to the degree desired with metal hydroxides, acetates, and similar salts.

The second route to ionomers involves modification of a preformed polymer. Sulfonation of EPDM, for example, permits the preparation of sulfonated EPDM with a content of sulfonic acid groups in proportion to the amount of sulfonating agent [9]. These reactions are conducted in solution, permitting the direct neutralization of the acid functionality to the desired level. The neutralized ionomer is isolated by conventional techniques, such as coagulation in a nonsolvent, solvent flashing, etc.

An alternate approach to modification of a preformed polymer involves a reaction conducted on a polymer melt, usually in an extruder [38]. The extruder sulfonation of EPDM has been described using the same sulfonating agents that are typically employed in solution. This continuous melt sulfonation was conducted on an oil-extended EPDM at temperatures of 90 °C to 100 °C. The rapid reaction with the unsaturation in EPDM led to conversions of 80% to 100% with residence times of about 6 to 12 min. Neutralization with metal stearates was conducted both in the extruder and on isolated EPDM–sulfonic acid product. Despite nonoptimal feed conditions, this extruder–reactor technique offers the advantages of shorter reaction times, elimination of solvent handling concerns, and simplication of polymer finishing steps. In addition, one patent has described the production of sulfo EPDM by alternate melt reaction schemes [39].

10B.4.1   Halato-Telechelic Ionomers

Previously [7], we described the synthesis of carboxylated elastomers and sulfonated ethylene–propylene terpolymers. Here we shall discuss telechelic polyisobutylene sulfonate ionomers, halato-telechelic carboxylate ionomers and sulfonated polypentenamer, and recent synthetic approaches to new ionomers.

10B.4.1.1 Telechelic Polyisobutylene Sulfonate Ionomers

Recent research on a new class of telechelic ionic polymers has been reported in a number of publications by Kennedy and Wilkes and their co-workers [40–46]. The synthesis procedure utilizes linear telechelic polyisobutylene diolefins and radial star triolefins. Sulfonation was carried out in hexane solution at room temperature with excess acetyl sulfate, generated in situ by the addition of sulfuric acid to aceatic anhydride. After the reaction, neutralization with ethanolic NaOH took place in tetrahydrofuran (THF) solution.

The quantitative nature of the terminal olefin sulfonation was determined by titration of the free acids and elemental sulfur analysis of the sodium ionomer. The success of these methods depends on there being no changes in polymer molecular weight upon sulfonation, since all molecular weights were measured prior to sulfonation. Assuming this to be the case, the functionality determinations, especially those measured by titration, agree with the expected values of 2 for the diolefins, and 3 for the triolefins.

10B.4.1.2 Halato-Telechelic Carboxylate Ionomers

The synthesis of low molecular weight difunctional carboxyl-terminated butadiene based polymers is well established and has been described in a recent review [47]. Anionic polymerization or free radical initiated polymerization processes are usually involved. The first route yields polymers of relatively narrow molecular weight distribution. However, in the molecular weight ranges that result in the best combination of properties (1500 to 6000), substantial amounts of organometallic catalyst are required. Free radical polymerization leads to broader molecular weight distributions, but being much cheaper it is generally preferred industrially. Chain transfer to solvent, which has an important effect on the final polymer functionality, is minimized by selection of appropriate solvents.

Teyssie and co-workers [48–55] have converted such carboxyl-terminated polymers to the salt forms, which they refer to as halato-telechelic polymers, by neutralizing with metal alkoxides in appropriate solvents, such as toluene. They point out the necessity for the quantitative removal of low molecular weight reaction products such as methanol to drive the reaction to completion and fully realize the inherent ionomeric properties of the polymers. In particular, it was found that ionomers prepared in this fashion produced gel at about 2% solids in nonpolar solvents. These gels were thermally reversible and the critical concentration for gel formation was related to chain molecular weight by the expression

$$C_{gel} = K M_n^{-0.5} \qquad (10B.1)$$

where K is a constant depending on the polymer backbone and solvent.

10B.4.2   Sulfonated Polypentenamers

In general, sulfonation of highly unsaturated polymer backbones leads to crosslinking through a series of side reactions that are not well understood. Thus it has proved impossible to sulfonate polybutadiene without attendant crosslinking. The only report of successful sulfonation of a polymer in this category remains that of the sulfonation of polypentenamer $[(CH_2)_3CH=CH]_x$ [56].

The reaction scheme may be summarized as follows:

$$[(CH_2)_3CH=CH]_x + SO_3 : O=P(OEt)_3$$
$$\rightarrow [(CH_2)_3CH=CH]_y[(CH_2)_3CH=CH]_z$$
$$\underset{SO_3^-Na^+}{\overset{|}{}}$$

Na salt of sulfonated polypentenamer

The reaction proceeds in chloroform at room temperature, and by this procedure it is possible to sulfonate above 20 mol% without crosslinking. This allows the preparation of water-soluble derivatives.

Although the reason for a crosslinking side reaction attendant on sulfonation of polybutadiene is not known, it may involve pendant vinyl groups. These tend to be much more reactive than either *cis* or *trans* double bonds in the backbone. The virtual absence of pendant vinyl groups in polypentenamer accounts for the possibility of sulfonating it without crosslinking. Further research is required, particularly mechanistic studies, to determine the precise sequence of reactions involved in the sulfonation process as well as the structures of the products.

10B.4.3   Copolymerization of Sulfonate Monomers

Nearly all of the ionomers based on carboxylates have been obtained by copolymerization of acrylic or methacrylic acid or related systems. Until recently, there have been only a few publications covering the synthesis and characterization of sulfonate ionomers by direct copolymerization of a sulfonate monomer. In part this is because of the difficulties of copolymerizing many polar sulfonate monomers with relatively nonpolar vinyl or diene monomers. The extreme insolubility of metal sulfonate monomers in solvents other than water has impeded the synthesis of such copolymers.

Recent publications [57–59] have shown that styrene sulfonates can be copolymerized with various monomers in conventional emulsion polymerizations. In this respect, the styrene derivative appears more suitable to successful copolymerization with nonpolar comonomers than do other sulfonate moieties. Recent articles by McGrath and co-workers [12] have shown that sodium styrene sulfonate is readily copolymerized with alkyl methacrylates in this way. The resulting copolymers display thermoplastic character (see Fig. 10B.5). Figure 10B.6 shows the stress–strain behavior of these copolymers. The substantial enhancement of tensile strength due to ionic group incorporation is especially evident at sulfonate levels of 5 mol%.

It was shown that in the series of methacrylate copolymers, the amount of ionic comonomer required to achieve a high level of ionic crosslinking was significantly higher

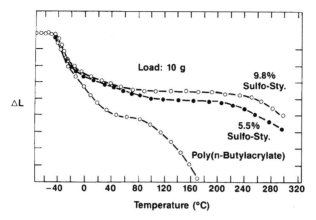

Figure 10B.5   Thermo mechanical analysis (TMA) penetration curves of n-butyl acrylate and sulfonated styrene copolymers (From Ref. [12])

than that observed for ionomers based on low polarity polymer backbones such as sulfonated EPDM (approx. 0.5 to 1 mol% metal sulfonate content). This difference can be attributed to less uniform distribution of ionic groups within the polymer backbone or the diminished strength of ionic association in a polymer matrix of increased polarity.

Other recent articles [57] describe analogous emulsion copolymerizations of styrene and sodium styrene sulfonate and have demonstrated some differences in the products, compared to those prepared by direct sulfonation of polystyrenes. While these systems are not elastomeric, the results suggest that emulsion copolymerization routes can lead to nonrandom incorporation of sulfonate groups in the polymer chain, clearly a result of the limited solubility of the very polar sulfonate monomer.

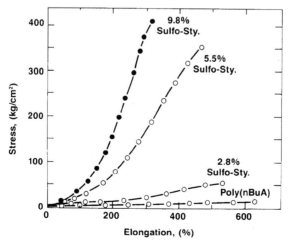

Figure 10B.6   Stress–strain behavior of *n*-butylacrylate and sulfonated styrene copolymers. (From Ref. [12]) (Note: To convert to MPa multiply kg/cm^2 by 0.0981)

More recently Agarwal and co-workers [13] have described the emulsion copolymerization of isoprene and sodium styrene sulfonate, and have shown such compositions to be excellent candidates for water-based pressure sensitive adhesives. These systems will be discussed later in this chapter.

10B.4.4   Block Copolymer Ionomers

The synthesis of several novel block copolymer ionomers has been described recently. Work by Weiss and co-workers [21] has been directed to the sulfonation of the styrene blocks in saturated triblock copolymers, for example, to poly(styrene–*block*-(ethylene–co-butylene)–*block*-styrene) by reaction with acetyl sulfate followed by conventional neutralization techniques (S–EB–S ionomers). The incorporation of ionic groups up to levels of 18 mol% (based on styrene content) was readily effected. Infrared analysis demonstrated that sulfonation occurred almost exclusively on the phenyl rings. Unlike the acid form of sulfonated EPDM, the S–EB–S acid derivatives exhibited no evidence of instability on storage. This difference was attributed to the greater stability of the aromatic sulfonic acid, as has been observed previously with highly sulfonated polystyrene. The S–EB–S ionomers exhibited improved thermal oxidative stability and were more hygroscopic than the unmodified polymers.

An alternate approach to block polymer carboxylate ionomers has been described by McGrath and co-workers [22]. In this synthesis di- and triblock copolymers of *tert*-butyl methacrylate and hexyl methacrylate were prepared by anionic polymerization. The products were readily hydrolyzed to remove the *tert*-butyl ester groups to give the acid group-containing polymers. Neutralization with KOH and CaOH provided the block ionomers. These systems, unlike conventional copolymerization approaches, offer well-defined ionomer structures in which the blocks can be precisely controlled. The precursor polymers exhibited a phase-mixed morphology, while the ionomers developed a two-phase morphology as determined by thermal, mechanical, and scattering behavior.

10B.4.5   Polyurethane Ionomers

While polyurethanes have been an important class of thermoplastic elastomers for many years, the modification of these systems with ionic functionalities has received relatively little attention until recently. Several publications have described ionic polyurethanes that show substantial enhancement of the mechanical properties compared to the base polyurethanes. A recent publication by Visser and Cooper [23] has compared the physical properties of carboxylated and sulfonated model polyurethane ionomers. These ionomers were prepared by using conventional urethane technology by reacting polytetramethylene glycol with tolylene diisocyanate to form the base polyurethane. These polymers were then reacted with sodium hydride. This adduct is then reacted with propane sultone or propiolactone to create the respective ionomer.

A comparison [23] of the sulfonate and carboxylate derivatives showed that the sulfonate ionomers gave higher tensile strengths. Differential scanning calorimetry (DSC) and dynamic mechanical thermal analysis (DMTA) results showed a higher degree of phase separation in

Typical structures are:

$$HO-\left[-CH_2-CH_2-CH_2-CH_2-O-\right]_n \cdot \cdot C-N-CH_3-N-C-\left[O-CH_2-CH_2-CH_2-CH_2-\right]_n \cdot \cdot OH$$

$$R = -(CH_2)_3-SO_3^-Na^+ \quad \text{or} \quad -(CH_2)_3CO_2^-Na^+$$

the sulfonated version than in the carboxylated ionomer. However, the Young's moduli were considerably higher for the carboxylated ionomers, showing the ionomer properties to be dependent on sample morphology. DMTA results indicate a higher crosslinking efficiency in the carboxylated ionomers, corresponding to their larger ionic aggregate sizes. This effect counterbalanced the lower acid strength of the carboxylate groups to give modulus values that were higher than expected. The moduli of all these ionomers were higher than predicted on the basis of ionic aggregates acting as crosslinks and fillers. This increased modulus was attributed to chain entanglements formed in the process of ionic aggregation.

10B.5   Recent Developments: Properties

10B.5.1   The Glass Transition Temperature T_g

In general, elastomeric halato-telechelic polymers exhibit glass transition temperatures that are insensitive to cation size and valence, degree of neutralization, and oligomer molecular weight. The situation is quite different with sulfonated polypentenamer. In a study of a series of sulfonated polypentenamers containing from 1.9 to 17.6 mol% pendant groups in the form of sodium salts, it was shown that the composition dependence of the T_g exhibits typical random copolymer behavior at low levels of sulfonation, but deviates significantly from this behavior above 10 mol% [60] (see Fig. 10B.7). Dynamic mechanical results indicated the presence of an ionic phase relaxation in addition to the T_g. This relaxation is present only in samples sulfonated above 10 mol% and is sensitive to the presence of polar impurities such as water. The most natural interpretation of this behavior is that above 10 mol%, microphase-separated clusters or ionic domains appear in significant concentration. Presumably the upturn in the T_g shown in Fig. 10B.7 is a result of such clusters acting as a combination of reinforcing fillers and highly functional crosslinks. This combined action results in a "bracing" effect on the chains of the matrix and hence elevates its T_g. In sulfonated polypentenamer and, by extension, in the halato-telechelic ionomer case, at sulfonation levels of less than 10 mol% the salt groups exist as ion pairs, quartets, or multiplets which are not phase separated from the matrix.

If we invoke the EHM model to explain the observed behavior, it might be postulated that the "regions of restricted mobility" around the multiplets begin to overlap significantly at the higher concentrations and that this in turn is responsible for the upturn in T_g.

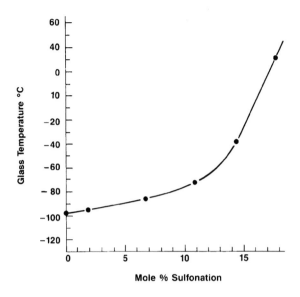

Figure 10B.7   Composition dependence of DSC T_gs for sulfonated polypentenamers. (From Ref. [60])

10B.5.2   Mechanical Properties

The properties of halato-telechelic elastomeric ionomers can be understood generally in terms of the formation of chain extension or pseudo-three-dimensional networks by the aggregation of the ion-pair groups on the chain ends. The degree of aggregation and the thermal stability of the aggregates is a function of concentration, type of cation and anion, and perhaps placement of the ionic groups on the chain. Kennedy and Wilkes [41–46] have carried out an extensive series of investigations with sulfonate terminated three-armed stars of polyisobutylene neutralized with calcium and potassium hydroxides. In experiments with oligomers of molecular weights below the critical molecular weight for entanglement formation of linear polyisobutylene (<9000), it was found that the simple rubber elasticity theory was capable of describing the stress–strain behavior at low to moderate elongations.

$$\sigma = \frac{\rho RT}{\overline{M}_c} \left(\lambda - \frac{1}{\lambda^2} \right) \tag{10B.2}$$

where σ is the stress, ρ the density, R the gas constant, T the absolute temperature, λ the extension ratio, and $\overline{M}_c$ the number average molecular weight between crosslinks.

Assuming only two ion pairs per ionic aggregate, $\overline{M}_{c2}$ would equal the average molecular weight of two arms, that is, $\overline{M}_c = 2/3\overline{M}_n$. For an oligomer in which $\overline{M}_n = 19,000$, $\overline{M}_c$ was found to be 6700 from Eq. (10B.2), which is quite close to the expected value of 6000. Above extension ratios of 5, stress induced crystallization occurred with a concomitant upturn in the stress–strain curve, and Eq. (10B.2) was no longer valid. The instantaneous permanent set is small, only 50% to 60% even at 700% elongation so that the networks possess considerable integrity.

In a later study, Kennedy and Wilkes [46] examined a series of oligomers with different molecular weights both above and below the critical entanglement molecular weight. The results generally amplified and confirmed those of the earlier study and it was concluded that three-arm star polyisobutylene ionomers of $\overline{M}$ between 11,000 and 34,000 possess high tensile properties together with low permanent set and hysteresis at ambient conditions. The presence of crystallinity at higher elongations enhances mechanical properties. These ionomers can be thermally formed above 150 °C. A general finding was that excess neutralizing agent increased the tensile properties with a maximum at about 100% excess agent.

The sulfonated polypentenamer ionomers have been studied by the dynamic mechanical relaxation technique [60]. The results provided strong evidence for the existence of microphase separated ionic aggregates or clusters above 10 mol% substitution. An ionic phase relaxation was present in samples above 10 mol% substitution and increased in magnitude as the degree of sulfonation increased. This relaxation was sensitive to low molecular weight polar impurities such as water and decreased substantially in temperature when the ionomers were saturated with water.

The effect of the cation on ionomer properties has been examined in several ionomer systems. In the case of sulfonated EPDM it was found [61] that physical properties such as tensile strength and melt flow were very dependent on the choice of cation as shown in Table 10B.3. Interestingly, the monovalent cations such as lithium and sodium exhibited stronger ionic association, as manifested by melt viscosity measurements, than divalent cations such as zinc and lead. The observations coupled with the excellent tensile properties exhibited by the zinc ionomer have contributed to the selection of zinc sulfo-EPDM as a primary ionic elastomer candidate for commercialization.

Table 10B.3 Effects of Various Cations on the Flow and Physical Properties of Sulfonated EPDM[a,b]

Metal	Apparent viscosity[c] (μPa-s[d]) rate (Hz)	Melt fracture at shear	Melt index (190 °C 3.3 MPa[e], 10 g/min)	Room temperature tensile strength (MPa[e])	Elongation (%)
Hg			disintegrated		
Mg	55.0	< 0.88	0	2.2	70
Ca	53.2	< 0.88	0	2.8	90
Co	52.3	< 0.88	0	8.1	290
Li	51.5	< 0.88	0	5.2	320
Ba	50.8	< 0.88	0	2.3	70
Na	50.6	< 0.88	0	6.6	350
Pb	32.8	88	0.1	11.6	480
Zn	12.0	147	0.75	10.2	400

[a] Based on Ref. [61]
[b] Sulfonate content: 31 meq/100 EPDM
[c] At 200 °C and 0.88 s^{-1}
[d] To convert μPa-s to centipoise, divide by 1000
[e] To convert MPa to psi, multiply by 145

The effects of these different cations on polymer flow and tensile properties indicate that ion pair association and the resulting network formation due to aggregation is more important than cation valency. Similar relationships were observed with carboxylate ionomers in Chapter 10A.

10B.6   Preferential Plasticization

The fact that the ionic phase relaxation is sensitive to low molecular weight polar impurities suggests the possibility that the relaxation of the ionic aggregates can be controlled by the deliberate addition of an appropriate polar diluent. Such an additive would affect only the ionic groups, leaving the properties of the matrix unchanged. Since the ionic aggregates act as physical crosslinks, it is clear that this approach could lead to the control of the temperature and, presumably, the shear rate, necessary to induce flow, and hence to the possible development of a thermoplastic elastomer with desirable processing characteristics.

In a study of the modification of ionic associations in sulfonated EPDM ionomers, it was found [62] that a crystalline additive such as zinc stearate can strongly affect material properties, in addition to being a highly effective preferential plasticizer for the ionic aggregates.

It was observed that zinc stearate is compatible with the ionomer even at high loadings (over 30% by weight) and that it enhances physical associations as reflected in mechanical properties and swelling characteristics. The morphological structure of the zinc stearate thus dispersed in the ionomer was found to be small microphase separated crystallites less than 500 nm (5000 Å) in diameter. These entities act as reinforcing fillers below their melting points and greatly enhance the flow properties of the ionomer above their melting points. This behavior is thermally reversible.

Figure 10B.8 summarizes some possibilities for varying the strengths of ionic interactions by utilizing different cations and preferential plasticizers. These modulus–temperature curves are a dramatic illustration of the versatility of the ionomer approach to controlling properties of thermoplastic elastomers.

The details of the mechanism of preferential plasticization of the ionic phase remain unclear. If the EHM model is correct, it will be necessary to reinterpret the origins of the effect. It is not obvious how a preferential plasticizer such as zinc stearate would act to enhance the mobility of chains located in the region of restricted mobility surrounding the multiplets as envisaged by the EHM model.

10B.7   Ionic Interactions in Polymer Blends

It is well known that most mixtures of high molecular weight polymers are incompatible in the thermodynamic sense due to the very small increase in their conformational entropy upon mixing. Compatibility, meaning sufficient interaction between the blend components to

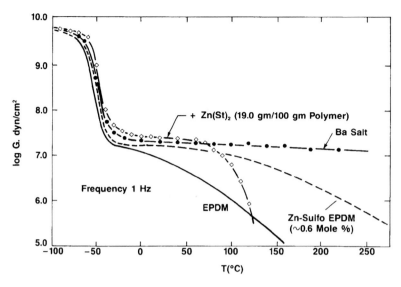

Figure 10B.8 Temperature dependence of G' at 1 Hz for sulfonated EPDM ionomers neutralized with various cations and containing preferential plasticizers as noted (From Ref. [76])

produce useful properties, may be achieved by a number of different routes including the addition of compatibilizing agents such as block copolymers, the use of chemical techniques such as grafting, etc. (see Chapter 7). A route to compatibilization involving ionomers has been described recently in a series of articles by Eisenberg and co-workers [14–16]. This involves the incorporation of specifically interacting acidic groups in one polymer with basic groups in the other to form ion pairs. Specifically, the blend of poly(styrene–co-styrene sulfonic acid) (PSSA) and poly(ethyl acrylate–co-4 vinylpyridine) (PEAVP) has been investigated. It was found that a functional group content of approximately 4 mol% is required for compatibilization of this blend. In this case compatibility was assessed by examining the dynamic mechanical relaxation behavior in the glass to rubber (primary) relaxation region. It was found that, above the 4 mol% functional group content mentioned previously, only a single tan δ or G'' peak associated with microbrownian motion accompanying the glass transition could be observed. Below this level of substituents, two peaks in these functions were clearly resolved. An increase in functional group content above 4 mol% had the effect of increasing the temperature of the glass–rubber relaxation in a manner similar to that observed with chemical crosslinks.

It is unlikely that the blends studied by Eisenberg are thermodynamically miscible, although this possibility cannot be ruled out entirely on the basis of the results available. They are certainly compatible in the operational sense and are of obvious scientific and technological interest.

The use of ionic interactions between different polymer chains to produce new materials with interesting properties dates back to at least the work of Michaels and Miekka [63]. These workers prepared materials they referred to as "polysalts" by mixing solutions of acidic polyelectrolytes, such as poly(acrylic acid) with basic polyelectrolytes, such as poly(vinylpyridine). The resulting materials were insoluble in all solvents although they would swell considerably in water. Their main area of application was as membranes for reverse osmosis. Later, Otocka, and Eirich [64] prepared similar materials and studied their mechanical properties. Because of their intractability, these materials have never been properly characterized.

Other specific interactions have also been utilized for compatibilization. Pearce and co-workers [65] have shown that the incorporation of short perfluorinated alcohol side chains onto polystyrene gave compatible blends with poly(ethylene oxide) (PEO) by hydrogen bonding. Hara and Eisenberg [66] have also demonstrated that poly(styrene–co-methacrylate) ionomers will interact with PEO via an ion–dipole mechanism to produce similar effects to those observed by Pearce. Although the first of these components is a thermoplastic, compatibilization by this route with a low T_g second component can lead to materials with T_gs below room temperature. This is a possible route to new elastomeric compositions.

During the past years a number of publications have described the enhanced miscibility of polymer blends through ionic interactions. For example, polymer complexes can be formed by the interaction of an amine containing polymer with an ionomer in which the cation is zinc or a transition metal [17, 18]. Typically, 4-vinylpyridine is the amine most widely employed because of its convenience in copolymerization with styrene and other vinyl monomers via free radical copolymerization. In fact, 2-vinylpyridine (as a comonomer with styrene) is almost completely ineffective in interacting with transition metal sulfonates [67]. Much of the published work involves interpolymer complexes formed through the specific interaction of transition-metal neutralized sulfonated EPDM, a thermoplastic elastomer, with thermoplastic glassy polymer, 4-vinylpyridine copolymers (especially styrene–co-4-vinylpyridine (PSVP) [17, 18, 67]. A variety of sulfonated EPDM polymers were prepared with different counter ions (zinc, magnesium, and sodium). A copolymer of styrene and 4-vinylpyridine was employed containing 8.5 mol% vinylpyridine (85 meq/100 g). The sulfonate content of the sulfonated EPDM was about 20 meq/100 g. As one study has shown, these polymers can be readily melt blended and either compression molded or injection molded. A range of blends of these component materials was prepared. Polystyrene was employed as a control, demonstrating the behavior of blends wherein no interactions occurred.

The work demonstrated that the PSVP blends provided a much stronger network than was the case with the polystyrene control. It was also clear that the zinc blends have properties that are considerably enhanced compared to the Mg- or Na-based ionomers. A wide variety of other cations have been investigated and it was found that for enhanced network strength, only zinc and transition metal cations are effective.

Melt rheological, morphological, and thermal analysis studies have all been conducted on these blends. The studies suggest optimal properties at a stoichiometry of 1 : 1 zinc to nitrogen. Compression-molded pads prepared from these systems have interesting properties. Those based on Zn sulfonated EPDM and PSVP can be virtually transparent, while under the same conditions those based on Na or Mg salts of sulfonated EPDM are nearly opaque.

Further investigations of the PSSA–PEAVP blends have also been carried out [68, 69]. Dynamic mechanical studies indicate that even at substitution levels as high as 10 mol% the compatible PSSA–PEAVP blends are not truly miscible on the molecular level but contain microphases. Fourier transform infrared spectroscopy (FTIR) on the PSSA–PEAVP blends showed that the sulfonic acid groups formed sulfonate anions and the pyridine groups formed pyridium cations, indicating the presence of coulombic bonds. Somewhat unexpectedly, it appears that all or nearly all of the acid–base groups interact even below substitution levels of 4 or 5 mol%, the degree of substitution identified by Eisenberg [14–16] as necessary for compatibility. Further FTIR results on zinc neutralized PSSA–PEAVP blends showed that the pyridine group coordinates to the zinc in the blend and that, when the zinc cations are hydrated, blending causes dehydration with the pyridine taking the place of water in the zinc coordination shell. Finally, viscoelastic measurements show the presence of a high-temperature loss peak in the PSSA–PEVP blends which was interpreted as indicating the presence of ionic aggregates in these blends. However, SAXS measurements show that the "ionic" peak present in ionomers is destroyed on blending. Available data are also interpretable on the basis of the EHM model for ionomer morphology. Clearly more extensive studies are needed to determine the actual structures present in the blends.

10B.8   Applications of Ionomeric Elastomers

10B.8.1   Thermoplastic Elastomers

Most ionomer applications exploit several characteristics that can be attributed either to ionic aggregation or to cluster formation, or the interaction of polar groups with ionic aggregates. Changes in physical properties caused by ionic aggregation in elastomeric systems or in polymer melts are most readily detected. Thus the marked enhancement in elastomeric green strength is a general characteristic of ionomer-based systems. The ionic aggregation is also apparent in increased melt viscosity. In the case of polyethylene-based metal carboxylate ionomers, the high melt viscosity is utilized in heat sealing. It also provides processing advantages during extrusion. Under some conditions, however, high melt viscosity is a limitation, for example, in injection molding. Other properties attributable to ionic aggregation include toughness, outstanding abrasion resistance, and oil resistance which is important in packaging applications.

   The interaction of various polar agents with the ionic groups and the ensuing property changes are unique to ionomer systems. This plasticization process is also important in membrane applications. A different application of ionic cluster plasticization involves the interaction of metal stearates with sulfonated EPDM to induce softening transitions. Plasticization is required to give processability to thermoplastic elastomers (TPEs) based on this technology. To produce TPEs, the sulfonated elastomer is compounded with mineral fillers, antioxidants, process oils, and/or polyolefins (polyethylene or polypropylene) using conventional elastomer blending equipment. The formulations can vary in hardness from soft (Shore A of 70) to quite hard semiplastic compositions. Typical processing conditions are similar to those employed for flexible poly(vinyl chloride). Many U.S. patents covering these compositions have been assigned to Exxon.

10B.8.2   Adhesives

One area of increasing activity for both sulfonate and carboxylate ionomers has been that of adhesives. The use of ethylene-based ionomers in many packaging applications is attributable to their excellent adhesive characteristics. Recent patent activity suggests the use of directly copolymerized carboxyl and sulfonate-containing polymers as water-based, pressure-sensitive adhesives (PSA). The presence of a suitable tackifier (an aliphatic petroleum resin) often improves the tackiness of these adhesive compositions by facilitating bond formation on contact. For this reason, a blend of an emulsified tackifier with an emulsion of a carboxylated styrene–butadiene copolymer is employed in PSA applications [70].

A recent publication [13] has highlighted the use of water-based copolymers of isoprene and sodium styrene sulfonate as PSAs. These emulsion copolymers are used in combination with emulsions of commercial hydrocarbon resins. Under proper control of polymer molecular weight and ionic content, these systems offer a good combination of cohesive strength and tackiness.

10B.8.3   Miscellaneous Applications

Many of the applications of elastomeric ionomers are similar to those found for styrenic block copolymers. For example, 5% to 10% of sulfonated EPDM is found to improve the properties of asphalt and thereby enhance its performance in roofing applications. The impact resistance of various thermoplastics such as nylon and Noryl® can be improved by incorporating sulfonated EPDM at modest levels. Finally, the use of ionomers as coatings to encapsulate materials as diverse as fertilizers and oxidizing chemicals has been described in patents [71]. These coated products can give controlled release of fertilizers and other agents.

10B.9   Conclusions

Although ionic interactions have been used to modify elastomer properties since at least the 1950s, the full potential of this technique has yet to be realized. Within the last 10 years, several important points pertaining to this subject have emerged.

Ionomeric associations dramatically modify polymer properties over a wide range of modulus, melt viscosity and transition temperatures. The reasons for this are only imperfectly understood because of a lack of knowledge of the degree of aggregation of ionic groups in the polymers and how structural variables affect them.

Multiple synthetic approaches to ionomeric elastomers exist and are commercially viable. Some of these have been discussed here and in our earlier review [7].

A wide range of potential applications exists for ionomers. TPEs are one example. Others include foams, elastic fibers, polymer modifiers, coatings, and solution applications.

10B.10 Future Developments

One report [72] has summarized the technical and patent literature in the field of TPEs from 1979 to 1984. Ionomeric TPEs led the field in terms of published articles during this period. Similarly, ionomeric TPEs were the subject of more patents than any other TPE approach over the same period. The versatility of ionic crosslinking coupled with the wide variety of synthetic approaches which can be employed to achieve these systems serves to make ionomeric TPEs an attractive research area. The ability to moderate the degree of ionic crosslinking by the strength of the ionic crosslink, the number of such interactions, and the use of external ionic plasticizers to control the nature of the resulting network offer unusual control over these systems. Several recent patents have suggested new uses for these ionomeric TPEs which vary from asphalt modification [73] to impact modification of engineering thermoplastics [74]. The use of Zn sulfonated EPDM as a waterproof, heat-sealable roofing membrane of exceptional tear strength has been described in a recent patent [75].

Based on these developments, research on ionomeric TPEs will be concerned with the synthesis of new ionomeric candidates, additional characterization of the morphology and flow behavior of available systems, and new applications that specifically exploit the unique characteristics of ionic crosslinks.

References

1. A. McAlevy, U.S. Patent 2,405,971 (August 29, 1946)
2. H.P. Brown and C.F. Gibbs, Presented at a Meeting of the ACS Rubber Division, September 1954, *Rubber Chem. Tech. 28*, 937 (1955)
3. H.P. Brown, *Rubber Chem. Tech. 30*, 1347 (1957)
4. H.P. Brown, Presented at a Meeting of the ACS Rubber Division, May 1963, *Rubber Chem. Tech. 36*, 931 (1963)
5. R.W. Rees, *Modern Plastics 42*, 209 (1964)
6. R.W. Rees and D.J. Vaughan, *Polym. Prepr. Am. Chem. Soc. Div. Polymer Chem. 6*, 287, 296 (1965)
7. W.J. MacKnight and R.D. Lundberg, *Rubber Chem. Tech. 57*(3), 652 (1984)
8. W.J. MacKnight and R.D. Lundberg, In *Thermoplastic Elastomers*, N.R. Legge, G. Holden and H.E. Schroeder (Eds.) Hanser (1987) Munich and New York
9. N.H. Canter, (to Exxon Research and Eng. Co.), U.S. Patent 3,642,728 (February 15, 1972)
10. W.J. MacKnight and T.R. Earnest, *J. Polym. Sci. Polym. Rev. 16*, 41 (1981)
11. J.P. Kennedy and R.F. Storey, *ACS Div. Org. Coat. Applied Polym. Sci. 46*, 182 (1982)
12. I. Yilgor, A. Packard, J. Eberle, E. Yilgor, R.D. Lundberg, and J.E. McGrath, *Polym. Prepr. ACS Div. Polym. Chem. 24*(2), 37 (1983)
13. P.K. Agarwal, F.C. Jaegisch, R.D. Lundberg, and V.L. Hughes, *ACS Polym. Mater. Sci. and Eng. Prep. 61*, 593 (1989)
14. P. Smith and A. Eisenberg, *J. Polym. Sci., Polym. Lett. Ed. 21*(3), 223 (1983)
15. Z.L. Zhou and A. Eisenberg, *J. Polym. Sci. Polym. Phys. Ed. 21*, 595 (1983)
16. A. Natansohn, R. Murali, and A. Eisenberg, *Chemtech 418*, July 1990
17. R.D. Lundberg, D.G. Peiffer, and R.R. Phillips, (to Exxon Res. and Eng. Co.), U.S. Patent 4,480,063 (October, 1984)

18. D.G. Peiffer, I. Duvdevani, P.K. Agarwal, and R.D. Lundberg, *J. Polym. Sci. Polym. Letters Ed. 24*, 581 (1986)
19. A. Sen and R.A. Weiss, *Polym. Preprints 28*(2), 220 (1987)
20. R.A. Florence, J.R. Campbell, and R.E. Williams, (to General Electric Co.), U.S. Patent 4,496,705 (January 29, 1985)
21. R.A. Weiss, A. Sen, C.L. Willis, and L.A. Pottick, *Polymer 32*(10), 1867 (1991)
22. C.D. Deporter, L.N. Venkateschanaran, G.A. York, G.L. Wilkes, and J.E. McGrath, *Polymer Preprints 30*(1), 201 (1989)
23. S.A. Visser and S.L. Cooper, *Macromolecules 24*, 2576 (1991)
24a. A. Eisenberg, *Macromolecules 3*, 147 (1970)
24b. A. Eisenberg, B. Hird, and R.B. Moore, *Macromolecules 23*, 4098 (1990)
25. W. Forsman, *Macromolecules 15*, 1032 (1982)
26. T.R. Earnest, J.S. Higgins, D.L. Handlin, and W.J. MacKnight, *Macromolules 14*, 192 (1981)
27. R.A. Register and S.L. Cooper, *Macromolecules 23*(11), 2978 (1990)
28. W.J. MacKnight and T.R. Earnest, Jr., *J. Polym. Sci. Macromol. Rev. 16*, 41 (1981)
29. W.J. MacKnight, W.P. Taggart, and R.S. Stein, *J. Polym. Sci. Polym. Symp. No. 45*, 113 (1974)
30. E.J. Roche, R.S. Stein, and W.J. MacKnight, *J. Polym. Sci. Polym. Phys. Ed. 18*, 1035 (1980)
31. M. Fujimura, T. Hashimoto, and H. Kawai, *Macromolecules 15*, 136 (1982)
32. D.J. Yarusso and S.L. Cooper, *Macromolecules 16*, 1871 (1983)
33. G. Fournet, *Acta Crystallzr. 4*, 293 (1951)
34. E.J. Roche, R.S. Stein, and W.J. MacKnight, *J. Polym. Sci. Phys. Ed. 18*, 1035 (1980)
35. M. Pineri, R. Dupliessix, S. Gauthier, and A. Eisenberg, *Advances in Chem. Series 187*, (1980) American Chemical Society, Washington, D.C., p. 283
36. T.R. Earnest, Jr., J.S. Higgins, and W.J. MacKnight, *Macromolecules 15*, 1390 (1982)
37. D.L. Handlin, W.J. MacKnight, and E.L. Thomas, *Macromolecules 14*, 795 (1980)
38. R. Siadat, R.D. Lundberg, and R.W. Lenz, *Polym. Eng. Sci. 20*8, 530 (1980)
39. R.D. Lundberg, H.S. Makoski, J. Bock, and T. Zawadski, (to Exxon Research and Eng. Co.) U.S. Patent 4,157,432 (June 5, 1979)
40. J.P. Kennedy and R.F. Storey, *Organ. Coat. Appl. Polym. Sci. Proc. 46*, 182 (1982)
41. Y. Mohajer, D. Tyagi, G.L. Wilkes, R.F. Storey, and J.P. Kennedy, *Polym. Bull. 8*, 47 (1982)
42. S. Bagrodia, Y. Mohajer, G.L. Wilkes, R.F. Storey, and J.P. Kennedy, *Polym. Bull. 8*, 281 (1982)
43. S. Bagrodia, Y. Mohajer, G.L. Wilkes, R.F. Storey, and J.P. Kennedy, *Polym. Bull. 9*, 174 (1983)
44. S. Bagrodia, G.L. Wilkes, and J.P. Kennedy, *J. Rheol. 28*, 474 (1983)
45. Y. Mohajer, S. Bagrodia, G.L. Wilkes, R.F. Storey, and J.P. Kennedy, *J. Appl. Polym. Sci. 29*, 1943 (1984)
46. G. Tant, G.L. Wilkes, and J.P. Kennedy, *Polym. Preprints 26*(1), 32 (1985)
47. D.N. Schulz, J.C. Sandra, and B.G. Willoughby, In *Anionic Polymerization: Kinetics, Mechanisms, and Synthesis* J.E. McGrath (Ed.) (1981) *ACS Symposium Series 166*, p. 427
48. G. Broze, R. Jerome, and P. Teyssie, *Macromolecules 14*, 224 (1981)
49. G. Broze, R. Jerome, and P. Teyssie, *Macromolecules 15*, 920 (1982)
50. G. Broze, R. Jerome, and P. Teyssie, *Macromolecules 15*, 1300 (1982)
51. G. Broze, R. Jerome, P. Teyssie, and C. Marco, *Macromolecules 16*, 996 (1983)
52. G. Broze, R. Jerome, and P. Teyssie, *J. Polym. Sci. Phys. Ed. 21*, 2205 (1983)
53. G. Broze, R. Jerome, P. Teyssie, and C. Marco, *Macromolecules 16*, 1771 (1983)
54. G. Broze, R. Jerome, and P. Teyssie, *J. Polym. Sci. Lett. 21*, 237 (1983)
55. R. Jerome, J. Horrion, R. Fayt, and P. Teyssie, *Macromolecules 17*, 2447 (1984)
56. D. Rahrig, W.J. MacKnight, and R.W. Lenz, *Macromolecules 12*, 195 (1979)
57. R.A. Weiss, R.D. Lundberg, and S.R. Turner, *J. Polym. Sci. Polym. Chem. Ed. 23*, 525, 535, 540 (1985)
58. B. Siadat, B. Oster, and R.W. Lenz, *J. Applied Polym. Sci. 26*, 1027 (1981)
59. R.A. Weiss, R.D. Lundberg, and A. Werner, *J. Polym. Sci. Polym. Chem. Ed. 18*, 3427 (1980)
60. D. Rahrig and W.J. MacKnight, *Adv. Chem. Series No. 187*, A. Eisenberg (Ed.) (1980) American Chemical Society, Washington, D.C.
61. H.S. Makowski, R.D. Lundberg, L. Westerman, and J. Bock, *Adv. Chem. Series 187*, 3 (1980)
62. (a) I. Duvdevani, R.D. Lundberg, C. Wood-Cordova, and G.L. Wilkes, *ACS Symposium Series 301* A. Eiserberg and F.E. Bailey, Eds) 185 (1986). (b) H.S. Makowski and R.D. Lundberg, *Adv. Chem. Soc. 187*, 37 (1980)

63. A.S. Michaels and R.G. Miekka, *J. Phys. Chem. 65*, 1765 (1961)
64. E.P. Otocka and F.R. Eirich, *J. Polym. Sci. A2 6*, 921 (1968)
65. S.P. Ting, B.J. Bulkin, and E.M. Pearce, *J. Polym. Sci. Polym. Chem. Ed. 19*, 451 (1981)
66. M. Hara and A. Eisenberg, *Macromolecules 17*, 1335 (1984)
67. R.D. Lundberg and P.K. Agarwal, *Indian U. Technol. 31*, 400, April 3, 1993
68. E.P. Douglas, K. Sakurai, and W.J. MacKnight, *Macromolecules 24*, 6776 (1991)
69. E.P. Douglas, K. Sakurai, and W.J. MacKnight, *Macromolecules 25*, 4506 (1992)
70. S.G. Takemoto and O.J. Morrison, (to Avery Int. Corp.), U.S. Patent 4,189,919 (February 19, 1980)
71. I. Duvdevani, P.V. Menalestas, E.N. Drake, and W.A. Thaler, (to Exxon Res. and Eng. Co.), U.S. Patent 4,701,204 (October 20, 1987)
72. N.R. Legge, Paper No. 73, 127th Meeting of Rubber Div., Los Angeles, April, 1985, *Elastomerics 117*(10), 19 (1985)
73. J.A. Cogliano, (to W.R. Grace & Co.), U.S. Patent 4,524,156 (June 18, 1985)
74. J.R. Campbell, P.M. Conroy, and R.A. Florence, (to General Electric Co.), PCT Int. Appl. WO 85 01,056 (March 14, 1985)
75. A.U. Paeglis, (to Uniroyal Inc.), U.S. Patent 4,480,062 (October 30, 1984)
76. P.K. Agarwal, H.S. Makowski, and R.D. Lundberg, *Macromolecules 13*, 1679 (1980)

11.1   Introduction

The commercial introduction of styrenic block copolymer thermoplastic elastomers (TPEs) occurred two decades ago. The first polymers were linear triblock copolymers with polystyrene hard blocks and polybutadiene elastomer blocks (S–B–S) or polyisoprene elastomer blocks (S–I–S), both formed by anionic polymerization. Ten years later, a second generation of linear triblock copolymers, in which the polybutadiene center blocks were hydrogenated, was brought to the market.

The hydrogenated block copolymers that we will discuss are linear triblock copolymers with terminal polystyrene blocks and elastomeric center blocks. The elastomer block consists of a copolymer of ethylene and butylene (S–EB–S) which is derived from a butadiene precursor by hydrogenation. These blocks, at sufficiently high molecular weight, have a sharp compositional boundary; the blocks are pure polystyrene or pure elastomer. Since the polymeric blocks are immiscible they phase separate into domains of polystyrene in a matrix of the elastomer. At less than about 50% polystyrene, the domains are dispersed.

The concept behind all TPEs is to provide in the same polymer the mechanical performance of a vulcanized rubber and the facile processing of thermoplastics. This is achieved over a limited range of service conditions in the TPEs that we know today. Indeed, it is the dream and the goal of researchers in this area to widen the window of service conditions to eventually equal that of vulcanized rubbers. It was this goal that drove the development of hydrogenated block copolymers. The experience with ethylene-propylene rubber (EPR) and ethylene–propylene–diene monomer (EPDM) rubbers clearly showed the advantages of saturated polymers in providing resistance to degradation from exposure to high temperatures (in air) and to UV radiation. Lack of resistance to degradation was a clear limitation of the unsaturated block copolymers and hydrogenation of these polymers gave the same measure of resistance to degradation as found in the olefin rubbers.

Clearly, the improvements expected from hydrogenation were achieved. Of even greater importance, however, it was found that the interrelationships governing phase separation were so strongly changed by hydrogenation that a different quality of block copolymer resulted, one having dramatically different properties. The unique rheology of these polymers, together with their high temperature stability, inevitably led to a new class of blended compounds, consisting of interpenetrating co-continuous polymer phases, the thermoplastic interpenetrating polymer networks (IPNs) [1, 2]. The ability to form co-continuous morphologies is a result of the network structure of the block copolymers in the melt and the consequent yield phenomena in a flow field during mixing. The network structure "templates" the surrounding phase while the yield stress prevents spontaneous demixing.

Styrene has been the chosen building block for the hard phase of block copolymers. Although it is one of the most readily available, anionically polymerizable monomers, its special utility in hydrogenated block copolymers arises from other factors to be discussed in detail. The chosen building block for the elastomer segment of the commercial hydrogenated block copolymers has been ethylene–butylene, derived from butadiene, rather than ethylene–propylene, derived from isoprene. Originally, this choice was made primarily on the basis of the traditional factors of cost and availability of the monomers, although there are other decision factors involved.

The first part of this chapter discusses in detail the chemical and physical structure of hydrogenated block copolymers, with emphasis on the differences from their unsaturated analogues and the properties that result. The second part deals with the role of such polymers in the formation of thermoplastic IPNs and the properties of such blends. An experimental section is provided at the end, in which the materials and the experimental methods and testing procedures are outlined briefly.

11.2   Hydrogenated Diene Block Copolymers

11.2.1   Chemical Structure

It is possible to hydrogenate styrenic block copolymers completely, including the aromatic unsaturation of the polystyrene blocks. This leads to a certain class of polymers that we will discuss shortly. However, it is usually the case that only the elastomer block is hydrogenated. Most of the art of the commercial process involves both this oftentimes difficult selective hydrogenation and also controlling the precursor block copolymer microstructure.

The hydrogenation of polyisoprene polymerized through the 1,4 carbons yields an olefin type rubber, an ethylene–propylene copolymer with perfectly alternating ethylene and propylene sequences. These polymers are completely free of polyethylenelike crystallinity and have a low T_g, about $-55\,^{\circ}\mathrm{C}$ to $-60\,^{\circ}\mathrm{C}$.

Isoprene monomer

$$\underset{\displaystyle C=C-\overset{\displaystyle C}{\underset{\displaystyle |}{C}}=C}{}$$

1,4-Polyisoprene–3,4-polyisoprene $[C-C\overset{C}{\underset{|}{=}}C-C]$ $[C-C]$ (with C branch and $=C$)

1:1-Ethylene-propylene–3-methylbutene $[C-C-\overset{C}{\underset{|}{C}}-C]$ $[C-C]$ (with two C branches)

Each isoprene monomer addition to the growing chain incorporates two methylene and one propylene sequence (or one ethylene and one propylene). Although some of the isoprene can also polymerize through the 3,4 carbons, the resulting hydrogenated structure, which has an isopropenyl side group, does not allow any crystallinity nor does it substantially increase T_g. In practice, in making block copolymers with ethylene–propylene blocks, the 3,4 addition reaction usually is suppressed as much as possible. The important point is that incorporation of each monomer unit enchains four "backbone" carbon atoms and includes a propylene unit or a branched monomer unit. Runs of polyethylene are impossible, regardless of the type of addition, and the chain is truly a rubbery polymer.

When butadiene is the rubber block monomer, the situation is somewhat more complex:

Butadiene monomer $C=C-C=C$

$$
\begin{array}{c}
C \\
\parallel \\
C \\
\mid
\end{array}
$$

1,4–1,2-Polybutadiene $[C-C=C-C]$ $[C-C]$

$$
\begin{array}{c}
C \\
\mid \\
C \\
\mid
\end{array}
$$

Ethylene–butylene $[C-C-C-C]$ $[C-C]$

In the synthesis of polybutadiene, polymerization can proceed through the 1,4 or the 1,2 double bonds. The random balance of 1,4 and 1,2 addition during polymerization can be controlled very readily in the range of about 8% to 70% 1,2 addition. Hydrogenation of these chains then yields either a polyethylene sequence from the 1,4 addition or a polybutylene sequence from the 1,2 addition. Each monomer can enchain in the backbone either four (a tetramethylene sequence) or two (an α-butylene sequence) carbon atoms. Hydrogenated pure 1,4-polybutadiene would be indistinguishable from linear high-density polyethylene, whereas hydrogenated pure 1,2-polybutadiene would yield atactic poly 1-butene. The random mixtures resulting from hydrogenation of anionic butadiene polymers produce elastomeric ethylene–butylene copolymers. (Such ethylene–butylene copolymer chains are not identical in molecular structure to those formed from random polymerization of ethylene and butylene, because of the absence of certain forbidden sequences. This arises from the enchainment of two ethylene units for each 1,4 addition. However, common physical properties apparently are not affected by this subtle difference.)

The elastomer segment composition in these block copolymers must be designed to adequately suppress polyethylene crystallinity, by interrupting polyethylenelike sequences with butylene monomer units. However, the T_g of polybutylene is around $-18\,°C$, high for a good quality elastomer, so that too much enrichment with butylene will substantially raise T_g. This balance is a requirement that is well known to manufacturers of Ziegler polymerized EPR or EPDM rubbers. The purpose is to obtain a saturated olefin elastomer block with the lowest possible T_g and the best elastomeric characteristics. Thus, the optimal composition will minimize the increase in T_g arising from butylene monomer incorporation and also minimize crystallinity by interrupting polyethylene sequences.

As the composition becomes rich in tetramethylene sequences arising from the hydro-genated polybutadiene 1,4 additions, the probability of runs of polyethylenelike sequences becomes higher and the crystallinity shows an increase displayed by the left-hand curve in Fig. 11.1. This curve matches a calculated relationship assuming that 20 methylene sequences, which would be produced by five sequential 1,4 additions to the chain, are capable of giving polyethylene crystallites at ambient temperature. Crystallization reduces the amorphous phase volume and reinforces the elastomer, overlaying and masking the T_g relationship on the left-hand side of the composition diagram. When crystallinity is almost completely suppressed, in the vicinity of 50% 1,2 addition, T_g is strongly increased by further butylene enchainment.

BLOCK COPOLYMERS OF STYRENE AND EB

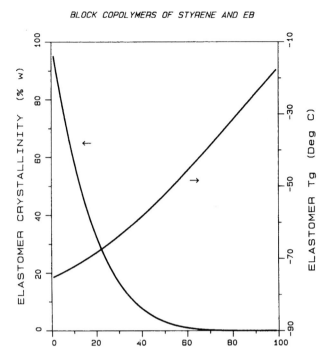

Figure 11.1   The effect of butylene (1,2 addition in the precursor) concentration on the crystallinity and T_g of the elastomer phase in S–EB–S polymers

11.2.2   Microphase Separation in Block Copolymers

The phase separation of block copolymers both in solution and in the solid state has been developed from thermodynamic principles by a number of authors [3–5]. All start from the thermodynamic equation of mixing for polymer pairs and involve summations of two main terms, the enthalpy of mixing and the change in entropy.

$$\Delta G = \Delta H - T\Delta S \tag{11.1}$$

Phase separation will occur when the free energy of mixing (ΔG) is positive. The enthalpy (ΔH) is equal to the heat of mixing and is related to the Flory–Huggins interaction parameter (X_{ab}), which in turn is a function of the differen in solubility parameters ($\delta_a - \delta_b$) between the block segments and the molecular weight and density (M_a, ρ) of the "a" segments.

$$\Delta H = f(\chi_{ab}) \tag{11.2}$$

$$\chi_{ab} = \frac{(\delta_a - \delta_b)^2 M_a}{\rho_a RT} \tag{11.3}$$

PHASE-MIXING IN BLOCK COPOLYMERS

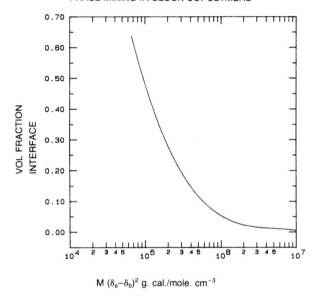

$M (\delta_a - \delta_b)^2$ g. cal./mole. cm^{-3}

Figure 11.2   The volume fraction of polymer homogeneously mixed at the interface as a function of the product of molecular weight and the squared difference in solubility parameter in (A–B) block copolymers. Drawn from relationship derived by Meier

The change in entropy is due mainly to the reduction in entropy from constrained A–B block junctions. One can view the equation describing microphase separation in block copolymers as one that describes the conditions under which phase separation will occur or alternatively as an expression of the energy level that must be overcome to dissociate the blocks (i.e. to provide homogeneous mixing of the segments). Phase separation leads to a domain structure consisting of areas of pure A block composition, areas of pure B block composition, and an area of mixed composition that we call the interface region or interface volume. If the polymer architecture allows ties between domain areas (such as a triblock A–B–A copolymer) a phase-separated network structure results.

The interface constitutes a volume that is quantitatively dependent on the driving force for phase separation. It is very important because it is a region of gradient composition which can actually lower the interfacial tension between different polymer pairs [6]. Figure 11.2 shows the relationship between the interface volume and the interaction parameter between blocks (M being the molecular weight and δ the solubility parameter in the figure) [7]. The figure shows that unless the polymer pairs are sufficiently different chemically, or molecular weight is sufficiently high, a very substantial volume of the polymer will be homogeneously mixed.

Figure 11.3 shows a map of solubility parameters for the most common pairs of polymers that make up anionically polymerized block copolymers. Both ordinate and abscissa are scaled in units of solubility parameter; polymers used as hard-blocks or end-block are plotted on the abscissa while polymers used as the elastomer-block are plotted on the ordinate. Values of T_g, and in one case (polyethylene) T_m, are indicated on the labels. Diagonal lines indicate the squared difference between ordinate and abscissa values and

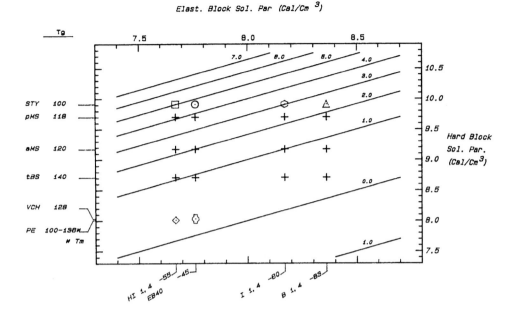

Figure 11.3 Map of hydrocarbon block copolymers. Solubility parameters are calculated from group contributions for the rubbery state, corrected to 25 °C. Diagonal lines indicate locus of squared difference in solubility parameter. ⊙ S–EB–S; ☐ S–EP–S; ◔ S–I–S; △ S–B–S; ◇ HY(B–I–B); ⬡ V

therefore describe the Flory–Huggins interaction parameter between polymer pairs that are located at intersections of the plotted polymers on the map at constant molecular weight. Solubility parameters are calculated by the method of group contributions [8]. Polymers that are either commercially manufactured or extensively synthesized are indicated by the crosses and symbols at the intersection of the respective polymers. The end-block polymers are based on polystyrene, alkyl-substituted polystyrenes, poly(α-methylstyrene) (αMS), poly(p-methylstyrene) (pMS) and poly($tert$-butylstyrene) (tBS), hydrogenated polystyrene (polyvinylcyclohexene, VCH) or polyethylene, which results from the hydrogenation of predominantly 1,4 polybutadiene and has a range of melting temperatures (T_m) depending on the level of 1,2 addition in the precursor. The elastomer blocks are 1,4 polyisoprene (I 1,4) and its hydrogenated polymer (HI 1,4); 1,4 polybutadiene (B 1,4) (typically with 5% to 10% 1,2 content), and hydrogenated polybutadiene of about 40% 1,2 content (EB$_{40}$).

The criteria for microphase separation into the domain network structure, the interface volume, and the consequent dissociation energy of the block copolymers are dependent on the interaction parameter, that is, on the squared difference in solubility parameter. Further, the processing rheology and the mechanical property set, especially time-dependent and temperature-dependent properties, also will be very dependent on this dissociation energy, which is the level of energy holding the polymer material in a network structure. So, to judge the criticality of this parameter, we can reference all of the block copolymers on this map to the S–B–S polymer shown. This S–B–S is easily processed, shows a network structure at reasonable molecular weights, and has high strength and reasonable creep rate at 23 °C, however, it has excessive creep at 50 °C. We can, therefore, say that the interaction parameter represented by this copolymer is a minimum value for useful block copolymers of ordinary molecular weight. Polymers that have squared difference in solubility parameter lower than that of S–B–S would have insufficient driving force for separation and therefore a greatly inferior property set. This is primarily why the alkyl-substituted polystyrene polymers, which have higher T_g values than polystyrene, are not used in commercial block copolymers. They would exceed the minimum interaction parameter only in the case where the elastomer block was hydrogenated. The advantages of stronger phase separation far outweigh the advantages of higher hard-block T_g. Thus polystyrene remains the choice for any amorphous hydrocarbon block copolymer. This last fact is clearly demonstrated in the case of the fully hydrogenated VCH–EB–VCH polymer. The interaction parameter is so severely reduced by hydrogenation that at only slightly elevated temperatures the polymer loses all strength and appears to be homogeneously mixed at ordinary melt temperatures.

The polymer labeled Hy(B–I–B) is a special case. In this polymer the end-block is a hydrogenated 1,4-polybutadiene and therefore is a "linear" polyethylene. The elastomer block is polyisoprene and on hydrogenation becomes a 1 : 1 ethylene–propylene copolymer. If the quality and molecular weight of the polyethylene end-blocks is such that rapid crystallization can occur, phase separation is the result of polyethylene crystallizing out of the homogeneously mixed segments. Crystallinity in these polymers can be measured by differential scanning calorimetry (DSC) although the crystallites cannot be seen by light or electron microscopy [9]. The reason these and similar polymers are not made commercially is because the polyethylene made this way is not very "linear," that is, the precursor 1,4-polybutadiene has a substantial 1,2 content that breaks up the crystallizable segments. Current anionic polymerization technology does not allow reduction of 1,2 content below about 5%. Also, without adequate phase separation in the melt, crystalline domains are not well developed and therefore are not as strong as spherulite phases in ordinary high-crystallinity polyethylene.

Thus, the principal commercial offering of hydrogenated polymers is S–EB–S, a polymer with polystyrene end-blocks, and ethylene–butylene rubber blocks. In this polymer the interaction parameter is at least two-and-one-half times the critical base case for S–B–S because of the shift to lower solubility parameter of the 1,4- and 1,2-polybutadiene segments on hydrogenation. The effect of this shift on the property set of S–EB–S compared to S–B–S is quite dramatic. To illustrate this shift, properties of two S–EB–S polymers (I and II) are now compared to those of an S–B–S. Further details of all three polymers are given in section 11.4.

The hydrogenated polymers typified by S–EB–S I have such a high association energy that they will not spontaneously disassociate above the polystyrene T_g, nor will they disassociate in response to a rather strong mechanical stress field. Figure 11.4 shows the apparent melt viscosity of such a polymer as a function of shear rate and temperature from 200 °C to 300 °C. The polymer melt viscosity is extremely non-Newtonian, resembling that of a solid (a solid would have a slope of -1 in this plot), with a slope of 0.98 at the lowest temperature. At 260 °C the slope has decreased to only about -0.89. This is a result of the intact network structure which persists in the so-called "melt" and prevents the development of dissipative flow in the shear field. At least up to 260 °C, flow is not temperature activated, an apparent contradiction in terms for a thermoplastic material. This same type of behavior is shown by covalently crosslinked rubbers and by some types of ionomers.

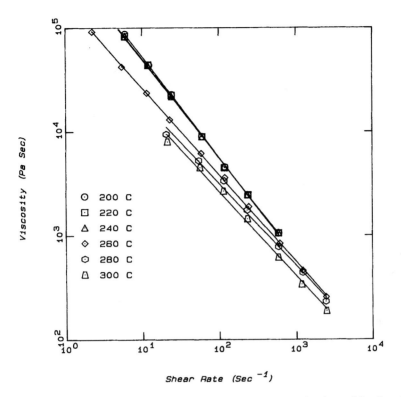

Figure 11.4   Melt viscosity of S–EB–S I hydrogenated block copolymer. The slope of the viscosity/shear-rate curve is nearly 1.0 up to 260 °C

11.2.3   Morphology and Strength

The equilibrium morphology of styrenic block copolymers, where the "A" phase volume fraction is lower than that of the "B" phase, consists of phase-separated "A" domains in the form of spheres, cylinders, or lamellae dispersed in a matrix of the "B" phase (Fig. 11.5). The geometric form will be that which minimizes the surface free energy between phases. The typical morphology of the polymers studied here is dispersed cylinders. If the polymer architecture is triblock or multiblock, the mechanism of network structure is bridging between domains by molecules whose "A" blocks enter different "A" domains. This creates a barrier to shear flow just as covalent crosslinking does. The mechanisms of failure and flow in a stress field, then, are very closely related, arising from the pulling-out of bridging chain "A" blocks, and are of comparable energy. In the "melt" state this energy is low, whereas in the "solid" state it is comparable to that of covalently crosslinked systems. The creation of sheared interfaces results in the formation of new units located in the stress field.

The tensile behavior of the three polymers is shown in Fig. 11.6. Both S–EB–S polymers (only one of which is shown) are stronger and have higher modulus than that of the S–B–S. This again is attributable to the higher association energy and the absence of substantial interface volume in the S–EB–S. The elongation of the S–EB–S polymers is lower because

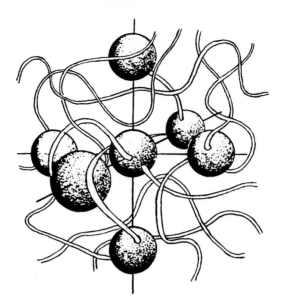

Figure 11.5   Spatial domain and network structure for the case of spherical domain geometry. A network, incapable of flow, is created when a few interdomain crosslinks are formed

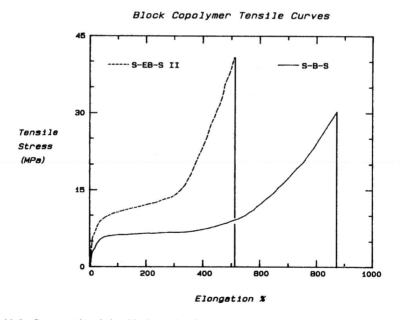

Figure 11.6   Stress–strain relationship in tension for saturated and unsaturated block copolymers

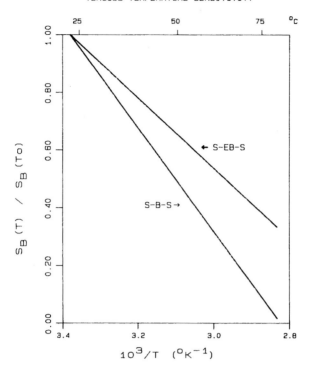

Figure 11.7   Arrhenius plot of tensile strength of S–EB–S II and S–B–S. Breaking stress is normalized to the value at 23 °C

the contour length of the S–EB–S is lower than that of the S–B–S (40% 1,2 vs. 8% 1,2) and therefore stress transfer to the dispersed domain structure occurs at lower strain. In line with this, the higher association energy is obvious in Fig. 11.7. Arrhenius plot of the tensile strength normalized to 23 °C. The rate of tensile loss in the S–EB–S with an increase in temperature is far less than that of the S–B–S analogue polymers. As shown in Fig. 11.8, the rate of stress relaxation increases with temperature to a far lesser degree in the S–EB–S materials and is even comparable at this level of strain to that of vulcanized polymers.

11.2.4   Dynamic Mechanical Properties

One of the problems common to both Ziegler-polymerized EPR or EPDM polymers and the S–EB–S polymers is the effect of polyethylene crystallinity on the elastic properties of the elastomer (i.e., resilience, rebound, recovery, etc.). To overcome this problem it is necessary that almost all methylene runs that otherwise would be large enough to join the polyethylene crystal must be interrupted. As was shown in Fig. 11.1, a balance between reduction of crystallinity and increase in T_g is achieved at a 1,2 content of about 35%. Figure 11.9 shows

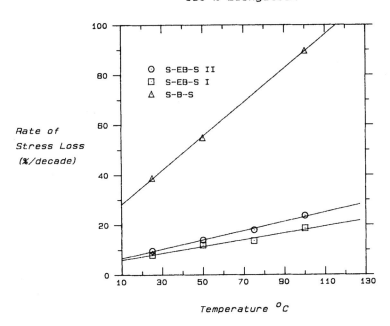

Figure 11.8   Rate of stress–relaxation as a function of temperature

that the dynamic hysteresis is minimized at about this same level. The curve was derived from a large amount of data that is not reported here but used somewhat different techniques. Values of dynamic hysteresis measured for the S–B–S polymer and the two S–EB–S polymers are shown in Table 11.1.

Conventional vulcanized natural rubber gum compounds have values in this test of less than 5% while clay-filled SBR compounds have values of about 25% to 30%, so these polymers are quite "rubbery" by comparison to standard rubbers.

The dynamic mechanical spectra of the three polymers are shown in Figs. 11.10, 11.11, and 11.12. The first difference that we focus on lies with the rubber block. This peak in the S–B–S sample (Fig. 11.12) has an intensity of 1.4 tan δ and a width of about 40 °C centered at about −84 °C. On the other hand, the EB peaks in the S–EB–S samples (Figs. 11.10 and 11.11) have intensities of about 0.3 and a width of 70 °C or more and are centered at −45 °C with a marked shoulder in the loss peak. This indicates a broader distribution of relaxation times, likely due to the existence of amorphous runs of ethylene and butylene and a small amount of crystallinity.

The modulus level in the rubbery plateau region is dependent primarily on the hard-block (polystyrene) concentration, being highest for the S–EB–S I sample. Since material in the interface will have transitions between those for pure A and pure B, the level of losses in the rubbery plateau region will be increased by the volume in the interface [10]. These data show that the S–B–S sample has the higher level of losses in this region (tan $\delta \sim 0.06$ compared to

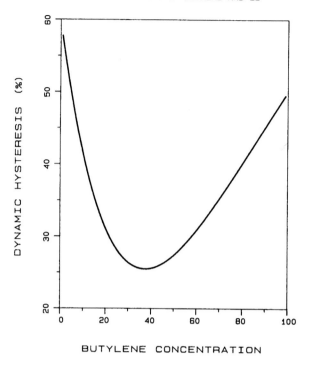

Figure 11.9   Dynamic hysteresis as a function of butylene concentration. Taken from process development data, continuous function shown is an empirical model

tan $\delta \sim 0.04$ for the two S–B–S polymers). The S–B–S polymer also has lower modulus than its hydrogenated analogue because of the decreased volume of pure A phase which determines the dynamic modulus level in this region.

In all three materials the loss peak associated with the polystyrene phase appears to be located at about 110 °C (somewhat higher for the S–EB–S I sample, which has higher molecular weight). However, from the modulus curve one can see that the onset of this transition occurs at about 75 °C in the S–B–S material and not until about 95 °C in S–EB–S II,

Table 11.1 Hysteresis of Block Copolymers

Sample	Dynamic hysteresis (23 °C, 150% elongation)
S–B–S	8.8
S–EB–S I	9.6
S–EB–S II	15.5

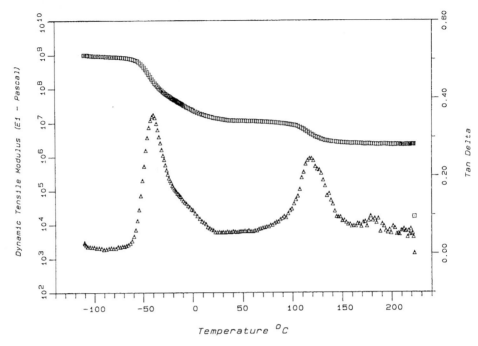

Figure 11.10   Dynamic mechanical spectrum at 11 Hz for S–EB–S I. Modulus (squares) and tan δ (triangles) as a function of temperature

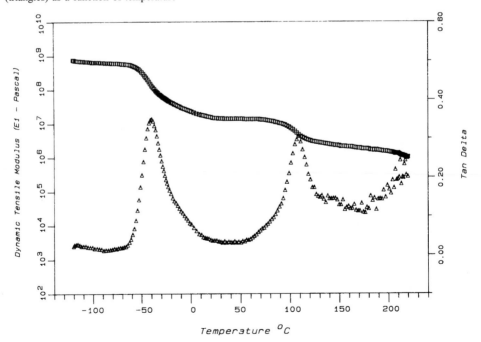

Figure 11.11   Dynamic mechanical spectrum at 11 Hz for S–EB–S II. Modulus (squares) and tan δ (triangles) as a function of temperature

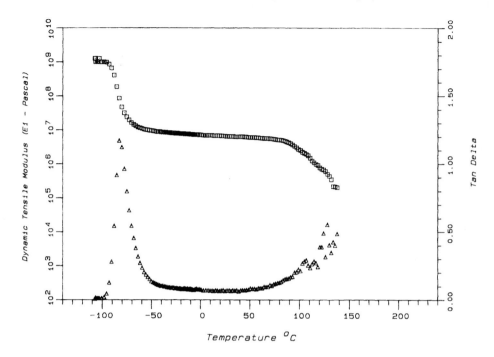

Figure 11.12   Dynamic mechanical spectrum at 11 Hz for S–B–S. Modulus (squares) and tan delta (triangles) as a function of temperature

and 100 °C in the S–EB–S I polymer. In addition, the plateau region above the transition is highest and flattest in the S–EB–S I sample, and does not even exist in the S–B–S sample. These phenomena are the result of the much stronger phase separation in the S–EB–S materials; recall Fig. 11.3. The volume of pure A phase is highest and the interface volumes lowest in the S–EB–S samples and phase separation above the end-block transition temperature does not occur at the strain levels of this test. There is a measurable yield stress above the styrene T_g that is related to the interaction parameter between the block segments and in both S–EB–S samples is high enough to maintain the integrity of the material.

The lack of any dissipative processes in the "melt" regimen suggests that the S–EB–S polymers are rather intractable. The melt viscosity as a function of shear stress is shown in Fig. 11.13 for the three samples studied here. The S–B–S sample on the right is typical of well-developed flow in a thermoplastic polymer. The analogous S–EB–S II polymer fails to reach a comparable level of flow even at 100 °C higher temperature although it does show some flow processes. In the higher molecular weight S–EB–S I materials, there is extremely little flow over its melt temperature range. This "melt" behavior is very similar to that of covalently crosslinked rubbers. Both the dynamic modulus in the plateau region above the styrene T_g from the dynamic data and the shear yield stress from the capillary measurements are of the same magnitude as that of a vulcanized, particulate-filled styrene-co-butadiene random copolymer SBR material.

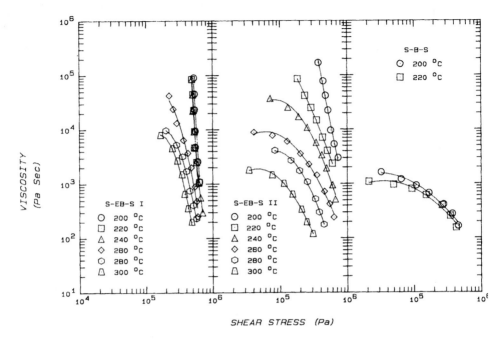

Figure 11.13   Corrected melt viscosity as a function of shear stress and temperature for the three block copolymers studied

11.3   Thermoplastic Interpenetrating Polymer Network Formation and Properties

11.3.1   Concept of Thermoplastic Interpenetrating Polymer Networks

IPNs have received a great deal of attention since they were first proposed by Klempner, Frisch, and Frisch in 1971 [11]. The rationale behind the concept of IPNs was that normally immiscible polymers could be locked together in a catenated structure by covalent cross-linking once an interpenetration of entangled chain segments was achieved. Crosslinking would provide a kinetic barrier to gross phase separation or reversion to the lower free energy, demixed condition. The ideal state was visualized as two infinite networks, everywhere interlocked but having no mutual chemical attachment.

    In the period since the IPN concept was introduced, a great many conventions and much specific terminology have developed dealing with the emerging technology. For the most part, these conventions are perceived from the point of view of the practitioner who is attempting to produce the structure. There is a very methodical, descriptive terminology that differentiates IPNs on the basis of the sequence of steps intended to achieve the interdispersed networks [12, 13]. Although thermoplastic IPNs have a great deal of conceptual commonality with thermosetting IPNs, at present they are synthesized in an entirely different way [14]. Therefore, it is difficult for thermoplastic IPNs to fit this increasingly accepted terminology convention that is being applied to thermoset IPNs.

It must be granted that chemical or "covalent" crosslinking is not an absolute requirement in the production of a network structure. Consider a material consisting of a polymer that has gelled in a solvent under some set of conditions. If one can note in a gel a resistance to some kind of disruptive field, a macroscopic stress field for instance, one then presumes that this resistance arises from a network structure. In the gel there is spatial connectivity of every element of the polymer phase; the polymer is continuous throughout the macroscopic volume. Spatial connectivity also applies to the solvent. The mechanism of stabilizing the gel after the instant of its formation need not be interchain chemical crosslinking. Such gels just as well could be locked into place through a reversible, physical mechanism and, in fact, quite commonly are. Block polymers of certain architecture form networks with some measurable level of association, just as crosslinking forms networks with a measurable level of association. Both types of network will dissociate if this level of energy is exceeded [7].

The conceptual structure evoked by the original IPN theory is that of co-continuous polymer networks or interdispersed gels, interpenetrating on some finite scale of mixing. In one theoretical extreme, the scale of mixing is small enough to be at the level of thermodynamic solubility of polymer chain segments. Interpenetration then could be traced only along the backbone of individual polymer chains. However, if there is association of like chains at a level greater than molecular scale, then the components begin to take on characteristics familiar to their species and the gel network is an interpenetration of polymer phases. In this case, interpenetration could be traced by a number of possible routes within any element of the network. Crosslinking within these phases can proceed but it constitutes an ordinary internal network, quite apart from the macro-network that constitutes that particular IPN.

The existence of identifiable phases clearly prevails in thermoplastic IPNs, although there may be some doubt in certain thermoset IPNs. The overall structure is rendered metastable by the formation, within one or more phases, of a persistent, micro-network structure, continuous within the macroscopic phase volume in all dimensions. This interval network, whether arising from covalent bonding or reversible physical interaction, indeed is a vital element of the structure but it does not define the IPN. It is the final, persistent, macro-topological structure of the phases that defines an IPN, regardless of the chemical or physical route by which it was created.

We define IPNs as equilibrium blends of two or more polymers in which at least two of the components have three-dimensional spatial continuity. Implicit in this definition, especially for thermoplastic IPNs, is the notion that each component is a polymer phase, generally with its own internal networklike structure from which its properties arise. In a binary mixture, the surface of each of the phases is an exact topological negative of the other, that is, they are "antitropic." It is obvious in that case that only one of the two phases need be spatially fixed to completely stabilize the entire framework. This definition, of course, is a morphological description of physical structure, not of the route by which an experimenter chose to create it.

Nature is replete with examples of materials that possess bicontinuous structure (our definition of IPN), from sandstone, open-celled sponges, and various naturally occurring biological membranes, to the polymer blend structures that result from spinodal decomposition above a lower critical solution temperature. Many writers, artists, and scientists have envisioned stylistic IPNlike frameworks, for example, the co-eating apple worms of George Gamow, "Double Planetoid" by Maurits Escher and the "Olympic Gels" of P. deGennes.

Scriven has very elegantly described space-filling topologies that are equilibrium bicontinuous structures and that are illustrated by the models of periodic minimum-surface structures of Schwarz and Neovius [15–17]. The requirements of bicontinuous structure are spatial (mathematical) connectivity, positive genus (multiply connected), with each subvolume intersecting (touching) each outer surface in more than one place. Scriven proposes that such bicontinuous structures can exist as an equilibrium morphology in water–oil emulsions that contain an amphiphile at the point in the phase diagram where inversion occurs.

We can now postulate that blends of two polymers that contain no additional phases, fall into three major morphological types depending on the scale of interdispersion and the continuity of the phases. With binary mixtures of polymers A and B one can have:

1. Miscible blends or combinations where the scale of mixing is at the magnitude of the molecular dimensions. The components lose their individual identity and characteristics of a new hybrid material are expressed.
2. Matrix-disperse blends where either A is dispersed in a matrix of B or the opposite case, B dispersed in A. The components retain their individual identity but the properties of the matrix are predominantly expressed.
3. Co-continuous blends or IPNs where each of the two polymers has three-dimensional continuity. The components retain their individual identities and thus the properties of both are fully expressed.

As one considers variations in the morphology of the individual polymeric phases of an IPN, as we will point out later, it becomes apparent that there are a number of morphological variants that are more subtle than phase connectivity. Both thermodynamic and kinetic factors are involved in determining which specific morphology will occur in a certain mixture. In binary thermoplastic IPN systems, in most cases, one can form the co-continuous structure at least as a transient structure, through the application of correct methods of mixing. Interfacial tension between the incompatible polymer components drives the system toward minimum surface free energy, that is, to phase growth. It is then the function of network structure within one or both continuous phases to kinetically inhibit retraction. This inhibition may be achieved by chemical or physical crosslinking. The possibilities include covalent bonding, ionic or phase association, domain formation, crystallization, and vitrification. All thermoplastic IPNs contain an internal network in at least one of the co-continuous polymer phases. Ideally, perhaps, for the ultimate in stability and properties, covalent bonds would have to be broken for phase disengagement to occur, or alternatively some process of comparable energy dissipation would have to take place.

11.3.2   Formation of the Interpenetrating Polymer Network

Intensive mixing of viscous incompatible polymer pairs in the melt can result in the development of an IPN structure in the shear field. The most efficient mixing of two components into a finely divided interdispersed structure can be achieved when the viscosities and the volume fractions of the two components are equal (Fig. 11.14). This situation, termed "isoviscous" mixing, maximizes coupling of adjacent fluid elements and assures an even distribution of the imposed shear field during mixing. Equal volume fractions maximizes the opportunity for maintaining connectivity since neither component is present in a minor amount. As the viscosities of the components diverge, the efficiency of shear stress transfer across an element of the lower viscosity phase is progressively reduced, requiring a reduction in its volume fraction to compensate for the reduced stress coupling. This mechanism of IPN formation describes the structure in the melt state, in the mixing shear field, and does not require the network characteristics we have illustrated for the S–EB–S I material. Such structures may survive to varying degrees after the shear field is removed and while the melt is being transferred prior to solidification. Rapid quenching of an IPN melt, prior to retraction, will preserve the structure by crystallization or vitrification. In the quiescent melt, however, the morphology will be transient in the absence of a kinetic barrier to flow such as would be provided by the block copolymer.

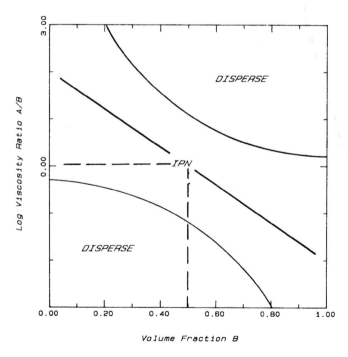

Figure 11.14   IPN mixing relationship showing the line along which co-continuity is easily obtained with two components that have similar viscosities. The permitted region is considerably expanded when block copolymers are used

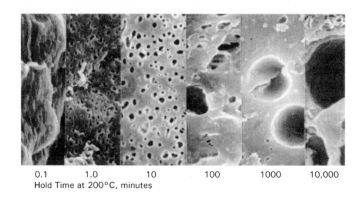

0.1   1.0   10   100   1000   10,000
Hold Time at 200°C, minutes

Figure 11.15   Phase growth of a polypropylene (70)/polybutylene (30) IPN blend. SEM at 5000 × shows the increase in phase size as time at 200 °C increases to 10,000 min. (1 week)

An example of a transient, unstabilized IPN was seen in blend BL1, a blend of polypropylene and polybutylene, two similar, linear polymers. On mixing, these polymers readily formed an extremely fine "celled" IPN that was preserved by quenching the material immediately after it cleared the extruder die. The small size, on the order of 0.1 μm, attests to the facility of interdispersing components of similar solubility parameter. When held quiescent in the melt (200 °C) this blend underwent gradual but extensive phase growth (Figs. 11.15 and 11.16), while retaining continuity of the phases, revealed by extraction and electron microscopy examination. The chemical similarity of the two polymers results in a low interfacial tension, the predominant driving force for phase growth or phase "retraction." Phase retraction in this system was an extremely slow process, with the phase size increasing about 100-fold over the period of a week. This IPN blend is sufficiently stable to retain fine cell size under practical melt fabrication and conditions. We conclude that low interfacial tension facilitates IPN formation and reduces the rate of retraction.

When polypropylene is blended with the S–EB–S I polymer, as in IPN–B2, a structure on the order of 0.2 μm is obtained. This structure shows no phase retraction or growth even after 1000 h at 200 °C (Fig. 11.17a). In this case the low interfacial tension is assisted by the network characteristics of the S–EB–S polymer, which provides the kinetic barrier to phase retraction. The same stabilization is shown when only 10% of the S–EB–S I is added to IPN–B1, resulting in a stable ternary IPN mixture. When the polypropylene level is increased to 50% as in IPN–B3 (Figure 11.17b), the characteristic cell size of the mixture is somewhat higher, 0.5 to 1 μm. As we go to polymers that are less chemically similar to the S–EB–S polymer, the binary IPNs from these polymers have increasing phase size. This is shown in the IPN–B4 (Fig. 11.17c), a blend of S–EB–S with nylon 12, and in IPN–B5 (Fig. 11.17d), a blend of S–EB–S with polyethersulfone.

The structure that can be seen in these photographs is typical of the thermoplastic IPN structure. It is more chaotic than any perceived minimal surface that might be present, for instance, in phases that are a result of a spinodal growth mechanism above the lower critical solution temperature. Complementary phases, which are deduced by SEM are depicted in the model in Fig. 11.18. This model is a conception resulting from the careful examination of a great number of IPN structures by SEM. The model shows both concave and convex

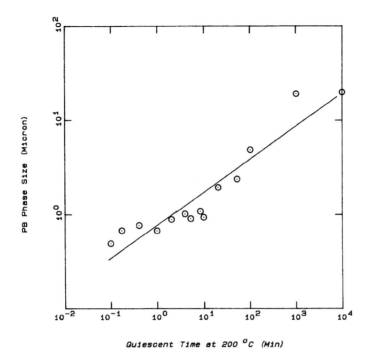

Figure 11.16   Phase size measured as the IPN of Fig. 11.15 is held in the melt

surfaces and interlocking and continuous phases. In a binary IPN, the co-continuous phases have some topological characteristic differences; one phase, usually the phase having network structure in the melt, seems to be primarily convex, forming a "skeletal" structure, while the second phase then occupies the space around that phase and has a "matrix" characteristic. This skeletal and matrix phase character is depicted in Fig. 11.19. The skeletal phase characteristic emphasized in this element is the rounded convex surface which maintains a "limblike" connectivity to other elements and constitutes a template for the matrix phase. The matrix is the surrounding phase, which takes the opposite, concave surface at the interface. These structural identities are easily seen in Fig. 11.20a, showing a deeply extracted nylon/S–EB–S IPN residue. The void areas clearly show the form of the skeletal S–EB–S phase and the antitropic nylon matrix. These characteristics are also seen in the SEM of IPN-B6 (Fig. 11.20b), a polypropylene/S–EB–S IPN which was etched with an oxygen plasma and viewed at a 120 degree angle. The small nodes or nodules at the etched surface are "skeletal" S–EB–S continuous phase. (Since polypropylene etches faster, the residue in the unshielded area is the S–EB–S.)

Figure 11.21 shows the cell size of binary blends as a function of the squared solubility parameter difference (the thermoplastic resin and the EB block). As the difference increases, or as interfacial tension increases, the cell size of the IPN shows a very rapid increase. The higher the interfacial tension, the greater will be the work required to increase the interfacial area between the phases. (These blends were not made with the high molecular weight S–EB–S polymer and result from ordinary viscous mixing.)

Figure 11.17a   Fracture surface SEM of extracted sample of an IPN blend containing 22% polypropylene and 78% S–EB–S I block copolymer. The S–EB–S has been extracted (2000 ×)

Figure 11.17b   Fracture surface SEM of extracted sample of an IPN blend containing 50% polypropylene and 50% S–EB–S I block copolymer. The S–EB–S has been extracted (2000 ×)

Figure 11.17c   Fracture surface SEM of extracted sample of an IPN blend containing 25% NY-LON 11 and 75% S–EB–S I block copolymer. The S–EB–S has been extracted (2000 ×)

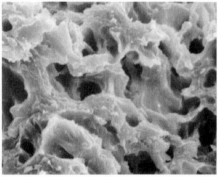

Figure 11.17d   Fracture surface SEM of extracted sample of an IPN blend containing 50% polyethersulfone and 50% S–EB–S I block copolymer. The S–EB–S has been extracted (2000 ×)

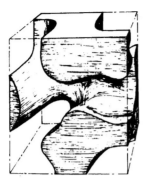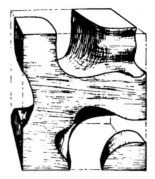

Figure 11.18   Model of the IPN structure based on SEMs of a large number of blends. Interlocking halves are separated and both convex and concave surfaces can be observed. Model is depicted at about 25,000 ×

As we mentioned earlier, the S–EB–S I polymer shows no measurable true shear flow and no apparent temperature activation of flow and is capable of templating the IPN structure. This polymer, in its gellike state, fragments in the blending device only along shear planes when mixing with an incomplete resin, and easily forms the skeletal continuous structure while the thermoplastic resin then takes up the antitropic continuous structure. This can take place by reversibly removing a few terminal blocks from domains, while retaining a "flow unit" structure of a size distribution that is consistent with the shear field.

The cohesive strength of the intact block polymer network structure prevents the breakup of the block polymer phase into droplets that would form a disperse phase; the reformation of the continuous phase then can occur by the restoration of the same number of chain segments that were breached in forming the macroscopic flow unit. This mechanism is analogous to dynamic vulcanization except that in the case of the block copolymer templated IPN the particles have the possibility to reconnect through the reversible network structure. This is the point in which S–EB–S diverges from all other network polymers, which merely serve to stabilize the IPN once formed by ordinary viscous mixing. The result is freedom from the constraints of the viscosity/composition ratio relationship and the interfacial tension

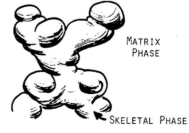

MATRIX
PHASE

Figure 11.19   Model of the skeletal element of the IPN. The components with the higher surface tension or the higher viscosity become the skeletal component with convex surface. The "antitropic" phase is the co-continuous matrix phase

SKELETAL PHASE

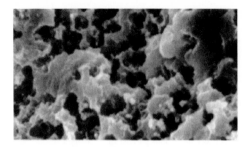

Figure 11.20a   SEM of a nylon/S–EB–S IPN mixture. The skeletal structure of the extracted rubber can easily be seen in the nylon residue

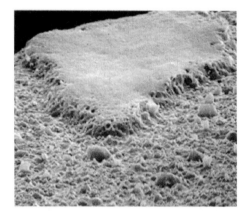

Figure 11.20b   View of SEM of an oxygen–plasma-etched sample of polypropylene and S–EB–S. Area at the top of the micrograph was shielded from the plasma. Since the polypropylene etches faster in the plasma, protuberances in the unshielded area are the S–EB–S skeletal phase

relationship. High molecular weight S–EB–S polymers consistently form fine phase size IPNs independently of these factors.

We have seen that factors important to IPN structure formation and stability are the viscosity ratio, compositional ratio, and solubility parameter difference. These control the interfacial tension and network structure. Other factors of importance are the absolute melt viscoelastic characteristics, viscosity and melt strength, and the degree of mixing or the energy field input in the mixing device. Figure 11.22 shows the effect of interfacial tension, of shear rate in the mixing field, and of quiesent phase retraction time on IPN phase size (in the case of isoviscous mixing) and the lack of effect (in the case of S–EB–S templating). High shear-rate in the mixing device will reduce the cell size up to a limit of the ability of the polymer pairs to resist degradation and to the practical limit of even the most intense mixing devices of about 2000 to 3000 s^{-1}. In addition to factors that are related to the shear field and

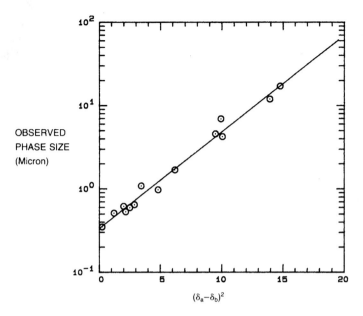

Figure 11.21   Observed phase size as a function of the square of the (D) solubility parameter difference between the resin IPN component and the elastomer block of the S–EB–S component

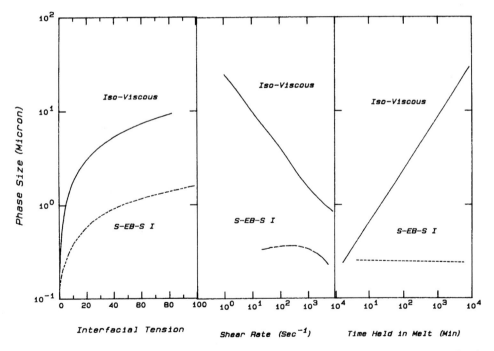

Figure 11.22   Factors that affect phase size in IPN structures, the interfacial tension, shear rate, and melt residence time. The response is different for the isoviscous case and the case where high molecular weight S–EB–S is used as the IPN former

the relationship between the IPN components, the individual components of the IPN mixture must resist breakup in elongational flow as the mixture is carried forward in the mixing extruder so as to maintain the connectivity that is developed in the shear field [18].

11.3.3   Mechanical Properties of Interpenetrating Polymer Networks

The benefit of a co-continuous structure vs. a dispersed structure is that the properties of a co-continuous structure generally can be described by an additive relationship. This effect for a polycarbonate/S–EB–S blend is shown in Figs. 11.23 and 11.24. In the first case there is only a hydrodynamic effect of the volume of polycarbonate used on the modulus. At the S–EB–S polystyrene glass transition temperature the small increase in modulus disappears. In the second, co-continuous, case, the modulus is increased in proportion to the polycarbonate volume fraction and thus the blend is still load-bearing up to the T_g of the polycarbonate rather than up to that of the polystyrene. Therefore, with co-continuous resin phases that are capable of withstanding high temperatures, the S–EB–S compounds are able also to provide stiffness and strength at temperatures well above the T_g of polystyrene.

In the IPN blends with co-continuous polymer phases, the components express their individual characteristics to the extent of the fraction of the total volume occupied by that phase. There is a load-path continuity that accompanies the structural phase co-continuity. In terms of intuitive continuum mechanical models, the IPNs resemble most closely a parallel combination of elements such as is employed for uniaxial, long fiber reinforcement in composites. The most important points in the application of any model relating properties of IPNs to the structure are the independence of the contribution of the components and the three-dimensional spatial connectivity of the elements.

The upper bound solutions in most continuum mechanical models (e.g., composite relationships developed by Kerner, Uemura, and Takayanagi; Nielson, Sato, and Furukawa; Coran and Patel; or Hapin and Tsai) resolve to simple linear additive relationships for the case of perfect adhesion and perfect connectivity in one dimension [19–24]. This is usually referred to as the "rule of mixtures" where strain is constant in each element. Lower bound relationships describe a series arrangement of elements where stress is constant in each element. The upper and lower bounds in blend moduli are given from the component moduli by:

$$\text{Upper bound}: \quad M_c = M_a \theta_a + M_b(1 - \theta_a) \tag{11.4}$$

$$\text{Lower bound}: \quad M_c = 1/(\theta_a/M_a + (1 - \theta_a)/M_b), \tag{11.5}$$

$$M_c = \text{modulus of the blend,}$$
$$\theta_a = \text{volume fraction of component A,}$$
$$M_a = \text{modulus of component A, and}$$
$$M_b = \text{modulus of component B.}$$

The IPN "unit cell," consisting of binary combinations with parallel elements with perfect adhesion, under a macroscopic stress field provides three load paths, one each through each component phase, and one via a series combination of the two materials. The latter load path, as a series element, is a consequence of the three-dimensional co-continuity of the IPN

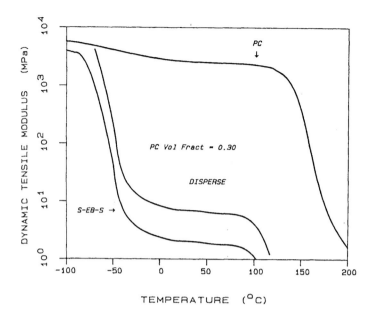

Figure 11.23   Dynamic modulus/temperature spectrum for S–EB–S I materials, polycarbonate (PC), and a blend of the two in which the PC is a particulate dispersed phase. The blend was made from powdered constituents far below the isoviscous temperature. The T_g of the polystyrene end-blocks in the S–EB–S I is at 110 °C

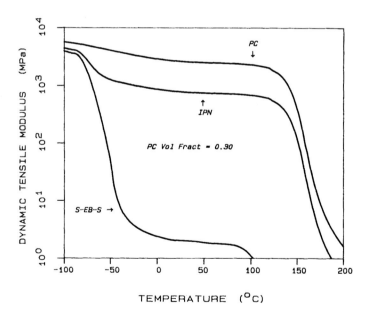

Figure 11.24   Dynamic modulus/temperature spectrum for S–EB–S I materials, polycarbonate (PC), and a blend of the two in which the PC is a IPN co-continuous phase. The blend was made from powdered constituents above the isoviscous temperature and with a long residence time in the melt

system. A somewhat more flexible model for this situation is a logarithmic rule of mixtures (the linear "rule of mixtures" is a special case of such a model with an exponent of 1.0). This is proposed by a number of workers to model the complex dielectric behavior of composites. Proposed by Looyenga with exponent weighting of 0.5 and later by Davies (with a weighting of 0.2) it is postulated to be a more rational model than the rule of mixtures for composites with three-dimensional phase continuity [25]. The model given below is shown plotted for a series of weighting exponents in Fig. 11.25.

$$M_c{}^K = M_a{}^K \theta_a + M_b K (1 - \theta_a) \tag{11.6}$$

Figure 11.26 shows a number of binary mixtures of S–EB–S polymers with various resins (listed in Section 11.4) at several levels. The measured blend modulus is expressed as a modulus ratio compared to the modulus of the S–EB–S I block copolymer, which is a component of all these blends. Each of these resins has about 350 times the modulus of the S–EB–S I polymer. The measured modulus data are shown plotted on the same grid as in Figs. 11.25 and 11.26, again showing the various logarithmic models. The data are located within a band in the upper region of this plot below the rule of mixture. They are described by weighting exponents in the range of 0.2 to 0.6 in the logarithmic mixture equation. It appears as if no single exponent is applicable to all the IPN systems.

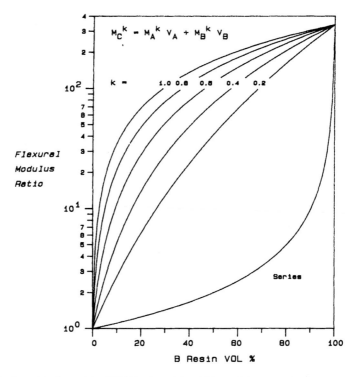

Figure 11.25   Logarithmic model of IPN blend modulus. This modulus is expressed as a ratio of the blend modulus to that of the softest component. Exponents in the model are given for various cases

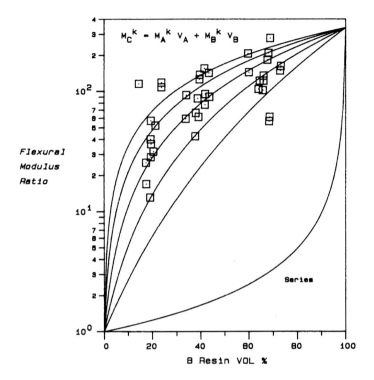

Figure 11.26   Binary blends of resins described in Section 11.4. Resin grades with modulus ratio (to that of the S–EB–S) of about 350 were used in this comparison. Data fall in the range of the logarithmic model with exponents in the range 0.2 to 0.6. Values of the exponent (k) are as shown in Fig. 11.25

In Fig. 11.27, the dynamic moduli measured at 205 °C for a series of binary mixtures of S–EB–S and PBT are shown. The lower solid curve is the Davies equation with exponent of 0.2. The data appear to fall in a range between this equation and the upper bound, with an exponent of about 0.6. Figure 11.28 shows the dynamic modulus of binary mixtures of S–EB–S and polycarbonate. These data are fit with a logarithmic model with an exponent of 0.5. Noteworthy in these data is the complete masking of the polystyrene glass transition of the S–EB–S by the polycarbonate continuous phase. Such mechanical property response can provide quite outstanding high-temperature integrity in blends at moderate stress levels with only moderate amounts of the high-temperature resin. For instance, amorphous resins can significantly raise the heat distortion temperature of thermoplastic crystalline resins.

Figure 11.29 shows the effect of temperature on the dynamic modulus of a number of IPN blend compositions, two vulcanized rubbers and a plasticized poly(vinyl chloride) (PVC). In every IPN blend the high-temperature resin, either polypropylene or polybutylene terephthalate (PBT) masks the polystyrene T_g of the S–EB–S and provides structural integrity up to the resin melt temperature. In the case of the blend containing both polypropylene and PBT, modulus level about the polypropylene melt temperature is equivalent to that of the vulcanized tread rubber. In this ternary IPN blend, the modulus level can be adjusted up or down by the polypropylene content while the PBT provides some resistance to deformation up to a temperature of 225 °C, where it also melts. The IPN blends show a true rubbery

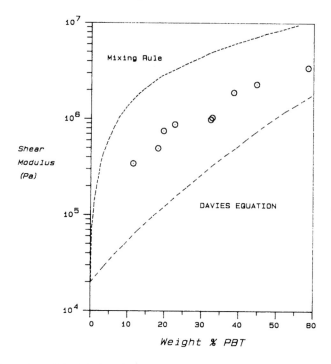

Figure 11.27   Dynamic modulus (obtained from a rheometrics mechanical blends spectrometer) of S–EB–S/ PBT blends in shear at 205 °C (a common bake oven temperature for curing painted high-performance materials). The data appear to be fit by an exponent in the logarithmic model of 0.5

plateau, not quite as flat as that of the vulcanized rubbers but much better than that of the PVC. They also provide the same level of integrity up to the same failure temperatures shown by the vulcanized rubbers. The Natural Rubber (NR) gum rubber degrades by losing rigidity at about 150 °C while the SBR begins to crosslink and embrittle at about the same temperature.

The relationships discussed for describing the modulus at very small deformations appear to adequately fit the co-continuous IPN blend modulus. The same rationale for modeling such high strain properties as tensile strength would be applicable only if the components had identical tensile response, a very unlikely situation. A useful analogy is found, however, in models of the tensile failure of macroscopic blends of textile fibers [26]. If we propose an IPN blend of two components or "fibers," one with higher breaking strain than the other, the fiber model proposes that if these fibers are clamped in parallel in a tensile fixture and pulled, then at the point of breaking elongation of the first fiber, the load supported by both fibers will be transferred to the remaining fiber. If this load represents a stress in excess of the breaking stress for the remaining fiber, the remaining fiber will also break, short of its breaking strain. This analysis presumes that the response of each fiber is independent of the other, and that there is no adhesion between fibers. The model is illustrated in Fig. 11.30. The fiber pairs are in uniaxial tension on the right and the stress–strain response is shown in the curve on the top left. As the strain causes breaking in fiber A (at elongation X_1) the load is

Since the original studies of anionic block copolymerization in the 1950s [1, 2] a variety of new polymerization methods (e.g. condensation, Ziegler–Natta, etc.) have contributed to an expanding number of block copolymer classes (e.g., A–B–C) and novel architectures (e.g., graft block). While some of these developments have resulted in important new materials (e.g., polyurethanes), anionic polymerization remains the only viable method for producing monodisperse block copolymers with well-defined architectures. Because current theories deal almost exclusively with model $(A–B)_n x$ type materials, we have restricted our attention in this chapter to studies based solely on this class of anionically polymerized block copolymers (see Fig. 12.1).

The phase behavior of undiluted (bulk) $(A–B)_n x$ block copolymers is determined by three experimentally controllable factors: the overall degree of polymerization N, architectural constraints characterized by n and the composition f (overall volume fraction of the A component), and the A–B segment–segment (Flory-Huggins) interaction parameter χ. (In the present review we use the terms monomer and segment interchangeably to imply statistical segment [3].) The first two factors are regulated through the polymerization stoichiometry and affect the translational and configurational entropy, while the magnitude of (the largely enthalpic) χ is determined by the selection of the A–B monomer pair. We note that for all the materials considered in this chapter, the interaction parameter has the temperature dependence $\chi \approx \alpha T^{-1} + \beta$, where $\alpha > 0$ and β are constants for given values of f and n. Since the $n = 1$ case has received the most comprehensive theoretical treatments and because the above factors qualitatively influence phase behavior independent of n, our introductory remarks and much of this text focuses on diblock copolymers.

At equilibrium, a dense collection of monodisperse diblock copolymer chains will be arranged in minimum free energy configurations. Increasing the energy parameter χ (i.e., lowering the temperature) favors a reduction in A–B monomer contacts. If N is sufficiently large, this may be accomplished with some loss of translational and configurational entropy by local compositional ordering as illustrated in Fig. 12.2 for the symmetric case $f = 0.5$. (Here we note that block crystallization, which can also occur, lies outside the scope of this chapter; we consider only amorphous copolymers). Such local segregation is often referred to as *microphase separation*; *macroscopic* phase separation is impossible in a single-component block copolymer. Alternatively, if either χ or N is decreased enough, the entropic factors will dominate, leading to a compositionally disordered phase. Since the entropic and enthalpic

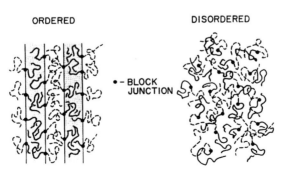

Figure 12.2   Schematic representation of order and disorder in a symmetric ($f = 1/2$) diblock copolymer showing lamellar order

contributions to the free energy density scale respectively as N^{-1} and χ, it is the product χN that dictates the block copolymer phase state. For $f = 0.5$ the transition between the ordered and disordered states occurs when $\chi N / n \sim 10$, as discussed below.

Two limiting regimens have been postulated to exist in the diblock copolymer phase diagram, as illustrated in Fig. 12.3. For $\chi N \ll 1$, a copolymer melt is disordered and the A–B interactions sufficiently weak that the individual chain statistics are unperturbed (i.e., Gaussian). The connectivity of the two blocks and the incompressibility of the melt, however, lead to a correlation hole [3, 4] that is manifested in scattering measurements as a peak corresponding to a fluctuation length scale $D \sim R_g \sim aN^{1/2}$. (Here R_g is the copolymer radius of gyration and a is a characteristic segment length.) As χN is increased to be O [10], a delicate balance between energetic and entropic factors produces a disorder-to-order phase transition. It has been suggested that in the vicinity of this transition, the A–B interactions are sufficiently weak that the individual copolymers remain largely unperturbed, the micro-domain period scales as $N^{1/2}$, and the ordered composition profile is approximately sinusoidal. We shall refer to such a regimen as the weak segregation limit (WSL). Current order–disorder transition (ODT) theories are based on this WSL assumption because it greatly simplifies calculations, although strict adherence of experimental systems to the WSL postulates is still not established. The second limiting regimen of phase behavior is referred to as the strong segregation limit (SSL) and corresponds to the situation of $\chi N \gg 10$. In this regimen, narrow interfaces of width [5] $a\chi^{-1/2}$ separate well-developed, nearly pure A and B microdomains. The interaction energy associated with A–B contacts is localized in these interfacial regions; the system would like to minimize the total area of such interface, but must do so under the constraint of incompressibility and with the entropic penalty of extended chain configurations [5, 6]. These opposing forces lead to perturbed chain configurations and microdomain dimensions (periods) that scale as $D \sim aN^{2/3}\chi^{1/6}$.

Since most theories and experiments dealing with block copolymer phase behavior can be categorized as either WSL or SSL, we have organized the present review under these general headings. However, as will be discussed further, this classification scheme may break

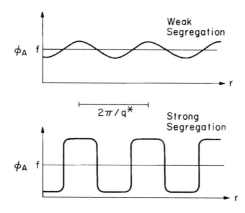

Figure 12.3   Comparison of the one-dimensional composition profiles characterizing the weak (WSL) and strong (SSL) segregation limits. ϕ_A and f refer to the local and stoichiometric (i.e., macroscopic) A-block volume fractions, respectively

down in the ODT region. Regardless, it proves most convenient to treat the transition region in the WSL section.

The preparation of model (undiluted) block copolymers for the purpose of studying the ODT [also referred to as the microphase separation transition (MST)] is complicated by the limited range in χN afforded by experimentally accessible temperatures. To overcome this difficulty, researchers often add modest amounts of a neutral solvent to the bulk material, thereby diluting the A–B contacts. Here a neutral solvent is defined as one that shows no preference for either block type. In general, such concentration solutions behave much like the bulk materials, with χ being replaced by an effective interaction parameter that is proportional to the copolymer concentration. Thus, we have included this restricted group of block copolymer solution studies in the present article; semidilute and dilute copolymer solutions fall outside the scope of our review.

This review is organized into four sections, each dealing with theory and experiment. In the first and second sections we cover recent developments in the strong and weak segregation limits, respectively. Limitations in size constrain us to significant developments that have occurred within roughly the past decade. Prior advances in block copolymers are documented in earlier reviews [7–12]. We further restrict ourselves to issues related to block copolymer thermodynamics, leaving a vast literature on copolymer dynamics to a future reviewer. However, in some instances dynamical properties and measurements are inextricably coupled with thermodynamic properties. In these circumstances we have not attempted to separate the issues. Our final topic involves block copolymer surfaces, which are addressed in the third section. Recent experimental and theoretical advances in this emerging area have added a new dimension to the field of block copolymer thermodynamics. A discussion and outlook section is devoted to assessing recent progress in this field and to speculating on future directions.

12.2   Strong Segregation Limit

12.2.1   Experiment

Until roughly a decade ago transmission electron microscopy (TEM) was the preeminent experimental technique for block copolymer structure. The combination of relatively large monodisperse microstructures and efficient heavy-metal staining techniques (e.g., osmium tetraoxide) produced truely spectacular electron micrographs of ordered phases in polystyrene–polydiene block copolymers. Five ordered phases were identified in the strong segregation limit. Two types of spherical and cylindrical microstructures, as well as a lamellar morphology (see Fig. 12.4), were shown to exist within well-defined composition ranges, in close agreement with Helfand's theoretical predictions [5]. During the past decade, our analytical capabilities have been greatly enhanced by the development of small-angle scattering techniques that complement advances in TEM and provide access to new thermodynamic and fluctuation quantities. The topics covered in this section reflect these recent developments.

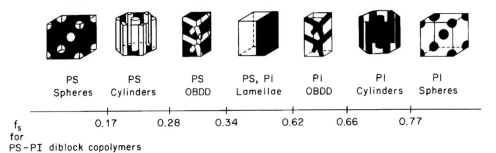

PS	PS	PS	PS, PI	PI	PI	PI
Spheres	Cylinders	OBDD	Lamellae	OBDD	Cylinders	Spheres

f_s
for
PS-PI diblock copolymers

Figure 12.4   Strong segregation limit (SSL) equilibrium morphologies for $(A-B)_n$ type block copolymers. The order–order transition compositions shown apply to polystyrene–polyisoprene diblock copolymers where ϕ_s corresponds to the polystyrene volume fraction

Hashimoto and co-workers have played a leading role in the application of small-angle X-ray scattering to the investigation of block copolymer thermodynamics. In a seminal series of publications [13–18] these authors used Small Angle X-ray Scattering (SAXS) and TEM to explore microdomain size and packing, and interfacial mixing in a model set of polystyrene–polyisoprene diblock copolymers. The quantitative measurement of the interfacial thickness between microphases represented an important development, particularly in the light of theoretical predictions of Helfand and Wasserman [5] regarding this parameter. Following procedures developed by Ruland [19], Vonk [20], and others [21], Hashimoto et al. evaluated the large scattering wavevector region of their SAXS data (the Porod regimen) and determined an interfacial thickness $t = 20 \pm 5$ Å independent microstructure (lamellar or spherical) and molecular weight. A subsequent small-angle neutron scattering (SANS) study of polystyrene–polybutadiene diblock copolymers by Bates et al. [22] produced essentially the same conclusions regarding interfacial thickness. [These authors also determined body-centered-cubic packing of spherical polybutadiene domains [23] in agreement with the weak segregation prediction of Leibler (see Section 12.3.1). However, this packing symmetry may not be universal since Richards and Thomason [24] have deduced a face-centered-cubic symmetry for polystyrene–polyisoprene diblock copolymers containing polystyrene spherical microdomains, also based on SANS measurements.] In independent SAXS measurements on polystyrene–polybutadiene diblock and triblock copolymers Roe et al. [25] found $10 \lesssim t \lesssim 17$ Å. [It should be noted that these authors report interfacial thickness measurements for $T > T_{ODT}$ (which we define as being in the WSL), suggesting the presence of microdomain structure within the disordered state.] All these small-angle scattering results for polystyrene–polydiene block copolymers are in reasonably good agreement with the prediction of Helfand and Wasserman [5], $t \approx \sqrt{2/3} a \chi^{-1/2} \cong 23$ Å (see Section 12.2.2), where a is the statistical segment length.

Although small-angle scattering experiments are capable of establishing a precise characteristic interfacial thickness, they are not able to discriminate between different mathematical expressions for the interfacial profiles because of limitations set by background and domain scattering. Spontak et al. [26] have attempted to address this deficiency by measuring the interfacial composition profile directly using TEM. This method requires extremely thin osmium-stained sections, which raises questions regarding stain-induced local segregation and swelling. Nevertheless, Spontak et al. report interfacial composition profiles with slightly greater characteristic thicknesses, $t \approx 26$ Å, than were found by the small-angle scattering Porod analysis.

One of the most exciting discoveries within the strong segregation regimen in recent years is the ordered bicontinuous double diamond (OBDD) morphology, depicted in Fig. 12.4 along with the five originally recognized equilibrium morphologies. A representative electron micrograph of this morphology is shown in Fig. 12.5. To our knowledge the earliest published TEM picture of the OBDD phase was obtained from a polystyrene–polydiene star-block copolymer and reported by Aggarwal [27] as the "wagon wheel" morphology. Subsequently, Thomas and co-workers [28, 29] initiated a research program aimed at ellucidating the solid-state morphology of well-defined star-block copolymers prepared by L.J. Fetters [30]. For a narrow range of experimental conditions including arm number, molecular weight, and composition ($\approx$26% by volume polystyrene) Thomas et al. found a novel bicontinuous morphology that did not conform to the well-known ordered phases. Soon thereafter came their dramatic identification of the now well-known OBDD phase [31], consisting of two continuous interpenetrating diamond (tetragonally coordinated) networks of polystyrene rods embedded in a continuous polyisoprene matrix. Evidence for this structure was provided by the combined use of SAXS and TEM. Small-angle scattering reflections from the ordered lattice were instrumental in identifying the exact lattice type and space group, which was then correlated with numerous tilted TEM images based on computer-generated two-dimensional crystallographic projections.

Following these publications on star-block copolymers, Hasegawa et al. [32] reported an equivalent ordered phase, denoted the "tetrapod-network structure" in polystyrene–polyisoprene diblock copolymers, thus demonstrating the occurrence of the OBDD phase in simple linear block architectures. [Because these diblock copolymers contained polystyrene as the major component (62% to 66% by volume) the two interpenetrating networks were composed of polyisoprene, that is, the "tetrapod-network structure" is actually an inverted version of the original OBDD phase identified by Thomas et al.] Both research groups

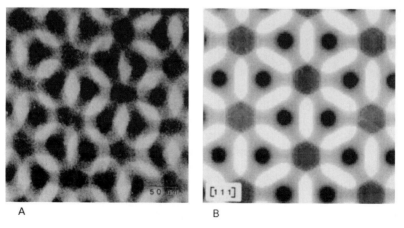

A   B

Figure 12.5 (A) Representative TEM of the ordered bicontinuous double diamond (OBDD) morphology. This image was obtained from an osmium-stained polystyrene–polyisoprene block copolymer. (B) A computer-simulated projection of the OBDD structure. (Provided by E.L. Thomas)

evaluated the stability of this morphology by preferential solvent casting and annealing experiments. For a narrow range of compositions, which varies slightly with chain architecture, the OBDD phase appears to be the equilibrium state; for polystyrene–polyisoprene diblock copolymers this occurs for polystyrene volume fractions of 0.28 to 0.34 (polystyrene double diamond structure) and 0.62 to 0.66 (polyisoprene double diamond structure) as indicated in Fig. 12.4 [32, 33].

Most recently, Thomas and co-workers have explored the underlying driving forces for the formation of the OBDD microstructure, demonstrating that it belongs to a class of geometrical structures characterized by constant mean curvature (CMC) surfaces [34]. Several TEM images of new block copolymer morphologies were also presented and shown to belong to a family of periodic area-minimizing surfaces that have attracted the attention of mathematicians for over a century. These elegant studies have revealed a beautiful physical manifestation of abstract mathematical concepts, made possible through the careful control of the three molecular parameters described in the Introduction.

Central to the theory of strong segregation in block copolymers is the concept of an extended block conformation, made necessary by the combined constraints of block joint localization at a narrow interface and an overall uniform density (see following section). The effects of extended chain conformation are most readily observed in the molecular weight dependence of the periodic lattice spacing and domain dimensions, $D \sim N^\delta$, where $\delta \approx 2/3$ in the SSL vs. $\delta = 1/2$ for unperturbed chain statistics (assumed) in the WSL. With the advent of SANS, the direct experimental determination of block chain statistics is now possible.

Richards and Thomason [35] reported the first SANS measurements on mixtures of partially labeled and unlabeled polystyrene–polyisoprene diblock copolymers. The SANS pattern contained two scattering contributions, one deriving from single-block scattering from within the polystyrene spherical domains (containing 4% deuterated polystyrene blocks) and a second associated with domain scattering. Richards and Thomason estimated the domain scattering contribution based on the SANS pattern obtained from an unlabeled sample, and subtracted it from the total scattering intensity to arrive at an estimate for the polystyrene block dimensions within the spherical domains. Unfortunately, this subtraction method is extremely sensitive to small differences in molecular weight and composition between the labeled and unlabeled polymers, which limits the practical application of the method.

Hadziioannou et al. [36] published the first SANS study of block chain dimensions in a lamellar microstructure. By mixing small amounts of partially deuterated polystyrene–polyisoprene block copolymer (perdeuterated polystyrene block) with unlabeled material of equal molecular weight, these researchers created scattering contrast within the polystyrene domains that could be used to determine the average block conformation within the lamellae. However, as with the previous case, for the concentrations of labeled polymer used by Hadziioannou et al. there is a strong scattering component due to interdomain interference that dominates the scattering intensity when the neutron beam is directed parallel to the plane of the lamellae. For unoriented (i.e., "polycrystalline") samples this interference effect nearly completely obscures single-block scattering. Hadziioannou et al. partially rectified this problem by shear-orienting their material, thereby providing a purely perpendicular incident neutral beam geometry. This eliminates the interlamellar scattering component, and allows a direct determination of chain dimensions parallel to the lamellae. Perpendicular dimensions, however, could not be measured. Their study indicated a significant lateral contraction of the block chain dimensions parallel to the lamellae, relative to the unconstrained size.

The topic of single chain scattering in undiluted block copolymers was treated theoretically by Jashan and Summerfield [37], and Koberstein [38] in the early 1980s. These authors recognized that by mixing specified amounts of partially labeled and unlabeled chains the contrast factor (i.e., the scattering length density for neutrons) could be matched between microphases, thus eliminating domain scattering. However, within a microdomain isotopic labeling would give rise to single block scattering. This contrast matching technique was first demonstrated by Bates et al. [39] in a polystyrene–polybutadiene diblock copolymer system in which spherical microdomains were prepared with 16% perdeuterated and 84% hydrogenous polybutadiene blocks. The method proved to be quite effective at eliminating domain scattering, and for low molecular weight polybutadiene blocks revealed an overall radius of gyration, in agreement with unperturbed dimensions. However, as the molecular weight was increased, the apparent block dimensions became unreasonably large; this behavior can now be attributed to small deviations from the exact contrast matching condition or to fluctuation effects (see below).

The most recent studies of block chain statistics have been made use of the contrast matching technique with the lamellar morphology. This combination affords access to both the lateral (parallel) and perpendicular components of the chain conformation within the microdomain space. SANS studies by Hasegawa et al. [40, 41] and Matsushita et al. [42] demonstrate that even at the theoretical contrast matching condition some residual domain scattering remains because of slight composition fluctuations within the isotopically mixed microphase; increasing molecular weight exacerbated this problem as inadvertently found by Bates et al. [39]. Nevertheless, contrast matching reduces the interference effects by up to two orders of magnitude, thereby facilitating the determination of the perpendicular block dimension. Hasegawa et al. [40, 41] report polystyrene block chain dimensions parallel and perpendicular to the lamellae to be approximately 70% and 160% of the unperturbed dimensions, respectively, for nearly symmetry polystyrene–polyisoprene diblock copolymers when $M_w = (7.5 - 10) \times 10^4$ g/mol.

12.2.2   Theory

By the middle of the 1970s, the physical principles that govern the microdomain period and the selection of ordered phases in the SSL had been well established by pioneering studies of Meier [43], Leary and Williams [44], and Helfand and Wasserman [5, 45]. Most notably, Helfand and Wasserman developed a self-consistent field theory that permits quantitative calculations of free energies, composition profiles, and chain conformations. They identified the three principal contributions to the free energy in the regimen $\chi N \gg 10$ as arising from: contact enthalpy in the narrow interfaces between nearly pure A and B microdomains, entropy loss associated with extended chain configurations to ensure incompressibility (i.e., stretching free energy), and confinement entropy due to localization of the block copolymer joints (covalent bonding sites between blocks) to the interfacial regions. Helfand and Wasserman showed that these narrow interfaces have characteristic thickness $a\chi^{-1/2}$ and by *numerical* solutions of the self-consistent field equations proposed that the microdomain period scales (asymptotically for $N \to \infty$) as $D \sim aN^{\delta}\chi^{\mu}$, with $\delta \approx 9/14 \approx 0.643$ and $\mu \approx 1/7 \approx 0.143$. In this asymptotic limit the confinement entropy of the junction is negligible compared with the stretching entropy. Helfand and Wasserman also developed numerical techniques for calculating the phase diagram in the SSL, and located the (virtually temperature independent) compositions that delimit the thermodynamic stability of spheres, cylinders, and lamellae. These compositions are in good agreement with experimental determinations of the phase boundaries, although we note that the OBDD phase (having not yet been discovered) was not included in the free energy competition.

While the Helfand–Wasserman theory is believed to contain all the proper physical ingredients necessary to describe the SSL, its practical application has been hindered because of the requirement of rather difficult numerical analysis. The theoretical advances of the past decade have been focused on developing *analytical* methods for estimating the free energy in the asymptotic limit $\chi N \to \infty$. Probably the most influential of these modern studies was an article by Semenov [6], which addressed a diblock melt in the SSL. Semenov argued that because the copolymers are strongly stretched in this regimen (see previous section), the required configurational integrals are dominated by the classical extremum of the energy functional (Hamiltonian). The extremum path corresponds to the most probable configuration of a copolymer block as it extends from an interface into a microdomain and experiences the chemical potential field produced by the surrounding molecules. Moreover, Semenov showed that the differential equation describing this path (which resembles the equation of motion of a classical particle) is analytically soluble, even when the chemical potential is determined self-consistently to maintain constant density of copolymer segments. This solution indicates that the copolymers are stretched nonuniformly (along their contours) as they enter into the microdomains and predicts that the chain ends are distributed at excess in the domain interiors. The classical mechanical analogy identified by Semenov has been further clarified by Milner, Witten, and Cats [46, 47] and was extensively exploited by these authors to treat related problems of grafted polymer brushes and surfactant interfaces. To appreciate the reduction in complexity afforded by this method, we note that the Helfand–Wasserman approach corresponds to the solution of a time-dependent problem in quantum mechanics, while the Semenov–Milner–Witten–Cates approach requires only the solution to the classical limit of this problem. Of course, the latter approach is legitimate only under conditions of strong chain stretching.

It is somewhat surprising that in spite of the significant chain deformations predicted for the SSL, the domain (stretching) contribution to the free energy per chain was found by Semenov to have the same scaling as for a Gaussian chain, namely $F_{\text{domain}}/kT \sim D^2/a^2N$, where D is the domain period. It is only the constant prefactor omitted from this expression that reflects the nonuniform stretching and distribution of ends. In the asymptotic limit $\chi N \to \infty$, the domain of free energy is balanced by the interfacial energy, which (per chain) is given by [5, 6, 45] $F_{\text{interface}}/kT \sim \gamma\sigma \sim Na\chi^{1/2}/D$. Here we have inserted well-known results for the interfacial tension, $\gamma \sim \chi^{1/2}a^{-2}$, and for the area per chain, $\sigma \sim Na^3/D$. By balancing the two free energy contributions, Semenov's prediction for the domain period is recovered, $D \sim aN^{2/3}\chi^{1/6}$. This result is believed to be asymptotically correct for large incompatibility, as the fluctuations about the classical path are $O(N^{1/2})$ and thus can be neglected for $N \to \infty$. The slight differences of these exponents from those of Helfand and Wasserman can be traced to the Helfand–Wasserman numerical estimate of F_{domain}. In particular, these authors found $F_{\text{domain}}/kT \sim (D/aN^{1/2})^{2.5}$, based on numerical calculations that extended only to $D/aN^{1/2} \approx 3$. We suspect that if the numerics had been carried out to $D/aN^{1/2} \gg 1$, the asymptotic scaling of Semenov would have been obtained. Finally, we note that Semenov's theory also provides estimates of the compositions that delimit the various ordered phases. As with the Helfand–Wasserman theory, these compositions are predicted to be temperature independent.

Another notable SSL theory of the past decade is that attributed to Ohta and Kawasaki [48]. Like the Helfand–Wasserman and Semenov theories the approach of Ohta and Kawasaki was field-theoretic, although Ohta and Kawasaki employed only a single scalar field describing the composition patterns. As a result, the physics associated with a nonuniform placement of chain ends is lost in this approach. Moreover, Ohta and Kawasaki used a random phase approximation for the free energy functional that is rigorously valid only for very weak compositional inhomogeneities, even though the inhomogeneities that characterize the SSL are $O(1)$ in volume fraction. In spite of this, Ohta and Kawasaki were able to obtain predictions for the domain periods and phase diagram that are qualitatively similar to those of Helfand–Wasserman and Semenov. They have also been able to reduce the calculation of the microdomain free energies to a purely geometrical problem.

Finally, we mention the SSL theory of Anderson and Thomas [49], which is the only theory thus far to contend with the OBDD phase. These authors modified the Ohta–Kawasaki approach to treat $(A–B)_n x$ star block copolymers. Although they found good agreement of the theoretical OBDD lattice parameters with experiment, Anderson and Thomas were not able to predict thermodynamic stability of the OBDD phase in the composition window where it is experimentally observed. Treatment of the OBDD phase using the methods of Helfand–Wasserman or Semenov has yet to be carried out.

12.3   Weak Segregation Limit

12.3.1   Theory

While the theoretical advances in the SSL were catalyzed by pioneering experiments involving the synthesis and characterization of model copolymers, developments in the WSL were strongly influenced by a seminal theoretical paper of Leibler [4]. Leibler considered the case of a monodisperse A–B diblock copolymer melt with degree of polymerization N, composition f, and equal monomer volumes and statistical segment lengths. For such a system, Leibler constructed a Landau expansion of the free energy to fourth order in a compositional order parameter field, $\psi(\mathbf{r}) = \langle \delta\phi_A(\mathbf{r}) \rangle$, where $\delta\phi_A(\mathbf{r}) = \phi_A(\mathbf{r}) - f$ is the fluctuation in microscopic volume fraction of A monomers at position $\mathbf{r}$. Such expansions prove useful in the vicinity of a second-order or weak first-order phase transition, where the amplitude of the order parameter field remains small in the low temperature phase [50]. Indeed, the ODT of symmetric ($f = 0.5$) or weakly asymmetric block copolymers is such a transition. Leibler's calculation was particularly remarkable, because he provided microscopic expressions for the (Landau) expansion coefficients as functions of the incompatibility, χN, and the copolymer composition. These coefficients were computed by means of the random phase approximation, introduced for polymer melt applications by de Gennes [3, 51].

By retaining only the leading harmonics in a Fourier representation of the various ordered-phase composition patterns, Leibler was able to exploit the Landau expansion and map out the phase diagram of a diblock copolymer melt near the ODT. (This approximation of neglecting higher harmonics in the description of the composition patterns can be rigorously justified only for the case of a second-order phase transition, but is expected to remain quantitative in weak first-order situations. Note also that the OBDD phase was not explicitly included in the free energy competition.) The phase diagram so obtained, in the parameter space of χN and f, is shown in Fig. 12.6a. The Landau theory predicts a critical point at $(\chi N)_c = 10.5$, $f_c = 0.5$, where a compositionally symmetric diblock melt is expected to undergo a second-order phase transition from the disordered to the lamellar phase. At such a transition, the amplitude of the lamellar pattern grows continuously from zero on lowering the temperature (i.e., increasing χN). The lattice constant (period) of the lamellar phase is predicted to be $D \approx 3.23 R_g \sim N^{1/2}$ at the symmetric ODT, consistent with the WSL assumption that the copolymers are only weakly perturbed by the inhomogeneous composition field. For asymmetric diblock copolymers, $f \neq 0.5$, the Landau theory predicts a weak first-order transition from the disordered phase to the BCC spherical phase. In contrast to the situation in the SSL, it is important to note that the Landau theory predicts first-order transitions between solid phases that can be accessed by changing *temperature*.

Besides the phase diagram shown in Fig. 12.6a, Leibler [4] provided an expression for the disordered phase structure factor, $S(q) = \langle \delta\phi_A(\mathbf{q})\delta\phi_A(-\mathbf{q})\rangle$, given by

$$S^{-1}(q) = N^{-1}F(x, f) - 2\chi \tag{12.1}$$

where $x \equiv q^2 R_g^2$ and $F(x,f)$ is a dimensionless function of wavenumber and composition

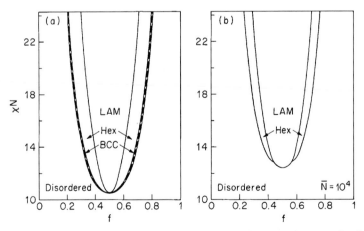

Figure 12.6   Theoretical phase diagrams for diblock copolymers in the weak segregation limit: (a) mean-field theory [4]; (b) fluctuation theory with $\bar{N} = 10^4$ [62]. LAM, Hex, and BCC correspond to lamellar, hexagonal (cylindrical morphology), and body-centered-cubic (spherical morphology) symmetries, and the dashed curve represents the mean-field (Landau) stability limit. (Reproduced from Ref. [63])

that is related to certain (Debye) correlation functions of a Gaussian diblock copolymer [4]. The most characteristic feature of this expression is the prediction of a Lorentzian peak at $x = x^*(f) \sim O(1)$, where F is minimum; the peak intensity diverges at the classical spinodal given by the condition $F(x^*, f) - 2(\chi N)_s = 0$. In the Landau theory, the spinodal and critical point coincide for $f = 0.5$. It should be noted that expressions similar to Eq. (12.1) were also derived by LeGrand and LeGrand [52] and by de Gennes [51].

The above expression for $S(q)$ has facilitated the interpretation of numerous X-ray and neutron scattering measurements on partially labeled model diblock copolymers. Similar expressions for more complex block copolymer architectures, such as graft and star copolymers, have been derived by Olvera de la Cruz and Sanchez [53] and by Benoit and Hadziioannou [54]. Other workers [55–58] have extended the Leibler expression to incorporate polydispersity effects, which are often important in practical applications. The latter extension is generally performed by averaging the Debye correlation functions that constitute $F(x,f)$ with an appropriate molecular weight distribution function, for example, the Schultz–Zimm function.

It was recognized by Leibler [4] that the Landau theory, which predicts mean-field critical behavior [59], is inadequate in the vicinity of the $f = 0.5$ critical point discussed above. Brazovskii [60] had previously demonstrated that such critical points, predicted by mean-field theories for systems exhibiting transitions between isotropic and striped (i.e., lamellar) phases, are suppressed by large-amplitude order parameter fluctuations. By means of a self-consistent Hartree approximation, Brazovskii showed that the mean-field critical point is replaced by a weak first-order phase transition, induced by fluctuations. It should be emphasized that such *fluctuation-induced first-order phase transitions* [61] have been predicted to occur in a variety of other physical systems, such as liquid crystals and driven (nonequilibrium) fluids. Experiments to test these predictions, however, have been quite limited.

Fredrickson and Helfand [62] extended Brazovskii's Hartree method of analysis to the A–B diblock copolymer melt considered by Leibler. They found that the fluctuation corrections are controlled by a Ginzburg parameter [59] $\bar{N}$, proportional to the copolymer molecular weight, defined by $\bar{N} = 6^3(R_g^3\rho_c)^2$, where ρ_c is the number density of copolymers in the melt. For fixed incompatibility χN, but $N \to \infty$, Fredrickson and Helfand found that Leibler's mean-field predictions are asymptotically approached. For finite $\bar{N}$, however, the fluctuation corrections impose both qualitative and quantitative changes in the phase diagram (Fig. 12.6b) and scattering behavior. In particular, the Hartree approximation leads to a suppression of the symmetric critical point at $(\chi N)_c = 10.5$, which is replaced by a weak first-order transition at (a lower temperature) $(\chi N)_{ODT} = 10.5 + 41.0\bar{N}^{-1/3}$. The amplitude of the lamellar composition pattern is predicted to be $O(\bar{N}^{-1/6})$ at the ODT. Because $\bar{N}$ is of order 10^3 to 10^4 for the typical experimental sample, these fluctuation corrections can be substantial. The changes in the phase diagram for asymmetric diblocks are even more dramatic, as is indicated in Fig. 12.6b for $\bar{N} = 10^4$. An important distinction with the mean-field diagram (Fig. 12.6a) is that the lamellar and the hexagonal phases are accessible at the ODT in the Hartree approximation for $f \neq 0.5$. However, the mean-field prediction of first-order transitions between ordered phases that can be accessed by changing temperature is preserved in the Hartree approximation.

The fluctuations manifest in the Fredrickson–Helfand theory also impact the structure factor of the disordered and ordered phases. For the disordered phase, the Hartree approximation for $S(q)$ has the same wavenumber dependence as in Eq. (12.1), but the bare Flory parameter χ is renormalized by composition fluctuations to an effective interaction parameter χ_{eff}. This renormalized parameter depends on temperature, composition, and molecular weight and is related to the bare parameter by

$$\chi_{\text{eff}}N = \chi N - \frac{C(f)}{2\bar{N}^{1/2}}[F(x^*, f) - 2\chi_{\text{eff}}N]^{-1/2} \tag{12.2}$$

where $C(f)$ is a composition-dependent coefficient. For a symmetric melt, the peak intensity $S(q^*)$ attains a maximum value that is $O(N\bar{N}^{1/3})$ at the ODT. In contrast, the mean-field theory gives rise to a divergent peak intensity at the symmetric ODT. The Hartree approximation also predicts an isotropic scattering component (in addition to the Bragg peaks) for the weakly ordered phases [62, 63]. This fluctuation component is also $O(N\bar{N}^{1/3})$, but is lower in intensity than the pretransitional disordered-phase component.

The Hartree approximation leads to an interesting physical picture of a symmetric diblock copolymer melt in the vicinity of the ODT [63]. While the Landau theory gives statistical weight only to the extremum composition field configurations in the ordered and disordered phases, namely the uniform and perfectly ordered configurations shown in Fig. 12.7a, the Hartree approximation also weights configurations such as those shown in Fig. 12.7b. The latter configurations have superimposed upon the extremum configurations isotropic composition fluctuations that have a preferred wavelength, $2\pi/q^*$, but random directions and phases. The Fredrickson–Helfand theory [62] suggests that the root-mean-squared amplitude of these fluctuations is $O(\bar{N}^{-1/6})$ and is thus comparable to the amplitude of the long-range-ordered lamellar component. It is interesting to note that the typical *equilibrium* composition field configurations in a disordered diblock melt (which fluctuate in time) are reminiscent of the transient *nonequilibrium* patterns encountered during the intermediate and late stages of spinodal decomposition [64].

A recent theoretical study of Semenov [65] suggests that the asymmetric wings of the phase diagram, that is $f \ll 0.5$ and $1 - f \ll 0.5$, could be much more complicated than is predicted by the Landau or Hartree theories. By using techniques discussed above for the SSL [6], Semenov argues that the free energy of formation of spherical micelles changes sign at an incompatibility (χN)M that is lower than the Landau or Hartree $(\chi N)_{\text{ODT}}$. Such micelles are localized, large-amplitude fluctuations in an (otherwise disordered) asymmetric diblock copolymer melt that are not accessible by the perturbative WSL approach described in the present section. Because of the favorable energetics of micelle formation near $(\chi N)_{\text{M}}$, Semenov postulates the existence of a spherical micellar phase that becomes more concentrated in micelles as χN is increased. By computing the interaction energy between two micelles, he further predicts a phase transition at which the micelles order into a microlattice with fcc symmetry. If Semenov's arguments are correct, this transition constitutes the ODT for asymmetric copolymers. Subsequent first-order structural transitions into hexagonal and bcc phases are also predicted by the theory.

At the time of this writing, the Hartree approximation has been explored for triblock copolymers [66], but not for more complex block copolymer architectures, such as stars. However, the extension to concentrated and semidilute diblock copolymer solutions with a neutral (nonselective) solvent has recently been carried out [67, 68]. It has frequently been assumed [69] that copolymer solutions have a uniform distribution of the nonselective solvent in the ordered microphases. This suggests that in the concentrated regimen, where swelling

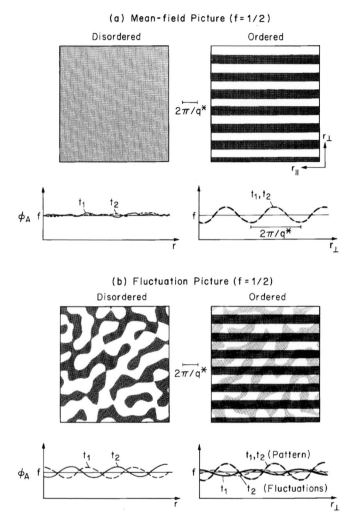

Figure 12.7   Instantaneous real-space composition patterns in the weak segregation limit: (a) mean-field theory, and (b) fluctuation theory. ϕ_A vs. r depicts the expected time dependence ($t_1 \neq t_2$) of each morphology, where ϕ_A is the local volume fraction of A segments. Recent SANS results support the fluctuation picture near the order–disorder transition (see Fig. 12.8). (Reproduced from Ref. [63])

effects are absent, the phase diagram of a copolymer solution is simply obtained by rescaling χ to $\phi\chi$ in the melt phase diagram, where ϕ is the copolymer volume fraction. Fredrickson and Leibler [67] have shown that this "dilution approximation" method [69, 70] neglects several aspects of the physics of such solutions. In particular, these authors demonstrated that even in the WSL there is a tendency for a neutral, good solvent to accumulate at the interfaces of the microdomains. This nonuniform placement of the solvent screens the unfavorable A–B monomer contacts, but does so with a translational entropy price. Screening occurs until that entropy cost exactly compensates the loss of contact enthalpy. For a good solvent, this compensation produces a periodic solvent composition profile with an amplitude that is N^{-1}

smaller than the amplitude of the A–B composition profile. As the solvent quality is decreased, the two order parameter fields have comparable amplitudes in the weakly ordered microphases and a tricritical point is encountered [67]. Another aspect of neutral copolymer solutions that distinguishes them from molten copolymers, is that the ODT is associated with a two-phase region in which disordered solvent-rich and ordered solvent-poor phases coexist. This region is very narrow for good solvents, but broadens as the solvent quality is decreased. Finally, we note that by reformulating the melt Hartree approximation in terms of concentration blobs, the important case of semidilute neutral copolymer solutions can be treated [67, 68].

12.3.2   Experiment

As discussed in the previous section, bulk block copolymers can be brought into the weak segregation regimen by decreasing either χ or N. The former is generally accomplished by selecting structurally similar monomers. For example, block copolymers ($f \sim 1/2$) prepared from styrene and α-methylstyrene remain disordered (i.e., homogeneous) at around $180\,^{\circ}\mathrm{C}$ for $M_w \lesssim 5 \cdot 10^5$ g/mol [71–73]. In contrast, symmetric polystyrene–polydiene block copolymers exhibit an order–disorder transition at about this temperature when $10^4 \lesssim M_w \lesssim 2 \cdot 10^4$ g/mol, depending on the polydiene type (e.g., polybutadiene or polyisoprene) and microstructure [74, 75]. Since these first reported, and in practice limiting cases, only a handful of block copolymers have been investigated in the WSL.

Pioneering studies by Cohen and co-workers [76, 78] on polydiene–polydiene diblock copolymers were conducted at about the time that the first WSL theory was developed. These investigations demonstrated WSL behavior in 1,4-polyisoprene-1,4-polybutadiene [76, 77] and 1,4-polybutadiene-1,2-polybutadiene [78] block copolymers which qualitatively supported Leibler's mean-field predictions, and gave impetus for subsequent research based on this class of materials. Since then, WSL investigations have been conducted with hydrogenated polystyrene–polydiene [75], polystyrene–poly(p-methylstyrene) [79], poly-(ethylene–propylene)–poly(ethylethylene) [63, 80–82], and polystyrene–polymethylmeth-acrylate [57, 83, 84] block copolymers. In addition, modest amounts of neutral solvents ($\lesssim 60\%$) have been added to polystyrene–polydiene materials to decrease the order–disorder transition temperature, thus bringing the WSL into the experimentally accessible temperature range [69, 85–87].

Within the weak segregation regimen the most significant feature is the order–disorder transition (ODT). Identification of the ODT temperature, denoted T_{ODT}, is often complicated by the weak first-order character of this phase transition, and the presence of significant composition fluctuations above and below T_{ODT}. Earlier studies established phase behavior based on the calorimetric or dynamic mechanical evaluation of the glass transition; a single glass transition is indicative of homogeneity (i.e., disorder), while two glass transitions signal microphase separation (i.e., order) [71–73, 75–78]. Although these techniques remain useful screening methods [80, 88], they are incapable of quantitatively establishing T_{ODT}.

The ordering of a block copolymer is accompanied by gross changes in the low-frequency rheological properties as first shown by Chung et al. [89], and Pico and Williams [90] for a poly(styrene–butadiene–styrene) (S–B–S) triblock copolymer, and plasticized S–

B–S, respectively. This behavior is characterized by the transition from a terminal dynamic mechanical response for $T > T_{ODT}$ [e.g., $G' \sim \omega^2$ and $G'' \sim \omega$ for $\omega \to 0$ where G' and G'' are the dynamic elastic and loss moduli [91], respectively] to a nonterminal response for $T < T_{ODT}$. At sufficiently low frequencies, G', and to a lesser extent G'', drop discontinuously as the temperature is raised through the first-order ODT. This discontinuity provides a quantitative means of identifying T_{ODT}; typical rheometer temperature control affords approximately $1\,°C$ precision in the determination of T_{ODT}. This technique has been demonstrated and exploited by several research groups studying both diblock [82, 92] and triblock [93–97] copolymers, and in our judgment represents the most efficient and accurate method for establishing T_{ODT}.

In addition to these abrupt changes in the limiting low-frequency ($\omega \to 0$) rheological properties at the ODT, composition fluctuations near the phase transition (see Fig. 12.7) lead to significant departures from thermorheological simplicity [82, 92]. A complete discussion of this dynamical behavior falls outside the scope of this review of block copolymer thermodynamics. Nevertheless, the continuous development of thermorheological complexity, particularly in the disordered state, is a direct manifestation of the fluctuation (i.e., thermodynamic) effects discussed previously and described below, and should not be confused with the discontinuous changes that mark the ODT [93].

As with the strong segregation limit, SAXS and SANS are very powerful and important experimental tools for investigating block copolymers in the weak segregation limit. The choice of X-rays or neutrons as incident radiation is dictated primarily by the choice of polymers, which determine the contrast factor. Nonpolar systems governed by relatively large χ parameters such as polystyrene–polydiene generally exhibit a sizeable electron density difference between components which provides good X-ray contrast. Accordingly, these materials are frequently studied by SAXS. Increasing block compatability by selecting structurally similar polymers such as isomers [78, 88] greatly reduces or eliminates X-ray contrast, making the use of SAXS either difficult or impossible. In this situation deuterium labeling (e.g., deuterating one block) provides strong neutron contrast, making SANS the experimental method of choice. These considerations are particularly important in the WSL. An intrinsically weak contrast factor [i.e., similar pure component electron (for SAXS) or neutron scattering length (for SANS) densities] will be sensitive to small changes in the local specific volume. Therefore, any inhomogeneously distributed (i.e., composition dependent) excess volume of mixing will modify this factor. If the local composition pattern changes with temperature, the contrast factor will vary with temperature. This effect will be most severe where the composition profile is most temperature dependent, that is, near the ODT. Such a spurious temperature dependence of the scattering intensity would preclude the quantitative evaluation of theory.

Shortly after publication of Leibler's landmark WSL theory, Roe et al. [25] demonstrated the existence of a broad temperature-dependent SAXS peak in a polystyrene–polybutadiene diblock copolymer above T_{ODT}. This, and other similar experiments [69], were found to be qualitatively consistent with Eq. (12.1), that is, decreasing temperature in the disordered state led to an increase in the peak scattering intensity. Quantitative assessments of Eq. (12.1) were first reported by Bates [56, 98] based on SANS measurements on partially deuterated 1,4-polybutadiene–1,2-polybutadiene diblock copolymers, and Mori et al. [74], who studied polystyrene–polyisoprene diblock copolymers by SAXS. These investigations demonstrated consistency between the measured and predicted disordered state structure factor for $f \sim 1/2$, and for the first time provided estimates of $\chi(T)$ obtained by fitting Eq. (12.1) to temperature-dependent small-angle scattering data, as originally proposed by Leibler [4]. Although the scattering results for the symmetric diblock copolymers produced results that were in good agreement with theory, Bates and Hartney [56] found a significant discrepancy between Eq. (12.1) and SANS data obtained from a series of asymmetric ($f \cong 0.25$) disordered 1,4-polybutadiene–1,2-polybutadiene samples. Along with the expected scattering peak at wavevector q^* these materials produced a significant forward scattering component that increased in intensity with increasing N. Bates and Hartney [56] speculated that this feature could derive from (domainlike) entities present in the disordered melt. The recent prediction by Semenov [65] regarding micelle formation in the disordered state (see previous section) is consistent with these observations. However, at the present time these measurements have not been confirmed and the issue of micelle formation in the disordered state remains unresolved.

As discussed in the previous section, Fredrickson and Helfand [62] have recently incorporated fluctuation effects into Leibler's original weak segregation mean-field theory, arriving at the phase diagram illustrated in Fig. 12.6 ($\bar{N} = 10^4$). Several experimentally testable differences between the mean-field and fluctuation theories can be identified. As the ODT is approached in the disordered phase both theories anticipate a rapid increase in the peak scattering intensity $I(q^*) \sim S(q^*)$ as indicated by Eq. (12.1). However, for the general case where $\chi = \alpha T^{-1} + \beta$ (see Introduction) the mean-field and fluctuation theories differ significantly in the predicted temperature dependence of $I(q^*)$; the former predicts $I^{-1}(q^*)$ to be linear in T^{-1} whereas fluctuation effects produce a nonlinear relationship between these parameters [see Eq. (12.2)].

Beginning with Roe and co-workers [25] the order–disorder transition has been examined in a variety of polystyrene–polydiene diblock [25, 76, 86], triblock [95], and star-block [86, 87] copolymers, and hydrogenated polystyrene–polybutadiene diblock copolymers [75], by small-angle X-ray scattering. These studies have relied on SAXS data obtained as a function of temperature for determining T_{ODT}. In general the ODT has been correlated with the temperature where a deviation from linearity in a plot of $I^{-1}(q^*)$ vs. T^{-1} is observed, which assumes mean-field behavior [25, 75, 95]. Alternatively, Hashimoto et al. [86, 87] have relied on the temperature dependence of the scattering peak position q^* in fixing T_{ODT}. [Previously these authors reported $q^* \sim T^0$ for $T > T_{\mathrm{ODT}}$ and $q^* \sim T^{1/3}$ for $T < T_{\mathrm{ODT}}$ [69]. Neglecting the intrinsic polymer coil thermal expansivity, the WSL and SSL theories predict $q^* \sim T^0$ and $q^* \sim T^{1/6}$, respectively (see Theory sections). In all these SAXS studies the ODT appears as a continuous transition as evidenced by an unbroken $I(q^*, T)$, contrary to the prediction of a first-order transition by both mean-field ($f \neq 1/2$) and fluctuation theories. None of these publications corroborate the assignment of T_{ODT} with rheological evidence of the phase transition.

Recently Bates and co-workers [80] reported the preparation of fully saturated hydrocarbon diblock copolymers in the WSL. A series of monodisperse, $f \cong 0.55$, poly(ethylene-propylene)–poly(ethylethylene) samples were studied rheologically [82] and by SANS [63, 81]. Owing to the chemical similarity between blocks this system exhibits an ODT at roughly five times the molecular weight of an equal composition polystyrene–polydiene material. For example for $M_{\mathrm{w}} = 57,400$ g/mol Bates et al. [63, 80–82] find $T_{\mathrm{ODT}} \approx 100\,^\circ\mathrm{C}$. This higher molecular weight brings these polymers well into the rheologically entangled state at the ODT which facilitates determining T_{ODT} [82]. In principle higher molecular weight polymers are also better candidates for evaluating the statistical mechanical WSL theories which are premised on a large N. Bates et al. [63] have shown an exact correspondence between the temperature at which the rheological properties in a poly(ethylene-propylene)–poly(ethylethylene) sample are discontinuous and where $I(q^*)$ exhibits a subtle (20%) discontinuity. These results conclusively demonstrate the first-order character of the ODT and underscore the value of dynamic mechanical analysis in establishing T_{ODT}. However, contrary to Hashimoto et al. [86, 87], Bates and co-workers [63] report $q^*(T)$ to be unaffected by the ODT.

A full evaluation of the poly(ethylene-propylene)–poly(ethylethylene) SANS results revealed the first clear evidence of composition fluctuations near the ODT. In the disordered state the principal scattering reflection could be quantitatively fit with the theoretical structure factor [Eq. (12.1)]. In addition, a shoulder was apparent at $q \cong 2q^*$ that became more prominent as the temperature was lowered toward T_{ODT}. This feature is not accounted for by current theory and suggests that the disordered phase may possess more "structure" (i.e., large composition gradients) than has previously been assumed. Overall, the disordered state SANS structure factor closely resembles the structure factor characterizing the final stage of spinodal decomposition in a symmetric binary polymer mixture [99], which is the basis for the real-space morphology of the fluctuating disordered phase depicted in Fig. 12.7. As shown in Fig. 12.8, $I^{-1}(q^*)$ is clearly nonlinear in T^{-1} over the entire temperature range examined, which extends 56 °C above T_{ODT}. A quantitative comparison of the mean-field and fluctuation theory predictions is also shown in Fig. 12.8. {It should be noted that these calculations were made without adjustable parameters [63]; N, f, and $\chi(T)$ were determined independently [82]}. This comparison confirms the predicted significance of fluctuation effects in the disordered state near the ODT and rules out the use of a mean-field assumption in evaluating $I(q^*, T)$ in these regions of the block copolymer phase diagram.

Bates et al. [63] also investigated the ordered state of a poly(ethylene-propylene)–poly(ethylethylene) specimen by using SANS. To facilitate these measurements a sample was shear-oriented based on the principles established by Mathis et al. [100]. Scattering experiments revealed a lamellar morphology which persisted up to T_{ODT}, confirming the fluctuation theory prediction that for slightly asymmetric compositions (here $f = 0.55$) the lamellar ordered phase should lead directly to the disordered phase (see Fig. 12.6b); mean-field theory predicts a lamellar-hexagonal-(body centered cubic)-disordered sequence of phase transitions (Fig. 12.6a). Fluctuations were also evidence in the two-dimensional SANS pattern (obtained from the oriented specimen) as the temperature approached T_{ODT}, in agreement with theory. Overall, the fluctuation theory is remarkably consistent with the (limited) experimental data [poly(ethylene-propylene)–poly(ethylethylene), $f = 0.55$] available for comparison at the time of this writing, leading us to speculate that the real-space morphologies near the ODT resemble those depicted in Fig. 12.7b; for $T \gg T_{ODT}$ and $T \ll T_{ODT}$ the classical pictures (Fig. 12.7a) are recovered.

Near the ODT the (polycrystalline) ordered and disordered states differ primarily in the extent of coherence of compositional order. Although this difference would be immediately apparent in direct images (see Fig. 12.7b), the associated small-angle scattering patterns barely reflect the phase transition [63]. Such long-range morphological features are best studied by direct methods such as TEM. Unfortunately, most model systems in the WSL are not readily studied by this technique. All the block copolymers considered in this review require staining prior to quantitative TEM analysis. Near the ODT, selective staining may seriously affect phase behavior, obviating use of the method. Samples that have chemically

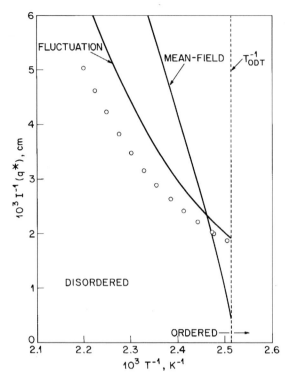

Figure 12.8   Reciprocal SANS peak intensity vs. inverse temperature for a disordered model poly(ethylene-propylene)–poly(ethylene) diblock copolymer near T_{ODT}. The mean-field and fluctuation curves have been calculated (no adjustable parameters) using the Landau [4] and fluctuation [62] theories, respectively, as described in Ref. [63]

similar blocks, such as poly(ethylene-propylene)–poly(ethylethylene) [80], are difficult if not impossible to selectively stain; a notable exception is the contrasting technique exploited by Cohen et al. [76–78] in examining polydiene–polydiene block copolymers. Polystyrene–polydiene block copolymers are easily selectively stained and thus represent the most attractive candidates for TEM analysis (see Section 12.2.1). In fact, several provocative electron micrographs obtained from polystyrene–polyisoprene–polystyrene triblock copolymers quenched from near the ODT (T_{ODT} was well above the polystyrene glass transition temperature) to room temperature seem to exhibit disordered morphologies similar to that shown in Fig. 12.7b [96, 101]. The formation of a glassy component probably reduces the effects of stain-induced changes in miscibility. However, it is difficult to assess the impact of such large temperature changes above T_g on the nonequilibrium state of the sample. This method might be improved by placing T_{ODT} slightly above T_g for polystyrene, through adjustment of N, which would dramatically increase the system response time relative to the temperature quench time. Under such conditions, a quantitative analysis of demonstrably "frozen" structures could be conducted.

12.4   Surface Behavior

12.4.1   Experiment

Block copolymer surface properties have been a topic of great interest for several decades owing in part to applications ranging from the formulation of adhesives to lubricating surfaces. Until recently, surface characterization tools have been limited to indirect methods such as X-ray photoelectron spectroscopy (XPS) and contact-angle wetting experiments. The past few years have witnessed vigorous growth in the area of surface analysis, particularly regarding block copolymer surfaces, because of the development of several new direct quantitative techniques, most noteably dynamic secondary ion mass spectroscopy (SIMS), and neutron reflectometry.

Early investigations of the wetting properties of undiluted block copolymers, which dealt mainly with semicrystalline materials, indicated a significant degree of surface enrichment in one block component [102]. This phenomenon was particularly evident in block copolymers containing polydimethylsiloxane which exhibited low energy surfaces, consistent with the surface properties of polydimethylsiloxone homopolymer. Surface enrichment of polystyrene in polystyrene–poly(ethylene oxide) diblock copolymers was also quantified by Thomas and O'Malley [103] using XPS measurements. Recently, Green et al. [104] employed XPS to evaluate the surface properties of a series of polystyrene–polymethylmethacrylate diblock copolymers. They report a molecular weight dependent surface preference for polystyrene. Although these studies clearly demonstrate the existence of a preferred surface component, which is easily rationalized based on the pure component surface tensions, the analytical methods discussed thus far are incapable of quantitatively delineating the actual surface profile or topology.

Direct evidence of surface segregation in a polystyrene–polyisoprene diblock copolymer was obtained using TEM by Hasegawa and Hashimoto [105]. Solvent cast specimens were stained with osmium tetraoxide, embedded in epoxy, and ultramicrotomed in preparation for microscopic examination. Regions of the ordered material (SSL) were photographed with lamellae arranged perpendicular and parallel to the free surface. In both cases a thin polyisoprene layer, approximately half a bulk lamellar dimension thick, existed at the polymer surface. This technique is suitable for observing strongly segregated surfaces but is not likely to be effective in the weak segregation regimen where staining could affect the system morphology.

SIMS is an alternative direct profiling technique that has recently has been applied to the investigation of block copolymer surfaces. This method relies on the intensity of secondary ions (e.g., $^1H^+$, $^2H^+$, $^{12}C^+$, etc.) emitted as a function of time during primary ion beam sputtering of the surface. A detailed SIMS analysis of a polystyrene–polymethyl methacrylate diblock copolymer film by Coulon et al. [106] illustrated how effectively this method images periodic lamellar structures that were formed parallel to the block copolymer free surface on annealing at elevated temperatures. These authors estimate an instrument resolution of about 125 Å. Another popular ion beam technique that finds application in polymer surface studies, forward recoil spectroscopy (FRS), is generally not appropriate for block copolymer surface problems because of current resolution limitations ($\gtrsim$ 200 Å) [107].

The most recently developed surface analytical technique with direct applicability to block copolymers is neutron reflectometry. Here a collimated neutron beam produced by

In our opinion the major unresolved issues in block copolymer thermodynamics are associated with the order–disorder transition, which we have included under the weak segregation limit heading. Foremost among these problems is the notion of the WSL itself. The WSL assumption originated with Leibler's mean-field treatment [4], and was retained by Fredrickson and Helfand [62] when fluctuation effects were incorporated into the theory. Our binary classification scheme of WSL and SSL remains a convenient and obvious one when dealing with currently available theories. However, categorizing the experimental studies near the ODT under these two headings is not so straightforward. Clearly there exist limits at sufficiently large and small values of χN where block copolymers conform to the SSL and WSL assumptions. Nevertheless it must be recognized that at present there is no theoretical estimate of the size or location of the intermediate region between these limiting behaviors. WSL theories assume a sinusoidal composition profile in the ordered phase, and unperturbed Gaussion coil behavior. Whether this accurately reflects the true situation in the transition region is simply not known. An important challenge for theorists is to develop a comprehensive theory that deals with the crossover from the WSL to the SSL.

Although recent SANS experiments [63, 81] support the qualitative (and to some extent, quantitative) predictions regarding composition fluctuations near the ODT, several troubling inconsistencies challenge the underlying WSL assumption. The observation of a shoulder at approximately twice the principal SANS reflection at temperatures above T_{ODT} by Bates et al. [63] cannot be explained by the perturbative methods implicit in the WSL theories. In fact, similarities between the disordered state structure factor and the final stage spinodal decomposition scattering pattern suggest a rather strongly segregated, albeit disordered, state. This recent finding is reminiscent of the SAXS results reported by Roe et al. [25] nearly a decade ago, where narrow (≈ 20 Å) interfacial thicknesses were determined above T_{ODT}. Stronger than predicted segregation in the disordered state might also account for the anomalous low-frequency rheological response found near the ODT [82, 92, 111]. Off-symmetry SANS measurements also suggest the existence of "domainlike" entities within the disordered melt [56], which have been predicted by Semenov [65] in the limit of small (or large) f.

These experimental findings near the order–disorder transition raise serious questions concerning the weak segregation limit assumption, and may invalidate our classification scheme. We believe the resolution of this issue will require both new theoretical approaches and additional quantitative experiments.

Finally, we have included block copolymer surfaces in this review. Only very recently have quantitative experimental techniques (e.g., SIMS and neutron reflectometry) been introduced to this field. Although identifying subjects that hold unusual promise is risky, we feel quite certain that developments in this area will come rapidly, and will have wide scientific and technological impact.

12.6   Update

Since this article was first published in 1990 there have been several significant developments in the field of block copolymer thermodynamics, both experimental and theoretical. Although we are not able to describe these findings in detail, an abbreviated summary, along with a list of current publications, is presented here as a supplement to our original review. This brief update, and the associated list of publications, is neither comprehensive nor balanced. Instead, it reflects changes in the field most related to the reprinted article.

12.6.1   Experiment

During the past 5 years there have been dramatic changes in our perception of block copolymer phase behavior. Three new ordered microstructures have been discovered: hexagonally modulated lamellae (HML) [112, 113], hexagonally perforated layers (HPLs) [112–115], and bicontinuous $Ia\bar{3}d$ [115–117] (also referred to as Gyroid) [117]. These are depicted in Fig. 12.10. Apparently, the occurrence of these new morphologies is controlled by the overall molecular weight (i.e., N or $\bar{N}$) in addition to the classical parameters χN and f indicating a new type of nonuniversality for diblock copolymer melts [118]. Recent experiment with a variety of materials demonstrate that the bicontinuous $Ia\bar{3}d$ phase disappears entirely as $\bar{N} \to \infty$ [118]. As $\bar{N}$ decreases, the window of χN space where it is observed increases, although there are no documented examples to date where it extends into the strong segregation limit. The HPL phase has been identified in a variety of undiluted diblock copolymers [112, 113, 115], and in diblock-homopolymer mixtures [114]. Topologically, the HPL and $Ia\bar{3}d$ bicontinuous microstructures are related, both being formed from flat tripod connectors as illustrated by Förster et al. [115]. However, the depictions in Fig. 12.10 should not be interpreted too seriously since these new phases are found near the ODT, where the distinction between separate microdomains, separated by sharp interfaces, is rather fanciful. What has been established by small-angle scattering and electron microscopy is the symmetry of each phase; the actual spatial distribution of polymer segments as a function of χN at a particular value of f is not well established except for lamellae [119, 120].

Discovery of the $Ia\bar{3}d$ bicontinuous phase has necessitated a reexamination of the OBDD morphology. An extensive investigation [115,118] of the polystyrene–polyisoprene diblock copolymer system near the ODT has led to a phase diagram that contains the $Ia\bar{3}d$ state in narrow channels located at $0.36 \lesssim f_{PI} \lesssim 0.39$ and $0.65 \lesssim f_{PI} \lesssim 0.69$; the OBDD phase was not found. Moreover, the bicontinuous phase is reported to be localized near the ODT,

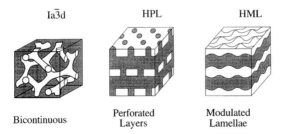

Figure 12.10   Three new microstructures that have been discovered since the original publication of this review. The previously reported OBDD phase has not been identified in recent experiments, and its existence is now questionable as explained in Section 12.6

extending no more than 100 °C below T_{ODT}, rather than the accepted channel of OBDD in the SSL [121, 122]. Hajduk et al. [123] have recently performed SAXS experiments on some of the original polystyrene–polyisoprene specimens that were reported to be OBDD [31, 33, 34] and conclude that this was an erroneous assignment; an $Ia\bar{3}d$ space group was actually indicated. Förster et al. [115] also question the occurrence of a bicontinuous phase in SSL polystyrene–polyisoprene specimens, suggesting instead that this phase may exist only in a metastable state far below the ODT. At the time of this writing this issue has not been definitively resolved. Nevertheless, it is the authors' opinion that the OBDD phase does not occur in diblock copolymer melts. However, addition of homopolymer, or mixing diblocks, could induce OBDD formation, analogous to what is found in surfactant (e.g., lipid)–water systems.

12.6.2   Theory

On the theoretical side, progress since 1990 has been more limited. Perhaps most exciting was the demonstration by Matsen and Schick [124] that the bicontinuous $Ia\bar{3}d$ phase (gyroid) is a stable equilibrium morphology for diblock copolymers within mean-field theory. Matsen and Schick used self-consistent field theory (SCFT), similar to that pioneered for block copolymers by Helfand and Wasserman [45a,b] and Hong and Noolandi [70], but with an important modification. Rather than attempt to self-consistently construct the Wigner-Seitz cell of a complex phase, for example, the $Ia\bar{3}d$, from three-dimensional numerical solutions of the SCFT equations, Matsen and Schick represented each microphase in the form of a generalized Fourier series satisfying the requisite symmetry operations. While this procedure can be implemented only for the WSL (a very large number of basis functions is required to describe a strongly segregated microphase), it dramatically reduces the computational requirements of solving the SCFT equations and allows for the study of arbitrarily complex microphases.

Matsen and Schick [124] find that Leibler's mean-field phase diagram for diblock copolymers is preserved very near $\chi N = 10.5$ (with only the "classical" phases present— lamellae, cylinders, and spheres), but that the gyroid phase emerges at a triple point ($\chi N = 11.1$, $f = 0.45$, 0.55) along the lamellae–cylinder coexistence curve. The channel of stability for the gyroid phase that opened up between lamellae and cylinders was not observed to close over the range of χN that Matsen and Schick could investigate: $11.1 < \chi N < 20$. Experimentally [118], however, the stability of the gyroid phase is restricted to temperatures near the ODT. Recent theoretical work in support of this is provided by Olmsted and Milner [125], who extended the SSL analysis of Semenov [6] to bicontinuous microphases such as the OBDD and $Ia\bar{3}d$ (gyroid). They find no evidence for the stability of phases other than the classical ones in the asymptotic SSL of $\chi N \to \infty$.

While the Matsen–Schick prediction of a stability window for the gyroid phase is an exciting development, it also raises some puzzling issues. Since the calculation is based on a variant of mean-field theory, we expect experimental systems to adhere strictly to the Matsen–Schick phase diagram only in the limit of $\bar{N} \to \infty$. For finite $\bar{N}$, fluctuation effects of the type considered by Fredrickson and Helfand [62] should modify the topology of the phase diagram near the ODT. Indeed, this is very likely the explanation for the experimental fact that the gyroid phase can be directly accessed at the ODT [118], whereas Matsen and Schick located it fully within the ordered region. Nevertheless, the experimental trends with $\bar{N}$ noted above are disturbing. The absence of the gyroid phase in the largest $\bar{N}$ samples [118], which should be the most "mean-field like," appears to be inconsistent with the Matsen–Schick calculations. Fluctuation effects could possibly conspire to enhance the window of gyroid stability as $\bar{N}$ is decreased, as observed experimentally, but these corrections to mean-field theory should vanish for $\bar{N} \to \infty$. Thus, some other physical mechanism must be responsible for suppressing the gyroid phase in the large $\bar{N}$ experimental systems. At present, it is not clear what this mechanism might be.

Overall, significant progress both in terms of theory and experiment has occurred over the past 5 years of research in this exciting and active area of polymer science. We anticipate comparable advances in block copolymer thermodynamics in the years to come.

Acknowledgment

We are indebted to our collaborators E. Helfand, R. Larson, L. Leibler, and J. Rosedale for their contributions to our understanding of the issues covered in this review.

References

1. M. Szwarc, M. Levy, and R. Milkovich, *J. Am. Chem. Soc. 78*, 2656 (1956)
2. S. Schlick and M. Levy, *J. Phys. Chem. 64*, 883 (1960)
3. P.G. deGennes, *Scaling Concepts in Polymer Physics* (1979) Cornell University Press, Ithaca, NY
4. L. Leibler, *Macromolecules 13*, 1602 (1980)
5. E. Helfand and Z.R. Wasserman, See Ref. [7], p. 99
6. A.N. Semenov, *Soviet Physics JETP 61*, 733 (1985)
7. I. Goodman (Ed.), *Developments in Block Copolymers—1* (1982) Applied Science, New York
8. I. Goodman (Ed.), *Developments in Block Copolymers—2* (1985) Applied Science, New York
9. D.J. Meier (Ed.), *Block Copolymers: Science and Technology* (1983) MMI Press by Harwood Academic Publ., New York
10. D.C. Allport and W.H. Jones (Eds.), *Block Copolymers* (1973) New York: Wiley
11. S.L. Aggarwal (Ed.), *Block Copolymers*, Plenum Press, New York
12. J.J. Burke and V. Weiss (Eds.), *Block and Graft Copolymers* (1973) Syracuse University Press, New York
13. T. Hashimoto, K. Nagatoshi, A. Todo, H. Hasegawa, and H. Kawai, *Macromolecules 7*, 364 (1974)
14. T. Hashimoto, A. Todo, H. Itoi and H. Kawai, *Macromolecules 10*, 377 (1977)
15. A. Todo, H. Uno, K. Miyoshi, T. Hashimoto, and H. Kawai, *Polym. Eng. Sci. 17*, 527 (1977)
16. T. Hashimoto, M. Shibayama, and H. Kawai, *Macromolecules 13*, 1237 (1980)
17. T. Hashimoto, M. Fujimura, and H. Kawai, *Macromolecules 13*, 1660 (1980)
18. M. Fujimura, H. Hashimoto, K. Kurahashi, T. Hashimoto, and H. Kawai, *Macromolecules 14*, 1196 (1981)
19. W. Ruland, *J. Appl. Crystallogr. 4*, 70 (1971)
20. C.G. Vonk, *J. Appl. Crystallogr. 6*, 81 (1973)
21. J. Koberstein, B. Morra, and R.S. Stein, *J. Appl. Crystallogr. 13*, 34 (1980)
22. F.S. Bates, C.V. Berney, and R.E. Cohen, *Macromolecules 16*, 1101 (1983)
23. F.S. Bates, R. E. Cohen, and C.V. Berney, *Macromolecules 15*, 584 (1982)
24. R.W. Richards and J.L. Thomason, *Macromolecules 16*, 982 (1983)

25. R.J. Roe, M. Fishkis, and J.C. Chang, *Macromolecules 14*, 1091 (1981)
26. R.J. Spontak, M.C. Williams, and D.A. Agard, *Macromolecules 21*, 1377 (1988)
27. S.L. Aggarwal, *Polymer 17*, 938 (1976)
28. D.B. Alward, D.J. Kinning, E.L. Thomas, and L.J. Fetters, *Macromolecules 19*, 215 (1986)
29. D.J. Kinning, D.B. Alward, E.L. Thomas, L.J. Fetters, and D.J. Handlin, *Macromolecules 19*, 1288 (1986)
30. L.J. Fetters, See Ref. [9], p. 17
31. E.L. Thomas, D.B. Alward, D.J. Kinning, D.L. Handlin, and L.J. Fetters, *Macromolecules 19*, 2197 (1986)
32. H. Hasegawa, H. Tanaka, K. Yamasaki, and T. Hashimoto, *Macromolecules 20*, 1651 (1987)
33. D.S. Herman, D.J. Kinning, E.L. Thomas, and L.J. Fetters, *Macromolecules 20*, 2940 (1987)
34. E.L. Thomas, D.M. Anderson, C.S. Henkee, and D. Hoffman, *Nature 334*, 598 (1988)
35. R.W. Richards and J.L. Thomason, *Polymer 22*, 581 (1981)
36. G. Hadziioannou, C. Picot, A. Skoulios, M.-L. Ionescu, A. Mathis, R. Dupplessix, Y. Gallot, and J.-P. Lingelser, *Macromolecules 15*, 263 (1982)
37. S.N. Jahshan and G.C. Summerfield, *J. Polym. Sci. Polym. Phys. Ed. 20*, 593 (1980)
38. J.T. Koberstein, *J. Polym. Sci. Polym. Phys. Ed. 20*, 593 (1982)
39. F.S. Bats, C.V. Berney, R.E. Cohen, and G.D. Wignall, *Polymer 24*, 519 (1983)
40. H. Hasegawa, T. Hashimoto, H. Kawai, T.P. Lodge, E.J. Amis, C.J. Glinka, and C.C. Han, *Macromolecules 18*, 67 (1985)
41. H. Hasegawa, H. Tanaka, T. Hashimoto, and C.C. Han, *Macromolecules 20*, 2120 (1987)
42. Y. Matsushita, Y. Nakao, R. Saguchi, K. Mori, H. Choshi, Y. Murago, I. Noda, M. Nagasawa, T. Chang, C.J. Glinka, and C.C. Han, *Macromolecules 21*, 1802 (1988)
43. D.J. Meier, *J. Polym. Sci. Pt. C 26*, 81 (1969)
44. D. Leary and M. Williams, *J. Polym. Sci. Pt. B 8*, 335 (1970)
45. E. Helfand, *Macromolecules 8*, 552 (1975); E. Helfand and Z.R. Wasserman, *Macromolecules 9*, 879 (1976)
46. S.T. Milner, T.A. Witten, and M.E. Cates, *Europhys. Lett. 5*, 413 (1988); *Macromolecules 21*, 2610 (1988)
47. S.T. Milner, T.A. Witten, and M.E. Cates, *Macromolecules 22*, 853 (1988); S.T. Milner and T.A. Witten, *J. Phys. Paris 19*, 1951 (1988)
48. T. Ohta and K. Kawasaki, *Macromolecules 19*, 2621 (1986); K. Kawasaki, T. Ohta, and M. Kohrogni, *Macromolecules 21*, 2972 (1988)
49. D.M. Anderson and E.L. Thomas, *Macromolecules 21*, 3221 (1988)
50. J.-C. Toledano and P. Toledano, *The Landau Theory of Phase Transitions* (1987) World Scientific, Teaneck, NJ
51. P.G. de Gennes, *Farad. Discus. Chem. Soc. 68*, 96 (1979)
52. A.D. LeGrand and D.G. LeGrand, *Macromolecules 12*, 450 (1979)
53. M. Olvera de la Cruz and I.C. Sanchez, *Macromolecules 19*, 2501 (1986)
54. H. Benoit and G. Hadziioannou, *Macromolecules 21*, 1449 (1988)
55. L. Leibler and H. Benoit, *Polymer 22*, 195 (1981)
56. F.S. Bates and M.A. Hartney, *Macromolecules 18*, 2478 (1985); F.S. Bates and M.A. Hartney, *Macromolecules 19*, 2892 (1986)
57. H. Benoit, W. Wu, M. Benmouna, B. Mozer, B. Bauer, and A. Lapp, *Macromolecules 18*, 986 (1985); L. Ionescu, C. Picot, M. Duval, R. Duplessix, and H. Benoit, *J. Polym. Sci. Polym. Phys. Ed. 19*, 1019 (1981)
58. K. Mori, H. Tanaka, H. Hasegawa, and T. Hashimoto, *Polymer 30*, 1389 (1989)
59. S.K. Ma, *Modern Theory of Critical Phenomena*. Benjamin/Cummings, Reading, MA
60. S.A. Brazovskii, *Soviet Physics JETP 41*, 85 (1975)
61. K. Binder, *Rep. Prog. Phys. 50*, 783 (1987)
62. G.H. Fredrickson and E. Helfand, *J. Chem. Phys. 87*, 697 (1987)
63. F.S. Bates, J.H. Rosedale, and G.H. Fredrickson, *J. Chem. Phys. 92*, 6255 (1990)
64. J.D. Gunton, M. San Miguel, and P.S. Sahni, *Phase Transitions and Critical Phenomena*, C. Domb and J.L. Lebowitz (Eds.) (1983) *8*, 267. Academic Press, New York
65. A.N. Semenov, *Macromolecules 22*, 2849 (1989)

66. A.M. Mayes and M. Olvera de la Cruz, *Macromolecules 24*, 3975 (1991)
67. G.H. Fredrickson and L. Leibler, *Macromolecules 22*, 1238 (1989)
68. M. Olvera de la Cruz, *J. Chem. Phys. 90*, 1995 (1989)
69. T. Hashijmoto, M. Shibayama, and H. Kawai, *Macromolecules 16*, 1093 (1983); id. *16*, 1427; id. *16*, 1434
70. K.M. Hong and J. Noolandi, *Macromolecules 16*, 1083 (1983)
71. M. Baer, *J. Polym. Sci. A 2*, 417 (1964)
72. L.M. Robeson, M. Matzner, L.J. Fetters, and J.E. McGrath, *Recent Advances in Polymer Blends, Grafts and Blocks.* L.H. Sperling (Ed.) (1974) Plenum Press, New York
73. S. Krause, D.J. Dunn, A. Seyed-Mozzaffari, and A.M. Biswas, *Macromolecules 10*, 786 (1977)
74. K. Mori, H. Hasegawa, and T. Hashimoto, *Polymer J. Jpn. 17*, 799 (1985)
75. J.N. Owens, I.S. Gancarz, J.T. Koberstein, and T.P. Russell, *Macromolecules 22*, 3380 (1989)
76. A.R. Ramos and R.E. Cohen, *Polym. Eng. Sci. 17*, 639 (1977)
77. R.E. Cohen and A.R. Ramos, *Macromolecules 12*, 131 (1979)
78. R.E. Cohen and D.E. Wilfong, *Macromolecules 15*, 370 (1982)
79. E.W. Fischer and W.G. Jung, *Makromol. Chem. Macromol. Symp. 26*, 179; W.G. Jung and E.W. Fischer, id. *16*, 281
80. F.S. Bates, J.H. Rosedale, H.E. Bair, and T.P. Russell, *Macromolecules 22*, 2557 (1989)
81. F.S. Bates, J.H. Rosedale, G.H. Fredrickson, and C.J. Glinka, *Phys. Rev. Lett. 61*, 2229 (1988)
82. J.H. Rosedale and F.S. Bates, *Macromolecules 23*, 2329 (1990)
83. S.H. Anastasiadis, T.P. Russell, S.K. Satija, and C.F. Majkrzak, *Phys. Rev. Lett. 62*, 1852 (1989)
84. P.F. Green, T.P. Russell, R. Jerome, and M. Granville, *Macromolecules 21*, 3266 (1988)
85. L.J. Fetters, R.W. Richards, and E.L. Thomas, *Polymer 18*, 2252 (1987)
86. T. Hashimoto, Y. Ijichi, and L.J. Fetters, *J. Chem. Phys. 89*, 2463 (1988)
87. Y. Ijichi, T. Hashimoto, and L.J. Fetters, *Macromolecules 22*, 2817 (1989)
88. F.S. Bates, H.E. Bair, and M.A. Hartney, *Macromolecules 17*, 1987 (1984)
89. C.I. Chung and J.C. Gale, *J. Polym. Sci. Polym. Phys. Ed. 14*, 1149; C.I. Chung and M.I. Lin, 1978, id. *16*, 545
90. E.R. Pico and M.C. Williams, *Nature 259*, 388 (1976)
91. J.D. Ferry, *Viscoelastic Properties of Polymers*, 3rd edit. (1980) John Wiley & Sons, New York
92. F.S. Bates, *Macromolecules 17*, 2607 (1984)
93. C.D. Han, J. Kim, and J.K. Kim, *J. Polym. Sci. Polym. Phys. Ed. 25*, 1741 (1987)
94. C.D. Han, J. Kim, and J.K. Kim, *Macromolecules 22*, 383 (1989)
95. C.D. Han, D.M. Baek, and J.K. Kim, *Macromolecules 23*, 561 (1990)
96. J.J. Widmaier and G.C. Meyer, *J. Polym. Sci. Phys. Ed. 18*, 2217 (1980)
97. E.V. Gouinlock and R.S. Porter, *Polym. Eng. Sci. 17*, 535 (1977)
98. F.S. Bates, *Macromolecules 18*, 525 (1985)
99. F.S. Bates and P. Wiltzius, *J. Chem. Phys. 91*, 3258 (1989)
100. A. Mathis, G. Hadziioannou, and A. Skoulios, *Polym. Eng. Sci. 17*, 570; 1979. *Colloid Polym. Sci. 257*, 136 (1979)
101. G. Hadziioannou and A. Skoulios, *Macromolecules 15*, 258 (1982)
102. M.J. Owen, See Ref. [9], p. 129
103. H.R. Thomas and J.J. O'Malley, *Macromolecules 12*, 323 (1979)
104. P.F. Green, T.M. Christensen, T.P. Russell, and R. Jerome, *Macromolecules 22*, 2189 (1989)
105. H. Hasegawa and T. Hashimoto, *Macromolecules 18*, 589 (1985)
106. G. Coulon, T.P. Russell, V.R. Deline, and P.F. Green, *Macromolecules 22*, 2581 (1989)
107. U.K. Chaturvedi, V. Steiner, O. Zak, G. Krausch, and J. Klein, *Phys. Rev. Lett. 63*, 616 (1989)
108. S.H. Anastasiadis, T.P. Russell, S.K. Satija, and C.F. Majkrzak, *J. Chem. Phys. 92*, 5667 (1990)
109. A. Menale, T.P. Russell, S.H. Anastasiadis, S.K. Satija, and C.F. Majkrzak, *Phys. Rev. Lett. 68*, 67 (1992)
110. G.H. Fredrickson, *Macromolecules 10*, 2535 (1987)
111. G.H. Fredrickson and E. Helfand, *J. Chem. Phys. 89*, 5890 (1988)
112. I.W. Hamley, K.A. Koppi, J.H. Rosedale, F.S. Bates, K. Almdal, and K. Mortensen, *Macromolecules 26*, 5959 (1993)
113. I.W. Hamley, M.D. Gehlsen, A.K. Khandpur, K.A. Koppi, J.H. Rosedale, M.F. Schulz, F.S. Bates, K. Almdal, and K. Mortensen, *J. Phys. II France 4*, 2161 (1994)

Abbreviations Used

Ac, PAc	acenaphtylene, polyacenaphtylene
Bd, PBd	butadiene, polybutadiene rubber
CR	polychloroprene rubber
ClSO$_2$PE	chlorosulfonated polyethylene
EPR	ethylene–propylene copolymer rubber
EPDM	ethylene–propylene–diene monomer rubber
IB, PIB	isobutylene, polyisobutylene
IIR, X-IIR	isobutylene–isoprene copolymer (butyl rubber), halobutyl rubber (X = Cl or Br)
Ind, PInd	indene, polyindene
IP, PIP	isoprene, polyisoprene
αMeSt, PαMeSt	α-methylstyrene, poly(α-methylstyrene)
MMA, PMMA	methyl methacrylate and poly(methyl methacrylate)
pClCH$_2$St and PpClCH$_2$St	p-chloromethylstyrene and poly(p-chloromethyl styrene)
pClSt, PpClSt	p-chlorostyrene, poly(p-chlorostyrene)
pFSt, PpFSt	p-fluorostyrene, poly(p-fluorostyrene)
St, PSt	styrene, polystyrene
b	block
co	copolymer (random) or statistical
cy	cyclized
g	graft
tr	trans

13.1   Introduction

Carbocationic polymerization is relatively a latecomer to the world of thermoplastic elastomers (TPEs). Nonetheless, within a short time this chemistry has enriched the science and technology of TPEs in so many ways that a comprehensive review, such as presented in this volume, would be incomplete without a discussion of TPEs prepared by carbocationic polymerizations.

This saga commences in the late 1960s, when the demonstrated utility of TPEs prepared by other techniques, mainly styrenic TPEs synthesized by anionic polymerizations, prompted research aimed at the preparation of similar structures/materials proposed by carbocationic processes. An important motivation for these investigations was the desire to replace the unsaturated rubbers, polybutadiene or polyisoprene, used in styrenic TPEs (Kraton®, Thermoplastic Elastomers, in particular) with the equally inexpensive but saturated (and consequently oxidatively and chemically resistant) polyisobutylene rubber as the elastomeric moiety.

Among the notable early successes toward cationically prepared TPEs were the discovery of the first generation of designed graft copolymers comprising a polyisobutylene backbone carrying polystyrene branches and similar rubbery–glassy sequential polymers (see Section 2.2). These early discoveries were possible because of a sufficient understanding of the initiation step of carbocationic polymerizations induced by active organic chlorides (RCl) in conjunction with certain alkylaluminum coinitiators (Et_2AlCl) ("controlled initiation"):

$$RCl + Et_2AlCl \longrightarrow R^\oplus + Et_2AlCl_2^\ominus \xrightarrow{+M} RM^\oplus \xrightarrow{+nM} R \rightsquigarrow M^\oplus \qquad (13.1)$$

However, after a flurry of activity by both academic and industrial investigators (see for example, [1, 2]), research in cationic grafting decelerated, mainly because at this time the other elementary steps (i.e., chain transfer, termination) could not be controlled and this lack of understanding of the mechanisms frustrated efforts to assemble the needed structures exhibiting superior physical properties and processing advantages.

Materials science of polymers—and the field of TPEs is certainly part of this discipline— is largely applied macromolecular engineering and is predicated upon a detailed under- standing of the elementary events of polymerization reactions. In hindsight, it is abundantly evident that the recent resurgence of carbocationic polymerization research, notably in block copolymers, occurred exactly because of the increased insight into the elementary processes, that is, initiation, propagation, chain transfer, and reversible termination, which in turn led to controlled rates, and products with designed molecular weights and molecular weight distribution.

This chapter concerns three large families of TPEs of which certain members can be prepared only by carbocationic polymerization techniques: graft copolymers (including bigrafts and graft blocks), block copolymers, and ionomers. All these products arose because of our detailed insight into the mechanism of carbocationic polymerizations: synthesis of grafts became possible because of controlled initiation (Section 13.2), synthesis of the various blocks because of the discovery of living (more precisely quasiliving) carbocationic polymerizations (Section 13.3), and synthesis of the ionomers because some of the fundamentals of ionomers in general have been specifically adapted to olefin-telechelic polyisobutylenes (Section 13.4).

13.2   Thermoplastic Elastomer Graft Copolymers

13.2.1   Introduction: Controlled Initiation and Structural Considerations

As mentioned in the Introduction, the first cationically prepared TPEs were graft copolymers of rubbery backbones carrying glassy branches. The development of this family of materials was based on the discovery that, although conventional Friedel–Crafts acids (e.g., $AlCl_3$, BF_3, $TiCl_4$) instantaneously induce olefin polymerizations in the presence of traces of ubiquitous moisture, certain other Friedel–Crafts acids, such as BCl_3, Et_2AlCl, Et_3Al, are inactive under the same conditions (i.e., they can be contacted with olefins without inducing polymerizations). However, these "inactive" acids will instantaneously coinitiate cationic polymerizations on the purposeful addition of active halogen compounds, for example, tertiary, allylic, or benzylic chlorides [see Eq. (13.1), where RCl = tertiary or allylic or benzylic Cl]. If such active chlorides are part of a preformed polymer, the polymerizations can start *only* at the active halide sites and the new chains that arise will become covalently attached to the polymer; in other words, they will become branches of the new graft. Thus if $\sim C^* \sim$ = tertiary, allylic, or benzylic carbon; M = cationically polymerizable monomer:

$$\underset{\underset{\text{Cl}}{|}}{\overset{*}{\sim\!\!\sim C \sim\!\!\sim}} + Et_2AlCl \longrightarrow \left[\underset{\oplus}{\overset{*}{\sim\!\!\sim C \sim\!\!\sim}} Et_2AlCl_2^{\ominus} \right] \xrightarrow{\ +\, M\ }$$

macrocation

$$\left[\underset{\underset{M^{\oplus}}{|}}{\overset{*}{\sim\!\!\sim C \sim}} Et_2AlCl_2^{\ominus} \right] \xrightarrow{\ +\, nM\ } \underset{\underset{M\!\sim\! M^{\oplus}}{|}}{\overset{*}{\sim\!\!\sim C \sim\!\!\sim}} \qquad\qquad (13.2)$$
$$Et_2AlCl_2^{\ominus}$$

growing branch
new graft copolymer

Since ubiquitous moisture does not initiate olefin polymerizations in conjunction with these Friedel–Crafts acids under these conditions, the products that arise will not be adulterated by homopolymer due to initiation although homopolymer can still form by chain transfer [2]. This principle of "controlled initiation" has led to a large number of interesting graft copolymers and Section 13.2.2 summarizes the chemistry and some properties of those grafts that exhibit TPE characteristics. References [2] and [3] concern a detailed introduction to controlled initiation in the context of cationic grafting processes, and Ref. [4] provides an updated discussion of the general principles.

For a graft copolymer, particularly a styrenic graft, to exhibit acceptable TPE properties, it should comprise a rubbery backbone carrying at least two but preferably many more than two (say ~ 10 or more) incompatible glassy branches. The molecular weight of the rubbery moiety should be fairly high ($> 50,000$ g/mol) and that of the branches at least medium (> 8000 g/mol). The molecular weight of the rubbery segment between two glassy branches ($M_{c,\text{rubber}}$) should preferably be higher than ~ 5000 g/mol; otherwise the TPE will tend to be brittle. In contrast, graft copolymers comprising a glassy backbone carrying rubbery branches are much less suitable as TPEs. The sketches in Fig. 13.1 illustrate these requirements. The more uniform the molecular weights of both the rubbery and glassy segments, in other words, the narrower the molecular weight distributions of the segments (both $M_{c,\text{rubber}}$) and $M_{n,\text{rubber}}$; and $M_{n,\text{glass}}$), the sharper will be the phase separation between the incompatible rubbery and glassy microdomains. Thus better overall mechanical properties (i.e., tensile strength, modulus, elongation) will arise. All these conclusions of course closely parallel the well-known teachings of glassy–rubbery–glassy triblock TPEs.

The early graft copolymers synthesized by controlled initiation approached but did not quite attain some of these ideal requirements: for example, the molecular weights of the backbones and branches were not uniform (i.e., $M_w/M_n \gg 1.0$) because they did not arise by living polymerization. The M_c between the glassy branches was also not uniform but statistically random, because the active halogens were introduced into the rubbery backbones

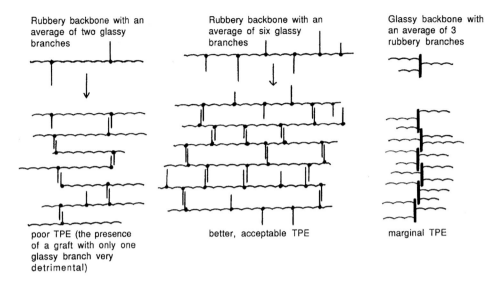

Figure 13.1   Scheme of graft structures: Useful TPE arises only with rubbery backbone with multiple glassy branches. (〰〰〰), rubbery segment; (——), glassy segment

by halogenations in solution and these are random processes. Thus dangling end-segments remained at the end of the backbones; these do not contribute to the load-bearing capability of the final TPE but act as useless diluents (see Fig. 13.1). Contemporary macromolecular engineering may reduce, indeed eliminate, these and other design shortcomings; however, studies have not yet been carried out in this direction.

13.2.2   Rubbery Backbone–Glassy Branches

In line with the principles discussed in Section 2.1, a large number and variety of TPEs comprising a rubbery backbone carrying glassy branches have been prepared by carbocationic techniques [4].

Table 13.1 summarizes the products described and lists references. The grafts are subdivided into four groups: grafts with one kind of glassy branches, with two kinds of glassy branches (bigrafts), or with one kind of rubbery and one kind of glassy branches (bigrafts), and grafts that carry a block as the branch wherein the block contains a glassy sequence (graft blocks). The subsequent sections highlight some representative TPEs. Obviously, at this early stage of development, the largest group of TPE grafts are in the first, simplest, group.

13.2.2.1   Rubbery Backbone Carrying One Kind of Glassy Branch

13.2.2.1.1 Backbone: EPR or EPDM   Probably the simplest cationically prepared TPEs are grafts of commercially available EPR backbones carrying multiple PSt branches [5]. They can be readily prepared by employing lightly chlorinated (1 to 5 wt% Cl) EPR as the initiator, in conjunction with various alkylaluminum compounds (Me$_3$Al, Et$_3$Al, Et$_2$AlCl) as coinitiators, for the polymerization of styrene (St):

$$
\begin{array}{c}
\qquad\qquad\quad CH_3 \qquad\qquad\qquad\qquad\qquad CH_3 \quad CH_3 \\
\qquad\qquad\quad | \qquad\qquad\qquad\qquad\qquad\quad\; | \qquad\quad | \\
\sim\!\!\sim\!\!CH_2CH_2CH_2CHCH_2CH_2CH_2CH_2CH_2CHCH_2CHCH_2\sim\!\!\sim
\end{array}
$$

$$\downarrow Cl_2$$

$$
\begin{array}{c}
\qquad\qquad\quad CH_3 \qquad\qquad\qquad\qquad\qquad CH_3 \quad CH_3 \\
\qquad\qquad\quad | \qquad\qquad\qquad\qquad\qquad\quad\; | \qquad\quad | \\
\sim\!\!\sim\!\!CH_2CH_2CH_2CHCH_2CH_2CH_2CH_2CH_2CHCH_2CCH_2\sim\!\!\sim \\
\qquad\qquad\qquad\qquad\qquad\qquad\qquad\qquad\qquad\qquad | \\
\qquad\qquad\qquad\qquad\qquad\qquad\qquad\qquad\qquad\quad Cl
\end{array}
$$

$$\downarrow +Et_2AlCl$$

$$
\begin{array}{c}
\qquad\qquad\quad CH_3 \qquad\qquad\qquad\qquad\qquad CH_3 \quad CH_3 \\
\qquad\qquad\quad | \qquad\qquad\qquad\qquad\qquad\quad\; | \qquad\quad | \\
\sim\!\!\sim\!\!CH_2CH_2CH_2CHCH_2CH_2CH_2CH_2CH_2CHCH_2\overset{\oplus}{C}CH_2\sim\!\!\sim \\
\\
\qquad\qquad\qquad\qquad\qquad\qquad\qquad\qquad\qquad\quad Et_2AlCl_2^{\ominus}
\end{array}
$$

$$\downarrow +Styrene$$

$$
\begin{array}{c}
\sim\!\!\sim\!\! EP\ Rubber \sim\!\!\sim \\
| \\
Polystyrene
\end{array}
$$

References [5] and [6] provide detailed synthesis, characterization, and physical–mechanical property information. In regard to the synthesis, the effect of chlorine content, nature of the alkylaluminum, grafting time, solvent polarity, temperature, styrene concentration, etc. have been described. Under suitable conditions, grafting efficiencies close to $\sim 90\%$ have been achieved. As shown in Fig. 13.2, the moduli of solution cast EPM–g–PSt films increase with PSt content. Figure 13.3 shows the effect of PSt content on the stress–strain properties of a series of grafts prepared with Et$_3$Al [6]. As with styrenic TPEs, the products change from elastomers to tough plastics with increasing PSt content.

EPDM–g–PSts have also been prepared as precursors of EPDM–g–PSt–g–PαMeSt bigrafts and their TPE characteristics demonstrated (see Section 13.2.2.1).

An attempt by French workers to graft Ind from chlorinated EPR by essentially the same synthetic principles gave unsatisfactory products due to various side reactions (crosslinking, gelation) and the lead was discontinued [7]. In contrast, the use of the tBuCl/Et$_2$AlCl initiating system in the presence of Ind and unchlorinated EPDM yielded TPEs with respectable mechanical properties [7]. The grafting was proposed to occur by the attack of growing PInd$^{\oplus}$ cations on the unsaturated sites of EPDM (grafting onto) or by allylic H$^{\ominus}$ abstraction by the tBu$^{\oplus}$ cation followed by graft initiation (grafting from). Table 13.2 shows the available mechanical property data [7].

EPDM–g–PSt and –PαMeSt grafts have also been prepared by the macromonomer technique, specifically by using cationically prepared cyclopentadienyl (Cp)-headed PSt and

Table 13.1 TPE Graft Copolymers: Rubbery Backbone–Glassy Branches

	Reference
With one kind of glassy branch	
EPM–g–PSt[a]	[5, 6]
EPM–g–PInd[a]	[7]
EPDM–g–PSt[a]	[8]
EPDM–g–PαMeSt[a]	[8]
PBd–g–PSt	[9]
PBd–g–αMeSt[a]	[10]
PBD–g–PInd	[7]
IIR–g–PSt[a,b]	[15, 16]
IIR–g–PαMeSt[a]	[17]
IIR–g–PInd[a]	[7, 16]
IIR–g–PAc	[16]
IIR–g–(Ind–co–αMeSt)	[7]
IIR–g–PpClSt[a]	[18]
ClSO₂PE–g–PSt	[20]
CR–g–PαMeSt	[17]
P(IB–co–pClCH₂St)–g–PSt	[24, 25]
P(IB–co–pClCH₂St)–g–PαMeSt	[26]
P(IB–co–pClCH₂St)–g–PInd	[24]
P(Bd–co–pClCH₂St)–g–PSt	[27]
With two kinds of glassy branches (bigrafts)	
EPDM–g–PSt–g–PαMeSt[a]	[28–30]
With one glassy and one rubber branch (bigraft)	
EPDM–g–PSt–g–PIB[a]	[28, 30]
EPDM–g–PαMeSt–g–PIB[a]	[33]
With a block branch (graft Block)	
CR–g–(PIB–b–PαMeSt)	[31]

[a] Lightly chlorinated backbone used
[b] Lightly brominated backbone used

PαMeSt (Cp-PSt, Cp-PαMeSt) [8]. The grafts were obtained by terpolymerization ethylene + propylene + Cp-PSt (or Cp-PαMeSt) using Ziegler–Natta catalysts. These materials are expected to be TPEs; however, the needed characterization research is yet to be performed.

An attempt has been made to prepare EPR– and EPDM–g–PSts in the melt in the presence of AlCl₃ in a laboratory-scale internal mixer [32]. Although extensive (40% to 50%) grafting occurred, side reactions were significant and the properties of the products have not been characterized.

13.2.2.1.2 Backbone: PBd   PBd–g–PSt and –PInd were prepared by grafting St or Ind onto PBd by the use of tBuCl/Et₂AlCl and TiCl₄, respectively [7, 9]. The effects of synthesis variables on graft composition, grafting efficiency, and microdomain morphology have been studied [9]. Regrettably, mechanical property information is unavailable on these potentially useful TPEs.

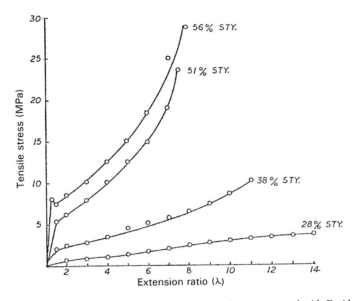

Figure 13.2   Stress–strain curves of EPM–*g*–polystyrene copolymers prepared with Et₃Al

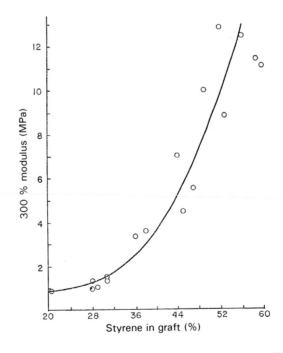

Figure 13.3   Effect of polystyrene content of the EPI–*g*–polystyrene system on 300% modulus

Table 13.2 Mechanical Properties of EPDM–*g*–PInd

Rubber	M_n	EPDM/PInd	$M_{n,graft}$	$M_{n,PInd}$	Temperature (°C)	Tensile strength (MPa)	Elongation (%)
E550-2504[a]	55,000	67/33	67,000	40,000	20	5.3	576
					60	2.1	364
					80	1.0	144
E550-2504[a]	55,000	54/45	130,000	40,000	20	12.2	250
					60	6.2	168
					80	8.4	67
Dutral-38[b]	66,000	46/54	256,000	56,500	20	16.3	395
					60	6.2	368
					80	2.2	112

[a] Ethylidene norbornene content: 3.5%
[b] Ethylidene norbornene content: 12%
Source: Ref. [7]
To convert MPa to psi, multiply by 145

PBd–*g*–PαMeSt was readily synthesized by grafting αMeSt from chlorinated PBd by the use of Me$_3$Al, Et$_3$Al, and Et$_2$AlCl coinitiators [10]. The focus of this research was synthesis and characterization and little property data are available.

13.2.2.1.3 Backbone: IIR   Halobutyl rubbers (X-IIR, where X = Cl or Br) are chlorinated or brominated copolymers of isobutylene and small amounts (0.6 to 2.5 mol%) of isoprene [11, 12]. Because of their outstanding combination of air retention, chemical and oxidative resistance, and mechanical properties, large quantities of these rubbers are used by the automotive industries in tire inner liners, etc. In contrast to the well-known incompatibility of butyl rubber (IIR), the few but critical percentages of halogens in halobutyl rubbers provide, surprisingly, a degree of beneficial compatibility with other hydrocarbon rubbers used in tires, in addition to desirable cure versatility and cure safety [13]. The halogenation of IIR yields allylic halogens in the chain [14]:

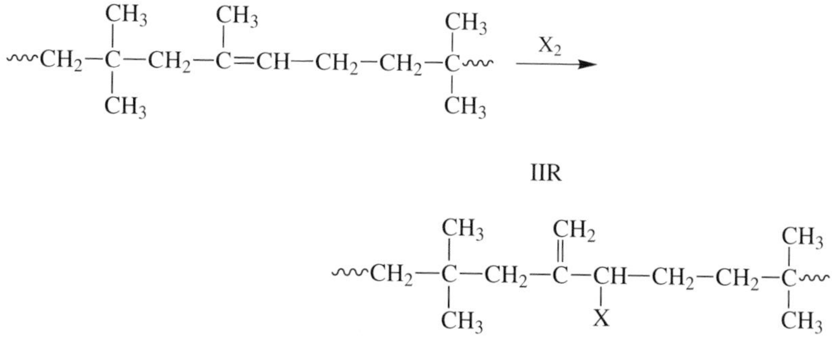

IIR

X – IIR(X = CL, Br)

The cationic grafting of glassy PSt branches from Cl-IIR occurs readily by the use of certain alkylaluminum compounds (Me₃Al, Et₃Al, Et₂AlCl) and TPEs of good overall quality can be obtained:

$$Cl-IIR \xrightarrow[Et_3Al]{St} \wedge\wedge CH_2-\underset{\underset{CH_3}{|}}{\overset{\overset{CH_3}{|}}{C}}-CH_2-\underset{\underset{CH=CH}{}}{\overset{\overset{St-St-St\wedge\wedge}{|}}{\overset{CH_2}{|}}{C}}=CH-CH_2-CH_2-\underset{\underset{CH_3}{|}}{\overset{\overset{CH_3}{|}}{C}}\wedge\wedge$$

For a discussion of this structure see [15].

Surprisingly, the grafting from Br-IIR under similar conditions leads to much less satisfactory products [15].

Quantitative proof for grafting of St from Cl-IIR together with the absence of Cl-IIR degradation during grafting has been obtained (agreement between experimental and theoretical molecular weights calculated from composition data) [15]. The effects of experimental variables (e.g., temperature, solvent polarity, styrene concentration, nature and concentration of alkylaluminums, halogen in X-IIR) on the composition and micro-architecture of IIR–g–PSt have been investigated and it has been shown that under select conditions one can obtain close to 100% grafting efficiencies and good quality TPEs. Brief hot-milling helps to improve the rheological properties of IIR–g–PSt [2].

Figures 13.4 and 13.5 show the effect of PSt content on the 300% modulus and tensile strength, respectively, of representative IIR–g–PSts [16]. Additional information substantiating the TPE characteristics of these interesting graft copolymers is available [16].

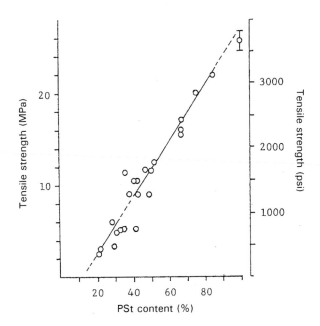

Figure 13.4   Effect of PSt content on tensile strength of chlorobutyl rubber–g–PSt prepared with Et₂AlCl coinitiator

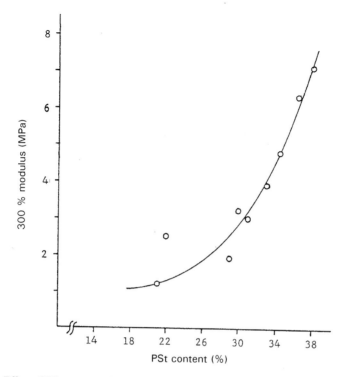

Figure 13.5   Effect of PSt cont-ent of Cl-butyl rubber–g–PSt on 300% modulus

The same technique used to prepare IIR–g–PSt grafts has also been used for the synthesis of IIR–g–PInd and –PAc grafts [16]. Figure 13.6 shows some characteristic stress–strain data relative to these new TPEs.

IIR–g–PαMeSt has been synthesized at high (up to 96% to 97%) grafting efficiencies by grafting αMeSt from Cl-IIR with SnCl$_4$ or BCl$_3$ coinitiators in the presence of the proton trap 2,6-di-*tert*-butylpyridine [17]. Unfortunately, mechanical property data for this potentially intriguing TPE are not available.

French investigators prepared IIR–g–PInd and IIR–g–(PInd–*co*–αMeSt) (i.e., grafts carrying branches of Ind–αMeSt copolymer) by starting the polymerization of Ind or Ind–αMeSt mixtures with Cl-IIR in conjunction with Et$_2$AlCl under various conditions [7]. While the mechanistic considerations are somewhat flawed (on account of the incorrect structure of the Cl-IIR) the mechanical properties, particularly the high-temperature strength properties of the products, are remarkable. Table 13.3 summarizes available information [7].

13.2.2.1.4 Backbone: ClSO$_2$PE   Rubbery chlorosulfonated polyethylene (ClSO$_2$PE) is a commercially available material (Hypalon$^®$) produced by free radical chlorination of low-density polyethylene in CCl$_4$ solution in the presence of SO$_2$. The substituents disrupt the crystallinity of the starting polyethylene and introduce into the chain a variety of chlorines (i.e., $\sim$3% primary, 90% secondary, 3% tertiary, and 4% sulfonyl) [19]. The cationic grafting of St from a ClSO$_2$PE (Hypalon-30$^®$ from DuPont, M$_n$ = 47,000 g/mol) containing $\sim$42% Cl and $\sim$0.9% S has been described [20]. This grafting most likely involves the *tert*-Cl sites.

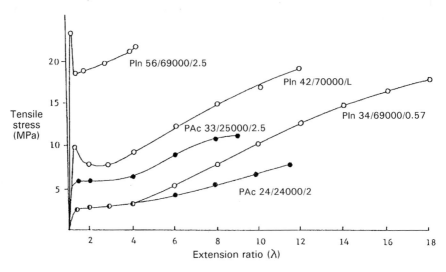

Figure 13.6   Stress–strain curves of Cl-butyl rubber–*g*–PIn and Cl-butyl rubber–*g*–PAc. (The three numbers identifying the curves give, respectively, the % glassy polymer content of the graft/M_n of the glassy branches/the branch-backbone ratio)

Backbone degradation accompanied grafting even under the most favorable conditions (Et$_2$AlCl coinitiator, CH$_2$Cl$_2$ solvent, -50 °C).

Pure grafts were not obtained and the materials tested most likely contained some ungrafted ClSO$_2$PE and PSt. Solution cast films were hazy, however, and considerable stress-whitening occurred. Table 13.4 summarizes some mechanical properties of two products: A = 77% PSt, M_n = 130,000 g/mol, gel content = 5%; and B = 45% PSt, M_n = 79,000 g/mol, gel content = 6%.

13.2.2.1.5 Backbone: CR   The repeat units present in CR (neoprene from DuPont) are as follows [21–23]:

$$-CH_2-\overset{\overset{\textstyle Cl}{|}}{C}=CH-CH_2- \qquad CH_2-\overset{\overset{\textstyle Cl}{|}}{\underset{\underset{\textstyle CH=CH_2}{|}}{C}}- \qquad \overset{\textstyle -CH_2-C-}{\underset{\textstyle HC-CH_2Cl}{\|}} \quad and \quad \overset{\textstyle -CH_2-CH}{\underset{\textstyle \underset{CH_2}{\overset{\|}{C-C}}}{|}}$$

major   minor

Because of the existence of the small but critical amounts of allylic chlorines, CR is a satisfactory cationic initiator in the presence of suitable Friedel–Crafts acids. Indeed, this rubber has been used to initiate the polymerization of αMeSt in conjunction with SnCl$_4$ or BCl$_3$ in the presence of the proton trap 2,6-di-*tert*-butylpyridine and the process has led to close to 100% grafting efficiencies [17]. Regrettably, physical property data have not been published.

Table 13.3 Characteristics and Mechanical Properties of IIR–g–PInd and IIR–g–P(Ind-co-MeSt)

	Graft				Testing	
Cl-IIR (%)	PInd (%)	PαMeSt (%)	$M_{n,graft}$	Temperature (°C)	Tensile strength (MPa)	Elongation (%)
80	20	—	238,000	Room	3.8	543
70	30	—	202,000	25	6.3	252
				60	3.0	110
				80	1.7	63
60	40	—	384,000	20	9.8	203
				60	4.0	40
				80	3.0	27
60	30	10	190,000	Room	11.1	188
60	23	17	220,000	25	13.0	380
				80	6.5	390
				120	2.8	180

Source: Ref. [7]
To convert MPa to psi, multiply by 145

13.2.2.1.6 Backbone: P(IB–co–pClCH₂St)   Statistical IB–pClCH$_2$St copolymers were prepared and used to provide backbones for PSt, PInd, and PαMeSt branches [24–26]; however, the mechanical properties of these potential TPEs have not been investigated.

13.2.2.1.7 Backbone: P(Bd–co–pClCH₂St)   The graft comprising a statistical Bd–pClCH$_2$St copolymer carrying PSt branches was claimed but insufficient data have been provided [27] for a meaningful property assessment.

13.2.2.2   Rubbery Backbone with Two Different Kinds of Glassy Branches (Bigrafts)

13.2.2.2.1 EPDM–g–PSt–g–PαMeSt   Advances in macromolecular engineering have led to the synthesis of bigrafts, that is, polymers in which a common backbone carries two

Table 13.4 Mechanical Properties of ClSO₂PE–g–PSt

	A*		B*	
Cross-head speed (cm/min)	5	0.5	5	0.5
Rate of strain (% min)	200	20	200	20
Tensile strength (MPa)				
Engineering stress	21.5	17.6	9.3	7.6
True stress	21.8	20.1	23.0	21.6
Elongation (%)	1.2	15	146	185
Modulus (100%)	—	—	125	100
Permanent set (%)	—	5	40	80

Source: Ref. [20]
To convert MPa to psi, multiply by 145
* See text for characterization data

different kinds of grafts [28, 29]. Bigrafts consisting of a rubbery backbone, for example, EPDM, having two kinds of glassy branches, for example, PSt and PαMeSt, are of great potential significance for TPEs. For example, the EPDM–g–PSt–g–PαMeSt, consisting of 63% EPDM, 19% PSt, and 18% PαMeSt, exhibits not three but only two T_gs: one at −45 °C characteristic of EPDM and one at 139 °C due to coalesced (molecularly compatible?) PSt/PαMeSt domains [30]. The latter transition is quite distant from both the T_g of PSt ($\sim$ 100 °C) and that of PαMeSt ($\sim$ 173 °C). The exact position of the high-temperature transition may be controllable by the relative concentration of the two glassy components. Thus, it appears that the high T_g of such bigrafts could be fine-tuned by controlling the relative concentration of the glassy components, PSt and PαMeSt.

Stress–strain properties of EPDM–g–PSt–g–PαMeSt indicate TPE characteristics. As shown by the data in Figs. 13.7 and 13.8, the grafting of a second relatively higher T_g segment PαMeSt to a TPE already containing a glassy segment PSt enhances tensile strength and moduli [28, 30].

13.2.2.3   Rubbery Backbone with One Glassy and One Rubbery Branch (Bigrafts)

13.2.2.3.1 EPDM–g–PSt–g–PIB and EPDM–g–PαMeSt–g–PIB   Bigrafts carrying one glassy and one rubbery branch (e.g., EPDM–g–PSt–g–PIB) have been prepared [28, 30, 33] and demonstrated to exhibit TPE character [30]. For example, a bigraft of EPDM = 44.5%, PSt = 32.4%, and PIB = 23.1% showed 23 MPa (3,300 psi) tensile strength, 7.3 MPa (1,060 psi) 300% modulus, and 915% elongation [30].

13.2.2.4   Rubbery Backbone with a Block Copolymer Branch

13.2.2.4.1 CR–g–(PIB–b–PαMeSt)   A series of graft blocks, consisting of a rubbery backbone and grafted branches of a diblock copolymer (one segment of which is glassy, the other rubbery), have been prepared [31]:

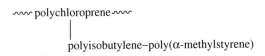

Although these products are expected to exhibit TPE characteristics, this has not been demonstrated.

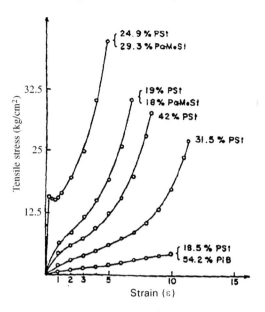

Figure 13.7   Stress–strain properties of representative monograft and bigraft copolymers. The numbers indicate the composition of the branches of the graft

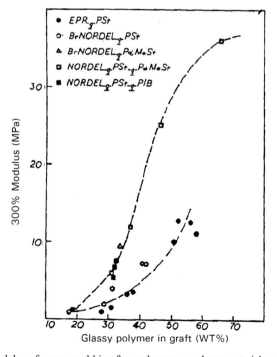

Figure 13.8   300% modulus of mono- and bigraft copolymers vs. glassy material content

13.3   Thermoplastic Elastomer Block Copolymers

13.3.1   Introduction: Living Polymerization and Sequential Monomer Addition

Long before the discovery of living carbocationic polymerizations and that of the introduction of the sequential monomer addition technique for the preparation of triblock TPEs, efforts were made to synthesize glassy–rubbery–glassy block copolymers by cationic techniques. Motivation for this research, of course, came from the spectacular academic and commercial successes of the rapidly expanding field of styrenic TPEs (notably the Kraton® family marketed by Shell) prepared by anionic polymerizations and from the realization that similar cationically prepared glassy–rubbery–glassy structures should have better overall properties than the anionically polymerized products.

Thus, Kennedy et al. prepared PαMeSt–b–PIB–b–PαMeSt by initiating bidirectional polymerization of αMeSt, first by using the telechelic tCl-PIB-Clt/Et$_2$AlCl initiating system [34] and later by the tCl-PIB–Clt/SnCl$_4$ combination in the presence of the proton trap 2,6-di-$tert$-butylpyridine [35, 36]. While the blocking efficiencies were quite high in both instances, the products were of broad molecular weight distributions which is quite undesirable for good TPE properties. Fodor et al. [37] synthesized PSt–b–PIB–b–PSt by slowly and continuously condensing IB gas to a bifunctional initiating system (p-dicumyl chloride/TiCl$_4$) stirred in a solvent at $-90\,°$C, and after the PIB has reached the desirable molecular weight, added St. In addition to the target triblock, however, side reactions led to PIB–b–PSt diblocks which compromised the properties of the TPE. In contrast to these early products prepared before the discovery of living carbocationic polymerization (see [3]), excellent TPEs have been obtained by the latter technique.

The group of anionically prepared TPEs that inspired these early researchers were PSt–b–PBd–b–Pst and Pst–b–PIP–b–PSt and the hydrogenated derivative of the former: PSt–b–poly(tetramethylene–co-1-butene)–b–PSt. These commercially available TPEs are preferentially synthesized by first preparing by sequential monomer addition to a PSt–b–PBd$_{1/2}^{\ominus}$Li$^{\oplus}$ or PSt–b–Pi$_{1/2}^{\ominus}$Li$^{\oplus}$ moiety and subsequently coupling two such fragments by difunctional coupling agents such as CH$_2$Br$_2$ to give the final TPE triblock (see Chapter 3):

$$\text{St} \xrightarrow{\text{secBuLi}} \text{St}^{\ominus} \xrightarrow{\text{1/2Bd}} \text{St–}b\text{–Bd}_{1/2}^{\ominus} \xrightarrow{\text{CH}_2\text{Br}_2} \text{PSt–}b\text{–PBd–}b\text{–PSt}$$

A major disadvantage of this process is the strict stoichiometric requirement in the coupling step: if the ratio $-$PBd$_{1/2}^{\ominus}$/CH$_2$Br$_2$ is not exactly 2.0 (a very difficult requirement to meet), the final product will contain PSt–b–PBd$_{1/2}$ diblock contaminants and these will seriously reduce the ultimate strength properties of the triblock. The resistance to oxidation and ozone of PSt–b–PBd–b–PSt can be reduced by selectively hydrogenating the rubbery midsegment, which is a random copolymer of cis- and $trans$-1,4- and 1,2-Bd repeat units:

$$\begin{array}{c} \text{—CH}_2\text{—CH=CH—CH}_2\text{—CH}_2\text{—CH—} \\ | \\ \text{CH=CH}_2 \end{array} \xrightarrow{\text{H}_2} \begin{array}{c} \text{—CH}_2\text{—CH}_2\text{—CH}_2\text{—CH}_2\text{—CH}_2\text{—CH—} \\ | \\ \text{CH}_2\text{—CH}_3 \end{array}$$

Corresponding polymers with a PIP midblock can be similarly be produced and hydrogenated. Hydrogenation, however, complicates the overall process and increases the cost.

Another fundamental weakness of these TPEs is the relatively low T_g of the PSt glassy segments ($<100\,°C$).

The following analysis indicates how cationically prepared styrenic TPEs could eliminate these process and product deficiencies:

1. The use of PIB as the rubbery midsegment would eliminate the need for hydrogenation. Hydrogenation of the PBd midsegment removes the unsaturation present in the original polymer but it creates tertiary Hs that are still vulnerable to oxidation. In contrast, PIB is saturated and contains only relatively stable primary and secondary Hs(CH_3- and $-CH_2-$ groups). IB can be polymerized only by cationic means and PIB-based rubbers ($T_g \sim -73\,°C$) are low-cost commodities whose molecular weight can be readily controlled.

2. The use of bifunctional cationic initiators (X-PIB-X) would produce biliving species ($^{\oplus}$PIB$^{\oplus}$) that can induce the polymerization of St to produce triblock; thus coupling would be eliminated. Indeed many efficient soluble bifunctional (and even trifunctional) initiators for cationic polymerizations have been described leading to PIB bi- and trications, $^{\oplus}$PIB$^{\oplus}$ and $^{\oplus}$PIB$^{\oplus}_{\oplus}$, which in turn can initiate the polymerization of St or St derivatives. In contrast, hydrocarbon-soluble bifunctional initiators are quite difficult to prepare; however, hydrocarbon solubility is mandatory because only in this medium can the needed PBd enchainment be obtained which, on hydrogenation, produces the rubbery poly(ethylene–co-1-butene) midsegment (see equation above).

3. TPEs with relatively high use temperatures could be made by employing St derivatives, such as pClSt and Ind, which readily undergo controlled cationic (but not anionic!) polymerization. The T_g of these PSt derivatives are much higher than that of PSt. Indeed, the T_gs of the glassy segments can be "fine tuned" by randomly copolymerizing cationically responsive St derivatives (see Section 13.3.2.2).

The road toward truly high-quality styrenic TPEs by cationic routes was opened by three discoveries made in rapid succession: (1) the living carbocationic polymerization of IB [38, 39]; (2) the living carbocationic polymerization of St and St derivatives [40–42]; and perhaps most importantly, (3) the finding that living PIB di- and trications $^{\oplus}$PIB$^{\oplus}$, $^{\oplus}$PIB$^{\oplus}_{\oplus}$ readily induce the living polymerization of St and St derivatives [43, 44]. These findings set the stage for the current development of PIB-based styrenic TPEs to be presented in the next section.

13.3.2 PIB-Based Glassy–Rubbery–Glassy Linear Triblocks and Radial Blocks of a Rubbery Core and Glassy Branches

Table 13.5 lists PIB-based linear- and radial-block TPEs prepared to date, together with references; Fig. 13.9 shows repeat units and T_gs of the glassy segments prepared, and Fig. 13.10 shows the bi- and trifunctional initiators described to date.

13.3.2.1 Styrenics

13.3.2.1.1 PSt–b–PIB–b–PSt and Radial(PIB–b–PSt)₃   PSt–b–PIB–*b*–PSt, conceived and developed as the cationic equivalent to anionically prepared PSt–*b*–PBd–*b*–PSt, was the first

Table 13.5 PIB-Based TPEs: Linear and Three-Arm Radial Blocks

	Reference
Linear and Three-Arm Star Triblocks with Glassy Homopolymers	
PSt–*b*–PIB–*b*–PSt	[43–49]
Radial(PIB–*b*–PSt)₃	[43, 44, 47]
PαMeSt–*b*–PIB–*b*–PαMeSt	[51, 52, 54]
PpMeSt–*b*–PIB–*b*–PpMeSt	[43, 46, 50, 51]
PptBuSt–*b*–PIB–*b*–PptBuSt	[43, 51, 55]
PInd–*b*–PIB–*b*–PInd	[43, 51, 55, 56, 70]
Radial(PIB–*b*–PInd)₃	[43, 51]
PAc–*b*–PIB–*b*–PAc	[60]
PpClSt–*b*–PIB–*b*–PpClSt	[43, 61, 62]
Radial(PIB–*b*–PpClSt)₃	[61, 62]
PpFSt–*b*–PIB–*b*–PpFSt	[65]
(*tr*-1,4-PIP)–*b*–PIB–*b*–(*tr*-1,4-PIP)	[66, 67]
cyPIP–*b*–PIB–*b*–cyPIP	[66]
PMMA–*b*–PIB–*b*–PMMA	[71, 72, 75]
i/sPMMA–*b*–PIB–*b*–s/iPMMA	[73]
Linear Triblocks With Glassy Copolymer Blocks	
P(*pt*BuSt–*co*–Ind)–*b*–PIB–*b*–P(*pt*BuSt–*co*–Ind)	[43, 51]
P(St–*co*–*p*ClSt)–*b*–PIB–*b*–P(St–*co*–*p*ClSt)	[43]
P(*p*MeSt–*co*–Ind)–*b*–PIB–*b*–P(*p*MeSt–*co*–Ind)	[69]
P(*p*MeSt–*co*–*p*ClSt)–*b*–PIB–*b*–P(*p*MeSt–*co*–*p*ClSt)	[43]
Linear and Three-Arm Star Ionomers	
$X^{\oplus}SO_3^{\ominus}$-PIB-$SO_3^{\ominus}X^{\oplus}$ and Radial (PIB-$SO_3^{\ominus}X^{\oplus}$)₃	[76, 76a, 77–82]
$(X^{\oplus}SO_3^{\ominus}$-$)_n$-PIB-$SO_3^{\ominus}X^{\oplus})_n$ and Radial [PIB-$(SO_3^{\ominus}X^{\oplus})_n$]₃	[83]

and is still the best-investigated triblock TPE synthesized by living cationic polymerization using sequential monomer addition. The synthesis, investigated on three continents [43–48], invariably involves a bifunctional initiator (see Fig. 13.10), $TiCl_4$ (or BCl_3) coinitiator, and the use of a moderately polar solvent mixture at −70 to −90 °C. Sequential monomer addition occurs in two steps: (1) the living polymerization of IB to a desirable molecular weight $^{\oplus}PIB^{\oplus}$ (M_n = 50,000 to 150,000 g/mol, $M_w/M_n \approx 1.1$) followed by (2) St addition to the $^{\oplus}PIB^{\oplus}$ charge and growing the triblock to a predetermined length ($M_{n, PSt}$ = 5000 to 25,000 g/mol, M_w/M_n of triblock ≈ 1.1). As expected, the triblock exhibits two T_gs and a high degree of microphase separation [44, 45, 57].

Great strides have been made in improving the physical–mechanical properties of PSt–*b*–PIB–*b*–PSt relative to the first triblocks described [43, 44]. The best overall properties to date have been published by Faust and co-workers [49]. Figure 13.11 shows the stress–strain curve of a PSt–*b*–PIB–*b*–PSt with M_n = 14,600–*b*–86,900–*b*–14,600, M_w/M_n (PIB) = 1.07 and M_w/M_n (triblock) = 1.12 [49]. Representative triblocks exhibited 23 to 26 MPa (3,300 to 3,800 psi) tensile strengths, which is similar to those of commercially available, anionically prepared styrenic TPEs. The tensile values are independent of $M_{n, PIB}$ in the 40,000 to 160,000 g/mol range. Tensile strengths start to rise at $M_{n, PSt} \approx 5000$, the point where the PIB–PSt domains start to phase separate, and reach a plateau in the 20 to 25 MPa (2,900 to 3,600 psi) range when $M_{n, PSt}$ is greater than ∼ 15,000 g/mol.

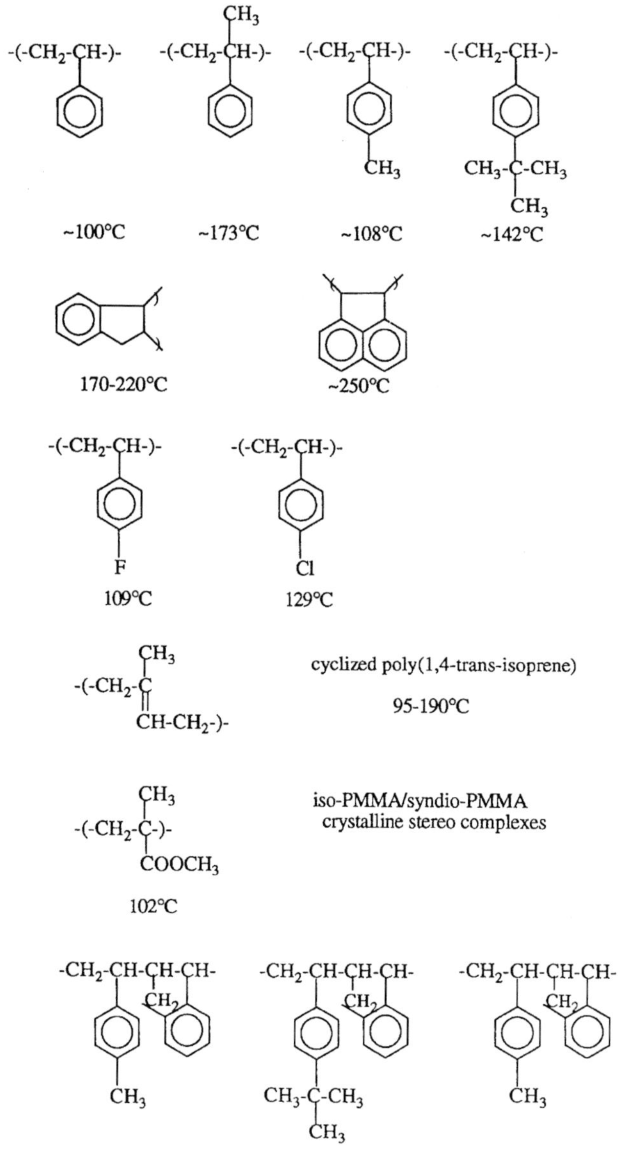

-(-CH₂-CH-)-   -(-CH₂-CH-)-   -(-CH₂-CH-)-   -(-CH₂-CH-)-

~100°C   ~173°C   ~108°C   ~142°C

170-220°C   ~250°C

-(-CH₂-CH-)-   -(-CH₂-CH-)-

F   Cl

109°C   129°C

$$-(-CH_2-\underset{\underset{CH-CH_2-)-}{\overset{\overset{CH_3}{|}}{C}}}$$

cyclized poly(1,4-trans-isoprene)

95-190°C

$$-(-CH_2-\underset{\underset{COOCH_3}{|}}{\overset{\overset{CH_3}{|}}{C}}-)-$$

iso-PMMA/syndio-PMMA
crystalline stereo complexes

102°C

-CH₂-CH-CH-CH-   -CH₂-CH-CH-CH-   -CH₂-CH-CH-CH-
CH₂   CH₂   CH₂

CH₃   CH₃-C-CH₃   CH₃
       CH₃

Figure 13.9   Repeat units of glassy segments and their T_gs

Koshimura and Sato [45] showed that these polymers have excellent damping characteristics (similar to IIR over a wide frequency range) and outstanding gas barrier properties. These PIB-based TPEs are tear resistant, quite soft (Shore A2 hardness increases from 25 to 87 as PSt content increases from 12 to 52 wt%), and exhibit good melt flow and resistance to shear degradation [45, 47].

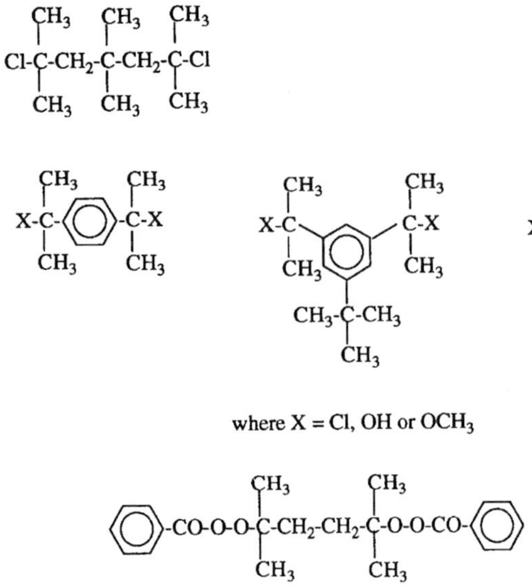

$$CH_3 \quad CH_3 \quad CH_3$$
$$Cl\text{-}C\text{-}CH_2\text{-}C\text{-}CH_2\text{-}C\text{-}Cl$$
$$CH_3 \quad CH_3 \quad CH_3$$

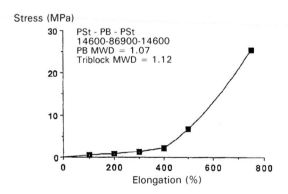

where X = Cl, OH or OCH$_3$

Figure 13.10   Bi- and trifunctional initiators for the synthesis of linear and three-arm radial blocks (coinitiator: TiCl$_4$)

Three-arm star radial blocks (radial(PIB–*b*–PSt)$_3$], where the radial core arises from a trifunctional aromatic initiator residue, have also been prepared and studied [43, 44, 47]. Motivation for the synthesis of radial blocks arises because of anticipated superior rheological characteristics at molecular weights equal to those of the linear block isomers. Regrettably, however, the needed rheological information has not yet been published.

13.3.2.1.2 PαMeSt–b–PIB–b–PαMeSt   The incentive for the preparation of this triblock is the relatively high T_g of the PαMeSt segment ($\sim$173 °C [52]) and low cost of the monomers. Also, because of its relatively low ceiling temperature [53] PαMeSt is rather difficult to

Figure 13.11   Stress–strain curve for the PSt–PIB–PSt triblock copolymer

synthesize by living anionic polymerization. However, efforts to prepare PαMeSt–b–PIB–b–PαMeSt TPEs by cationic means were unsuccessful [51] or met with only limited success [54].

13.3.2.1.3 PpMeSt–b–PIB–b–PpMeSt   This TPE (T_g of PpMeSt ~ 108 °C) has been synthesized by the same authors and by the same method used for the preparation of the PSt–b–PIB–b–PSt parent. However, only a limited amount of information is available in regard to physical–mechanical properties [43, 46, 50, 51]. In one example, a PpMeSt–b–PIB–b–PpMeSt of M$_n$ = 18,000–b–78,000–b–18,000 exhibited 14 MPa (2000 psi) tensile strength, 9 (1,300 psi) MPa 300% modulus, 420% elongation, and 53 Shore A hardness [50].

13.3.2.1.4 PptBuSt–b–PIB–b–PptBuSt   The motivation for the synthesis of this triblock was the relatively high T_g (~142 °C) of the outer glassy segment and the commercial availability of the monomer. The procedure for the preparation of this TPE was similar to that used for the PSt–b–PIB–b–PSt parent [43, 51, 55]. The triblock showed two T_gs (PIB = − 65 °C, PptBuSt = 144 °C), and its tensile properties and hardness were largely determined by the PptBuSt content. A representative triblock of M$_n$ ~ 100,000 g/mol, with M$_w$/M$_n$ = 1.2 and 40 wt% PptBuSt, exhibited 16 MPa (2,300 psi) tensile stress, 5 MPa (720 psi) 300% modulus, 550% elongation, and 76 Shore A hardness. The stress–strain behavior of this family of materials was typical of elastomers [55].

13.3.2.1.5 PInd–b–PIB–b–PInd and Radial(PIB–b–PInd)₃   Ind, a potentially very inexpensive monomer obtainable from coal tar, is ready converted to PInd (T_g ~ 220 °C) by cationic means [1, 2]. The living cationic polymerization has been investigated [57, 59] and the knowledge generated was used to prepare PInd–b–PIB–b–PInd triblocks [43, 51, 55]. The synthesis procedure was virtually identical to that used with St and St-derivatives, that is, addition of predetermined amounts of Ind to ⊕PIB⊕ dications of predetermined molecular weight (see Section 3.1). Representative TPEs showed microphase separation into PIB and PInd domains by T_g (−65 °C and 209 °C) and electron microscopy, and again the stress–strain behavior was typical of elastomers [56]. A PInd–b–PIB–b–PInd of M$_n$ = 93,500 g/mol (comprising M$_{n, PIB}$ = 54,000 and M$_{n, PInd}$ = 19,600) exhibited 20 MPa (3000 psi) tensile strength, 17 MPa (2000 psi) 300% modulus, 400% elongation, 60 Shore A hardness, and 15% tensile set. Owing to its saturated nature this TPE exhibits good thermostability (95% weight retention up to 346 °C in N$_2$ and 335 °C in air). Melt viscosities were similar to Kraton® TPEs [56].

13.3.2.1.6 PAc–b–PIB–b–PAc   In view of the very high T_g of PAc (~ 250 °C) efforts have been made for the synthesis of PIB-based TPEs with this St derivative (see Fig. 13.9 for formula). While the living cationic polymerization of Ac could not be achieved, some circumstantial evidence has been presented for the synthesis of PAc–b–PIB–b–PAc carrying very short PAc blocks (M$_n$ ~ 9000 g/mol) [60]. A triblock contaminated with diblock (perhaps also PIB and PAc) triblock containing 87 wt% PIB and 13 wt% PAc (M$_{n, PIB}$ ≈ 65,000) showed ~ 15 MPa (2,200 psi) tensile strength at ~ 800% elongation.

Because of the low PAc content the polymer was very soft: Shore 2A hardness was 37 to 40 [60].

13.3.2.1.7 PpClSt–b–PIB–b–PpClSt, Radial(PIB–b–PpClSt)₃ and PpFSt–b–PIB–b–
PpFSt PIB-based TPEs with chlorinated styrenic glassy segments could be of interest because of the combination of the relatively high T_g, polarity, and flame resistance of the hard block. In contrast to the living cationic polymerization of pClSt, the controlled anionic polymerization of this monomer is difficult even at $-78\,^{\circ}C$ owing to chlorine abstraction from the monomer by the counterion [64]. The cationic synthesis, characterization, and some properties of well-characterized linear P*p*ClSt–*b*–PIB–*b*–P*p*ClSt and three-arm star radial(PIB–*b*–P*p*ClSt)₃ have been described recently [61, 62]. Before the synthesis of terblock copolymers by sequential monomer addition, conditions for the living polymerization of *p*ClSt were developed [63]. Subsequently, the blocking of *p*ClSt from $^{\oplus}$PIB$^{\oplus}$ and $^{\oplus}$PIB$^{\oplus}$ was achieved and the properties of the resulting block TPEs were studied [62]. For example, a linear triblock of P*p*ClSt–*b*–PIB–*b*–P*p*ClSt (63/37 wt% overall composition containing a PIB midsegment of $M_n = 58,200$ ($M_w/M_n = 1.11$), exhibited 21 MPa (3000 psi) ultimate tensile strength, 15 MPa (2,200 psi) 300% modulus, 460% elongation, and 70 Shore A hardness. The stress–strain traces of various TPEs of this type were similar to those of typical rubbers.

The syntheses of P*p*FSt–*b*–PIB–*b*–P*p*FSt and radial(PIB–*b*–P*p*FSt)₃ have been accomplished under conditions essentially identical to those used for the preparation of the *p*Cl derivative. A linear block copolymer exhibited the expected TPE character [65]. In view of the presence of F in this product, it would be of interest to investigate the surface characteristics of this TPE.

13.3.2.2 T_g Control of Glassy Segments by Copolymerization of Styrenic Monomers

The T_g of a copolymer obtained by the random copolymerization of two monomers falls in between the T_gs of the homopolymers. In cationic polymerization the reactivity of various nonpolar or hydrocarbon substituted styrenic monomers is expected to be similar and therefore polymerization of mixed monomers should yield essentially random copolymers. For example, it has been demonstrated that, *p*MeSt and Ind give close to random copolymers [68] and that the T_g of the copolymers is determined by the relative *p*MeSt/Ind composition. Thus the T_g of such random copolymers can be "fine-tuned" between $\sim 115\,^{\circ}C$ (the T_g of P*p*MeSt) and $\sim 194\,^{\circ}C$ (that of PInd) [68].

This background knowledge has been used to prepare TPEs with glassy segments of predictable T_gs [69]. The synthesis started by preparing a bifunctional living $^{\oplus}$PIB$^{\oplus}$ of appropriate molecular weight ($M_n = 50,000$ to $70,000$) and narrow dispersity ($M_w/M_n = 1.1$), and adding to this dication mixtures of *p*MeSt + Ind. The second step was the copolymerization of these styrenic monomers until a preselected copolymer chain length was reached [69]. Table 13.6 summarizes synthesis conditions of four representative P(Ind–*co*–*p*MeSt)–*b*–PIB–*b*–P(Ind–co-*p*MeSt) together with characterization data and some mechanical properties [69]. The stress–strain traces were typical of those of high-quality TPEs. Similar results have also been published describing a TPE whose glassy segment was a P(*pt*BuSt–*co*–Ind) copolymer [51].

Table 13.6 Synthesis, Characterization, and Select Mechanical Properties of P(Ind-co-pMeSt)-b-PIB-b-P(Ind-co-pMeSt)

| | PIB midsegment[a] | | | | Triblock[b] | | | | Compositions | | | Mechanical properties | | | |
Expt.	Conversion (%)	M_n^c	M_w/M_n^c	I_{eff}^d, (%)	Conversion of Ind and pMeSt, (%)	M_n^c triblock	M_n^f outer block	M_w/M_n^c	Overall composition of triblock rubber/glass (wt/wt)	Composition of outer block Ind/pMeSt (wt/wt)	T_g, Rubber/glass (°C)	Tensile strength (MPa)	Modulus (300%) (MPa)	Elongation (%)	Hardness (Shore A2)
C1	100	71,600	1.07	98	84	105,300	16,900	1.25	68/32	82/18	−61/189	18.8	12.0	410	64
C2	95	66,200	1.07	100	89	101,900	17,900	1.39	65/35	61/39	−61/175	21.3	14.8	450	62
C3	99	70,200	1.08	101	100	115,000	22,400	1.34	61/39	41/59	−63/148	23.4	14.6	460	56
C4	106	79,700	1.10	93	100	160,900	40,600	1.54	47/53	56/44	−60/162	—	—	—	—
M1	—	—	—	—	—	116,000	19,200	1.24	67/33	0/100	−63/113	14.1	9.1	420	53
I1	—	—	—	—	—	83,200	13,100	1.29	70/30	100/0	—	17.0	12.1	400	55
I2	—	—	—	—	—	93,400	19,600	1.39	59/41	100/0	65/209	20.5	17.0	400	70

[a] Conditions of IB polymerization: DiCumCl = 1.5 mM, TiCl$_4$ = 20.0 mM, IB = 1.87 M, Et$_3$N = 1.5 mM, CH$_3$Cl/nC$_6$H$_{14}$ = 4/6 (vol/vol), V_0 = 150 mL, 2 hours, −80°C

[b] Conditions of block polymerization: to the above living ⊕PIB⊕ charge at −80°C were added, in CH$_3$Cl (12 mL)/nC$_6$H$_{14}$ (18 mL), the following comonomer mixtures. C1: Ind 0.06$_2$ mol/pMeSt 0.015 mol. C2: Ind 0.046 mol/pMeSt 0.030 mol. C3: Ind 0.046 mol/pMeSt 0.031 mol. C4: Ind 0.077 mol/pMeSt 0.076 mol

[c] By GPC (using PIB calibration)

[d] Calculated for $(W_p/\bar{M}_n)$/(mole of initiator)

[e] Calculated from overall composition by ^{1}H NMR spectroscopy assuming 100% blocking efficiency of Ind-co-pMeSt

[f] Calculated from $(\bar{M}_{nTriblock} - \bar{M}_{nPInd})/2$.

To convert MPa to psi, multiply by 145

Source: Ref. [69]

13.3.2.3   Cyclized Polyisoprene as Glassy Segment

A series of unusual TPEs has been made in which the glassy outer segments are cyclopoly-isoprene (cyPIP) [66]. The synthesis of cyPIP–b–PIB–b–cyPIP type TPEs involved three steps in one pot: (1) the living bifunctional polymerization of IB to $^{\oplus}$PIB$^{\oplus}$, (2) the addition of IP and formation of tr-1,4-PIP segments, and (3) the acid-induced cyclization of the latter to glassy cyPIP segments ($T_g = 95\,°C$ to $190\,°C$, depending on the extent of cyclization). The structure of cationically prepared cyPIP has not been fully elucidated [66]; among the glassy units may be structural elements of three or four fused rings, for example,

Although cyclization reduces the number of unsaturations in the glassy segment, the product exhibits rather strong UV absorption which is characteristic of conjugated double bonds [66].

The tensile strength of extensively cyclized products increases with time. This effect may be caused by progressive phase separation of the rubbery PIB and glassy cyPIP micro-domains. For example, the tensile strength and elongation of a product containing a PIB segment of $M_n = 55,000$ and 14.4 mol% cyPIP changed from ~ 0.5 MPa (70 psi) and $\sim 1250\%$ at 2 days to ~ 10 MPa (1400 psi) and $\sim 1100\%$ after 34 days of storage at room temperature [66].

13.3.2.4   PMMA-b–PIB-b–PMMA and Hard Segments of Stereocomplexes

A series of PIB-based TPEs having PMMA glassy segments have been prepared by cationic → anionic chain-end transformation. The multistep synthesis proceeded by first preparing a α,ω-ditoluyl-PIB, lithiating it, adding 1,1-diphenylethylene, and initiating the polymerization of MMA by the latter intermediate [72]. A series of PMMA–b–PIB–b–PMMAs with various compositions (molecular weights) have been synthesized, characterized (T_gs at $-61\,°C$ and $105\,°C$), and their physical–mechanical properties investigated [72]. The higher molecular weight products (e.g., those with segmental $M_n \approx 10,000$–$53,000$–$10,000$) exhibited good mechanical strength (~ 15 MPa or 2000 psi) at $\sim 600\%$ elongation [72]. The PMMA segments in these products were essentially syndiotactic PMMAs (sPMMA) with T_g somewhat higher than $100\,°C$.

Blends of sPMMA and isotactic PMMA (iPMMA) yield crystalline stereocomplexes in which the s- and iPMMA chains are in close contact resulting in a significant rigidification of the blend; the melting point of iPMMA–sPMMA stereocomplexes is in the $170\,°C$ to $220\,°C$ range (see [74] and references therein). With this in mind, efforts have been made to increase the use temperature of sPMMA–b–PIB–b–sPMMA by blending this triblock with iPMMA, in other words, by preparing PIB-based TPEs in which the outer segments were iPMMA–sPMMA stereoblock complexes [73].

Physical testing showed that stereocomplexation significantly enhanced the mechanical properties, and that experimental variables strongly affected the ultimate properties (e.g., segment molecular weight, complexation conditions, annealing, casting conditions, etc.) [73].

Very recently the synthesis of PMMA–*b*–PIB–*b*–PMMA by a combination of cationic polymerization (of IB) and group-transfer polymerization (of MMA) has been reported; however, the properties of the products were not investigated [75].

13.4   Thermoplastic Elastomer Ionomers

13.4.1   Introduction

Ionomers are hydrocarbon polymers containing a relatively small amount (typically < 15 wt%) of ionic groups. This small but critical amount of charged moieties exerts a profound effect on the physical–mechanical solution and thermal properties of these materials. The ionic groups may be situated randomly along the backbone or exclusively at the end of the hydrocarbon chains. Wherever these groups may be they segregate from the sea of hostile hydrocarbons into coulombic domains ("multiplets" or higher "clusters") and thus give rise to ionic "crosslinks" of these polymers. Significantly, the ionic crosslinks can be thermally relaxed to permit melt flow and this relaxation is reversible. The exact "melting out" temperature of these ionic crosslinks can be affected by a variety of factors, for example, the nature of the ionic groups ($-SO_3^{\ominus}X^{\oplus}$, $-COO^{\ominus}X^{\oplus}$), the presence of excess inorganic additives, etc. Thus these polymers may be regarded a class of TPEs useful for hot melt adhesives, coatings, films etc. Chapters 10A and 10B concern ionomers in general and emphasize their TPE characteristics.

PIB-based ionomers made by cationic processes have often been discussed in the recent scientific literature and the next section concerns a brief review focusing on their synthesis and mechanical properties.

13.4.2   PIB-Based Ionomers

Linear and three-arm star PIBs carrying $-SO_3^{\ominus}Mt^{\oplus}$ end-groups, first described in 1982 [75], were prepared by quantitative acetyl sulfonation of olefin di- or tritelechelic PIBs, followed by neutralization of the terminal $-SO_3^{\ominus}$ groups with various inorganic reagents. More recently, it has been found that the same ionomers can also be made from *t*Cl-telechelic PIB [76]; schematically, for the tritelechelic variety:

Since these ionomers carry the hard ionic $-SO_3^{\ominus}Mt^{\oplus}$ ($Mt^{\oplus} = Na^{\oplus}$, $K^{\oplus}$, $Zn^{2\oplus}$, $Ca^{2\oplus}$, $Cr^{3\oplus}$, etc.) groups exclusively at chain termini, dangling hydrocarbon polymer chain ends are absent. Thus virtually every change will contribute to the load-bearing capacity of the network ("endless" ionomer networks) [4]. The mechanical properties of these ionomers were found to be comparable to, and in some instances better than, those of EPDM-based ionomers with random $-SO_3^{\ominus}Mt^{\oplus}$ groups [78].

A variety of such telechelic ionomers were the subject of a number of investigations focusing on the mechanical properties of these ionomeric TPEs [77–82]. For example, ionomers with relatively low ion contents (< 3 mol%) but elastic arms probably contained only small multiplets (ion-pair dimers). Linear ionomers were mechanically weaker than analogous three-arm star products, most likely because the former contain only relatively weaker ionic crosslinks but the latter also contain permanent covalent crosslinking points (the center core of the three-arm star). Good elastomeric properties arose only within a well-defined range of molecular weights. Strain-induced crystallization has been observed in low molecular weight ($M_n \sim 9,000$ g/mol) three-arm star $Ca^{2\oplus}$ neutralized PIB ionomers [77]. Low molecular weight ($M_n = 8300$) products exhibited low elongation ($\sim 150\%$) while those with M_n in the 11,000 to 34,000 range showed high extentions (often $> 1000\%$), high ultimate tensile strengths, low permanent set, and low hysteresis [79]. They can be processed above $\sim 150\,^{\circ}C$.

The addition of excess neutralizing agents profoundly affected mechanical properties. For example, the addition of excess KOH or $Ca(OH)_2$ to an ionomer greatly enhanced its ultimate strength. Thus the addition of 125 parts of $Ca(OH)_2$ to a three-arm star sulfonated PIB ($M_n = 14,000$) increased the ultimate tensile strength from ~ 6 MPa (900 psi) for the starting polymer to 9 MPa (1300 psi) for the ionomer, both at $\sim 1000\%$ elongation [81]. The excess inorganic material is most likely located at the ionic sites. Ionomers neutralized by zinc acetate were found to have the lowest softening temperatures [82]. This additive is an efficient plasticizer of PIB-based ionomers and significantly reduces their melt viscosities.

Linear and triarm star radial PIB-based ionomers carrying a multiplicity of terminal $-SO_3^{\ominus}Mt^{\oplus}$ groups have recently been prepared [83]. They were synthesized by first preparing a PSt–*b*–PIB–*b*–PSt or radial(PIB–*b*–PSt)$_3$ with short PSt segments (i.e., $M_{n,\,PSt} \approx 4000$; $M_{n,\,PIB} = 24{,}600$ for the linear and 37,500 for the three-arm variety), and subsequently partially sulfonating the aromatic blocks. According to the preliminary report available [83], the mechanical properties of the sulfonated ionomers increased relative to those of the unsulfonated parents.

13.5   Summary

This survey concerns TPEs made by cationic polymerization, or, more precisely, TPEs that can be made *only* by cationic polymerization or whose preparation requires at least one critical cationic polymerization step. Three types of such TPEs have been identified: (1) grafts (including bigrafts and block grafts), having a rubbery backbone (PIB, EPDM) carrying preferably more than two glassy branches; (2) triblocks comprising a rubbery midsegment (PIB) flanked by two glassy outer segments; and (3) ionomers that are rubbery PIB strands physically "crosslinked" by $-SO_3^{\ominus}Mt^{\oplus}$ groups placed exclusively at the end of the rubbery chains.

The synthesis of the grafts is mainly based on the synthetic principle of controlled initiation which, particularly in the presence of proton traps (D*t*BP), leads to essentially uncontaminated grafts. The synthesis of the best triblocks is effected by living bi- or tridirectionally growing $^{\oplus}PIB^{\oplus}$ or $^{\oplus}PIB^{\oplus}$, except in those cases where the terminal $-SO_3^{\ominus}Mt^{\oplus}$ groups are introduced by acetyl sulfonation and subsequent neutralization.

Among the unique features of cationic polymerizations are the use of a large variety of inexpensive reactive monomers, leading to both rubbery polymers (IB) and glassy polymers (*p*ClSt, Ind, αMeSt, etc.), whose controlled living polymerization can be achieved *only* by this technique. Other advantages are in the use of inexpensive chemicals (monomers, coinitiators such as TiCl$_4$, solvents), continuous processes, and reasonably simple reaction conditions. Most cationic polymerizations are preferably carried out at relatively low reaction temperatures (e.g., $-30\,°C$ to $-90\,°C$) but these temperatures are routinely achieved during commercial production of IIR.

The understanding and consequent control of cationic polymerization has sufficiently advanced, particularly during the past decade, to push this technique to the forefront of macromolecular engineering, a skill that is absolutely needed for the design and synthesis of the necessarily complex TPE microarchitectures.

Acknowledgments

This chapter was prepared with financial assistance from the National Science Foundation (Grant 94-23202).

Note Added in proof (March, 1996).   Many excellent publications have appeared on carbocationically prepared TPEs, since this manuscript was completed. Among the most significant papers are those by: G. Kaszas, *Polym. Mat. Sci. Eng.*, *68*, 325 (1993), R. Faust, *Macromol. Symp.*, *85*, 295 (1994), D. Li and R. Faust, *Macromolecules*, *28*, 4893 (1995), B. Zaschke and J.P. Kennedy, *Macromolecules*, *28*, 4426 (1995), R. Storey and D.W. Baugh, *Polym. Prepr.*, *36*, 414 (1995), and Z. Fodor and R. Faust, *J.M.S.-Pure Appl. Chem.*, *A33*, 305 (1996).

Most of these papers concern the synthesis, characterization, and select properties of polyisobutylene-based TPEs.

References

1. J.P. Kennedy, *Cationic Polymerization of Olefins: A Critical Inventory* (1975) John Wiley & Sons, New York
2. J.P. Kennedy and E. Maréchal, *Cationic Polymerization* (1978) John Wiley & Sons, New York
3. J.P. Kennedy (Ed.), *J. Appl. Polym. Sci. Appl. Polym. Symp. 30*, 1–11 (1977)
4. J.P. Kennedy and B. Ivan, *Designed Polymers by Carbocationic Macromolecular Engineering: Theory and Practice* (1992) Hanser, Munich and New York
5. J.P. Kennedy and R.R. Smith, In *Recent Advances in Polymer Blends, Grafts and Blocks*. L.H. Sperling (Ed.) (1974) Plenum Press, New York
6. R.R. Smith, Ph.D. Thesis, The University of Akron, 1984
7. P. Sigwalt, A. Polton, and M. Miskovic, *J. Polym. Sci. Symp. No. 56*, 13 (1976)
8. A. Gadkari and M. Farona, *Polym. Bull. 17*, 229 (1987)
9. J.P. Kennedy and J.M. Delvaux, *Adv. Polym. Sci. 38*, 141 (1981)
10. R. Ambrose and J.J. Newell, *J. Polym. Sci. Polym. Chem. Ed. 17*, 2129 (1979)
11. J.P. Kennedy, In Part I of *Polymer Chemistry of Synthetic Elastomers*. Kennedy-Tornqvist (Ed.) (1968) Wiley–Interscience, New York
12. D.J. Buckley, *Rubber Chem. Tech. 32*, 1475 (1959)
13. Anon., Enjay Butyl HT, Brochure of the Enjay Chemical Co., 1961
14. I.J. Gardner and J.V. Fusco, Rubber Meeting of the American Chemical Society, Montreal, Canada, May 1987, Abstr. #90
15. J.P. Kennedy and J.J. Charles, *J. Appl. Polym. Sci. Appl. Polym. Symp. 30*, 119 (1977)
16. J.J. Charles, Ph.D. thesis, The University of Akron, 1983
17. J.P. Kennedy and S.C. Guhaniyogi, *J. Macromol. Sci. Chem. A18*, 103 (1982)
18. J.P. Kennedy and F.P. Baldwin, Belgian Patent 701,850 (1968)
19. A. Nersasian and D.E. Andersen, *J. Appl. Polym. Sci. 4*, 74 (1960)
20. J.P. Kennedy and D.M. Metzler, *J. Appl. Polym. Sci. Appl. Polym. Symp. 30*, 105 (1977)
21. M.M. Coleman, D.L. Tabb, and E.G. Brame, *Rubber Chem. Tech. 50*, 49 (1977)
22. M.M. Coleman and E.G. Brame, *Rubber Chem. Tech. 51*, 668 (1978)
23. R.C. Ferguson, *Anal. Chem. 36*, 2204 (1964)
24. B. Pary, M. Tardi, A. Polton, and P. Sigwalt, *Eur. Polym. J. 4*, 393 (1985)
25. M. Tazi, M. Tardi, A. Polton, and P. Sigwalt, *Eur. Polym. J. 22*, 451 (1986)
26. M. Tazi, M. Tardi, A. Polton, and P. Sigwalt, *Br. Polym. J. 19*, 369 (1987)
27. Z. Janovic and K. Saric, *Croat. Chem. Acta 48*, 49 (1976)
28. J.P. Kennedy, E.G. Melby, and A. Vidal, *J. Macromol. Sci. Chem. A9*, 833–847 (1975)
29. J.P. Kennedy and A. Vidal, *J. Polym. Sci. Polym. Chem. Ed. 13*, 1765 (1975)
30. J.P. Kennedy and A. Vidal, *J. Polym. Sci. Polym. Chem. Ed. 13*, 2269 (1975)
31. J.P. Kennedy and S.S. Plamthottham, *Polym. Bull. 7*, 337–344 (1982)
32. B. Pukansky, J.P. Kennedy, T. Kelen, and F. Tudos, *Polym. Bull. 6*, 327–334 (1982)
33. A. Vidal and J.P. Kennedy, *Polym. Lett. 14*, 489–491 (1976)
34. J.P. Kennedy and R.A. Smith, *J. Polym. Sci. Polym. Chem. Ed. 18*, 1539–1546 (1980)
35. J.P. Kennedy, S.C. Guhaniyogi, and L.R. Ross, *J. Macromol. Sci. Chem. A18*, 199–128 (1982)

36. J.P. Kennedy, S. Guhaniyogi, and L.R. Ross, *Org. Coat. Appl. Polym. 46*, 178–181 (1982)
37. Zs. Fodor, J.P. Kennedy, T. Kelen, and F. Tudos, *J. Macromol. Sci. Chem. A24*(7), 735–747 (1987)
38. J.P. Kennedy and R. Faust, (to The University of Akron), U.S. Paten 4,910,321
39. J.P. Kennedy and M.K. Mishra, (to The University of Akron), U.S. Patent 4,929,683
40. R. Faust and J.P. Kennedy, *Polym. Bull. 18*, 21–28 (1988)
41. R. Faust and J.P. Kennedy, *Polym. Bull. 18*, 29–34 (1988)
42. R. Faust and J.P. Kennedy, *Polym. Bull. 18*, 35–41 (1988)
43. J.P. Kennedy, J.E. Puskas, G. Kaszas, and W.G. Hager, (to The University of Akron), U.S. Patent 4,946,899
44. G. Kaszas, J.E. Puskas, J.P. Kennedy, and W.G. Hager, *J. Polym. Sci. Pt. A. Polym. Chem. 29*, 427–436 (1991)
45. K. Koshimura and H. Sato, *Polym. Bull. 29*, 705 (1992)
46. H. Everland, J. Kops, A. Nielsen, and B. Ivan, *Polym. Bull. 31*(2), 159 (1993)
47. G. Kaszas, *Polym. Mat. Sci. Eng. 68*, 375 (1993)
48. M. Gyor, Zs. Fodor, H.C. Wang, and R. Faust, *Polym. Preprints 34*, 562 (1993)
49. R. Faust, *Makromol. Chem. Macromol. Symp. 85*, 295 (1994)
50. T. Tsunogae and J.P. Kennedy, *Polym. Bull. 27*, 631–636 (1929)
51. J.E. Puskas, G. Kaszas, J.P. Kennedy, and W.G. Hager, *J. Polym. Sci. Pt. A. Polym. Chem. 30*(1), 41–49 (1992)
52. J.M. Cowie and P.M. Toporowski, *Eur. Polym. J. 4*, 621 (1968)
53. L.J. Fetters, In *Block and Graft Copolymerization, Vol. 1.* R.J. Ceresa (Ed.) (1973) John Wiley & Sons, New York
54. Y. Tsunogae and J.P. Kennedy, *J. Polym. Sci. Pt. A. Polym. Chem. 32*, 403 (1994)
55. J.P. Kennedy, N. Meguriya, and B. Keszler, *Macromolecules 24*(25), 6572–6577 (1991)
56. J.P. Kennedy, S. Midha, and Y. Tsunogae, *Macromolecules 26*, 429–435 (1993)
57. J.P. Kennedy, S. Midha, and B. Keszler, *Macromolecules 26*, 424–428 (1993)
58. L. Thomas, M. Tardi, A. Polton, and P. Sigwalt, *Macromolecules 25*, 5886 (1992)
59. L. Thomas, M. Tardi, A. Polton, and P. Sigwalt, *Macromolecules 26*, 4075 (1993)
60. Zs. Fodor and J.P. Kennedy, *Polym. Bull. 29*(6), 697–705 (1992)
61. J.P. Kennedy and J. Kurian, *Polym. Mat. Sci. Eng. 63*, 371–375 (1990)
62. J.P. Kennedy and J. Kurian, *J. Polym. Sci. Pt. A. Polym. Chem. 28*, 3725–3738 (1990)
63. J.P. Kennedy and J. Kurian, *Macromolecules 23*, 3736–3741 (1990)
64. J. Liutkus, M. Hatzakis, J. Shaw, and J. Paraszczak, *Polym. Eng. Sci. 43*(18), 1047 (1983)
65. J. Kurian, Ph.D. Thesis, The University of Akron, 1991
66. G. Kaszas, J.E. Puskas, and J.P. Kennedy, *J. Appl. Polym. Sci. 39*(1), 119–144 (1990)
67. J.E. Puskas, G. Kaszas, and J.P. Kennedy, *J. Macromol. Sci, Chem. A28*, 65–80 (1991)
68. Y. Tsunogae, I. Majoros, and J.P. Kennedy, *J. Macromol. Sci. Pure Appl. Chem. A30*(4), 253–267 (1993)
69. Y. Tsunogae and J.P. Kennedy, *J. Macromol. Sci. Pure Appl. Chem. A30*(4), 269–276 (1993)
70. J.P. Kennedy, B. Keszler, Y. Tsunogae, and S. Midha, *Polym. Prepr. 32*(1), 310 (1991)
71. J.P. Kennedy and J.L. Price, *Polym. Mat. Sci. Eng. 64*, 40–41 (1991)
72. J.P. Kennedy and J.L. Price, *ACS Symp. Ser. No. 496*, pp. 258–277 (1992)
73. J.P. Kennedy, J.L. Price, and K. Koshimura, *Macromolecules 24*(25), 6567–6571 (1991)
74. G. Helary, G. Belorgey, and T. Hogen-Esch, *Polymer 33*(9), 1953 (1992)
75. W.G. Ruth, W.J. Brittain, A.V. Lubnin, J. Kuang, and J.P. Kennedy, *Polym. Prepr. 34*(2), 584–585 (1993)
76. J.P. Kennedy and R.F. Storey, *Org. Coat. Appl. Polym. Sci. 46*, 192–185 (1982)
77. R.F. Storey and Y. Lee, *J. Polym. Sci. Pt. A. Polym. Chem. 29*, 317 (1991)
78. S. Bagrodia, Y. Mohajer, G.L., Wilkes, R.F. Storey, and J.P. Kennedy, *Polym. Bull. 8*, 281–286 (1982)
79. Y. Mohajer, D. Tyagi, G.L. Wilkes, R.F. Storey, and J.P. Kennedy, *Polym. Bull. 8*, 47–55 (1982)
80. S. Bagrodia, Y. Mohajer, G.L. Wilkes, R.F. Storey, and J.P. Kennedy, *Polym. Bull. 9*, 174–181 (1983)
81. S. Bagrodia, Y. Mohajer, G.L. Wilkes, R.F. Storey, and J.P. Kennedy, *Polym. Preprints 24*(2), 88–89 (1983)
82. Y. Mohajer, S. Bagrodia, G.L. Wilkes, R.F. Storey, and J.P. Kennedy, *J. Appl. Polym. Sci. 29*, 1943–1950 (1984)
83. S. Bagrodia, G.L. Wilkes, and J.P. Kennedy, *J. Appl. Polym. Sci. 30*, 2179–2184 (1985)
84. R.F. Storey, B.J. Chisolm, and Y. Lee, *Polym. Preprints 33*(2), 184 (1992)

14 Macromonomers as Precursors for Thermoplastic Elastomers

Roderic P. Quirk, William J. Brittain, and Gerald O. Schulz

14.1 Introduction. 396

14.2 Synthesis of Macromonomers. 398
    14.2.1 Anionic Polymerization 398
        14.2.1.1 Polystyrene Macromonomers. 398
        14.2.1.2 Poly(methyl methacrylate) Macromonomers. 404
        14.2.1.3 Poly(2-vinylpyridine) Macromonomers. 406
    14.2.2 Group Transfer Polymerization. 407
    14.2.3 Michael Insertion Polymerization 408
    14.2.4 Free Radical Polymerization 409

14.3 Homopolymerization of Macromonomers. 410
    14.3.1 Anionic Polymerization 410
    14.3.2 Group Transfer Polymerization. 411

14.4 Copolymerization of Macromonomers. 411
    14.4.1 Kinetics. 411
    14.4.2 Anionic Copolymerization. 413
    14.4.3 Model Graft Copolymer Synthesis. 414
    14.4.4 Free Radical Copolymerization. 416
        14.4.4.1 Effects of Macromonomer Reactivity. 417
        14.4.4.2 Effects of Copolymerization Process 419
        14.4.4.3 Effects on Graft Copolymer Properties. 422

14.5 Summary 425

References. 426

backbone chain and also for the synthesis of the macromonomer precursor which forms the grafted branches. Thus, a diverse assortment of elastomeric and hard segment components can be envisioned which, in turn, can lead to materials that exhibit a wide range of useful TPE properties.

14.2   Synthesis of Macromonomers

Living polymerizations provide the maximum degree of control for the synthesis of macromonomers with predictable, well-defined structures. New living polymerization systems have been developed that proceed via a variety of mechanistic types including anionic, cationic, radical, Ziegler–Natta, ring-opening metathesis, coordination, and group-transfer polymerization [19]. However, anionic polymerization is one of the most reliable and useful methods for the routine synthesis of macromonomers, especially those based on styrene and alkyl methacrylate backbones. These are of interest for the preparation of comb-type graft copolymers that exhibit the properties of thermoplastic elastomers.

14.2.1   Anionic Polymerization

14.2.1.1   Polystyrene Macromonomers

Living anionic polymerization provides an excellent methodology for the synthesis of macromonomers with low degrees of compositional heterogeneity [2, 20–22]. The absence of chain termination and chain transfer reactions provides polymers whose molecular weights are precisely predicted and controlled by the stoichiometry of the polymerization, as shown in Eq. (14.1) [20, 21].

$$M_n = (\text{g of monomer})/(\text{moles of initiator}) \qquad (14.1)$$

In general, these polymers can also be prepared with narrow molecular weight distributions; however, this depends on the relative rates of initiation (R_i) and propagation (R_p) [23]. In addition, the polydispersity (compositional heterogeneity in molecular weight) will decrease with increasing molecular weight in accord with the relationship developed by Flory [24a, 25] [Eq. (14.2)], where the second approximation is valid only for higher degrees of polymeriza-

$$X_w/X_n = 1 + [X_n/(X_n + 1)^2] \cong 1 + [1/X_n] \qquad (14.2)$$

tion. From Eq. (14.2), it is apparent that the polydispersity will be highest for the lower molecular weight macromonomers. To reduce the polydispersity, it is advisable to use a very reactive initiator such as *sec*-butyllithium rather than *n*-butyllithium [26]. Each initiator molecule initiates one polymer chain [Eq. (14.1)] by addition of monomer units. Since the initiator residue is located at the initiating (α) chain end of the polymers, a macromonomer can be prepared by utilizing an initiator that contains a polymerizable functional group as shown in Scheme 2, where $-X$ is a polymerizable functional group which must also be stable with respect to the anionic polymerization of monomer [11]. This functionalized initiator procedure

$$X-R-Li + nM_1 \longrightarrow X-R-[M_1]_n-Li \xrightarrow{H_2O} X-R-[M_1]_n-H$$

Scheme 2

is most effect for heterocyclic ring-opening polymerization where the propagating anion is generally not reactive with respect to addition to vinyl functionality such as a styryl or methacryloyl group [27].

Another attribute of living anionic polymerization that is important for the synthesis of macromonomers is that these polymers retain the carbanionic chain end when all of the monomer has been consumed [28]. Therefore, these carbanionic chain ends (e.g., P^-Li^+) can react with a variety of electrophilic reagents (e.g., X–Y) to generate polymers with functional end groups, –Y, as shown in Eq. (14.3) [29].

$$P^-Li^+ + X-Y \rightarrow P-Y + LiX \tag{14.3}$$

The concept of macromonomers as conceived and developed by Milkovich [1–4] is most clearly represented by the synthesis of a methacryloyl-terminated polystyrene as shown in Scheme 3. This synthesis illustrates some of the complexities and attributes of anionic macromonomer synthesis. First, poly(styryl)lithium (PsLi) can be synthesized readily with molecular weights ranging from less than 1×10^3 g/mol to the upper ranges of useful macromonomers (20 to 30×10^3 g/mol) [30]. However, poly(styryl) lithium is too reactive for direct reaction with methacryloyl chloride; vinyl addition competes with acylation [5]. The reactivity of the chain end can be attenuated by first end-capping with ethylene oxide to form the corresponding lithium alkoxide (Scheme 3). It is noteworthy that only one epoxide

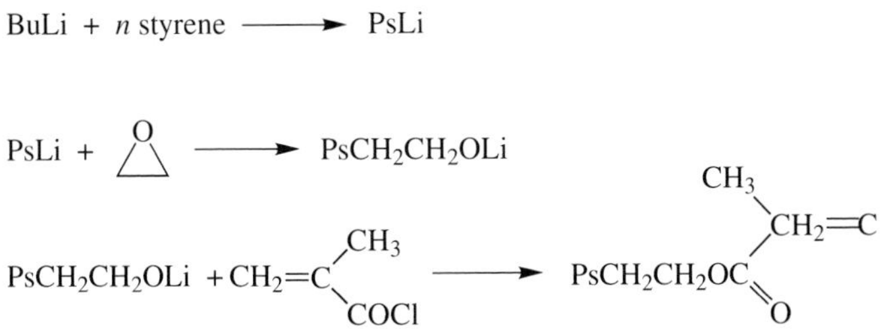

$$BuLi + n \text{ styrene} \longrightarrow PsLi$$

$$PsLi + \triangle O \longrightarrow PsCH_2CH_2OLi$$

$$PsCH_2CH_2OLi + CH_2{=}C\begin{smallmatrix}CH_3\\COCl\end{smallmatrix} \longrightarrow PsCH_2CH_2OC\begin{smallmatrix}CH_2{=}C{-}CH_3\\O\end{smallmatrix}$$

Scheme 3

unit adds to the chain end with lithium as the counterion [31]. The polymeric alkoxide was then reacted with methacryloyl chloride to form the methacrylate-functionalized macro-monomer. High functionality (>90%) can be obtained for these reactions even using large-scale equipment [3].

It is important to quantitatively determine the functionality of macromonomers. A common procedure is to subject the macromonomer to copolymerization with other monomers, using free-radical polymerization, and then determine the amount of "macromonomer" that does not undergo copolymerization by size-exclusion chromato-graphy (SEC); this method assumes that the amount of polymer remaining corresponds to the maximum amount of unfunctionalized macromonomer [2–4]. The vinyl functionality can be determined by titration with bromine [32] or mercuric acetate [33, 34] and this can be used to estimate M_n; comparison of this value with either M_n (SEC) or M_n (osmometry) provides an estimate of functionality. The ^{1}H NMR spectrum of the methacrylate-functionalized polystyrene exhibits resonances at $\delta = 5.4$ and 5.8 (5.9) ppm for the vinyl protons (=CH$_2$), $\delta = 3.75$ ppm for the oxymethylene group (−OCH$_2$), and at $\delta = 1.8$ ppm for the vinyl methyl group (=C−CH$_3$) [32, 34]. In general, high functionalities have been reported for the methacrylate-terminated polymers using these analytical methods.

This same methodology can be used to prepare diblock macromonomers as shown in Scheme 4. The living diblock copolymer polystyrene–*block*-poly(isoprenyl) lithium was first prepared by sequential monomer addition, followed by end-capping with ethylene oxide and then termination with methacryloyl chloride [35].

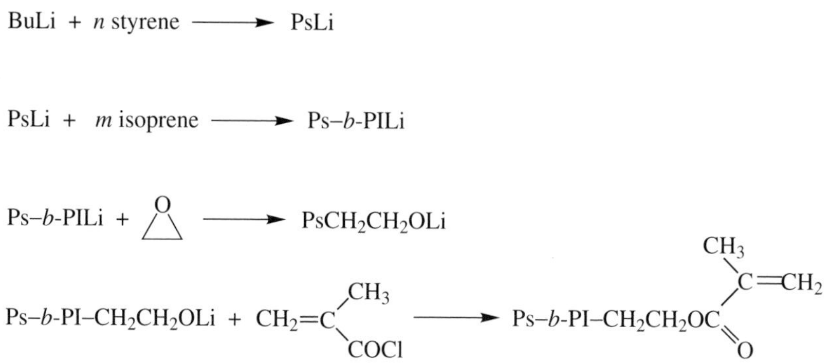

BuLi + n styrene ⟶ PsLi

PsLi + m isoprene ⟶ Ps–*b*-PILi

Ps–*b*-PILi + (ethylene oxide) ⟶ PsCH$_2$CH$_2$OLi

Ps–*b*-PI–CH$_2$CH$_2$OLi + CH$_2$=C(CH$_3$)(COCl) ⟶ Ps–*b*-PI–CH$_2$CH$_2$OC(=O)C(CH$_3$)=CH$_2$

Scheme 4

Macromonomer synthesis via the direct reaction of polymeric organolithium compounds with unsaturated alkyl halides is complicated by side reactions such as lithium–halogen exchange, coupling to form dimeric species and elimination reactions when a hydrogen is located in a vicinal position relative to the halogen [36]. For example, direct addition of a benzene solution of poly(styryl) lithium (M$_n$ = 6 × 10^3 g/mol) to a tetrahydrofuran (THF) solution of p-vinylbenzyl chloride (8 molar excess) at 0 °C yields only a 50% yield of the desired macromonomer; a 50% yield of dimer (SEC analysis) is also obtained as shown in Scheme 5 [37]. However, when THF (17 vol%) was added to poly(styryl) lithium before the addition to p-vinylbenzyl chloride (13 molar excess), the macromonomer was obtained in quantitative yield. The macromonomer functionality ($f = 0.9$ to 1.1) was determined by UV analysis utilizing the high molar extinction coefficient of the styryl unit ($\varepsilon = 1.64 \times 10^4$ liter/mol-cm) compared to polystyrene ($\varepsilon = 1.33 \times 10^2$ liter/mol-cm) at $\lambda = 250$ nm. Other recent studies have confirmed the efficiency of this procedure [38]; functionalities of 95% to 99% were determined using ^{1}H NMR by integrating the resonance for the methyl groups from the *sec*-butyl initiator fragment relative to the methylene group of the vinyl benzyl unit at the chain end (M$_n$ = 2 × 10^3 to 17.5 × 10^3 g/mol). Reactions of poly(styryl) lithium with p-vinylbenzyl chloride at −78 °C in a toluene–THF mixture provided macromonomers with functionalities of 90% by ^{1}H NMR analysis and 71% to 81% by UV analysis [39]. Macromonomer functionalities (UV analysis) of only 72% have been reported for termination reactions of polystyrene–*block*-poly(isoprenyl) lithium with p-vinylbenzyl chloride in a toluene–THF (100/1, vol/vol) mixture at −78 °C [40]. The analogous

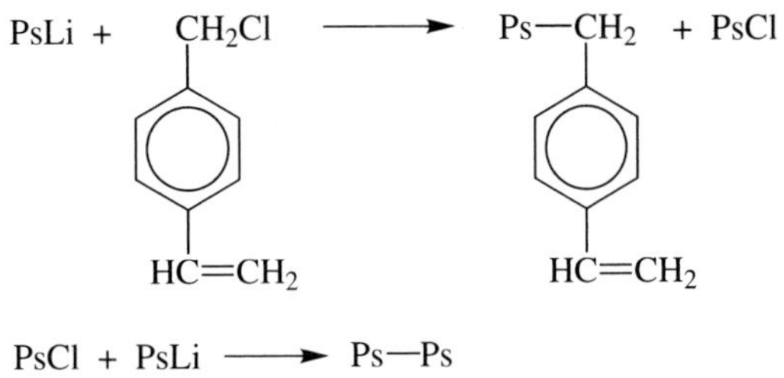

Scheme 5

functionalizations of poly(α-methylstyrene)–*block*-poly(isoprenyl) lithium were reported to be quantitative [40].

It has been reported that macromonomer synthesis can also be effected by termination with *p*-vinylbenzyl tosylate [41]. The direct reaction of poly(isoprenyl) lithium with *p*-vinylbenzyl tosylate in either benzene or a benzene–THF mixture produced the corresponding macromonomer with reported functionalities of 96% to 100% and no dimer formation.

An extension of the concept of deactivation of the chain end to reduce side reactions is the use of 1,1-diphenylethylene end-capping as shown in Scheme 6 [42]. Both polymeric and

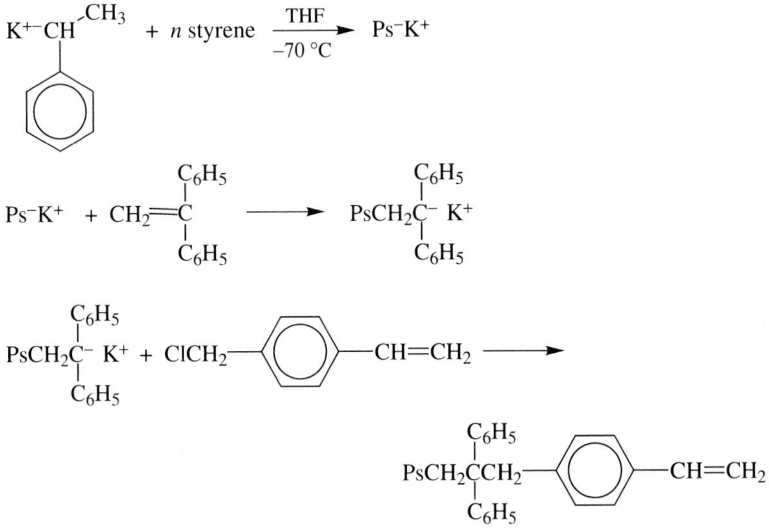

Scheme 6

simple organometallic compounds add quantitatively and relatively rapidly with 1,1-diphenylethylene to produce the corresponding 1,1-diphenylmethyl carbanion [43]. The chain end reactivity is attenuated to minimize addition reactions to the double bonds in the terminating agents. Since the desired addition reactions may not occur to the exclusion of addition to the double bonds or lithium–halogen exchange reactions that lead to coupled polymers, it is important to carry out adequate characterization to establish the end-group functionality of the macromonomers. This vinylbenzyl-functionalized polystyrene was titrated using mercuric acetate [33] for end-group analysis of double bonds. This estimate of number average molecular weight was in good agreement with M_n (SEC). It is noteworthy that the use of 1-phenylethyl potassium as initiator in THF at $-70\,^\circ$C resulted in polymers exhibiting somewhat broad molecular weight distributions ($M_w/M_n = 1.1$ to 1.15) [22].

Macromonomer syntheses by termination reactions with other alkyl halides have also been reported. For example, poly(styryl) lithium was terminated with allyl chloride [44]. Although direct characterization of the resulting macromonomer was not reported, copolymerization with a mixture of ethylene and propylene using vanadium Ziegler–Natta catalysts, for example, $VOCl_3/Et_3Al_2Cl_3$, incorporated the macromonomer into the resulting polymer with a reported efficiency of 80%. It has been reported that termination reactions of living carbanions with both vinyldimethylchlorosilane and vinyl(chloromethyl) dimethylsilane are quantitative in THF over a wide temperature range, although rather minimal characterization of functionality was provided [45].

An alternative procedure for the synthesis of polystyrene macromonomers is to use an initiator that has either a polymerizable functional group that is stable to the anionic polymerization conditions or that has a protected functional group. However, there are few examples of this type of synthesis, primarily because most unsaturated groups are not stable to polystyryl carbanions.

Waack and Doran [46, 47] have described the use of allyllithium and vinyllithium in THF as initiators for the anionic polymerization of styrene to produce the corresponding vinyl-substituted styrenes. Unfortunately, both of these initiators are relatively inefficient. Only 11% of allyllithium and only 5% of vinyllithium reacted with styrene; the result was that polystyrenes with molecular weights much higher than predicted by stoichiometry [see Eq. 14.1)] were obtained.

It would be expected that initiators with styryl units would not be useful for styrene polymerization because the styryl group would undergo competitive addition reactions with the growing carbanionic chain end. However, it has been reported that 4-vinylbenzyllithium can be used to initiate the polymerization of styrene without addition to the initiator double bond [48]. 4-Methylstyrene was metalated with lithium diisopropylamide as shown in Scheme 7. Unfortunately, the equilibrium metalation reaction does not proceed to form the

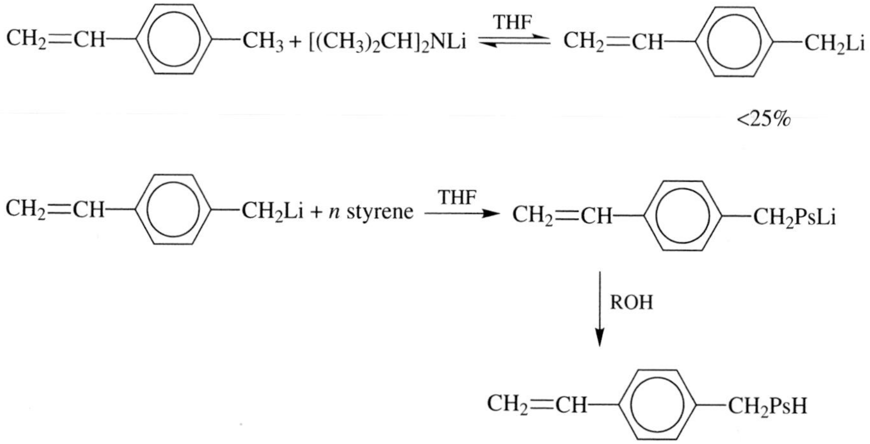

Scheme 7

desired 4-vinylbenzyllithium in very high yield; consequently, the residual 4-methylstyrene undergoes competing metalation during the polymerization reaction. However, it was reported that the resulting polystyrene macromonomer ($X_n < 15$) was monofunctional based on both UV-visible analysis ($\lambda = 296$ nm) and ^{1}H NMR analysis of double bond functionality. No evidence for 4-methylphenyl groups was observed by ^{1}H NMR. It should be noted that although 4-methylstyrene should be less reactive than styrene, it is not expected to be unreactive. Competitive addition to the styrene unit at the chain end and branching would be expected if higher molecular weight macromonomer synthesis was attempted. For example, the copolymerization parameter of styrene monomer in copolymerizations with 4-methylstyrene is 0.72 in benzene [49]. An analogous functionalized initiator was formed from 4-trimethylsilylmethylstyrene and lithium diisopropylamide. This polymerized 4-trimethylsilylmethylstyrene to form a complex type of macromonomer because of competing polymerization and metalation [50].

Termination reactions of living poly(butadienyl) lithium with dimethylfulvene have been reported to incorporate cyclopentadiene chain-end functionality as shown in Scheme 8 [51].

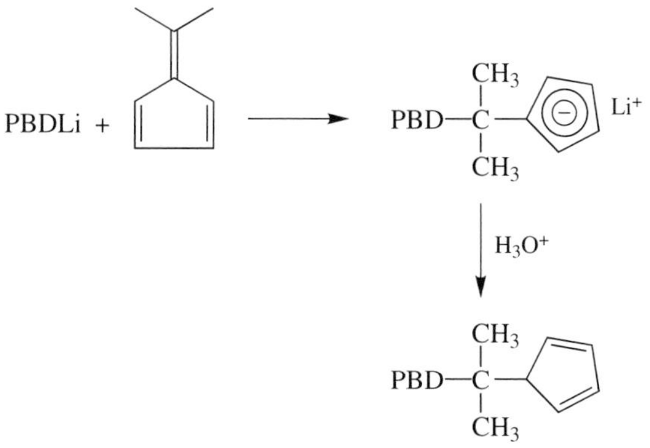

Scheme 8

The resulting cyclopentadienyl-functionalized polybutadiene was reacted with various dienophiles but no other characterization was reported.

Polystyrene macromonomers with terminal 1,1-diphenylethylene functionality have been prepared by the reaction of one equivalent of poly(styryl) lithium with two equivalents of 1,4-*bis*(1-phenylethenyl) benzene in the presence of small amounts of THF ([THF]/[Li] = 20) as shown in Scheme 9 [52]. The macromonomer was characterized UV spectroscopy utilizing the strong absorption of the 1,1-diphenylethylene group at 260 nm ($\varepsilon = 1.18 \times 10^4$) compared with the weak absorbance at this wavelength for polystyrene ($\varepsilon = 182$). In addition, for a macromonomer with $M_n = 5.4 \times 10^3$ g/mol, ^{1}H NMR spectroscopy could also be used to evaluate the chain-end functionality. The macromonomer exhibited characteristic resonances for the terminal diphenylalkyl methine proton and vinyl protons at $\delta = 3.5$ ppm and 5.4 ppm, respectively. Both of these methods indicate that the functionality is approximately 98% for addition reactions effected at 5 °C to 8 °C in benzene; less than 1.4% of the dimer adduct is obtained under these conditions.

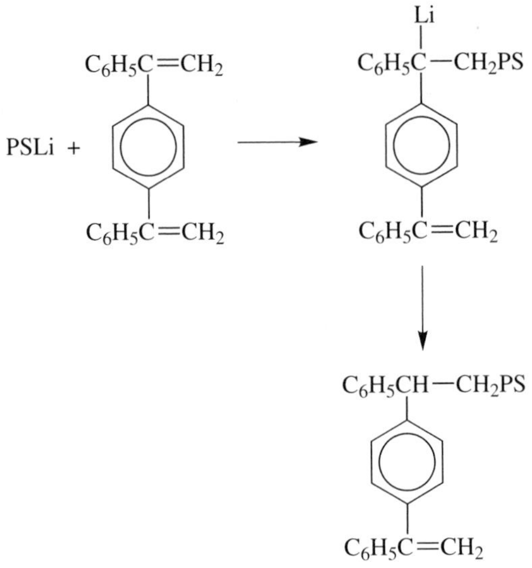

Scheme 9

14.2.1.2   Poly(methyl methacrylate) Macromonomers

Alkyl methacrylates are another important class of anionically polymerizable monomers that can be used to form high T_g macromonomers for synthesis of TPEs. However, unlike styrenic monomers, which can be polymerized anionically at room temperature and above in hydrocarbon solution, alkyl methacrylates must be polymerized at low temperatures. The controlled, living anionic polymerization of alkyl methacrylates can be effected using a less reactive, more hindered initiator such as the adduct of 1,1-diphenylethylene with butyllithium in THF at −78 °C as shown in Scheme 10 [53, 54]. Direct termination reactions at −78 °C with a twofold excess of *p*-vinylbenzyl iodide was reported to proceed in up to 92% completion in several hours as determined by UV and NMR analysis [54]. The bromo analog reacted similarly, but the chloride failed to react. The results with *p*-vinylbenzyl bromide have been confirmed and extended to the termination reaction with *p*-isopropenylbenzyl bromide [55]. The macromonomers were reportedly characterized by chemical titration and UV

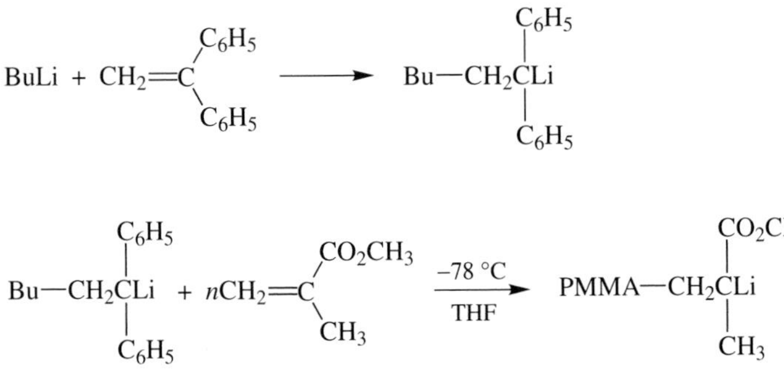

Scheme 10

analysis [55]. A similar macromonomer syntheses was reported for *tert*-butyl methacrylate using tritylsodium as initiator and *p*-vinylbenzyl chloride termination at $-78\,^{\circ}$C in THF [56]. The functionality was reported to be 1.12.

Termination of living anionic poly(methylmethacrylate) (PMMA) with allyl bromide at $-78\,^{\circ}$C produced a quantitative yield of the corresponding allyl-functionalized PMMA [53]. The functionality was identified and quantitatively determined by ^{1}H NMR using the vinyl absorptions at $\delta = 4.80$, 4.96, 5.12, 5.42, and 5.52 ppm. ^{1}H NMR analysis also indicated high functionality for the macromonomer prepared by termination of living anionic PMMA with 4-vinylbenzoyl chloride; the aromatic resonances are clearly observed at $\delta = 7.0$ to 7.25 ppm and the vinyl protons at $\delta = 5.4$ to 6.8 ppm [57].

Teyssié and co-workers [58] have found that the addition of LiCl to the *sec*-BuLi initiated anionic polymerization of *tert*-butyl acrylate (*t*BA) leads to a dramatic improvement in the control of molecular weight and molecular weight distribution. This modified initiator system has been used to make vinyl-terminated macromonomers of poly(*tert*-butyl acrylate) (P*t*BA). Gnanou and co-workers [59] terminated living P*t*BA with either 4-vinylbenzyl bromide or 3-(chlorodimethylsilyl) propyl methacrylate at $-25\,^{\circ}$C in THF. Based on UV analysis, they reported that the functionalization was nearly quantitative for macromonomers with M_n of 8000 and 15,000 g/mol. Teyssié and co-workers [60] synthesized ω-methacryloyl P*t*BA by reacting with living P*t*BA with benzaldehyde at $-78\,^{\circ}$C to generate an intermediate lithium alkoxide. This was subsequently reacted with methacryloyl chloride to afford a macromonomer with $>87\%$ functionality based on ^{1}H NMR; however, details of the characterization were not given. The same group used a series of terminating agents to prepare ω-styryl P*t*BA including 4-vinylbenzoyl chloride, 4-(chlorodimethylsilyl) styrene, and 4-(chlorodimethylsilyl) α-methylstyrene. ω-Styryl P*t*BA macromonomers with $M_n = 1500$ to 4500 g/mol were synthesized with functionalities of 0.91 to 0.97 based on ^{1}H NMR and UV.

Grignard reagents can also be used as initiators to prepare highly isotactic PMMA macromonomers [61]. The *tert*-butylmagnesium bromide-initiated polymerization of methyl methacrylate at $-78\,^{\circ}$C in THF proceeds slowly (24 h) to produce living isotactic PMMA. This reacts with *p*-vinylbenzyl bromide (but not with *p*-vinylbenzyl chloride) to form the corresponding macromonomer with 96% to 97% double bond functionalities (^{1}H NMR), but only in the presence of a polar additive such as hexamethylphosphoramide. For the preparation of highly syndiotactic PMMA macromonomers, a mixed initiator formed from *tert*-butyllithium and tri-*n*-butylaluminum ([Al]/[Li] ≥ 3) was used for polymerizations at $-78\,^{\circ}$C in toluene [61]. Termination with *p*-vinylbenzyl bromide in the presence of TMEDA provided the macromonomer in essentially quantitative yield.

Unsaturated Grignard reagents have also been used as initiators for anionic polymerization of methyl methacrylate [62]. PMMA macromonomers prepared using either *p*-vinylbenzyl-magnesium chloride or *m*-vinylbenzylmagnesium chloride at $-78\,^{\circ}$C to $-110\,^{\circ}$C in THF or toluene contained one vinylbenzyl group per chain as determined by ^{1}H NMR; functionalities greater than 1 were obtained at higher temperatures. However, the conversions were generally low ($<55\%$) and the molecular weight distributions were quite broad. Predominantly syndiotactic polymers were obtained in THF, while stereoblock polymers were formed in toluene. Similar results were reported using *o*-vinyl-benzylmagnesium chloride [63].

14.2.1.3   Poly(2-vinylpyridine) Macromonomers

The anionic polymerization of 2-vinylpyridine is complicated by addition reactions of the carbanionic chain ends to the pyridine rings, which result in branching reactions [64–66]. To obtain living polymers with controlled molecular weights and somewhat narrow molecular weight distributions ($M_w/M_n = 1.1$ to 1.4), a more hindered, less reactive initiator such as the adduct of 1,1-diphenylethylene with butyllithium has been used in THF with 15% dimethylformamide at $-80\,°C$ [67]. Termination with *p*-bromomethyl-*α*-methylstyrene proceeded rapidly to form the macromonomer "quantitatively"; however, no definitive characterization data were presented. It was necessary to first attenuate the chain end reactivity by addition of ethylene oxide (slow reaction even at $0\,°C$), followed by methacryloyl chloride to form the corresponding methacrylate-terminated macromonomer [67]. No characterization data were presented. An efficient, well-characterized macromo-nomer synthesis was based on the use of the lithium salt of 2-ethylpyridine as initiator and the slow addition of 2-vinylpyridine vapor at $-78\,°C$ in THF as shown in Scheme 11 [68]. An excess of *p*-vinylbenzyl chloride (0.52 to 2.58 molar excess) was added. This system provided macromonomers ($M_n = 2.2$ to 24×10^3 g/mol) with narrow molecular weight distributions ($M_w/M_n = 1.05$ to 1.08) and quantitative functionalities ($f = 97\%$ to 99%; NMR, UV, or copolymerization). An *α*,*ω*-bis(*p*-vinylbenzyl)poly(2-vinylpyridine) as also prepared analogously using lithium naphthalene as initiator [68].

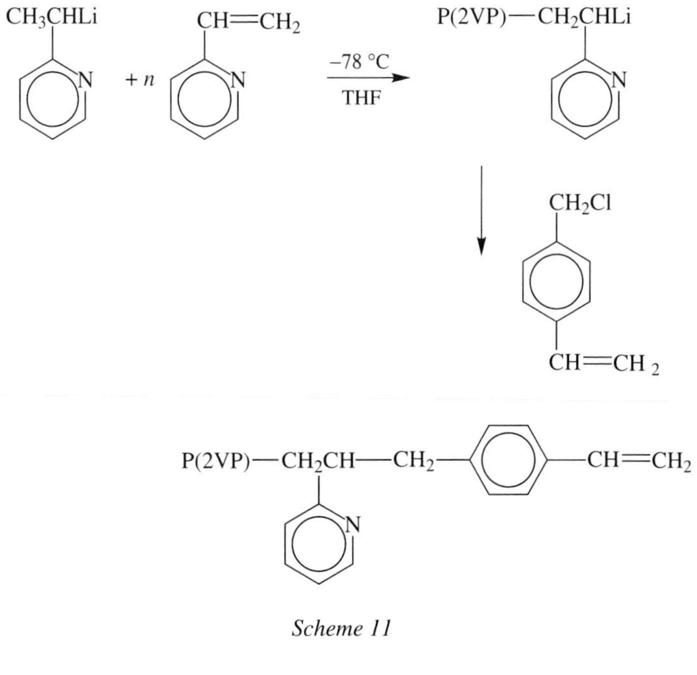

Scheme 11

14.2.2   Group Transfer Polymerization

Group transfer polymerization (GTP) is a useful synthetic method for the room temperature polymerization of methacrylates [69]. GTP can be used to control the molecular architecture of polymethacrylates including the synthesis of macromonomers. Macromonomers have been prepared via GTP using both termination and initiation techniques. Simms and Spinelli [70] prepared α-methacryloyl PMMA by initiating the polymerization of MMA with [[2-methyl-1-(trimethylsiloxy)ethoxy)-1-propenyl]oxy] trimethylsilane (1) (see Scheme 12). Quenching the living polymer with methanol in the presence of fluoride gave PMMA with a hydroxyl end group (3); reaction of the alcohol with methacryoyl chloride gave the corresponding macromonomer (4). McGrath and co-workers [71] reported a similar approach to the same macromonomer and in a later publication, Sheridan and McGrath [72] prepared ω-acryloyl PMMA by substituting acryloyl chloride in the last step of the synthesis. Radke and Müller [73] prepared ω-methacryloyl PMMA using three different synthetic procedures, including reaction of ω-hydroxy PMMA (3) with either methacryloyl chloride or methacrylic acid–dicyclohexylcarbodiimide. They also quenched the living polymer (2) with methanol to afford ω-trimethysiloxy PMMA which was reacted with methacryloyl fluoride.

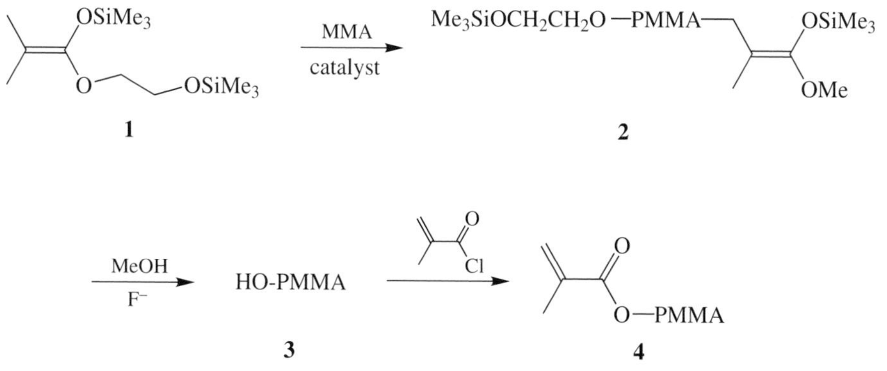

Scheme 12

Asami and co-workers [74] prepared ω-styryl PMMA by terminating the living PMMA polymer chain from GTP with 4-vinylbenzyl bromide or 4-vinylbenzyl tosylate. The highest functionality of the ω-styryl PMMA obtained by this termination method was 83%. The α-styryl PMMA was also synthesized with a functionalized initiator, *m,p*-vinylphenylketene methyl trimethylsilyl acetal (5) [Eq. (14.4)]. Preparation of the α-styryl macromonomer (6) by the initiation method produced materials with $M_n = 3600$ to $11,400$ g/mol and $M_w/M_n = 1.09$ to 1.33. Witkowski and Bandermann [75] synthesized α-styryl PMMA using initiator 5 with only *p*-vinyl substitution rather than a mixture of *m* and *p* isomers used in Asami's work.

Bandermann also prepared PMMA with a butadiene end group using initiator [Eq. (14.5)].

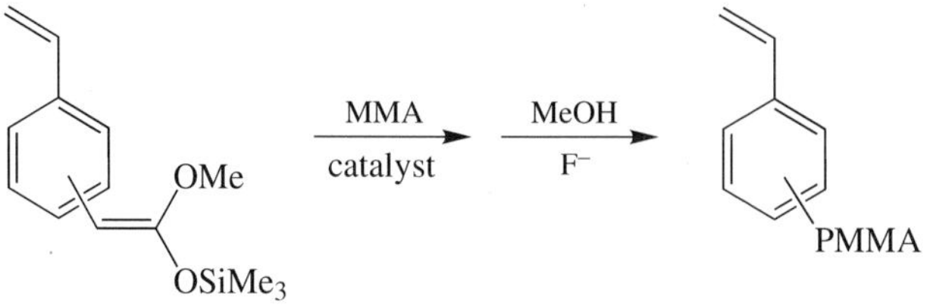

$$(14.4)$$

5   **6**

$$(14.5)$$

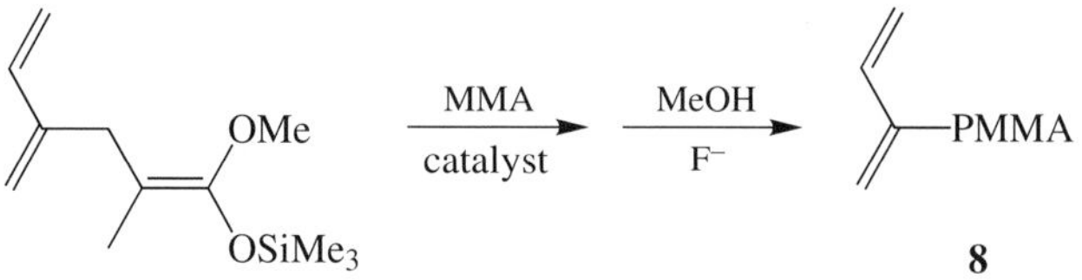

7   **8**

There are a number of other potentially useful routes to macromonomers using GTP by the termination method. New functionalization chemistry described by Quirk and Ren [76] leads to chelic PMMA chains which could be derivatized to afford macromonomers. Any method that gives ω-hydroxy PMMA could be used to make ω-methacryloyl PMMA by reaction with methacryloyl chloride.

14.2.3   Michael Insertion Polymerization

There are several other methods for the preparation of polyacrylate macromonomers, but they are less controlled than GTP. One method described by Lewis and Haggard [77] is the alkoxide-initiated oligomerization of acrylates, which can be categorized as a Michael insertion polymerization [Eq. (14.6)]. The possible disadvantage of this technique is a broad molecular weight distribution (typically, DP = 5 to 20) of oligomers; however, the simplicity of the method is a compensating advantage. The final outcome of this polymerization is a mixture of oligomers with terminal double bonds that result from alkoxide elimination from the initiating end of the oligomer chain. The average molecular weight of the oligomers can be controlled by the amount of alcohol present. This process was successfully used to oligomerize a series of acrylates. For the successful exploitation of this synthetic chemistry in the fabrication of TPEs, it would be necessary to use high T_g acrylates such as isobornyl acrylate.

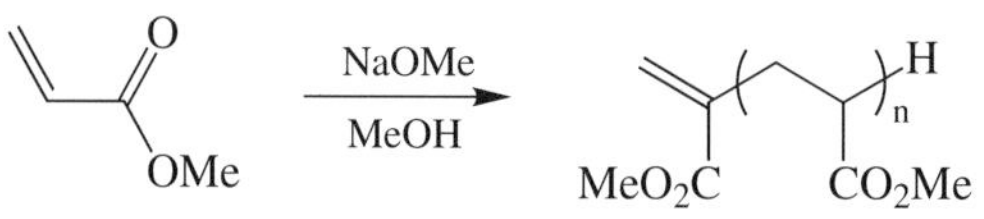

$$(14.6)$$

14.2.4   Free Radical Polymerization

The use of free radical polymerization to prepare macromonomers is difficult because of the requirements for a high degree of functionality and reasonably well-controlled molecular weight range. Procedures to control the molecular weight in free radical polymerization are generally unacceptable either because high temperatures or pressures are needed to carry out the polymerization reaction, or because chain transfer agents have objectionable odors or toxicity, or because of the adverse properties associated with a high incidence of initiator or chain transfer fragments. However, since some vinyl monomers cannot be prepared by living polymerization methods, it is necessary to resort to radical syntheses [11]. Even though the macromonomer chains cannot be prepared with the precise control of living polymerization procedures, macromonomers prepared by free radical techniques can be thoroughly characterized (average chain length, chain-length distribution, chain-end functionality) before copolymerization with another monomers. Thus, it is possible to generate comb-type graft copolymers with well-characterized graft branches.

A typical free radical method for the preparation of a PMMA or polystyrene macromonomers is shown in Scheme 13. The reaction is initiated by azobisisobutyronitrile

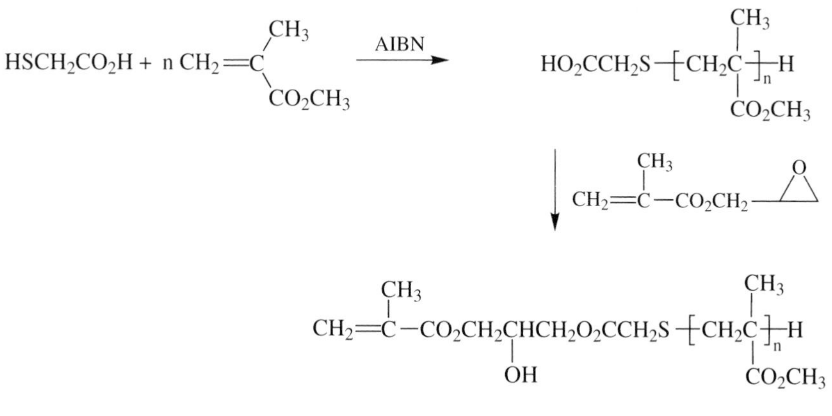

Scheme 13

(AIBN) and utilizes efficient chain-transfer agents such as thioglycolic acid [78–82] or iodoacetic acid [83] to introduce the carboxyl functional group at one of the macromonomer chain ends as a macromonomer precursor [5, 10, 11, 13, 84]. A reaction with glycidyl methacrylate generates the polymerizable methacryoyl end group. These macromonomers have M_n values of 1000-8000 g/mol and M_w/M_n values of 1.5 to 2.8.

A one-step synthesis of PMMA macromonomers is possible via free radical polymerization using cobalt(II) chelates as chain transfer agents [Eq. (14.7)] [85, 86]. The significant advantage of the cobalt(II) chelates is their high chain transfer constants, which means that very little catalyst is needed to produce low molecular weights and that the outcome of the process is a functionalized polymer with a terminal vinyl group (a consequence of the catalytic chain transfer cycle). PMMA macromonomer with functionalities ranging from 95% to 100% and $M_n = 1000-$

25,000 g/mol were reported using both solution and emulsion polymerization techniques [87].

$$\text{(structure)} \xrightarrow[\text{AIBN}]{\text{Co(II) Chelate}} \text{(polymer structure)} \tag{14.7}$$

14.3   Homopolymerization of Macromonomers

14.3.1   Anionic Polymerization

Macromonomers should be polymerizable by a variety of mechanisms, depending on the chain-end functional group. However, simply because of the low concentration of reactive groups per gram of macromonomer, one would expect slow rates of homopolymerization. Thus, for polymerization mechanisms in which chain termination reactions and chain transfer reactions can compete with propagation, these side reactions can limit conversion and molecular weight. The homopolymerization of macromonomers is of considerable interest from the standpoint of structure–property relationships, since a polymacromonomer can be considered as the ultimate branched polymer with one branch (or graft) per repeating unit as shown in Eq. (14.8) [7, 22, 34].

The anionic polymerization of styrene-terminated macromonomers provides a useful system for considering the scope and limitations of homopolymerization of macromonomers,

$$n\text{CH}_2{=}\underset{\underset{\text{P}}{|}}{\overset{\overset{\text{R}}{/}}{\text{C}}} \xrightarrow{\text{Initiator}} \left(\text{CH}_2{-}\underset{\underset{\text{P}}{|}}{\overset{\overset{\text{R}}{|}}{\text{C}}} \right)_n \tag{14.8}$$

since these should be living polymerizations and free of chain termination and chain transfer reactions. The anionic homopolymerization of a macromonomer is analogous to attempts to prepare polymers with molecular weight $>10^6$, that is, impurities must be at concentrations less than those of the chain ends [88, 89]. The butyllithium-initiated homopolymerization of a p-vinylbenzyl-functionalized polystyrene macromonomer ($M_n = 3 \times 10^3$ g/mol) in benzene at 40 °C gave a polymer with M_w (light-scattering) $= 97 \times 10^3$ g/mol [7]; the calculated number average molecular weight was 21×10^3 g/mol. Only a small amount of residual macromonomer was evident in the SEC trace of the polymacromonomer, which attests to the high functionality of the macromonomer and also to the ability to obtain high molecular weight polymacromonomers using alkyllithium-initiated polymerization. Analogous results were obtained for the low-temperature anionic polymerization of a methacrylate-functionalized polystyrene macromonomer [7]. Unfortunately, no homopolymerizations have been investigated for diblock macromonomers which could form thermoplastic elastomers of the type poly[(hard phase)–block-(soft phase)] methacrylate].

14.3.2   Group Transfer Polymerization

GTP is a useful synthetic process for the polymerization of methacrylates, and thus offers potential for the polymerization ω-methacryloyl macromonomers. Little work has been reported on the homo- or copolymerization of macromonomers using GTP. Asami and co-workers [90] studied the homopolymerization of a ω-methacryloyl polystyrene ($M_n = 3800$ g/mol) using bifluoride catalyst. Under a variety of experimental conditions, only oligomers (DP = 2 to 6) were obtained with macromonomer conversions of 8% to 99%. The best results were obtained at $-78\,°C$. Copolymerization of the polystyrene macro-monomer with MMA gave high conversions of macromonomer for molar ratios of MMA/ω-methacryloyl/polystyrene $\geqslant 27$ at $0\,°C$. Higher concentrations of macromonomer gave poor results. To our knowledge, there have not been any reports on the use of GTP to prepare TPEs using ω-methacryloyl macromonomers.

14.4   Copolymerization of Macromonomers

14.4.1   Kinetics

The principal interest in macromonomers arises from their ability to undergo copolymeriza-tion with backbone-forming monomers to form comb-type graft copolymers, as previously shown in Scheme 1. The scope of this method for the synthesis of a broad range of heterophase polymers is seemingly limitless. Macromonomers can be prepared from a wide variety of monomers utilizing appropriate polymerization mechanisms and some form of controlled, preferably living, polymerization methodology which provides a well-defined macromonomer in terms of molecular weight, molecular weight distribution, and polymeriz-able functionality. In principle, this broad range of macromonomers can be copolymerized with almost any monomer providing that a suitable mechanism is available which results in useful copolymerization parameters for the functional groups involved. Even with the restriction that the macromonomer contain at least one hard-phase (high T_g) block segment and that the backbone forming monomer provide a soft-phase (low T_g) backbone, the macromonomer method for synthesis of TPEs is undoubtedly the most versatile method available.

The kinetics of copolymerization of a macromonomer (M) with a comonomer (C) can be described in terms of the general kinetic equations for copolymerization, Eqs. (14.9) to (14.12),

$$\text{\sim\sim\sim M* + M} \xrightarrow{k_{MM}} \text{\sim\sim\sim M—M*} \tag{14.9}$$

$$\text{\sim\sim\sim M* + C} \xrightarrow{k_{MC}} \text{\sim\sim\sim M—C*} \tag{14.10}$$

$$\text{\sim\sim\sim C* + C} \xrightarrow{k_{CC}} \text{\sim\sim\sim C—C*} \tag{14.11}$$

$$\text{\sim\sim\sim C* + M} \xrightarrow{k_{CM}} \text{\sim\sim\sim C—M*} \tag{14.12}$$

in which the symbol * indicates the reactive chain end regardless of mechanism. By substituting the monomer reactivity ratios ($r_M = k_{MM}/k_{MC}$; $r_C = k_{CC}/k_{CM}$) and using the steady-state approximation for both types of reactive chain end, it is possible to derive the well-known instantaneous copolymer composition equation, [Eq. (14.13)], in which the subscripts "p" indicate the composition of the respective monomer or macromonomer in the polymer that has been formed at a given time [5, 91]. Thus, the copolymer composition depends on the monomer reactivity ratios (rates of homopolymerization divided by the rates of crossover to the other monomer) and the monomer feed ratios in the polymerization mixture.

$$\frac{d[C]_p}{d[M]_p} = \frac{[C](r_C[C] + [M])}{[M](r_M[M] + [C])} \qquad (14.13a)$$

$$\frac{d[C]_p}{d[M]_p} = \frac{1 + r_C([C]/[M])}{1 + r_M([M]/[C])} \qquad (14.13b)$$

Because of the high molecular weight of the macromonomer relative to the comonomer, the molar concentration of the macromonomer is quite low relative to the comonomer, that is, [C]/[M]>50 [92]. For this case, the following inequalities generally hold:

$$1 \gg r_M([M]/[C]) \quad \text{and} \quad r_C([C]/[M]) \gg 1$$

and the instantaneous copolymerization equation [Eq. (14.13)] reduces to a simpler form as shown in Eq. (14.14). Thus, macromonomer copolymerizations are predicted to depend on

$$\frac{d[C]_p}{d[M]_p} = r_C \frac{[C]}{[M]} \qquad (14.14)$$

the comonomer reactivity ratio and the molar concentration ratio of the comonomer to the macromonomer, but the copolymerization will not depend on the macromonomer reactivity ratio. From attempts to directly evaluate the macromonomer reactivity ratios as a function of macromonomer molecular weight, it has been concluded that meaningful values of r_M can be obtained only for macromonomers of relatively low molecular weight ($\leq 10^3$ g/mol) [93, 94]. In contrast, values of r_c can be determined with some degree of confidence from Eq. (14.14) providing that the number average molecular weight of the macromonomer is 10^4 g/mol or greater [93]. The problem is that it is difficult to obtain copolymer composition data over a wide range of copolymer compositions because of the disparities in molecular weights [94]. For example, the inequalities above contain not only the molar feed ratios, but also the monomer reactivity ratios. Thus, the inequalities may not hold for copolymerizations with low values of r_C and high values of r_M.

It is possible to integrate Eq. (14.14) from $t = 0$ to time t and obtain Eq. (14.15). r_C can then be calculated by plotting the data for comonomer conversion versus macromonomer conversion as indicated by the equation. This equation should be valid up to high conversion providing that the concentration of comonomer, C, is still large enough to neglect propagation by the sites derived from the macromonomer, M [42]. It has been concluded that the probability of finding MM diads along the chain is negligible if the molar fraction of macromonomer in the feed remains below 0.05 [5]. Thus, it is generally concluded that from

$$\log \frac{[C]_t}{[C]_0} = r_C \log \frac{[M]_t}{[M]_0} + Y \qquad (14.15)$$

the copolymerization equation, the amount of compositional heterogeneity obtained in the comb-type branch copolymer from macromonomer copolymerization will be acceptable for conversion up to 50% to 70%. The problem will be exacerbated if the comonomer reactivity ratio is greater than 1, which will tend to increase the amount of macromonomer in the feed as conversion increases. The opposite will be true if r_C is less than 1, that is, the concentration of macromonomer and the number of grafts will tend to decrease with conversion.

14.4.2   Anionic Copolymerization

Although the range of monomers that can be polymerized anionically is limited, anionic copolymerization of a structurally well-defined macromonomer with another monomer has the advantage of controlled polymerization with respect to the molecular weight and molecular weight distribution of the polymer backbone. Few anionic copolymerizations of macromonomers have been reported. The poly(styryl) lithium-initiated copolymerizations of p-vinylbenzyl-functionalized polystyrene macromonomers ($M_n = 2.7 \times 10^3$, 5.6×10^3, and 12.7×10^3 g/mol) with both styrene and p-methylstyrene in benzene at 25 °C were investigated [42]. For the macromonomer molecular weights investigated, the monomer reactivity ratios were not dependent on the molecular weight of the macromonomer in accord with the Flory equal reactivity principle [24b]. The observed monomer reactivity ratios (r_C) were 1.98 for styrene and 0.997 for p-methylstyrene. These results may not be general, however, because of the expected compatibility of the monomers with the macromonomers and the comb-type graft copolymer product. Thus, phase separation during the copolymerization would not be a problem. In other systems of interest for preparation of phase-separated TPEs in which the backbone chain and the graft chain are not miscible, phase separation is often a problem. This can be dependent on molecular weight of the macromonomer and can occur during the copolymerization [13].

The butyllithium-initiated copolymerization of p-vinylbenzyl-functionalized poly-styrene macromonomers with butadiene has been examined in cyclohexane [38]. Sodium *tert*-butoxide ([Na]/[Li] = 1) was added to these copolymerization mixtures to promote randomization in the copolymerization [95]. The proposed synthesis of polybutadiene–*graft*-polystyrene was investigated to explore the effects of the molecular weight of the graft chains and their distribution along the elastomeric backbone on the properties of these heterophase systems. The macromonomer molecular weights varied from $M_n = 1.6 \times 10^3$ to 38×10^3 g/mol and the predicted polybutadiene backbone molecular weight was designed to be $M_n = 250 \times 10^3$ g/mol.

The copolymerization data were analyzed using an integrated form of the macro-monomer copolymerization equation [Eq. (14.11)] and the results are shown in Table 14.1 in terms of $1/r_B$ which is equal to k_{BM}/k_{BB}, that is, the rate of crossover of poly(butadienyl) lithium chain end to macromonomer compared to the rate of addition of butadiene. This has been described as a measure of the reactivity of the macromonomers since it would be expected that the rate constant for homopolymerization of butadiene would be constant. It is noteworthy that for low conversions and low molecular weight macromonomers ($<10,000$ g/mol) a random copolymerization behavior ($r = 1$) for the poly(butadienyl) lithium chain end is observed below conversions of 25% to 30%. For a given macro-

Table 14.1 Values of $1/r_C$ as a Function of Conversion for the Copolymerization of *p*-Vinylbenzyl-Substituted Polystyrenes with Butadiene

Conversion of butadiene in %	$1/r_C$ for different macromonomer molecular weights and macromonomer feeds			
	$M_n = 5000^a$ $F = 25–33\%^b$	$M_n = 9000^a$ $F = 20–30\%^b$	$M_n = 9000^a$ $F = 40–50\%^b$	$M_n = 30,000^a$ $F = 25–33\%^b$
30	1.02	1.00	0.92	0.44
40	0.98	0.93	0.88	0.43
50	0.91	0.83	0.70	0.42
60	0.84	0.74	0.60	0.41
70	0.81	0.65	0.51	0.38
80	0.76	0.59	0.46	0.37
Conversion[c]	98.8	97.7	98.3	89.9

[a] Macromonomer molecular weights (g/mol)
[b] Amount of macromonomer in feed (wt%)
[c] Maximum conversion of macromonomer in the anionic copolymerization
Source: Ref. [38]

monomer, the macromonomer reactivity, that is, $1/r_B$, decreases with increasing butadiene conversion. In addition, this measure of macromonomer reactivity decreases with increasing molecular weight of the macromonomer for a given degree of conversion. This type of behavior would be expected for a system that exhibits phase separation during the copolymerization. Based on data for polystyrene–*block*-polybutadiene–*block*-polystyrene (S–B–S) copolymers and theoretical calculations, a critical minimum molecular weight for phase separation between 5000 and 10,000 g/mol is expected [96]. To the extent that phase separation occurs during formation of the graft copolymer, the macromonomer would tend to be located in a polystyrene-rich phase and the butadienyllithium chain end would be in a polybutadiene-rich phase. This effect would limit the accessibility of the macromonomer to the polymeric organolithium chain end. Another factor that will tend to promote phase separation with increasing molecular weight of the macromonomer is the quality of the solvent for polystyrene. Since cyclohexane is a theta solvent for polystyrene ($T_\theta \simeq 35\,°C$) [97], this would also tend to limit access of the macromonomer to the chain end as the molecular weight of the macromonomer increases.

14.4.3   Model Graft Copolymer Synthesis

The macromonomer-based copolymerization method for the synthesis of comb-type graft copolymers solves some of the classic problems in traditional grafting reactions: (1) the lack of control of the grafted branch molecular weight and molecular weight distribution and (2) the contamination of the graft copolymer with the homopolymers of the backbone and the grafting monomer [13]. However, this method does not control the number of graft branches per molecule or the distribution of graft branches along the polymer backbone. In principle, living anionic polymerization with a nonhomopolymerizable macromonomer can provide a method of preparing branched and graft polymers with control of all of the structural parameters and low degrees of compositional heterogeneity, as illustrated in Scheme 14. Thus, living polymerization will form the first (A) backbone segment, **9**, which will react with

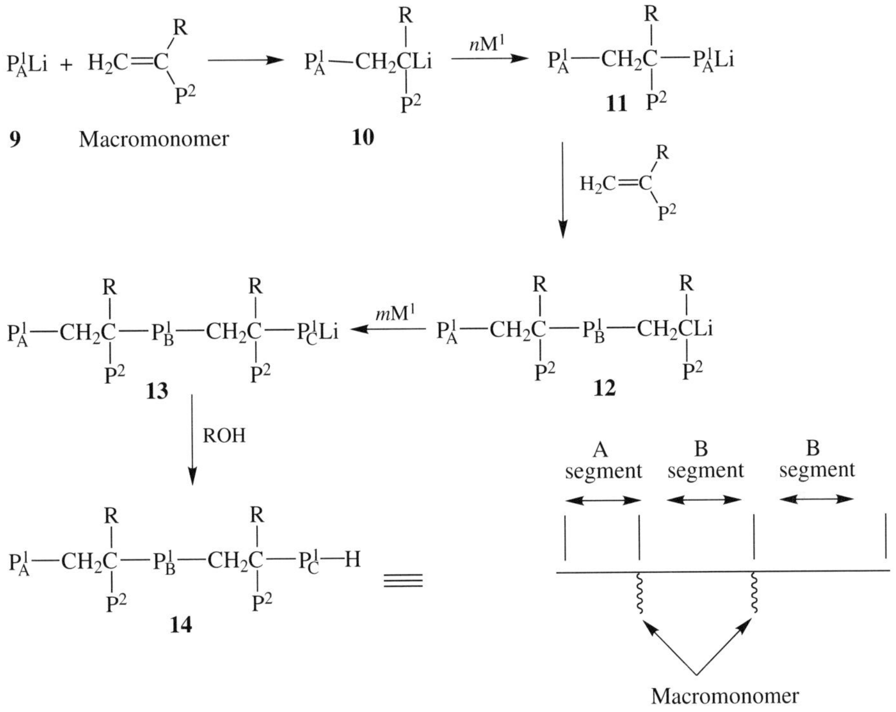

$$P_A^l Li + H_2C = C \overset{R}{\underset{P^2}{}} \longrightarrow P_A^l - CH_2 \overset{R}{\underset{P^2}{C}} Li \xrightarrow{nM^1} P_A^l - CH_2 \overset{R}{\underset{P^2}{C}} - P_A^l Li$$

9   Macromonomer   **10**   **11**

$$\downarrow \quad H_2C = C \overset{R}{\underset{P^2}{}}$$

$$P_A^l - CH_2 \overset{R}{\underset{P^2}{C}} - P_B^l - CH_2 \overset{R}{\underset{P^2}{C}} - P_C^l Li \xleftarrow{mM^1} P_A^l - CH_2 \overset{R}{\underset{P^2}{C}} - P_B^l - CH_2 \overset{R}{\underset{P^2}{C}} Li$$

13   **12**

$$\downarrow ROH$$

$$P_A^l - CH_2 \overset{R}{\underset{P^2}{C}} - P_B^l - CH_2 \overset{R}{\underset{P^2}{C}} - P_C^l - H$$

14

A segment   B segment   B segment

Macromonomer

Scheme 14

a nonhomopolymerizable macromonomer to add only one macromonomer unit to the chain end to form **10** and maintain the living nature of the polymerization. Addition of more backbone-forming monomer (nM^1) will then generate a new (B) backbone segment whose length will be defined by the ratio of the grams of monomer added to the moles of active chain end [see Eq. (14.1)]. The living polymer **11** can then react with another equivalent of nonhomopolymerizable macromonomer to place another graft branch at the precise point on the backbone which was defined by the B segment length. Addition of more backbone-forming monomer (mM^1) to the adduct **12** will then generate a new (C) backbone segment whose length will be defined by the ratio of the grams of monomer added to the moles of active chain end [see Eq. (14.1)]. At this point the whole sequence of macro-monomer/comonomer addition could be repeated to generate more graft branches at specific locations along the backbone or the living polymer could be terminated to form the precisely defined graft copolymer with two graft branches located at segment distances A and A + B along the polymer backbone (**14**). Although such a model graft copolymer synthesis has not been accomplished, a demonstration of the feasibility of this approach has been performed using a 1,1-diphenylethylene-functionalized polystyrene macromonomer to form a hetero three-armed, star-branched copolymer as shown in Scheme 15 [52].

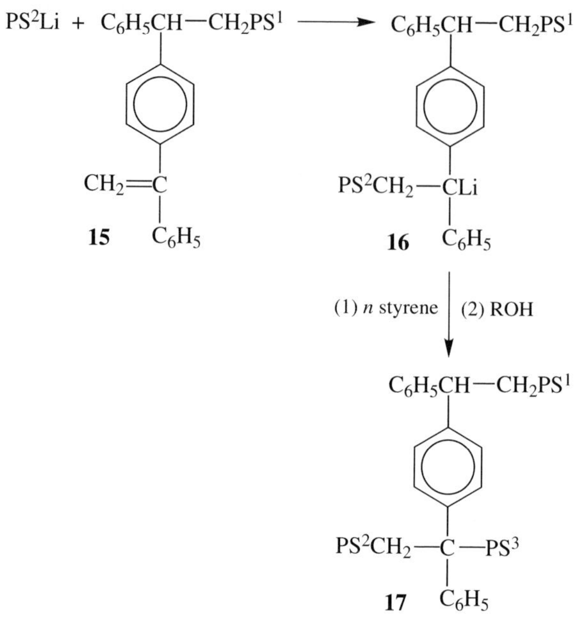

$$PS^2Li + C_6H_5CH-CH_2PS^1 \longrightarrow C_6H_5CH-CH_2PS^1$$

CH$_2$=C
15 C$_6$H$_5$

PS^{2}CH$_2$—CLi
16 C$_6$H$_5$

(1) n styrene | (2) ROH

C$_6$H$_5$CH—CH$_2$PS^1

PS^{2}CH$_2$—C—PS^3
17 C$_6$H$_5$

Scheme 15

For the synthesis of hetero three-armed, star-branched polymers, the first step involves the addition of a polymeric organolithium compound, for example, poly(styryl) lithium, with the 1,1-diphenylethylene-functionalized polystyrene macromonomer, **15**, to form the corresponding coupled product, **16**, a diphenylalkyllithium. For stoichiometric amounts of poly(styryl) lithium and macromonomer, it was found that the efficiency of this coupling reaction is >96%. This result also shows that the vinyl functionality of the macromonomer is >96%. Finally, the third arm was formed by addition of monomer (e.g., styrene) in the presence of THF to promote the crossover reaction. SEC analyses of the heteroarm star polymer product (**17**) showed that each of these steps proceeded efficiently to give the expected products; only relatively small amounts of nonstar product were observed. A narrow molecular weight distribution star product ($M_w/M_n = 1.02$) was easily obtained by one fractionation step. This methodology can be extended to other backbone-forming monomers, especially polydienes, to form model graft TPEs.

14.4.4   Free Radical Copolymerization

The principal application for macromonomers has been the formation of comb-type graft copolymers by free radical copolymerization with backbone-forming monomers [2, 5, 9]. Macromonomers capable of free radical copolymerization offer the advantages of application in commercial polymerization systems. To be effective and provide the desired properties for TPEs, impact plastics, and other blends and alloys, appropriate macromonomers and polymerization conditions must be selected. Furthermore, once the graft copolymer has been prepared, processing is needed to achieve the desired morphology and properties.

The macromonomer copolymerizations can be carried out by bulk, solution, emulsion, or aqueous suspension polymerization processes [2]. Solution polymerization in a good solvent for both macromonomer and backbone offers convenience for some applications, but limits molecular weight because of solution viscosity. Aqueous suspension polymerization using oil-soluble initiators offers high molecular weight polymers and easy recovery of the product. Suspension polymerization using appropriate surfactants and oil-soluble initiators has been used to prepare stable latices.

14.4.4.1   Effects on Macromonomer Reactivity

Macromonomers (M) terminated with methacrylate or other functionalities generally copolymerize with various monomers by copolymerization kinetics defined by monomer reactivity ratio r_C (k_{CC}/k_{CM}) of the comonomer C, as shown in Eq. (14.14). In this equation, it is seen that macromonomer (M) reactivity (k_{CM}) relative to comonomer polymerization (k_{CC}) is determined by the expression $1/r_C$. Numerous investigations of macromonomer reactivities have been published, including a recent review on copolymerizations by Meijs and Rizzardo [98]. In general, the intrinsic reactivity of a functional group should not depend on the molecular weight of the polymer to which it is attached, in accord with the Flory equal reactivity postulate [24b]. However, as noted by Flory [24c], application of this principle presupposes a homogeneous reaction.

Deviations of macromonomer reactivity from values predicted by conventional copoly-merization values for low molecular weight monomers have long been recognized, and attributed to phase separation between the macromonomer and the backbone polymer segments. Early investigations found macromonomer reactivity decreasing at higher monomer conversions, and attributed these effects to phase separation [3, 4]. Such effects are not observed in compatible systems such as the copolymerization of a methacrylate-terminated polystyrene macromonomer ($M_n = 13,000$) (25% to 45% macromonomer) with styrene monomer [4] as discussed in Section 14.4.2. For example, the styrene monomer reactivity ratio was determined to be 0.61 for the AIBN-initiated, bulk copolymerization [4]; the literature values for ethyl methacrylate copolymerization with styrene (C) range from $r_C = 0.55$ to 0.67 [99]. A methacrylate-terminated polystyrene macromonomer is too reactive for copolymerization with vinyl chloride (C) since the value of r_C was determined to be 0.05, again in good agreement with literature values for the low molecular weight comonomers [100]. In this system, compositional heterogeneity would be expected because of the rapid depletion of the macromonomer from the feed.

A copolymerization model proposed by Harwood [101] considered partitioning of monomers and solvents between the volume occupied by the growing polymer chain and the free solvent volume outside of the polymer coils. This "bootstrap" model describes an effective monomer concentration in the domain of a growing polymer radical. Apparent changes in the measured reactivity ratios from solvent effects are considered to arise from changes in monomer concentration in the vicinity of the growing chain radical, not from changes in reactivity.

As an extension of the "bootstrap" model, Percec and co-workers [102, 103] developed a micelle model to account for the observed solvent and concentration effects on the determined r_C. To account for increased macromonomer reactivity, $1/r_C$, with increased macromonomer concentration, they proposed a micellar model. This gave a heterogeneous copolymerization with variable concentrations of monomers inside and outside of the micelle with the growing active chain. This difference in monomer concentration in micelles compared to the concentration in the external phase was described by a partition coefficient, k. The experimental reactivity ratio, r_C, would therefore be a product of the true monomer reactivity ratio, r_C°, and a partition coefficient k, as shown in Eq. (14.16), where k is defined by Eq. (14.17).

$$r_C = r_C^\circ(k) \tag{14.16}$$

$$k = \frac{[C]_{chain}/[C]_{bulk}}{[M]_{chain}/[M]_{bulk}} \tag{14.17}$$

The micellar model (Fig. 14.2) is based on a system in which the monomer and the macromonomer differ in polarity, that is, polyphenylene oxide (PPO) macromonomer copolymerized with methyl methacrylate. The micelle formation was attributed to solubility differences of the PPO macromonomer and PMMA backbone polymer in toluene solution. It is well known that a solution of two incompatible polymers in a good solvent undergoes phase separation when a critical concentration is reached. Therefore phase separations resulting in apparent changes in macromonomer reactivity can be expected at higher monomer conversions in these macromonomer polymerization systems, especially in bulk or suspension polymerization systems. Thus, decreases in methacrylate-terminated polystyrene macromonomer reactivity in aqueous suspension copolymerizations with butyl acrylate were found at higher conversions [4]. r_C increased from 0.38 to 1.47 as butyl acrylate conversion increased from 10% to 79%. In any case, when the macromonomer and the comonomer have differing polarities, micellar or other forms of polymer phase separations are likely to occur at some stage in these macromonomer copolymerizations irrespective of polymer and solvent polarities.

These same methacrylate-terminated polystyrene macromonomers also exhibited lower reactivity with ethyl ($r_C = 0.7$) or butyl acrylate ($r_C = 0.8$) when copolymerizations were carried out in 1 : 1 benzene/DMSO solution [4]; for small molecule copolymerizations, these monomer reactivity ratios range from 0.13 to 0.47 [99]. Perhaps partitioning of highly polar DMSO into the micelles of growing acrylic backbone polymer radicals provides higher polarity in the polymer domain, which would tend to increase the concentration of acrylate monomer in this region and so reduce the effective polystyrene macromonomer concentration in the vicinity of the active free radical site. However, copolymerization of the same methacrylate-terminated polystyrene macromonomer with methyl methacrylate monomer

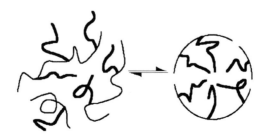

Figure 14.2   Schematic representation of reversible micelle formation for graft copolymers

Table 14.2 Effect of Penultimate Polystyrene Block Segments on the Macromonomer Conversion for Aqueous Suspension Polymerization with Styrene Monomer

Macromonomer[a]	Polystyrene block M_n	Wt% unreacted macromonomer	Percent macromonomer conversion
I(15)S(5)-MA	5000	0	100
I(15)S(2.5)-MA	2500	0	100
I(15)S(1.5)-MA	1500	1.0	97
I(15)S(1)-MA	1000	2.6	91
I(20)-MA[b]	—	7.4	76

[a] Numbers in parentheses indicate block $M_n \times 10^{-3}$. All isoprene blocks have $M_n = 15,000$ except where otherwise noted
[b] Isoprene block $M_n = 20,000$
Source: Ref. [35]

in DMSO/benzene exhibited approximately equal reactivity ($r_C \sim 1$), but with much scatter of data [4]. Nevertheless, it appears that choice of appropriate cosolvents can be used to avoid loss of reactivity by controlling the partitioning of macromonomer between the bulk solution and the vicinity of the growing polymer chain.

A means of overcoming phase separation effects is to prepare a multiblock macromonomer with a short compatible chain segment adjacent to the polymerizable end group. Compatibility of this segment with the backbone composition allows more favorable partitioning of the polymerizable end group of the macromonomer into the vicinity of the active polymerization site. A series of methacrylate-functionalized polystyrene–*block*-polyisoprene macromonomers (PS–*b*-PI-MA) were prepared by sequential monomer addition as shown in Scheme 4 [35]. Copolymerization of styrene with a PS–*b*-PI-MA macromonomer with 10,000 block polystyrene and 25,000 block polyisoprene segments in an aqueous suspension polymerization resulted in relatively high amounts (50%) of unreacted macromonomer. Possible causes for these low macromonomer conversions were believed to be due to phase separation or possible intramolecular degradative chain transfer to the polyisoprene segments when they were adjacent to the methacrylate group. Complete conversion was obtained when a macromonomer was prepared with a short polystyrene block adjacent to the polymerizable methacrylate group (PI–*b*-PS-MA). A polystyrene block molecular weight of about 2500 g/mol or greater adjacent to the methacryoyl group was found to be adequate in providing complete macromonomer conversions in these polymerizations. It was believed that these results, shown in Table 14.2, were achieved by favorable partitioning of the methacrylate group into the phase containing the growing polystyrene backbone radical.

14.4.4.2 Effects of Copolymerization Process

Effective free radical copolymerization of macromonomers in solution, bulk, or aqueous suspension polymerization requires consideration of the macromonomer properties, comonomers, solvents, and process. It is particularly helpful to consider these parameters in light of possible phase separation phenomena taking place during the polymerization, especially in the latter stages. As discussed above, micellar or other types of phase separations are likely to occur during polymerization, even in good solvents for both macromonomer graft and copolymer backbone. Similarly, polymerizations in bulk or in aqueous suspension would be expected to promote phase separation effects. Therefore a development program should consider the following effects governing the polymerization.

The macromonomer must have monofunctionality to avoid crosslinking, and the degree of functionalization should be high. As discussed previously in Section 14.2, in addition to spectroscopic methods, functionality can be readily determined by copolymerization in known systems, followed by SEC to determine the amount of unreacted macromonomer [2, 3]. Madruga and San Roman [104] found high levels of unreacted macromonomer in a study analyzing the copolymerization kinetics of butyl acrylate with a commercial methacrylate-terminated polystyrene macromonomer. Independent of the initial macromonomer concentrations, the weight fraction of unreacted macromonomer in the products ranged from 25% to 33%. This was attributed to the presence of unfunctionalized macromonomer. Their conclusions appear to be valid, since the solution polymerization recipe and process used should have resulted in essentially complete macromonomer conversion. This indicates the importance of reliable characterization of every batch of experimental or commercial macromonomer for functionality.

It is also usually advisable to employ macromonomer functional groups with relatively high reactivity in the polymerization system, that is, where $1/r_C$ is greater than unity. Where the macromonomer reactivity is equal or less than the comonomer, incomplete conversion of the macromonomer frequently results from phase separation or incomplete conversions of the comonomer. However, if the functional group is too reactive, then the macromonomer is incorporated predominantly in the initially formed graft copolymer, giving compositional heterogeneity as discussed for vinyl chloride copolymerizations [100]. Poly(methyl methacrylate) macromonomers were prepared and functionalized with acrylate groups in order to have equal reactivity in free radical copolymerizations with 2-ethylhexyl acrylate (2-EHA) [72]. The macromonomer functionalities were nearly 100%, as determined from the narrow polydispersity indices. The AIBN-initiated copolymerizations of these macromonomers with 2-EHA in ethyl acetate solution resulted in high yields of graft copolymers. Macromonomer conversions were high, but incomplete, and extraction was required to remove the unreacted macromonomer (ca. $10 \pm 5\%$). Thus, to minimize unreacted, nonvolatile macromonomer in the graft copolymer product, it is preferable to utilize a polymerizable functional group having greater reactivity than the comonomer.

The macromonomer should not undergo significant side reactions such as chain transfer or scission during the polymerization. Percec and Wang [105] found lower macromonomer reactivity ($1/r_C$) in copolymerizations initiated with AIBN than with peroxides. The greater apparent reactivity obtained with peroxide initiation could be attributed to grafting of monomer onto the poly(2,6-dimethyl-1,4-phenylene ether) macromonomer via peroxy radical attack. The SEC determination of r_C used for measurement of the reduction of the free macromonomer peak would not differentiate between macromonomer removal by grafting or by copolymerization.

Evidence for grafting reactions with an acrylic-terminated butadiene–styrene diblock macromonomer was also found in copolymerizations with methyl methacrylate for the development of new high impact strength dental base resins [106]. Although extraction studies using an unfunctionalized butadiene–styrene diblock copolymer (S–B) showed that a relatively small amount of grafting (12%) occurred in these benzoyl peroxide-initiated systems, only the acrylic-terminated S–B macromonomer achieved 100% incorporation and enhanced impact properties [106].

Easier processing graft copolymers with minimal gel were prepared by copolymerization of styrene with methacrylate-terminated, polystyrene–*block*-polyisoprene diblock macromonomers (PS–*b*-PI-MA) ($M_nPS = 10,000$ g/mol, $M_nPI = 25,000$ g/mol), using AIBN as initiator [35]. Benzene-insoluble gel was avoided when macromonomer contents were less than 30%, but gel formed at higher macromonomer levels. Evidence for grafting reactions was obtained by copolymerization of styrene with an unfunctionalized polyisoprene chain in the presence of AIBN.

Free radical copolymerization in solution should consider the solubility of both the starting macromonomer and the final backbone polymer. If the solvent is very poor for the backbone segment (but good for the macromonomer), then phase separation of the newly formed backbone occurs, and precipitation of this phase takes place. The macromonomer grafts are capable of stabilizing the particles as a nonaqueous dispersion (NAD), but the macromonomer may suffer incomplete conversion. When polymerization of polystyrene macromonomer with acrylonitrile was carried out in benzene, a poor solvent for polyacrylonitrile (PAN), conversion of macromonomer was incomplete [3]. Apparently the PAN phase arising from premature precipitation of growing acrylonitrile polymer backbone became a primary locus of polymerization, as described by the Percec micellar model [102, 103]. Complete polymerization of macromonomer took place when the copolymerization was carried out in DMF, a good solvent for both PAN and polystyrene macromonomer [3].

Polymerization in bulk or in aqueous suspension presents a more severe problem. Phase separation leads to incomplete macromonomer conversion compared to polymerization in a good solvent. Use of a water-insoluble cosolvent delays phase separation, and can provide high conversions. For example, the addition of benzene to a copolymerization of ethyl acrylate with a methacrylate-functionalized polystyrene macromonomer in aqueous suspension polymerization has been shown to dramatically reduce the amount of unreacted macromonomer [3].

In general, compatibility of the macromonomer in the comonomer phase and the delay of phase separation during copolymerization depend on the following factors [3]:

1. Comonomer backbone composition, and the relative compatibility with the macromonomer. Relative polarities and solubility parameters of backbone and macromonomer affect the degree of phase separation.
2. The macromonomer molecular weight. Increasing molecular weight decreases compatibility and promotes phase separation.
3. Level of macromonomer in the monomer feed composition. Low macromonomer levels resulted in increased unreacted macromonomer in the product because the macromonomer is excluded from the comonomer-backbone phase.
4. Unreacted macromonomer. The presence of too much unreacted macromonomer (>5%) can lead to opaque or hazy products.

As previously pointed out, if phase separation of the polymeric backbone and the macromonomer occurs during the polymerization, this will reduce the probability of macromonomer incorporation into the growing chain [3].

If pure graft copolymer is desired from an aqueous suspension polymerization process, new particle formation from initiation in the aqueous phase should not be allowed to take place. Unlike normal emulsion polymerization, water-insoluble macromonomers will not diffuse through the aqueous phase to copolymerize with comonomer in newly formed emulsion polymer particles. This problem can be avoided by employing oil-soluble initiators such as lauroyl peroxide, which generate no water-soluble initiator fragments. The resulting graft copolymer product can be recovered as filterable beads [3].

If a stable latex is desired, it can be prepared as a modified suspension polymerization system by preemulsifying the monomer solution of the oil-soluble initiator using certain surfactant systems [3]. Thus, polymerization is restricted to the emulsified monomer/macromonomer droplets, and particle nucleation is prevented.

14.4.4.3   Effects on Graft Copolymer Properties

The macromonomer graft polymers prepared by the various processes can exhibit properties similar to those of A–B–A triblock polymers when prepared from macromonomers having high functionalities and relatively narrow molecular weight distributions. Polystyrene macromonomer copolymers with low T_g acrylic monomers were TPEs having relatively good elastic recovery when macromonomer graft levels were 20% to 30%, as shown in Table 14.3. The graft polymers with the polystyrene graft levels of >30% exhibited yielding at low strains in the stress–strain curves. Increasing the hard block contents brought about increased yield strengths, and reduced elasticity. At 45% to 55% polystyrene contents the products were ductile thermoplastics with leathery properties [3].

In systems where phase separation is good, better elastomeric properties are obtained from copolymers having lower hard block contents. In Table 14.3 it is seen that the graft copolymers with 20% to 25% polystyrene had high elastic recovery and low permanent set. Higher polystyrene contents in the graft copolymers had increasing yielding and increasing permanent set. As in block copolymers, these effects result from stress softening from yielding of polystyrene domains and slippage of elastomer chains.

When macromonomer copolymerization is properly carried out to high monomer conversion, it is possible to prepare relatively pure graft polymers uncontaminated with

Table 14.3 Effect of Macromonomer Content on Tensile Properties of Polystyrene Macromonomer–Acrylic Graft Copolymer TPEs[a]

Wt% macromonomer	Yield strength (MPa)	Tensile strength (MPa)	Percentage elongation (%)	Permanent set (%)
20	—	6.3	750	5
25	0.9	9.8	790	5
30	2.6	10.4	810	40
35	3.8	13.3	560	50
45	11.5	14.6	400	130

[a] Graft copolymers were prepared in aqueous suspension from methacrylate-functionalized polystyrene macromonomer ($M_n = 11,000$ g/mol) and 1 : 1 ethyl acrylate–butyl acrylate (wt/wt)

To convert MPa to psi, multiply by 145

Sources: Refs. [3, 107]

Table 14.4 Tensile Properties of Polystyrene Macromonomer–Acrylic (30:70, wt/wt) Graft Copolymer TPEs[a]

Acrylic comonomer(s)	Polystyrene macromonomer M_n (g/mol)	Tensile strength (MPa)	Percentage elongation (%)	Permanent set (%)
BA	11,000	>7.1	>1000	25
BA	16,000	7.7	800	15
1:1,EA/BA	11,000	10.4	810	40
1:1,EA/BA	16,000	11.5	550	22
2:1,EA/BA	11,000	12.8	680	42
1:1,EA/BA[b]	16,000	13.0	410	43

[a] BA, butyl acrylate; EA, ethyl acrylate; acrylic monomer conversions are reported to be >95%
[b] 35 wt% macromonomer
To convert MPa to psi, multiply by 145
Source: Ref. [3]

homopolymers. This is important because TPE properties depend on a hard block molecular weight that is uniform and relatively low. Unreacted macromonomer, as homopolymer, should be minimal. This requirement is usually necessary for providing appropriate domain morphology. Furthermore, low molecular weight grafts are needed to minimize chain ends that are detrimental to physical properties.

Also important in the development of properties is the extent of phase separation and the compatibility of the macromonomer in the backbone phase. Partial compatibility of the low T_g matrix with the reinforcing domains will lower tensile strength and modulus. If the partially compatible domains yield under stress, this will reduce elastic recovery. Thus the compatibility of the comonomer should be considered together with macromonomer molecular weight in developing a TPE system. It is seen in Table 14.4 that tensile strengths of the polystyrene macromonomer graft copolymers increased with increasing polarity of the acrylic backbone (higher amounts of ethyl acrylate relative to butyl acrylate) because of improved phase separation. The butyl acrylate copolymers had lower tensile properties, which was attributed to plasticization of polystyrene domains with partially compatible poly(butyl acrylate) segments. Increasing the graft molecular weight also increased tensile strength and reduced permanent set because of reduced yielding of polystyrene domains.

When the molecular weight of the macromonomer is too low, there may be inadequate reinforcing by the hard block domains. In the case of polystyrene macromonomers, molecular weights less than 10,000 to 13,000 result in acrylate graft copolymers exhibiting lower tensile strengths from increased ductility of the hard block domains. Likewise, Xie and Zhou [108] found that graft copolymers formed from butyl acrylate and an MA-functionalized PMMA macromonomer (see Scheme 13) with $M_n = 4500$ g/mol provided only 3.3 MPa (480 psi) tensile strength, whereas copolymerization with a MMA macromonomer with $M_n = 13,700$ g/mol resulted in a tensile strength of 5.3 MPa (770 psi) (see Table 14.5). The fact that the permanent set was reduced by a factor of 2 can be attributed to the reduced yielding of PMMA domains expected for the higher molecular weight macromonomer. The tensile strength was reported to increase with increasing wt percentage of PMMA grafts;

Table 14.5 Tensile Properties of Methacrylate-Functionalized PMMA Macromonomer–*n*-Butyl Acrylate (25 : 75, wt/wt) Graft Copolymer TPEs[a]

PMMA–MA macromonomer M_n (g/mol)	Tensile strength (MPa)	Ultimate elongation (%)	Permanent set (%)
4500	3.3	620	30
10,100	4.8	565	23
13,700	5.3	495	17
16,000[b]	6.0	575	17
18,900	5.8	470	16
22,400	6.1	460	16

[a] Solution polymerization in benzene using AIBN
[b] 33.3 wt% PMMA macromonomer
To convert MPa to psi, multiply by 145
Source: Ref. [108]

optimum TPE properties were obtained with 33 wt% PMMA with $M_n = 16,000$ g/mol as shown by the fourth entry in Table 14.5.

Processing effects are very pronounced in these multiphase systems. Not only will processing conditions affect orientation of the domains, but shearing forces are needed to establish monolithic interconnecting domain structures. Furthermore, some shearing may be needed to break down gel in higher molecular weight graft copolymers. The need for shear processing to develop properties is seen in Table 14.6. These polystyrene macromonomer–acrylic copolymers were prepared by aqueous suspension polymerization and were recovered either as beads or coagulated from stable latices. It is apparent that brief milling was useful for the development and optimization of tensile strength and elongation in the molded specimens [3]. The hot milling induced flow of the polystyrene domains and formed the reinforcing domain morphology.

Table 14.6 Effects of Milling[a] on Tensile Properties of Graft Copolymers Formed from Copolymerization of Methacrylate-Functionalized Polystyrene Macromonomer with Acrylates

Polymer	Tensile strength (MPa)		Percent elongation	
	Unmilled	Milled	Unmilled	Milled
14-3[b]	3.8	9.9	70	540
14-12[c]	8.1	13.0	190	400
12-1[d]	5.0	10.0	740	940

[a] Copolymers were milled on a two-roll laboratory mill for 2 min at 143 °C to 150 °C
[b] Suspension copolymer of PS-MA ($M_n = 11,000$ g/mol)–butyl acrylate (40 : 60, wt/wt)
[c] Suspension copolymer of PS-MA ($M_n = 11,000$ g/mol)–butyl acrylate (50 : 50, wt/wt)
[d] Latex copolymer of PS-MA ($M_n = 13,000$ g/mol)–1 : 1 butyl acrylate/ethyl acrylate (35 : 50, wt/wt)
To convert MPa to psi, multiply by 145
Source: Ref. [3]

Films cast from the 12-1 latex in Table 14.6 were very weak because they apparently lacked a monolithic reinforcing network. On the other hand, when this polymer was cast as a film from toluene solution, the film was strong and elastomeric. As in the case of block copolymers, equilibrium domain morphology is established when these macromonomer copolymer films are prepared from good solvents.

Melt processing conditions and molding temperatures were found to affect the clarity and impact strength of blends of polystyrene with the graft copolymer formed from the copolymerization of styrene with a methacrylate-functionalized polystyrene–*block*-polyisoprene macromonomer (see Scheme 4) with styrene [PS(10)PI(25,000)-MA–styrene = 30 : 70, wt/wt] [35]. It was concluded that the observed rodlike morphology of the polyisoprene domains with radii of 200 Å was very susceptible to orientation from processing, and this affected clarity and impact properties [35]. Clarity and impact strength were improved when the molding temperature was increased from 300 °F to 400 °F. In specimens milled for only 1 min at 300 °F prior to compression molding, these polyisoprene rods were uniformly distributed and highly oriented. These specimens lacked impact strength, apparently because the 200 Å domains were too small to initiate the crazing and provide the energy dissipation needed for toughness. However, in specimens milled for 10 min the polyisoprene rods formed a network of the rodlike domains. This morphology apparently provided the coarser structure needed for craze initiation, and resulted in high impact strength. The highest impact strengths were observed with a macromonomer level of 25 wt%.

The preparation of impact-resistant polystyrene from blends of polystyrene macromonomer–butyl acrylate TPE with polystyrene resin demonstrated the versatility of these graft copolymers. When this macromonomer-based graft copolymer having 30 wt% polystyrene grafts was blended with the polystyrene melt on a 300 °F two-roll mill it imparted high impact strength to the polystyrene. Compression-molded specimens of these blends having 21% and 28% butyl acrylate rubber contents had notched Izod impact strengths of 270 J/m (5.0 ft lb/in.), with flexural moduli up to 145 MPa (200,000 psi). Transition electron micrographs of compression-molded samples showed 1 to 3 μm rubber particles occupying 40% to 50% of the blend volume. Most of the polystyrene macromonomer graft resided in the rubber phase as 200 Å domains, and provided the elastomeric reinforcement [107].

14.5   Summary

Macromonomers provide a useful method for the preparation of TPEs by copolymerization of soft-phase-backbone-forming monomers with hard-phase macromonomers to form comb-type graft copolymers. Macromonomers are readily prepared by living polymerizations methods of all mechanistic types either by using a functionalized initiator or by post polymerization, chain-end-functionalization reactions of the living chain ends. The advantages of the macromonomer method for forming TPEs are that both the macromonomer synthesis and the subsequent copolymerization process can be effected using any polymerization methodology. This flexibility has the potential to lead to the development of the widest possible variety of TPEs since the hard-phase-forming monomer and the soft-phase-forming monomers can be selected and polymerized by independent and different chemical processes. It has been demonstrated that variation of processing and molding conditions can exert dramatic effects on the ultimate properties of the TPEs derived from macromonomer copolymerizations.

References

1. Macromer® is a trademark of CPC International Inc. R. Milkovich and M.T. Chiang, U.S. Patent 3, 786, 116 (1974)
2. R. Milkovich, In *Anionic Polymerization. Kinetics and Mechanism, ACS Symposium Ser. No. 166*. J.E. McGrath (Ed.) (1981) American Chemical Society, Washington, D.C., p. 41
3. G.O. Schulz and R. Milkovich, *J. Appl. Polym. Sci. 27*, 4773 (1982)
4. G.O. Schulz and R. Milkovich, *J. Polym. Sci. Polym. Chem. Ed. 22*, 1633 (1984)
5. P.F. Rempp and E. Franta, *Adv. Polym. Sci. 58*, 1 (1984)
6. P. Rempp, P. Lutz, P. Masson, and E. Franta, *Makromol. Chem. Suppl. 8*, 3 (1984)
7. P. Rempp, P. Lutz, P. Masson, P. Chaumont, and E. Franta, *Makromol. Chem. Suppl. 13*, 47 (1985)
8. P. Rempp, E. Franta, P. Masson, and P. Lutz, *Prog. Colloid Polym. Sci. 72*, 112 (1986)
9. Y. Kawakami, In *Encyclopedia of Polymer Science and Engineering, Vol. 9*. J.I. Kroschwitz (Ed.) (1987) Wiley–Interscience, New York, p. 195
10. V. Percec, C. Pugh, O. Nuyken, and S.D. Pask, In *Comprehensive Polymer Science, Vol. 6, Polymer Reactions*. G.C. Eastmond, A. Ledwith, S. Russo, and P. Sigwalt (Eds) (1989) Pergamon Press, Elmsford, NY, p. 281
11. Y. Gnanou, *Ind. J. Technol. 31*, 317 (1993)
12. Basic Definitions of Terms Relating to Polymers, *Pure Appl. Chem. 40*, 482 (1974)
13. P. Dreyfuss and R. P. Quirk, In *Encyclopedia of Polymer Science and Engineering, Vol. 7*. J.I. Kroschwitz (Ed.) (1986) John Wiley & Sons, New York, p. 551
14. R.P. Quirk and Y. Wang, *Polym. Int. 31*, 51 (1993)
15. F.S. Bates and G.H. Fredrickson, *Annu. Rev. Phys. Chem. 41*, 525 (1990)
16. R.P. Quirk, D.J. Kinning, and L.F. Fetters, In *Comprehensive Polymer Science. Vol. 7. Specialty Polymers and Polymer Processing* (1989) Pergamon Press, Oxford, p. 1
17. G. Riess and G. Hurtrez, In *Encyclopedia of Polymer Science and Engineering, Vol. 2*. J.I. Kroschwitz (Ed.) (1985) John Wiley & Sons, New York, p. 324
18. G. Holden, E.T. Bishop, and N.R. Legge, *J. Polym. Sci. Pt. C 26*, 37 (1969)
19. R.P. Quirk and J. Kim, *Rubber Chem. Technol. 64*, 450 (1991)
20. M. Morton, *Anionic Polymerization: Principles and Practice* (1982) Academic Press, New York
21. S. Bywater, In *Encyclopedia of Polymer Science and Engineering, Vol. 2*. J.I. Kroschwitz (Ed.) (1985) John Wiley & Sons, New York, p. 1
22. P. Rempp and E. Franta, In *Recent Advances in Anionic Polymerization*. T.E. Hogen-Esch and J. Smid (Eds.) (1987) Elsevier, New York, p. 353
23. R.P. Quirk and B. Lee, *Polym. Int. 27*, 359 (1992)
24. P.J. Flory, *Principles of Polymer Chemistry* (1953) Cornell University Press, Ithaca, New York, (a) p. 338; (b) p. 75; (c) p. 78
25. J.F. Henderson and M. Szwarc, *J. Polym. Sci. Macromol. Rev. 3*, 317 (1968)
26. S. Bywater and D.J. Worsfold, *J. Organometal. Chem. 10*, 1 (1967)
27. R.P. Quirk and J. Kim, In *Ring-Opening Polymerization*. D.J. Brunelle (Ed.) (1993) Hanser, Munich and New York, p. 263
28. R.N. Young, R.P. Quirk, and L.J. Fetters, *Adv. Polym. Sci. 56*, 1 (1984)
29. R.P. Quirk, In *Comprehensive Polymer Science, First Supplement*. S.L. Aggarwal and S. Russo (Eds) (1992) Pergamon Press, Oxford, UK, p. 83
30. M. Morton and L.J. Fetters, *Rubber Chem. Technol. 48*, 359 (1975)

31. R.P. Quirk and J-J. Ma, *J. Polym. Sci. Polym. Chem. Ed. 26*, 2031 (1988)
32. K. Ito, Y. Masuda, T. Shintani, T. Kitano, and Y. Yamashita, *Polym. J. 15*, 443 (1983)
33. J.B. Johnson and J.P. Fletcher, *Anal. Chem. 31*, 1563 (1959)
34. P. Masson, E. Franta, and P. Rempp, *Makromol. Chem. Rapid Commun. 3*, 499 (1982)
35. G.O. Schulz and R. Milkovich, *Ind. Eng. Chem. Prod. Res. Dev. 25*, 148 (1986)
36. B.J. Wakefield, *The Chemistry of Organolithium Compounds* (1974) Pergamon Press, Oxford, UK, p. 144
37. R. Asami, M. Takaki, and H. Hanahata, *Macromolecules 16*, 628 (1983)
38. M. Arnold, W. Frank, and G. Reinhold, *Makromol. Chem. 192*, 285 (1991)
39. Y. Tsukahara, K. Tsutsumi, Y. Yamashita, and S. Shimada, *Macromolecules 23*, 5201 (1990)
40. K. Ishizu, K. Shimomura, and T. Fukutomi, *J. Polym. Sci. Polym. Chem. 29*, 923 (1991)
41. R. Asami and M. Takaki, *Makromol. Chem. Suppl. 12*, 163 (1985)
42. Y. Gnanou and P. Lutz, *Makromol. Chem. 190*, 577 (1989)
43. R.P. Quirk and J. Ren, *Macromolecules 1992*. J. Kahovec (Ed.) (1993) VSP, Netherlands, p. 133
44. J-J. Ma, D. Pang, and B. Huang, *J. Polym. Sci. Polym. Chem. 24*, 2853 (1986)
45. P. Chaumont, J. Herz, and P. Rempp, *Eur. Polym. J. 15*, 537 (1979)
46. R. Waack and M.A. Doran, *Polymer 2*, 365 (1961)
47. R. Waack and M.A. Doran, *J. Org. Chem. 32*, 3395 (1967)
48. Y. Nagasaki and T. Tsuruta, *Makromol. Chem. 187*, 1583 (1986)
49. J. Chen and L.J. Fetters, *Polym. Bull. 4*, 275 (1981)
50. Y. Nagasaki and T. Tsuruta, *Makromol. Chem. Rapid Commun. 10*, 403 (1989)
51. G. Jalics, *J. Polym. Sci. Polym. Chem. Ed. 15*, 1527 (1977)
52. R.P. Quirk and T. Yoo, *Polym. Bull. 31*, 29 (1993)
53. B.C. Anderson, G.D. Andrews, P. Arthur, Jr., H.W. Jacobson, L.R. Melby, A.J. Playtis, and W.H. Sharkey, *Macromolecules 14*, 1599 (1981)
54. G.D. Andrews and L.R. Melby, in *New Monomers and Polymers*. B.M. Culbertson and C.U. Pittman, Jr., (Eds.) (1984) Plenum Press, New York, p. 357
55. P. Lutz, P. Masson, G. Beinert, and P. Rempp, *Polym. Bull. 12*, 79 (1984)
56. K. Ishizu, K. Mitsutani, and T. Fukutomi, *J. Polym. Sci. Pt. C Polym Lett. 25*, 287 (1987)
57. S.D. Smith, *Polym. Preprints (Am. Chem. Soc. Div. Polym. Chem.) 29*(2), 48 (1988)
58. S.K. Varshney, C. Jacobs, J.P. Hautekeer, Ph. Bayard, R. Jérôme, R. Fayt, and Ph. Teyssié, *Macromolecules 24*, 4997 (1991)
59. K. Antolin, J-P. Lamps, P. Rempp, and Y. Gnanou, *Polymer 31*, 967 (1990)
60. S.K. Varshney, Ph. Bayard, C. Jacobs, R. Jérôme, R. Fayt, and Ph. Teyssié, *Macromolecules 25*, 5578 (1992)
61. K. Hatada, T. Kitayama, K. Ute, E. Masuda, T. Shinozaki, and M. Yamamoto, *Polym. Bull. 21*, 165 (1989)
62. K. Hatada, H. Nakanishi, K. Ute, and T. Kitayama, *Polym. J. 18*, 581 (1986)
63. K. Hatada, T. Shinozaki, K. Ute, and T. Kitayama, *Polym. Bull. 19*, 231 (1988)
64. A.R. Luxton, A. Quig, M.J. Delvaux, and L.J. Fetters, *Polymer 19*, 1320 (1978)
65. I.G. Krasnoselskaya and B.L. Erussalimsky, *Makromol. Chem. Rapid Commun. 6*, 191 (1985)
66. W. Toreki and T.E. Hogen-Esch, *Polym. Preprints (Am. Chem. Soc. Div. Polym. Chem.) 29*(2), 416 (1988)
67. P.R. Rao, P. Masson, P. Lutz, G. Beinert, and P. Rempp, *Polym. Bull. 11*, 115 (1984)
68. M. Takaki, R. Asami, S. Tanaka, H. Hayashi, and T. Hogen-Esch, *Macromolecules 19*, 2900 (1986)
69. O.W. Webster, W.R. Hertler, D.Y. Sogah, W.B. Farnham, and T.V. RajanBabu, *J. Am. Chem. Soc. 105*, 5706 (1983). W.J. Brittain, *Rubber Chem. Technol. 65*, 580 (1992)
70. J.A. Simms and H.J. Spinelli, *J. Coat. Technol. 59*, 125 (1987)
71. J.D. DeSimone, A.M. Hellstern, E.J. Siochi, S.D. Smith, T.C. Ward, P.M. Gallagher, V.J. Krukonis, and J.E. McGrath, *Makromol. Chem. Macromol. Symp. 32*, 21 (1990)
72. M.S. Sheridan and J.E. McGrath, *Makromol. Chem. Macromol. Symp. 64*, 85 (1992)
73. W. Radke and A.H.E. Müller, *Makromol. Chem. Macromol. Symp. 54/55*, 583 (1992)
74. R. Asami, Y. Kondo, and M. Takaki, *Polym. Preprints (Am. Chem. Soc. Div. Polym. Chem.) 27*(1), 186 (1986)
75. R. Witkowski and F. Bandermann, *Makromol. Chem. 190*, 2173 (1989)

76. R.P. Quirk and J. Ren, *Polym. Int. 32*, 205 (1993)
77. S.N. Lewis and R.A. Haggard, U.S. Patent 4, 158,736 (1979)
78. K. Ito, N. Usami, and Y. Yamashita, *Macromolecules 13*, 216 (1982)
79. Y. Yamashita, Y. Tsukahara, and H. Ito, *Polym. Bull. 7*, 289 (1982)
80. Y. Tamashita, Y. Tsukahara, K. Ito, K. Okada, and Y. Tajima, *Polym. Bull. 5*, 335 (1981)
81. Y. Chujo, T. Shishino, Y. Tsukahara, and Y. Yamashita, *Polym. J. 17*, 133 (1985)
82. Y. Chujo, K. Murai, Y. Yamashita, and Y. Okumura,, *Makromol. Chem. 186*, 1203 (1985)
83. Y. Tamashita, K. Ito, H. Mizuno, and K. Okada, *Polym. J. 14*, 255 (1982)
84. Y. Yamashita, *J. Appl. Polym. Sci. Appl. Polym. Symp. 36*, 193 (1981)
85. K.G. Suddaby, R.A. Sanayei, A. Rudin, and K.F. O'Driscoll, *J. Appl. Polym. Sci. 43*, 1565 (1991)
86. S.D. Ittel, A.A. Gridnev, B.B. Wayland, and M. Fryd, *Polym. Preprints (Am. Chem. Soc. Div. Polym. Chem.) 35*(1), 704 (1994)
87. A.H. Janowicz, U.S. Patent 5,028,677 (1991)
88. D. McIntyre, L.J. Fetters, and E. Slagowski, *Science 176*, 1041 (1972)
89. E.L. Slagowski, L.J. Fetters, and D. McIntyre, *Macromolecules 7*, 394 (1974)
90. R. Asami, M. Takaki, and Y. Moriyama, *Polym. Bull. 16*, 125 (1986)
91. F.R. Mayo and C. Walling, *Chem. Rev. 46*, 191 (1950)
92. V. Jaacks, *Makromol. Chem. 161*, 161 (1972)
93. G.G. Cameron and M.S. Chisholm, *Polymer 26*, 437 (1985)
94. J.P. Kennedy and C.Y. Lo, *Polym. Bull. 8*, 63 (1982)
95. C.F. Wofford and H.L. Hsieh, *J. Polym. Sci. Pt. A1 7*, 461 (1969)
96. D.J. Meier, *J. Polym. Sci. Pt C Polym. Symp. 26*, 81 (1969)
97. H.G. Elias, In *Polymer Handbook*, 3rd edit. J. Brandrup and E.H. Immergut (Eds) (1989) John Wiley & Sons, New York, p. VII/205
98. G.F. Meijs and E. Rizzardo, *J. Macromol. Sci. Rev. Macromol. Chem. Phys. C30*, 305 (1990)
99. R.Z. Greenley, In *Polymer Handbook*, 3rd edit. J. Brandrup and E.H. Immergut (Eds.) (1989) John Wiley & Sons, New York, p. II/153
100. G.O. Schulz and R. Milkovich, *Polym. Int. 33*, 141 (1994)
101. H.J. Harwood, *Makromol. Chem. Macromol. Symp. 10/11*, 331 (1987)
102. V. Percec, U. Epple, J.H. Wang, and H.A. Schneider, *Polym. Bull. 23*, 19 (1990)
103. V. Percec and J.H. Wang, *Makromol. Chem. Macromol. Symp. 54/55*, 561 (1992)
104. E.L. Madruga and J. San Roman, *Makromol. Chem. Rapid Commun. 12*, 319 (1991)
105. V. Percec and J.H. Wang, *Macromol. Rep. A28* (Suppl. 3), 221 (1991)
106. R.A. Rodford and M. Braden, *Biomaterials 13*, 726 (1992)
107. Unpublished results of G.O. Schulz
108. H. Xie and S. Zhou, *J. Macromol. Sci. Chem. A27*, 491 (1990)

15A Order–Disorder Transition in Block Polymers

Takeji Hashimoto

15A.1 Introduction.   430

15A.2 Nature of the Order–Disorder Transition of Block Copolymers   431

15A.3 Equilibrium Aspects of the Order–Disorder Transition   434

15A.4 Characterization of the Order–Disorder Transition by Scattering Techniques   437
    15A.4.1 Principles of the Method   437
    15A.4.2 Experiments.   437

15A.5 Changes of Spatial Concentration Fluctuations Accompanied by Order–Disorder
    Transitions   442
    15A.5.1 Disordered State   443
    15A.5.2 Ordered State and Transition.   446

15A.6 Kinetic Aspects of the Order–Disorder Transition   456

15A.7 Changes of Properties Accompanied by the Order–Disorder Transition   460

References.   460

15A.1   Introduction

The simplest (A–B) block copolymers such as polystyrene–*block*-polyisoprene (S–I) and polystyrene–*block*-polybutadiene (S–B) are amorphous and nonpolar. The only interactions among A–A, A–B, and B–B are London dispersive forces. The difference in the cohesive energy densities (or van der Waals interactions) between segments A and B results in the net repulsive interactions between A and B. These are described in terms of Flory–Huggins thermodynamic interaction parameters in the mean-field approximation [1].

In the "ordered state" such block copolymers form "microdomains" with a long-range order [2] having a spatial periodicity D of the order of the size of the polymer coil, as characterized by the radius of gyration of the block copolymer R_g. In the "disordered state" they form a homogeneous structure in which segments A and B are molecularly mixed [3]. The transition from the disordered state to the ordered state is a phenomenon known as "microphase separation," and the microdomain structures evolve as a consequence of the microphase separation. The transition from the ordered state to the disordered state is known as a "microphase dissolution."

The nature of the order–disorder transition of block copolymers has been explored quite extensively [3–12]. It is a fundamental problem from both industrial and academic viewpoints.

Industrially it is associated with processability and performances of block copolymers as thermoplastic elastomers (TPEs) [13–17], pressure-sensitive hot-melt adhesives [17, 18], viscosity stabilizers for oils, and so on [17]. At processing temperatures, typically 170 °C to 180 °C, some block polymers can be in a disordered state and in this case their melt viscosities are relatively low. This improves their processability and also the interfacial wetting between the adhesives and substrates.

On the other hand, if they are in an ordered state at processing temperature, they will exhibit high viscosity and quite remarkable non-Newtonian flow behavior [18–25], which causes difficulty in processing. At service temperatures, they are in an ordered state. The dispersed domains of polystyrene block chains act as fillers for the elastomer phase [13–17] and hence improve the performances of these elastomers. When small amounts of similar block polymers are added to an oil, they are molecularly dissolved in the oil at higher temperature. This dissolution increases the viscosity of the oils and hence counterbalances the normal loss of viscosity of oils with increase in temperature. When temperature is lowered, the block polymers form micelles or droplets. The transition from the molecularly dissolved state to the micellar state reduces the viscosity of the oils and thus counterbalances the normal increase in viscosity at lower temperature [17].

The order–disorder transition of block copolymers is important also from academic viewpoints, because it is related to structure and structure evolution (ordering) and dissolution (disordering) in a cooperative system, a fundamental problem in equilibrium and nonequilibrium statistical physics in the condensed state [26]. To predict the order–disorder transition of block copolymers, the physics developed for the cooperative phenomena in atomic or small molecular systems have to be generalized to properly take into account the connectivity between similar and dissimilar monomeric units, that is, the connectivity of monomers A (B) and A (B) block sequence and the connectivity between the block sequence A and the block sequence B at their ends.

15A.2   Nature of the Order–Disorder Transition of Block Polymers

The order–disorder transition of the block copolymers is a thermodynamic transition that is controlled by counterbalancing two physical factors, the energetics and the entropy. This is a common feature of all the phase transitions in atomic or small molecular systems, such as metallic alloys, lattice gas, and ferromagnetic materials [26].

Figure 15A.1 illustrates schematically the order–disorder transition in A–B diblock copolymers. In the ordered state A and B segments segregate themselves in A and B microdomains. However, because of molecular connectivity the segregation would not result in macroscopic phase separation into two coexisting macroscopic liquid phases as in the mixture of A and B but rather results in a regular periodic microdomain with a long-range order as schematically shown in Fig. 15A.1a (e.g., alternating lamellar microdomains [27]). On the other hand, in the disordered state A and B segments are molecularly mixed as depicted in Fig. 15A.1b.

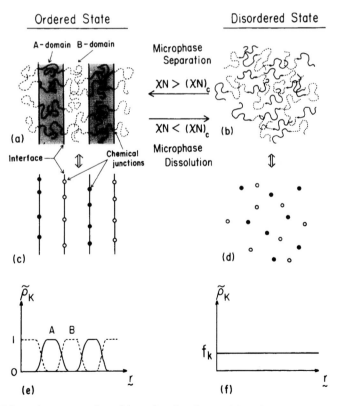

Figure 15A.1   Schematic representation of the order–disorder transition of an A–B diblock polymer. Parts (a) and (b) are molecular packing, (c) and (d) are the spatial distribution of the chemical junctions, and (e) and (f) are the spatial distribution of segmental density in the ordered and disordered states, respectively

The characteristics of the ordered state are symbolically depicted in Fig. 15.1c, where the chemical junctions between A and B block chains are confined to the narrow interfacial region drawn by a series of vertical straight lines. This region has a characteristic interface thickness of about 20 Å for the S–I block copolymers [28–30]. Generally, the average end-to-end vectors should orient normal to the interface. The solid and open circles differentiate the orientations of the end-to-end vectors of the block chains. The solid circles represent the junctions with their end-to-end vectors of A segments oriented right and conversely the open circles represent junctions with the A segments oriented left. Thus the block copolymers in the ordered state have the liquid crystalline characteristics, although the block copolymer molecules themselves are the flexible chains without any mesogenic groups. On the other hand, in the disordered state, the chemical junctions are randomly distributed in space and the orientations of the end-to-end vectors for A and B block chains are also random, as schematically illustrated in Fig. 15A.1d. It should be noted that the systems having the spherical and cylindrical microdomains in the ordered state have essentially identical characteristics to those described above.

The term *order–disorder transition* (ODT) has the same meaning as the terms *microphase separation* and *dissolution* conventionally used in the block copolymer literature or *microphase separation transition* (MST), as designated by Leibler [3]. In the case of A–B diblock copolymers, the conventional definition may be based on the following concept. The number of components (n) is equal to 2, and the number of independent variables (f) is also equal to 2 (e.g., temperature T and pressure P). Then in a phase-separated state, we have number of phases (α) equal to 2 according to the Gibb's phase rule; $f = n + 2 - \alpha$. The two phases are the A-rich and B-rich microdomains. This picture holds if our length scale of observation is smaller than R_g or smaller than the size of each domain D_A or D_B. However, for a large length scale, much larger than the domain repeat distance D of the microdomains, we have either a homogeneous structure that has a locally periodic variation of compositions inherent to the segregated A and B microdomains or one that has a thermally induced composition variation inherent to a single phase state. In other words, in this case our system has $n = 1$, since A and B are linked, and $f = 2$ and hence $\alpha = 1$. Thus we have physically a homogeneous state or phase with a *compositional order* (*hence designated ordered state or ordered phase*) in the segregation regimen shown in the left half of Fig. 15A.1 or that with a compositional disorder (hence designated disordered state or disordered phase) in the single-phase regimen shown in the right half of Fig. 15A.1. To our knowledge the term order–disorder transition appeared first in Ref. [31] in the block copolymer literature.

The relevant order parameter in such systems may be a reduced spatial segmental density profile $\tilde{\rho}_K$ (K = A or B) defined as

$$\tilde{\rho}_K(\mathbf{r}) = \rho_K(\mathbf{r})/\rho_{K0} \qquad (K = A \text{ or } B) \qquad (15A.1)$$

where $\rho_K(\mathbf{r})$ is the spatial segmental density profile for the K-segments and ρ_{K0} is that for the pure K-homopolymers. The densities $\tilde{\rho}_K$ satisfy

$$0 \leq \tilde{\rho}_K(\mathbf{r}) \leq 1 \qquad (15A.2)$$

and

$$\tilde{\rho}_A(\mathbf{r}) + \tilde{\rho}_B(\mathbf{r}) = 1 \qquad (15A.3)$$

The latter condition comes from incompressibility [32–34]. In the ordered state and in the strong segregation limit where $D \gg t_I$ (the characteristic interface thickness [28]) the order parameters $\tilde{\rho}_K$ vary as shown in Fig. 15A.1e, while in the disordered state they satisfy

$$\langle \tilde{\rho}_K \rangle_T = f_K \qquad \text{(disordered state)} \qquad (15A.4)$$

where

$$f_K = N_K/(N_A + N_B) = N_K/N \qquad (15A.5)$$

with N_K and N being the polymerization indices of K-block chain and the total block polymer, respectively.

The energetics originating from the repulsive interaction between A and B chains favor the ordered state since the interaction energy can be minimized by the segregation:

$$E_{AB} \sim k_B T \chi \tilde{\rho}_A(\mathbf{r}) \tilde{\rho}_B(\mathbf{r}) \qquad (15A.6)$$

where χ is the Flory–Huggins segmental interaction parameter between A and B, k_B is the Boltzmann constant, and T is absolute temperature. The segregation in the ordered state, however, gives a loss of entropy. There are two kinds of entropy loss: (1) the placement entropy (or translational entropy) associated with the spatial arrangements of the chemical junctions (the solid and open circles in Figs. 15A.1c and 15A.1d and (2) the conformational entropy associated with the number of possible states for given chain molecules A and B in the confined domain space [32, 50]. Entropy is obviously maximized in the disordered state. If the energetics outweigh the entropy, the systems attain the ordered state. On the other hand, if the entropy outweighs the energetics, they attain the disordered state. The equilibrium structure is the one that minimizes the free energy. The free energy F should be a functional of $\tilde{\rho}_A(\mathbf{r})$ or $\tilde{\rho}_B(\mathbf{r})$:

$$F = F\{\tilde{\rho}_K(\mathbf{r})\} \qquad (15A.7)$$

Let us now consider the order–disorder transition in block copolymers. In the ordered state, the size of the microdomains parallel and perpendicular to the interfaces, the shape of the domains (spheres, cylinders, lamellae, etc.) [27], the thickness of the interfacial region between two coexisting microdomains, and the mixing of unlike segments in each microdomain should all be characterized. It should be noted that a number of interesting ordered microdomain morphologies, other than the classical morphologies, have been reported recently [35]. In the disordered state, the thermal concentration fluctuations should be characterized. A spectrum of thermal concentrations fluctuations with wavelength $D = 2\pi/q$, where q is the wave number of a particular Fourier component of the fluctuations, can be generally analyzed from the elastic scattering profiles of neutrons, X-rays, and light. In the order–disorder transition, one should characterize the transition point, the kinetics, and the mechanism. In the bulk block copolymers the transition point is found to occur at

$$\chi N = (\chi N)_c \qquad (15A.8)$$

where $(\chi N)_c$ depends on f. If the system has a third component S (such as solvent, plasticizer, tackifying resin, etc.), the transition point is a function of polymer volume fraction Φ_P and χ_{KS} (K = A, B) where χ_{KS} is the χ-parameter between K and S [36–41, 50]. It is also found that $(\chi N)_c$ depends on the fraction of homopolymers A and/or B added to the A–B block polymers [42–46]. As for the mechanism, one can think about spinodal decomposition [47] and nucleation and growth [48] as in mixtures of homopolymers or small molecules and atoms [49].

15A.3   Equilibrium Aspects of the Order–Disorder Transition

Pioneering work on the prediction of the transition point were carried out by Meier [32, 50], Helfand [51], and Noolandi and Hong [52]. Leibler [3] has carried out Landau type analysis and developed most general theory on the transition point and phase diagram in the weak segregation limit in the context of the mean-field theory where $D \cong t_1$ and the deformation of polymer coils is not significant. He also predicted the thermal concentration fluctuations in the disordered state in the context of random phase approximation. Later on Hong and Noolandi [42] developed the theory of phase equilibria for the mixtures of A–B with A (or B). Fredrickson and Helfand [11] generalized the Leibler's Landau type mean-field theory to account for the effect of thermal fluctuations on the phase transition. They showed that the thermal fluctuations lower the mean-field critical temperature and hence suppress the two-phase region or increase $(\chi N)_c$. It is also shown that this effect changes the nature of the order–disorder transition from the second-order transition to a weakly first-order transition at $f = 0.5$. This effect, designated as the Brazovskii effect [53], was pointed out in Leibler's work [3]. We will review below Leibler's Landau type mean-field theory.

Figure 15A.2 shows the phase diagram predicted by Leibler [3] for the symmetric A–B diblock polymers where A and B have identical segmental densities ($\rho_{A0} = \rho_{B0}$), and statistical segment lengths $a_A = a_B = a$. He predicted the first-order transition for $f \neq 0.5$ and the second-order transition for $f = 0.5$. At a given χN, that is, at a given T and N inside the miscibility gap, he predicts morphological changes, as f increases from 0 to 1, going from spherical domains in body-centered cubic (bcc) lattice, hexagonally packed cylindrical domains, alternating lamellae, and their phase-inverted domain structures. At a given f ($\neq 0.5$), as χN increases the morphology changes from spherical domains to cylindrical domains and finally to lamellar domains. For an S–I block copolymer and a related copolymer system [54] thermoreversible morphological transitions were actually confirmed between the spherical domains and the cylindrical domains and also between the cylindrical domains and the lamellar domains.

For block copolymers the critical point occurs when $f = f_c = 1/2$ and $(\chi N)_c$ is equal to 10.5. In case of symmetric polymer–polymer mixture the critical point occurs also at the critical volume fraction of 1/2 but $(\chi N)_c$ has the value of 2. It should be noted here that for mixtures, N is defined as $N = N_A = N_B$. However, for the symmetrical block copolymer, N is defined as $N = N_A + N_B = 2N_A = 2N_B$.

Thus

$$(\chi N)_c = \begin{cases} 10.5 \text{ (block copolymers, } N = N_A + N_B) & (15A.9) \\ 2 \text{ (polymer–polymer mixtures, } N = N_A = N_B) & (15A.10) \end{cases}$$

Therefore the connectivity between A and B strongly affects the critical point $(\chi N)_c$ and enhances the miscibility of polymers A and B.

Leibler [3] also presented the theory for thermally induced concentration fluctuations of the block copolymers in the disordered state on the basis of random phase approximation. The

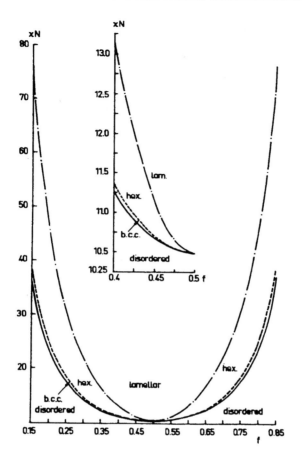

Figure 15A.2   Phase diagram of an A–B diblock polymer. [From Ref. [3], L. Leibler, *Macromolecules 13*, 1602 (1980), Copyright © 1980, American Chemical Society, with permission]

structure factor $\tilde{S}(q)$, which is a Fourier transform of the density–density correlation function, is given by

$$I(q) \sim \tilde{S}(q) = \left[\frac{S(q)}{W(q) - 2\chi} \right]^{-1} \qquad (15A.11)$$

where $S(q)$ and $W(q)$ are given in terms of S_{ij}, the Fourier transform of the density–density correlation functions of the "ideal block copolymer chains" whose properties are given by Gaussian statistics:

$$S(q) = \sum_{i,j=1}^{2} S_{ij}(q) \qquad (15A.12)$$

$$W(q) = S_{11}(q)S_{22}(q) - S_{12}^{2}(q) \qquad (15A.13)$$

$S_{ij}(q)$ are given in eqs. (IV-2) to (IV-4) in Ref. [3]. The scattered intensity $I(q)$ is proportional to the structure factor $\tilde{S}(q)$.

Figure 15A.3 shows the calculated structure factor $\tilde{S}(q)/N$ for the case of $N_A = N_B = N/2$ (i.e., $f = 1/2$) as a function of the reduced scattering vector qR_g where q is magnitude of the scattering vector defined as

$$q = (4\pi/\lambda)\sin(\theta) \tag{15A.14}$$

λ and 2θ are the wavelength and the scattering angle in the medium, respectively, and R_g is the radius of gyration of unperturbed chains

$$R_g{}^2 = Na^2/6 \tag{15A.15}$$

Two important conclusions may be derived from Fig. 15A.3. (1) The scattering profiles exhibited a single scattering maximum at $q = q_m$ even in the disordered state. (2) The q_m value depends on f and R_g,

$$q_m = q_m(f, R_g) \propto 1/R_g \qquad \text{(disordered state)} \tag{15A.16}$$

but is independent of the χ parameter and therefore temperature, except for a weak temperature dependence of R_g. Thus

$$q_m R_g \propto \chi^0 \propto T^0 \qquad \text{(disordered state)} \tag{15A.17}$$

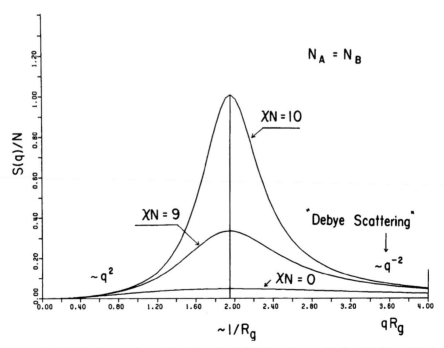

Figure 15A.3   Calculated scattering profiles for an A–B diblock polymer with $f = 1/2$ ($N_A = N_B = N/2$) and for three different values of χN on the basis of Leibler's theory or

$$D/R_g = 2\pi/(q_m R_g) \propto \chi^0 \propto T^0 \quad \text{(disordered state)} \quad (15A.18)$$

If the theory is correct, one can estimate R_g from the peak position of q_m and χ parameter by best-fitting the experimental and theoretical scattering profiles and hence extract single chain properties from the thermal concentration fluctuations in bulk. Existence of the scattering maximum is a consequence of the connectivity between A and B.

15A.4   Characterization of the Order–Disorder Transition by Scattering Techniques

The conventional techniques such as thermal analysis and volume analysis do not effectively characterize the order–disorder transition. The small heat of the transition may be related to the molecular connectivity, that is, a property attributed to the microdomains. Consequently the scattering technique seems to be the best technique for the characterization of the transition.

15A.4.1   Principles of the Method

If temperature dependence of χ is given by

$$\chi = A_1 + B_1/T \quad (15A.19)$$

as found in experimental results, then from Eqs. (15A.11) and (15A.19) the temperature dependence of the scattered intensity is given by

$$I^{-1}(q) \propto \frac{S(q)}{W(q)} - 2A_1 - \frac{2B_1}{T} \quad (15A.20)$$

and hence in the disordered state I^{-1} should linearly decrease with T^{-1}. It should be noted that $S(q)/W(q)$ is a function of f and R_g. Hence it weakly depends on T because of the weak dependence of R_g on T. Its temperature dependence is usually much weaker than that of χ. The deviations from linearity in the plot of I^{-1} vs. T^{-1} are a consequence of onset of the ordering or the microphase separation. In the disordered state the wavelength of the dominant mode of fluctuations D should satisfy Eq. (15A.18) and be weakly dependent on T. The increase of D as T decreases is again a consequence of the onset of the transition [7]. Thus the transition point can be doubly checked from the crossover behavior of I^{-1} vs. T^{-1} and D vs. T.

15A.4.2   Experiments

Figure 15A.4 shows typical small-angle X-ray scattering (SAXS) profiles from the block copolymer samples having ordered structure (HS-10) and disordered structure (HK-17) [55].

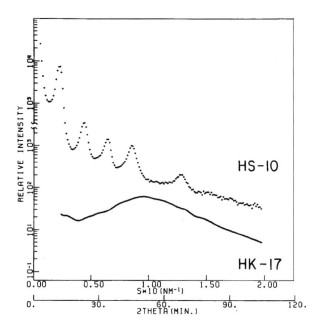

Figure 15A.4   Comparison of the SAXS profiles of HS-10 (ordered state) and HK-17 (disordered state). [From Ref. [55], Mori, Hasegawa, and Hashimoto, *Polym. J. 17*, 799 (1985), Copyright © 1985, The Society of Polymer Science, Japan, with permission]

Both HS-10 and HK-17 are the S–I diblock copolymers. HS-10 has number average molecular weight $M_n = 8.1 \times 10^4$, and $M_w/M_n = 1.13$, and the weight fraction of the polystyrene segment $w_S = 0.50$. For HK-17, $M_n = 8.5 \times 10^3$, $M_w/M_n = 1.25$, and $w_S = 0.50$.

The SAXS profiles were obtained for the as-cast films at room temperature. Although the profiles were plotted on a relative intensity scale, the intensity of HK-17 and HS-10 can be compared. HS-10 shows multiple-order scattering maxima from the single lamellar identity period D ($= 452$ Å)

$$2D \sin \theta = n\lambda \tag{15A.21}$$

or

$$sD = n \tag{15A.22}$$

where s is defined as

$$s = q/(2\pi) = (2 \sin \theta)\lambda \tag{15A.23}$$

On the other hand, HK-17 shows only the first-order scattering maximum, which is very broad and very weak compared with that of HS-10. The analyses based on Eqs. (15A.18) and (15A.20) will prove that the scattering from HK-17 is typical of the scattering from the disordered state.

Since f is about the same for the two polymers, the difference in the two SAXS profiles is attributed to an effect of the polymerization index N. For HK-17, N is small so that χN at room temperature $(\chi N)_{\mathrm{RT}}$ satisfies $(\chi N)_{\mathrm{RT}} < (\chi N)_{\mathrm{c}}$. Hence it is in the disordered state. However, for HS-10, N is large so that $(\chi N)_{\mathrm{RT}} > (\chi N)_{\mathrm{c}}$. Hence it is in the ordered state. The analysis shown in the next section indicates $(\chi N)_{\mathrm{RT}} = 10.1$ for HK-17 and 33.5 for HS-10.

The change of the SAXS profiles above and below the order–disorder transition are shown in Fig. 15A.5 for polystyrene–*block*–polybutadiene–*block*–polystyrene (S–B–S) triblock copolymer in dioctyl phthalate (DOP). The S–B–S used in the studies was the Shell copolymer Kraton® D1102 thermoplastic rubber. This has $w_{\mathrm{S}} = 0.28$ and block molecular weights of 10,000, 51,000, and 10,000. DOP is a neutral solvent for both polystyrene and polybutadiene and is added to lower the ODT temperature. The weight fraction of polymer in the solution was 0.6.

Figure 15A.6 shows the behavior of $D = 2\pi/q$ vs. T^{-1} and I^{-1} vs. T^{-1} at various temperatures [55]. Above 90 °C, the system is in the disordered state where I^{-1} changes linearly with T^{-1} and D is independent of T. Below 90 °C, the system is in the ordered state where the change of I^{-1} with T^{-1} deviates from the linear relationship and D increases with

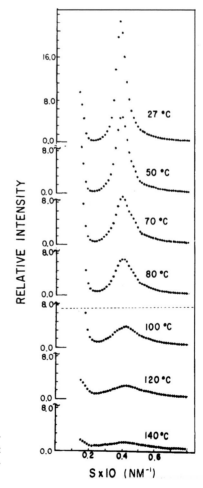

Figure 15A.5 Scattering intensity profiles for a 60 wt% S–B–S triblock polymer (TR1102) in dioctylphthalate (DOP) at various temperatures. The order–disorder transition temperature T_{c} is 90 °C (see Fig. 15A.6)

decreasing T because of a mechanism that will be clarified in the next section. The horizontal dotted line in Fig. 15A.5 indicates that the profiles above and below it are those for the ordered and disordered states, respectively.

The neutral solvent should decrease the effective interaction parameter χ_{eff} between the S and B segments,

$$\chi_{eff} = \chi \Phi_p \qquad (15A.24)$$

in the context of mean-field approximation [36, 41], where χ is the interaction parameter of the S and B segments in bulk. In this case in Eq. (15A.11) χ should be replaced by χ_{eff}, and the intensity I should be replaced by I/Φ_p for the concentrated solution in a neutral solvent [41]. The factor Φ_p in I/Φ_p is associated with the correction for the change of the scattering contrast.

$$I(q)/\Phi_p \sim \left[\frac{S(q)}{W(q)} - 2\chi\Phi_p \right]^{-1} \qquad (15A.25)$$

The effect of the solvent in lowering the order–disorder transition temperature can be estimated as follows:

$$\chi_c \Phi_p = \chi_{co} \qquad (15A.26)$$

or from Eq. (15A.19):

$$\left(A_1 + \frac{B_1}{T_c} \right) \Phi_p = A_1 + \frac{B_1}{T_{co}} \qquad (15A.27)$$

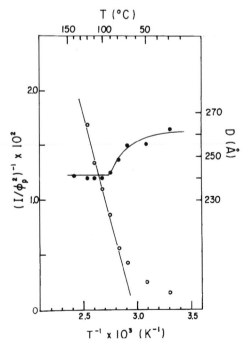

Figure 15A.6 Reciprocal scattered intensity I^{-1} (relative intensity) and the wavelength D of the dominant mode of fluctuations as a function of T^{-1} for the same sample as in Fig. 15A.5

where χ_{co} is the critical value of χ and T_{co} is the transition temperature when $\Phi_p = 1$, that is, for bulk block copolymers. Obviously the higher the concentration Φ_p, the higher the transition temperature T_c. Figure 15A.7 shows the temperature dependence of the order–disorder transition temperature measured from the crossover behaviors of the intensity $I^{-1}(q)$ with T^{-1} and D with T^{-1} for a solution of an S–B block copolymer in n-tetradecane [56]. The measured T_c linearly increases with increasing polymer concentration.

It is interesting to consider the transition temperature of the "tapered" block copolymers in which average mole fraction of a constituent monomer (e.g., A) continuously changes from unity to zero when it is scanned from one end of the block copolymer to the other end. Such tapered block polymers are synthesized by a simultaneous copolymerization of styrene and isoprene or styrene and butadiene monomers in certain reaction media using living anionic polymerization. Quantitative studies on morphology, viscoelastic properties, and polymer-ization mechanism indicated that they can be approximated by block copolymers with block chains of an A-rich random copolymer and a B-rich random copolymer [8, 57, 58]. A thermodynamic property such as the order–disorder transition also confirms qualitatively the validity of this approximation; the tapered block polymers have much lower order–disorder transition temperature T_c than the pure block polymers with corresponding N and f.

Table 15A.1 gives data on a tapered block copolymer of styrene and isoprene prepared by a simultaneous living anionic polymerization in benzene with a trace amount of tetrahy-drofuran. This polymer has $T_c = 150\,^\circ C$, much lower than that of the corresponding ideal block copolymer with comparable molecular weight and w_S, which was prepared by a sequential living anionic polymerization. T_c for the latter is over $220\,^\circ C$ [8]. This is because the effective interaction parameter χ_{eff} between styrene-rich random copolymer (designated as A-block here) and isoprene-rich random copolymer (designated as B-block here) is smaller than the χ value for the ideal S–I block [44]:

$$\chi_{eff} = \chi(f_{AS} - f_{BS})^2 = \chi(f_{AI} - f_{BI})^2 \tag{15A.28}$$

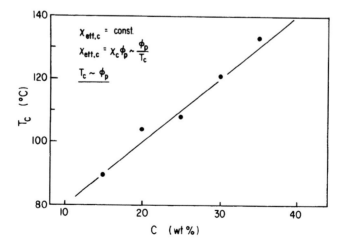

Figure 15A.7   Concentration dependence of the order–disorder transition temperature T_c for an S–B diblock polymer in n-tetradecane. [From Ref. [56], Hashimoto, et al., *Macromolecules* **19**, 754 (1986), American Chemical Society, with permission]

Table 15A.1 Tapered and Ideal Block Copolymers

Specimen	$\overline{M}_n \times 10^{-4}$	$\overline{M}_w / \overline{M}_n$	Wt% of styrene	Polymerization	T_c (°C)
Tapered block	4.3	1.05	47	Simultaneous polym. benzene/THF	150
Ideal block	4.9	1.13	45	Sequential polym. THF	> 220

Ideal block: pure S-block/pure I-block
Tapered block: "Styrene-rich" block/"isoprene-rich" block
Source: Ref. [8]

where f_{AS} and f_{AI} are the fractions of styrene and isoprene monomers in the A-block and f_{BS} and f_{BI} are the corresponding fractions in the B-block. Thus the T_c will be lowered by a factor associated with $(f_{AS} - f_{BS})^2$ or $(f_{AI} - f_{BI})^2$, that is,

$$\left(A_1 + \frac{B_1}{T_c} \right) (f_{AS} - f_{BS})^2 = A_1 + \frac{B_1}{T_{co}} \tag{15A.29}$$

where T_{co} is the T_c for the ideal block for which $f_{AS} = f_{BI} = 1$ and $f_{AI} = f_{BS} = 0$. For the block polymers with $f_{AS} = f_{BS}$, χ_{eff} vanishes and therefore the ordered state cannot exist.

15A.5 Changes of Spatial Concentration Fluctuations Accompanied by Order–Disorder Transitions

In this section we discuss the spatial concentration fluctuations $\psi(\mathbf{r})$ in the disordered state and ordered state. The quantity $\psi(\mathbf{r})$, which is called an "order-parameter," is a variable that is used to describe the thermodynamic state of the systems. The free energy of the systems is given by

$$F = F_0 + \Delta F\{\psi(\mathbf{r})\} \tag{15A.30}$$

where F_0 is the free energy of a uniform system and ΔF is the free energy functional of $\psi(\mathbf{r})$ associated with excess free energy due to fluctuations. The order parameter is defined as the local fluctuation of the reduced segmental density $\tilde{\rho}_A(r)$ from the average reduced segmental density f

$$\psi(\mathbf{r}) = \delta \tilde{\rho}_A(r) = \tilde{\rho}_A(r) - f \tag{15A.31}$$

The order parameter $\psi(\mathbf{r})$ is expanded into a Fourier series:

$$\psi(\mathbf{r}) = \sum_{\mathbf{q}} \psi_{\mathbf{q}} = \sum_{\mathbf{q}} \psi(\mathbf{q}) \exp(i\mathbf{q} \cdot \mathbf{r}) \tag{15A.32}$$

where $\mathbf{q}$ is the wave vector of the $\mathbf{q}$-Fourier component of $\psi(\mathbf{r})$.

The excess free energy due to the fluctuation ΔF can be also expanded in powers of the order parameter:

$$\Delta F = F - F_0$$
$$= \sum_q F(\mathbf{q})|\psi_q|^2 + \Delta F_3 + \Delta F_4 + \cdots \quad (15A.33)$$

where the first term of the right-hand side of Eq. (15A.33) is the second-order term, and ΔF_3, ΔF_4 etc. are the third-order and the higher-order terms.

In the disordered state, the generation and dissolution of the fluctuations are in dynamic equilibrium and the thermal average of $\psi(\mathbf{r})$ and its Fourier components $\psi(\mathbf{q})$ become zero

$$\langle \psi(\mathbf{r}) \rangle_T = 0 \quad (15A.34)$$

or

$$\langle \psi_q \rangle_T = 0 \quad (15A.35)$$

where $\langle\ \rangle_T$ denotes the thermal average.

From the scattering theory it is clear that the scattered intensity $I(\mathbf{q})$ at the scattering vector $\mathbf{q}$ is related to the intensity of Fourier components with the wave vector $\mathbf{q}$:

$$I(\mathbf{q}) \propto \langle |\psi_q|^2 \rangle_T \quad (15A.36)$$

If a system under consideration has isotropic concentration fluctuations, $\mathbf{q}$ should be replaced by $q = |\mathbf{q}|$. For the systems that can attain thermal equilibrium, Eq. (15A.36) can be calculated on the basis of the Boltzmann statistics [59] and by approximating $\Delta F \cong F(\mathbf{q})\psi_q^2$ in Eq. (15A.33),

$$\langle |\psi_q|^2 \rangle_T = \frac{\int \psi_q^2 \exp[-\psi_q^2 F(\mathbf{q})/k_B T]\, d\psi_q}{\int \exp[-\psi_q^2 F(\mathbf{q})/k_B T]\, d\psi_q} \quad (15A.37)$$
$$= \frac{k_B T}{2F(q)}$$

Thus the measurement of the scattering profile $I(q)$ is equivalent to the measurement of the susceptibility $F(\mathbf{q})^{-1}$ (i.e., the reciprocal of the free energy required to produce unit intensity of the fluctuations, $|\psi_q|^2 = 1$), as a function of q, or to the measurement of the spectral distribution of the fluctuations $\langle |\psi_q|^2 \rangle_T$. It is obvious that $\langle |\psi_q|^2 \rangle_T$ is not zero even for the disordered state, though $\langle |\psi_q| \rangle_T$ is zero.

15A.5.1   Disordered State

The scattering profiles in the disordered state were presented and discussed to some extent earlier. Here we further extend the discussion for the bulk S–I block polymer designated as HK-17. Figure 15A.8a shows the scattering profiles at various temperatures for this polymer [55].

The intensity profile at room temperature for this specimen is presented in Fig. 15A.4 using a semilogarithmic scale, whereas the profile in Fig. 15A.8 for higher temperatures is

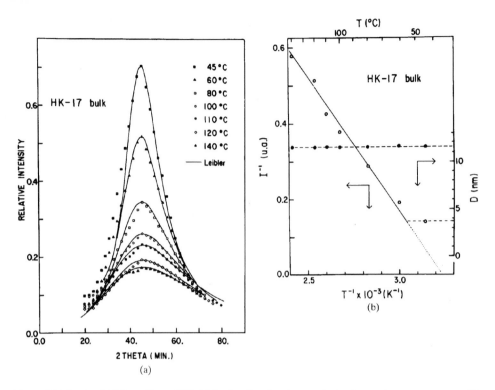

Figure 15A.8   (a) SAXS profiles of HK-17 in the disordered state obtained at various temperatures. The data points and solid lines refer to the experimental and theoretical results, respectively. (b) Reciprocal intensity I^{-1} and the wavelength D of the dominant mode of fluctuations as a function of T^{-1} for HK-17 in the disordered state. [From Ref. [55], K. Mori, H. Hasegawa, and T. Hashimoto, *Polym. J. 17*, 199 (1985), Copyright © 1985, Society of Polymer Science, Japan, with permission]

plotted on a linear scale. Figure 15A.8b shows the plots of I^{-1} vs. T^{-1} and the wavelength of the dominant mode of the fluctuations $D = 2\pi/q_m$ vs. T^{-1} to judge the state of order of the polymer. (The scattering vector q_m is the q at which the scattering intensity becomes maximum.)

There are a number of important observations that lead us to believe that the polymer sample is in the disordered state over the observed temperature range: (1) q_m is independent of temperature (Fig. 15A.8a) and hence the wavelength D is also independent of temperature (Fig. 15A.8b); (2) the reciprocal intensity I^{-1} linearly decreases with the reciprocal absolute temperature T^{-1} (the data point at 45 °C is an exception); and (3) the measured profiles are nicely fitted with the theoretical profiles (solid lines in Fig. 15A.8a) calculated from Eq. (15A.11). The weak temperature dependence D that is invoked by the weak temperature dependence of R_g [see Eq. (15A.16) or (15A.18)] cannot be discerned with this scale. The fit of the experimental profiles with Leibler's prediction becomes better than that shown in Fig. 15A.8a, when the molecular weight distribution is taken into account [38, 60, 61]. The deviation of the datum point of 45 °C from the linearity between I^{-1} vs. T^{-1} turned out to be

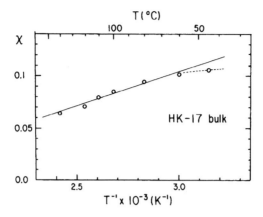

Figure 15A.9   Temperature dependence of the segmental interaction parameter χ determined by the best fits of the experimental and theoretical SAXS profiles for HK-17 in bulk and in the disordered state. [From Ref. [55], K. Mori, H. Hasegawa, and T. Hashimoto, *Polym. J.* 17, 799 (1985), Copyright © 1985, Society of Polymer Science, Japan, with permission]

caused by vitrification of the disordered sample. In other words, the thermal fluctuations in the disordered state are frozen in below its glass transition temperature T_g ($T_g \cong 51\,°C$):

$$I(q, T) \propto \langle |\psi_q|^2(T) \rangle_T \qquad \text{at } T > T_g$$
$$\propto \langle |\psi_q|^2(T_g) \rangle_T \qquad \text{at } T \leq T_g \qquad (15A.38)$$

Origin of the scattering maximum at $q = q_m$ is interpreted as a consequence of the fact that the connectivity between S and I block chains invokes the minimization of $F(q)$ at $q = q_m$. D and q_m are then the most probable wavelength and wave number of the fluctuations [62]. It should be noted that the mixtures of the polymers A and B have a minimum $F(q)$ at $q = 0$, since the smaller the value q, the smaller the gradient free energy [63] proportional to $(\nabla \psi(\mathbf{r}))^2$. Hence $I(q)$ has a maximum at $q = 0$ rather than at $q = q_m$.

Fitting the experimental points to theoretical curves as in Fig. 15A.8a allows us to estimate the temperature dependence of χ and R_g. Figure 15A.9 shows the temperature dependence of χ per monomer unit, estimated by a best-fitting procedure [55]. The χ value decreases with increasing temperature as shown in the figure, in which the straight line is given by

$$\chi = -0.0937 + 66/T \qquad (15A.39)$$

As the temperature is lowered, the χ parameter increases and hence the thermal concentration fluctuations increase. Consequently the scattering intensity increases and the scattering maximum becomes sharp as the temperature approaches the ODT temperature T_c. However, if $T_c < T_g$, then when the temperature is lowered below T_g, the system cannot attain thermal equilibrium, and the thermal concentration fluctuations are frozen in at $T = T_g$. This in turn gives rise to freezing in of the thermal concentration fluctuations and hence the χ value estimated by the fitting procedure is independent of temperature below T_g, as shown by the broken curve in Fig. 15A.9.

The χ values and temperature dependence of χ determined here are in reasonable agreement with those determined from the phase diagrams of corresponding, oligomeric mixtures or mixtures based on random copolymers [64]. However, the method presented here may be superior to others, because in the other methods the determination of the binodal line for the bulk specimen is usually difficult, and the achievement of the thermal equilibrium is also difficult and time-consuming. The measured cloud-point line is not necessarily identical to the binodal line. The effects of molecular weight distribution and asymmetry of monomer size on the estimated χ were investigated [60, 61], and χ thus estimated was found to depend on N and f [65].

15A.5.2   Ordered State and Transition

Here we consider the concentration fluctuations in the ordered state and their variation when the systems approach toward the disordered state or when the systems change from a strong segregation to a weak segregation regime.

Figure 15A.10 shows a typical transmission electron micrograph showing a long-range order (*superlattice*) of the microdomains for the alternating lamellar microdomains of S–I diblock polymer HY-12 cast from toluene solution (the block copolymer has $M_n = 5.24 \times 10^4$, $M_w/M_n = 1.16$, and $w_S = 0.52$). The transmission electron micrograph was obtained on the ultrathin section stained by osmium tetroxide, and the bright and dark portions correspond to the unstained polystyrene lamellae and the stained polyisoprene lamellae, respectively. Figure 15A.11 schematically represents the spatial concentration fluctuations in the direction normal to the lamellar interfaces where $\tilde{\rho}_K$ changes periodically with the spacing D, and extensive mixing of the unlike segments A and B occurs only in the narrow interfacial region with the characteristic interface thickness [28–30, 66, 67] $t_I \cong 20$ Å. D is of the order of radius of gyration R_g of the total block polymer and is much larger than t_I [66, 67]. Detailed SAXS analyses also indicate that the distribution of the domain identity

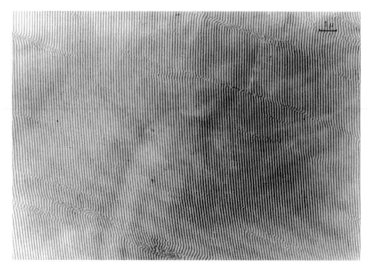

Figure 15A.10   Typical transmission electron micrograph of ultrathin section stained with OsO4 (S–I diblock polymer with $M_n = 5.24 \times 10^4$, $M_w/M_n = 1.16$ and $w_S = 0.52$)

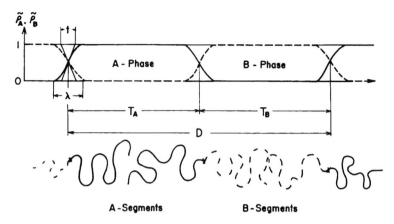

A-Segments   B-Segments

Figure 15A.11   Reduced segmental density profile $\tilde{\rho}_K$ (K = A, B) in the direction normal to the interface. t (corresponding to t_I in the text) and λ are the parameters characterizing the interface thickness. [From Ref. [66], T. Hashimoto, M. Shibayama, and H. Kawai, *Macromolecules 13*, 1237 (1980), Copyright © 1980, American Chemical Society, with permission]

period is much narrower than that of the molecular weight of the block polymers [68], for example,

$$D_W/D_N - 1 \ll M_W/M_N - 1 \tag{15A.40}$$

where D_W and D_N are the weight- and number-average domain identity periods.

The observations as described above are relevant to those for the strong segregation limit. The fact that D values are related to R_g values originates from the segregation effect and incompressibility. The two opposing physical factors, the interface energy and the free energy associated with the conformational and placement entropy, predict the equilibrium domain size, giving rise to the 2/3 power law on dependence of the domain size on N in the large χN limit [32, 50–52, 66–71]:

$$D_A \propto N_A^{2/3} \tag{15A.41}$$

$$D \propto (N_A + N_B)^{2/3} = N^{2/3} \tag{15A.42}$$

where D_A is the size of the A-domain, and D is the domain identity period [66, 67]. The 2/3 power law has been theoretically predicted by a number of workers [32, 50–52, 69, 70]. If the two block copolymers with different molecular weights are miscible at all compositions to form a single type of domain morphology, N_A and N are found to be replaced by number average polymerization indices $N_{A,n}$ and N_n, respectively [71, 72], when Eqs. (15A.41) and (15A.42) are applied to the mixtures.

Let us now consider what happens when repulsive interactions between A and B are weakened by raising the temperature or by adding neutral solvents for both A and B (i.e., by decreasing the polymer volume concentration Φ_p). In the strong segregation limit where T is low and Φ_p is high, the chains A and B segregate themselves into their respective domains, giving rise to the spatia segmental density profiles as shown in Fig. 15A.12a [7]. Owing to the repulsive interactions between chains A and B, they are stretched normal to the interface as

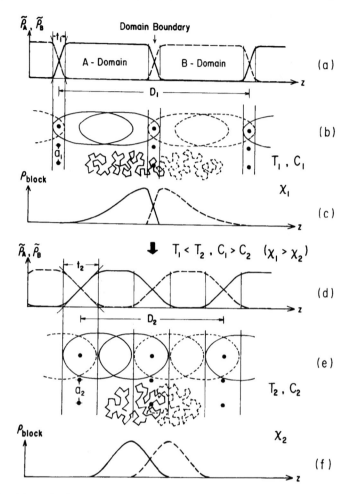

Figure 15A.12   Schematic diagram showing concentration and temperature dependence of the domain size (D_1 and D_2), average nearest-neighbour distance of the junctions along the interface (a_1 and a_2), and spatial segmental density profile of a given block chain in a direction normal to the interface. [From Ref. [7], T. Hashimoto, M. Shibayama, and H. Kawai, *Macromolecules 16*, 1093 (1983), Copyright © 1983, American Chemical Society, with permission]

shown in Fig. 15A.12b [7], where the solid and dashed ellipsoids of revolution stand schematically for the segmental clouds of the chains A and B, respectively. The segmental cloud of each chain heavily overlaps those of the same chains in neighboring polymers. This overlap is both lateral and longitudinal, ensuring a uniform overall segmental density everywhere in the domain space.

The stretching of polymers normal to the interface gives rise to the increase of R_g normal to the interface, which, coupled with the incompressibility, gives rise to an increase of D_1 and decrease of the interface-to-volume ratio S/V. The stretching results in the loss of the conformational entropy, and the reduction of S/V results in the loss of the placement entropy. These entropy losses tend to be compensated by the gain in reducing the interaction energies associated with S/V.

When T is raised or Φ_p is lowered, χ_{eff} between A and B decreases. Now some A-block chains are partially in the B domains, resulting in a larger value of the domain-boundary thickness t_2 and a smaller end-to-end distance. The latter, together with the incompressibility, results in a smaller domain size D_A and interdomain distance D_2, as shown in Figs. 15A.12d and 15A.12e. The segmental density distribution for a given block copolymer chain is skewed at low T and high Φ_p as shown in Fig. 15A.12c. It becomes a more symmetric Gaussian-type distribution at high T and low Φ_p, as shown in Fig. 15A.12f, resulting in increased conformational entropy of the block copolymer chains. The increased enthalpy of mixing caused by increasing S/V is outweighed by the increasing conformational entropy and placement entropy. It is important to note that the enthalpy of mixing decreases with increasing T and decreasing Φ_p because of the corresponding changes of χ_{eff} with T and Φ_p. This is why the smaller domain size is stable at higher T or lower Φ_p. This decrease of the domain size from D_1 and D_2 involves the increase of the average distance between the junctions from a_1 to a_2 as shown in Figs. 15.12b and 15.12e.

The physical insight gained above would suggest that the domain size, or the domain identity period D, is primarily determined by the segregation power between chains A and B, that is, by the effective interaction χ_{eff} between A and B:

$$D \cong D(\chi_{eff}) \tag{15A.43}$$

The greater the value χ_{eff}, the larger the value D. The value D depends also on the chain dimension R_g. However, the concentration and temperature dependence of R_g is much weaker than those of χ_{eff} in the concentration and temperature ranges discussed here [73]. For the neutrally good solvents and at higher polymer concentration it may be reasonable [41] to assume Eq. (15A.24). Moreover if the segmental interaction parameter χ in bulk is given by Eq. (15A.19) and further approximated by

$$\chi \cong B_1/T \tag{15A.44}$$

it follows from Eqs. (15A.24), (15A.43), and (15A.44) that

$$D \cong D(\Phi_p/T) \tag{15A.45}$$

If this is the case, the D value measured at various Φ_p and T should fall onto a single master curve when they are plotted as a function of Φ_p/T. The increase in T is equivalent to the decrease in Φ_p. More generally D values should be plotted as a function of χ_{eff} or $\chi\Phi_p$.

When T is raised and/or Φ_p is lowered, χ_{eff} decreases and approaches $(\chi_{eff})_c$. This is the weak segregation regime where D and the interface thickness become comparable or, in another words, the boundaries between A and B microdomains are diffuse, giving rise to a sinusoidal profile for $\tilde{\rho}_K(\mathbf{r})$. If T is raised further and/or Φ_p is lowered further, $\chi_{eff} < (\chi_{eff})_c$. The systems then become disordered, as discussed previously.

Figure 15A.13 shows temperature dependence of SAXS [7] from (a) 50 wt% and (b) 40 wt% toluene solution of the S–I block polymer designated L-2 with $M_n = 3.1 \times 10^4$, $M_w/M_n = 1.13$, and $w_S = 0.40$, and Fig. 15A.14 shows similar data from a 60 wt% solution [7]. Temperature dependence of SAXS from 15, 20, 25, 29, 47, and 70 wt% toluene or DOP solutions were also measured for the S–I block copolymer designated L-8. This has $M_n = 9.4 \times 10^4$ and $w_S = 0.50$. The large excess scattering intensity at small s (smaller than the value s of the scattering maximum) is observed in Figs. 15A.13 and 15A.14 when it is compared to the corresponding intensity in Fig. 15A.8a. This effect is ascribed to the

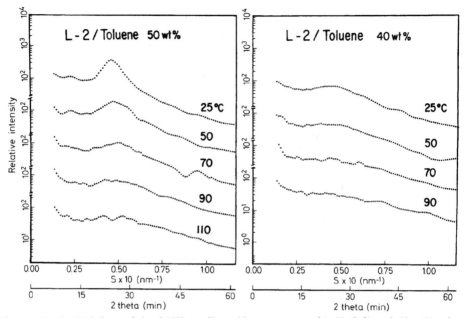

Figure 15A.13   Variations of the SAXS profiles with temperatures for 50 (left) and 40 wt% toluene solutions of L-2 (right). [From Ref. [7], T. Hashimoto, M. Shibayama, and H. Kawai, *Macromolecules 16*, 1093 (1983), Copyright © 1983, American Chemical Society, with permission]

concentration fluctuations of solvent and block copolymer molecules [41]. Figure 15A.15 represents the wavelengths of the dominant mode of fluctuations D of the block copolymer solutions at various Φ_p values as a function of T for (a) L-8 and (b) L-2, whereas Figs. 15A.16 and 15A.17 represent the plots $(I/\Phi_p^2)^{-1}$ with T^{-1} used to check the state of order for the solutions of these block copolymers [7]. In Figs. 15A.16 and 15A.17, the intensity I denotes $I(q)$ at $q = q_{max}$ in arbitrary units.

In Fig. 15A.15 the D values generally appear to first decrease with increasing temperature until they fall to a constant value that is independent of Φ_p but is dependent on molecular weight (cf. constant values $D \cong 36$ and 21 nm (360 and 210 Å), for L-8 and L-2, respectively). The decrease of D with increasing T for a given Φ_p and the decrease of D with decreasing Φ_p at a given T are the consequence of decreasing segregation power [see Eqs. (15A.24) and (15A.43) and Fig. 15A.12].

The region in which $D = 2\pi/q_m$ is independent of both T and Φ_p is the one where the block copolymers are in the disordered state. This point is confirmed further in Figs. 15A.16 and 15A.17. For example, for 15 and 20 wt% solutions of polymer L-8, I^{-1} linearly decreases with T^{-1} over the entire temperature range covered and for 25 wt% solution at temperatures higher than 100 °C (see Fig. 15A.16). These temperature and concentration ranges are identical to those where the relation

$$D = 2\pi/q_m \sim (\Phi_p/T)^0 \quad \text{(disordered state)} \quad (15A.46)$$

is observed in Fig. 15A.15a. For a 40% solution of polymer L-2 the linear decrease of I^{-1} with T^{-1} was observed over the entire temperature range covered and also at $T > 90$ °C for 50 wt% solutions. The relation of Eq. (15A.46) is again observed in Fig. 15A.15. From this

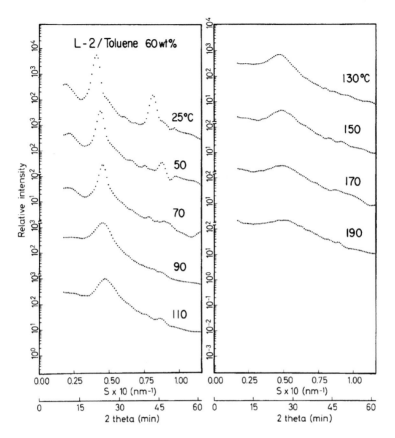

Figure 15A.14   Variations of the SAXS profiles with temperature for a 60 wt% toluene solution of L-2. [From Ref. [7], T. Hashimoto, M. Shibayama, and H. Kawai, *Macromolecules 16*, 1093 (1983), Copyright © 1983, American Chemical Society, with permission]

behavior we conclude that in Fig. 15A.13a the profiles below and above 90 °C are those from the ordered and disordered states, respectively, whereas all the profiles in Fig. 15A.13b are from the disordered state. Similarly the profiles in Fig. 15A.14 at $T \leq 150$ °C are from the ordered state. The comparison of the wavelengths D of L-8 and L-2 in the disordered state should give dependence of D with $N^{1/2}$, which, together with Eq. (15A.46), predicts following scaling rule in the disordered state:

$$D \propto N^{1/2}(\Phi_p/T)^0 \qquad \text{(disordered state)} \qquad (15A.47)$$

Figure 15A.18 shows a plot of D values for L-2 and L-8 at various values of Φ_p and T [7]. In Fig. 15A.18a logarithms of D/b are plotted as a function of logarithms Φ_p/T, and in Fig. 15A.18b logarithms of $D/bZ^{1/2}$ are plotted as a function of x/x_c, where b is the statistical segment length. Z is the polymerization index ($Z = N$) and x and x_c are defined as follows:

$$x = \Phi_p/T \qquad (15A.48)$$

$$x_c = (\Phi_p/T)_c \qquad (15A.49)$$

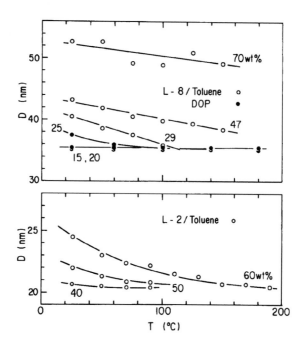

Figure 15A.15   Temperature dependence of the wavelength D of the dominant mode of the fluctuations for 15, 20, and 25 wt% DOP solutions and 29, 47, and 70 wt% toluene solutions of L-8 (a) and for 40, 50, and 60 wt% toluene solutions of L-2 (b). [From Ref. [7], T. Hashimoto, M. Shibayama, and H. Kawai, *Macromolecules 16*, 1093 (1983), Copyright © 1983, American Chemical Society, with permission]

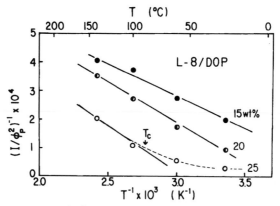

Figure 15A.16   Variations of $(I/\Phi_p^2)^{-1}$ with T^{-1} for 15, 20, and 25 wt% DOP solutions of L-8. T_c stands for the ODT temperature. [From Ref. [7], T. Hashimoto, M. Shibayama, and H. Kawai, *Macromolecules 16*, 1093 (1983), Copyright © 1983, American Chemical Society, with permission]

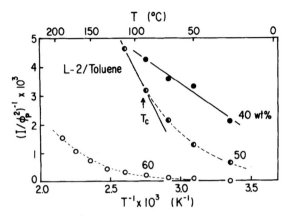

Figure 15A.17   Variation of $(I/\Phi_p^2)^{-1}$ with T^{-1} for 40, 50, and 60 wt% toluene solutions of L-2. T_c stands for the order–disorder transition temperature. [From Ref. [7], T. Hashimoto, M. Shibayama, and H. Kawai, *Macromolecules 16*, 1093 (1983), Copyright © 1983, American Chemical Society, with permission]

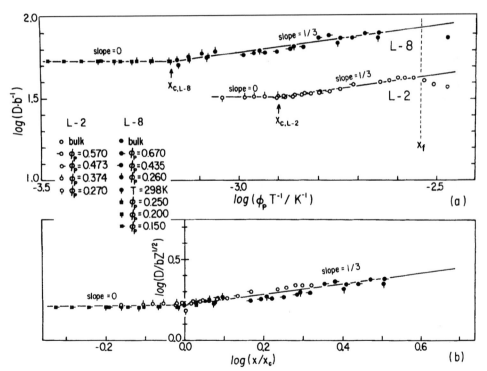

Figure 15A.18   Master curves (a) between D/b and Φ_p/T and (b) between $D/bZ^{1/2}$ and $x^* = x/x_c$ where $x = \Phi_p/T$ and $x_c = (\Phi_p/T)_c$. Z is the polymerization index which is equal to N in the text. Note that the data points at $x > x_f$ in (a) were neglected in the master curve shown in (b), because they are in the regimen of "kinetic control" rather than in the regimen of "thermodynamic control." [From Ref. [7], T. Hashimoto, M. Shibayama, and H. Kawai, *Macromolecules 16*, 1093 (1983), Copyright © 1983, American Chemical Society, with permission]

As shown in Fig. 15A.18a, for both polymers the D values measured at various values of Φ_p/T nicely fall onto a single master curve within experimental accuracy, when they are plotted as a function of Φ_p/T. D follows the relationship given by Eq. (15A.46) in the disordered state and in the ordered state:

$$D \propto (\Phi_p/T)^{1/3} \qquad \text{(ordered state)} \qquad (15\text{A}.50)$$

where thermal equilibrium is attainable, that is, at $x < x_f = (\Phi_p/T)_f$. The molecular weight dependence of D is found to be approximately identical to that in bulk block copolymer and hence

$$D \propto N^{2/3}(\Phi_p/T)^{1/3} \qquad \text{(ordered state)} \qquad (15\text{A}.51)$$

The value x_c corresponds to the ODT point. If $x > x_c$, the systems are ordered, and if $x < x_c$, they are disordered.

The molecular weight dependence of D and x can be rescaled by scaling D with the unperturbed chain dimension $bZ^{1/2}$ and x with x_c as in Fig. 15A.18b. Thus we obtain following scaling rule

$$D/(bN^{1/2}) \propto (x^*)^m, \qquad x^* = x/x_c \qquad (15\text{A}.52)$$

where

$$m = 1/3 \quad \text{for} \quad x^* > 1 \quad \text{(ordered state)} \qquad (15\text{A}.53)$$
$$m = 0 \quad \text{for} \quad x^* < 1 \quad \text{(disordered state)} \qquad (15\text{A}.54)$$

and

$$x_c \sim N^{-1/2} \qquad (15\text{A}.55)$$

Figure 15A.19 summarizes the temperature and concentration dependence of D in the disordered state and ordered state [7]. The crossover of D vs. x occurs at $x = x_c$ which has

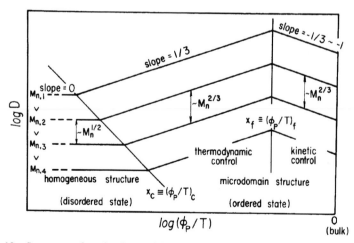

Figure 15A.19   Summary of molecular weight, concentration, and temperature dependence of the wavelength D. The variations of $\log D$ with $\log(\Phi_p/T)$ are plotted for four different molecular weights, M_{n1} to M_{n4}. [From Ref. [7], T. Hashimoto, M. Shibayama, and H. Kawai, *Macromolecules 16*, 1093 (1983), Copyright © 1983, American Chemical Society, with permission]

molecular weight dependence given by Eq. (15A.55). Thus the larger the molecular weight, the smaller the value of x_c and consequently the crossover takes place at lower concentrations and high temperatures. The reason for the unique exponent $-1/2$ has been unclear for a long time but our recent results indicate that N dependence of $\chi(\chi \sim N^{-1/2})$ is a possibility [65, 74]. At $x > x_f$ the system cannot attain thermodynamic equilibrium and the wavelength D is controlled by kinetics. Origin and effects of the nonequilibrium are discussed in detail elsewhere [75, 76].

When the degree of polymerization of block copolymers N is lowered, the critical point $x_c \equiv (\Phi_p)_c$ increases because the critical concentration $\Phi_{p,c}$ increases or T_c decreases. However, the vitrification point x_f does not change much with N. Therefore the range of x between x_c and x_f where block copolymers can attain thermal equilibrium in the ordered state becomes narrow. This causes the change of D with x to be small, as is well expected from the results shown in Fig. 15A.18a. Thus block copolymers having small N are expected to exhibit only a very small change of D with x across the order–disorder transition. This is equivalent to saying that $D = D(\chi_{eff} N)$ changes only a little with Φ_p or T, because $\chi_{eff} N$ for the copolymers with a small N changes very little over the range of Φ_p or T covered in experiments. This makes a determination of the ODT temperature from the plot of D vs. T difficult. As N is lowered, the Brazovskii effect becomes increasingly significant [11] and the plots of $I^{-1}(q_m)$, where $I(q_m)$ is the intensity $I(q)$ at $q = q_m$, vs. T^{-1} in the disordered state tend to have a prominent curvature [12, 77, 78]. This curvature makes it difficult to determine T_{ODT} from $I(q_m)^{-1}$ vs. T^{-1} data also.

In conclusion, a unique determination of T_{ODT} appears to be difficult from the scattering data of both $I(q_m)^{-1}$ vs. T^{-1} and D vs. T^{-1}. Owing to these difficulties, the change of low-frequency linear dynamic mechanical behavior with temperature has been proposed as the most reliable method for unequivocal determination of T_{ODT} [12, 77]. However, it has been reported recently [78] that despite the difficulties stated above, T_{ODT} can still be clearly determined from a high precision scattering experiment where the scattered intensity is measured with a small temperature increment or decrement in the temperature range centered on T_{ODT}. Such experiments may pinpoint T_c with accuracy less than $1\,°C$ from discontinuities [77–79] at T_{ODT} in the plots of $I(q_m)^{-1}$ vs. T^{-1} and σ_q^2 vs. T^{-1}, where σ_q is full width at half maximum of the first-order scattering maximum.

Figure 15A.20 sketches changes in various quantities such as D/R_g, $I(q_m)^{-1}$, and σ_q^2 with $1/T$ around T_c [99]. Figure 15A.20a shows a conventional picture based on the Leibler's Landau type mean-field theory, as described earlier. We define here T_{ODT} thus determined as $T_{c,conv}$ (a conventional concept of order–disorder transition temperature). Figure 15A.20b shows a new picture that takes into account the Brazovskii effect. $I(q_m)^{-1}$ and σ_q^2 shows a discontinuity at T_c but D/R_g shows almost no change at T_c. We define this T_c as $T_{c,new}$. Figure 15.20c shows a unified picture in which $T_{c,conv}$ in part (a) corresponds to a crossover temperature T_{MF}, and $T_{c,new}$ in part (b) corresponds to a true ODT temperature T_{ODT}. In the unified picture, at $T > T_{MF}$ the concentration fluctuations is characterized in terms of the Leibler's mean-field theory so that D/R_g is independent of $1/T$ and I_m^{-1} decreases linearly with $1/T$; at $T_{ODT} < T < T_{MF}$ the system is still in the disordered state but the concentration fluctuations deviate from those characterized by the mean-field theory, so that D/R_g increases with $1/T$, and $I(q_m)^{-1}$ and σ_q^2 are larger than the corresponding mean-field values owing to the strong thermal fluctuation effects. The plots of $I(q_m)^{-1}$ and σ_q^2 vs. $1/T$ show a remarkable curvature. At the true T_{ODT}, D/R_g shows almost no change, but $I(q_m)^{-1}$ and σ_q^2 show a

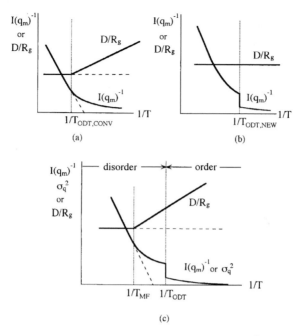

Figure 15A.20   Schematic diagram showing changes in D/R_g, $I(q_m)^{-1}$, and σ_q^2 across the ODT temperature T_c. (a) A conventional picture, (b) a new picture, and (c) a unified picture. σ_q is full-width at half maximum of the first-order scattering maximum from block copolymers. The remarkable curvature exists in the plot of $I(q_m)^{-1}$ vs. T^{-1} in the disordered state in (b), due to the Brazovskii effect [3, 11, 53], while this curvature is less significant and ignored in the schematic diagram (a)

discontinuous decrease. With increasing N, T_{ODT} approaches T_{MF} and the extent of the discontinuity in σ_q^2 and I_m^{-1} decreases. The earlier scattering analyses that led to the conventional picture (Fig. 15A.20a) seemed to overlook this discontinuity in I_m^{-1} and σ_q^2 at the true ODT temperature. The critical point T_c or x_c discussed in conjunction with Figs. 15A.15 to 15A.19 should correspond to the mean-field-to-non-mean field crossover point T_{MF} in the unified picture illustrated in Fig. 15A.20c).

15A.6   Kinetic Aspects of the Order–Disorder Transition

Here we briefly discuss kinetics of thermally induced disordering and ordering processes. The studies are important for a molecular design of processability and also to understand nonequilibrium statistical physics of polymers. This is a field to be explored extensively in the future.

When the thermodynamic driving force for the phase separation is suddenly removed by a temperature jump (T jump) above the ODT temperature T_c of the block copolymers, they undergo ODT (i.e., the disordering transition) according to a molecular mechanism as schematically illustrated in Fig. 15A.21 [56, 80–82]. A molecular process of the disordering

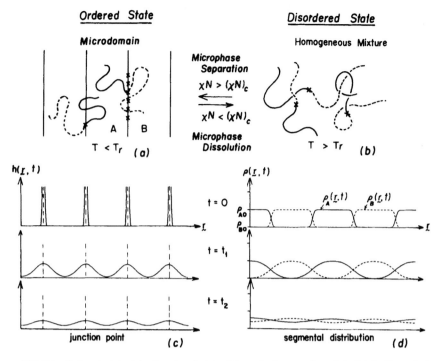

Figure 15A.21   Ordered state (a), disordered state (b), and change of spatial distribution of chemical junctions (c) and segments $\tilde{\rho}_K(\mathbf{r}, t)$ (K = A, B) (d) with time in the process of order-to-disorder transition (microphase dissolution). [From Ref. [56], T. Hashimoto, et al., *Macromolecules* **19**, 754 (1986), Copyright ©1986, American Chemical Society, with permission]

may be envisioned by reptation of chain molecules across the interfaces. The reptation is a curvilinear diffusion of polymer molecules along its own contour in bulk or in concentrated solution [83, 84].

After a T jump, each molecule undergoes mutual diffusion, and as a consequence the spatial distribution of the chemical junctions $h(\mathbf{r})$ becomes broader with time as shown in Fig. 15A.21c. As the reptation takes place the segments lose their memories with respect to their positions and orientations, resulting in mixing of the unlike segments, increased interface thickness, and eventually in an essentially uniform segmental density distribution characteristic for the disordered mixtures as shown in Fig. 15A.21d.

The ordering process may not necessarily be a reversed process of the disordering process. However, the reversed disordering process can be a possible process of the ordering [47, 81]. That is, analogously to the spinodal decomposition of mixtures [85–87], after a temperature drop (T drop) below T_c, concentration fluctuations of the junctions (Fig. 15A.21c) or the segments (Fig. 15A.21d) are built up throughout the entire sample space, and these fluctuations grow with time. Another possible process of the ordering may be envisioned as analogous to the nucleation and growth of mixtures; an ordered phase may be nucleated in the matrix of the disordered phase by a thermal activation process [48, 88]. Thus an intermediate structure is biphasic, comprising the ordered region and the disordered region, in contrast to the single phase structure in the ordering process analogous to the spinodal decomposition. As time goes on, the ordered phase will grow at the expense of the disordered phase to result eventually in a completely ordered phase as depicted in Fig. 15.21a.

To gain some insights into the physics underlying the disordering process, let us further simplify our treatment by assuming that T is raised sufficiently high above T_c, so that the thermodynamic interaction between A and B polymers does not significantly affect the diffusion [80]. Under this condition, the change of the segmental density profile with time $\tilde{\rho}_K(\mathbf{r}, t)$ should be governed by a Fickian mechanism:

$$\frac{\partial \tilde{\rho}_K(\mathbf{r}, t)}{\partial t} = D_c \nabla^2 \tilde{\rho}_K(\mathbf{r}, t) \qquad (K = A \text{ or } B) \qquad (15A.56)$$

where D_c is the diffusivity for the center-of-mass motion of the block copolymer as a whole, which should be a function of the self-diffusivities of the constituent polymers, $D_{c,K}$. Note that for simplification, Eq. (15A.56) ignores also the effect of the random thermal force [47, 89] on $\tilde{\rho}_K(\mathbf{r}, t)$.

Similarly we obtain

$$\frac{\partial h(\mathbf{r}, t)}{\partial t} = D_c \nabla^2 h(\mathbf{r}, t) \qquad (15A.57)$$

for the spatial distribution of the chemical junctions. By solving Eq. (15A.56) or Eq. (15A.57) under appropriate initial conditions as shown in Figs. 15A.21c and 15A.21d at $t = 0$, one obtains $\tilde{\rho}_K(\mathbf{r}, t)$ and $h(\mathbf{r}, t)$. From these one can calculate the observable properties as a function of time. For example, elastic scattering of X-rays, neutrons, and light as a function of time is given by

$$I(q, t; T) = I(q, t = 0; T) \exp\{-2R(q; T)t\} \qquad (15A.58)$$

$$R(q; T) = q^2 D_c(T) \qquad (15A.59)$$

Thus from the decay rate of the elastic scattered intensity one can determine D_c. The rate $R(q)$ is the decay rate for the q-Fourier component of the fluctuations. The time required for the decay of q-Fourier component of the fluctuations is given by

$$\tau_q = R(q)^{-1} = q^{-2} D_c^{-1} \qquad (15A.60)$$

One can estimate the time τ_{qm} required for the decay of the wavelength $D = 2\pi/q_m$ of the dominant fluctuations from Eq. (15A.60):

$$\tau_{qm} = q_m^{-2} D_c^{-1} = (D/2\pi)^2 D_c^{-1} \qquad (15A.61)$$

τ_{qm} is proportional to the mean-squared displacement, and therefore the larger the domain identity period D, the longer the time τ_{qm} required for the decay.

It should be noted that Eq. (15A.58) predicts $I(q, t = \infty; T) = 0$. This is not correct because the intensity $I(q, t = \infty; T)$ should be identical to the static scattered intensity at $T > T_c$ from the disordered state:

$$I(q, t = \infty; T) = I_s(q; T) \qquad (15A.62)$$

Equation (15A.59) may be modified to take this effect into account [82]:

$$I(q, t; T) = [I(q, t = 0; T) - I_s(q; T)] \exp[-2R(q; T)t] + I_s(q; T) \qquad (15A.63)$$

This correction corresponds to the one for the effect of random thermal noise [89] on the dynamics of the concentration pattern $\tilde{\rho}_K(\mathbf{r}, t)$. A treatment of the dynamics of the ordering and disordering processes, including that of the effect of finite χ on the dynamics of ordering and disordering processes, is developed [47] in the context of time-dependent Ginzburg–Landau formalism [26]. This treatment is a generalization of the simple treatment which yielded Eqs. (15A.58) and (15A.59). The generalization modifies $R(q; T)$ in Eq. (15A.59) as follows,

$$R(q; T) = q^2 \Lambda(q)/\tilde{S}(q) \qquad (15A.64)$$

where $\Lambda(q)$ is q-dependent Onsager transport coefficient of block copolymers and $\tilde{S}(q)$ is the structure factor given by Eq. (15A.11). The detailed formula for $\Lambda(q)$ was not given in the treatment but rather given later by Kawasaki and Sekimoto [90]. $\Lambda(q)$ in the limit of $q = 0$ depends on collective diffusivity D_{app} which in turn depends on D_c and the χ-dependent thermodynamic driving force for the disordering. $1/\tilde{S}(q)$ has a maximum at $q = q_m$ owing to the connectivity between A and B in block copolymers. This demands that a certain Fourier component of the fluctuations with wave number $q_m \cong 1/R_g$ has the maximum growth rate and these fluctuations become a dominant mode in the early stage of the ordering process. The q-dependence of the Onsager kinetic coefficient ignored in the treatment of Ref. [47] was elaborated by Kawasaki and Sekimoto [90].

The disordering process was investigated for S–B diblock polymer solutions with n-tetradecane (C14) by a repetitive T jump method described in detail elsewhere [82]. The block polymer has $M_n = 5.2 \times 10^4$ and $w_S = 0.30$ and its 35 wt% polymer solution has ODT at $T_c = 130\,°C$. SAXS curves were obtained at 2-s intervals after the T jump from room temperature to $160\,°C$. The scattered intensity decayed with time, indicating that the microdomains are dissolved at a time scale of the order of a minute or less for this particular system. The time-resolved SAXS studies were conducted as a function of polymer concentration and temperature [56, 82]. The collective diffusivity (D_{app}) of the order of 10^{-14} to 10^{-13} cm^2/s was estimated by this technique. The time required for the disordering is short, despite the very small D_{app}. This is because the distance $D = 2\pi/q_m$ required for the diffusion is very small. The concentration and temperature dependence of D_{app} are described in detail elsewhere [56]. D_{app} is expected to be a strong function of polymerization index N as well as f and type of the solvent.

The ordering process was studied by the time-resolved SAXS technique for a 55 wt% solution of an S–B–S triblock polymer in dipentene, the S–B–S having $M_n = 5.8 \times 10^4$ and $w_S = 0.48$ and dipentene being a neutrally good solvent [47, 81]. The solution has $T_c = 160\,°C$ and the ordering was activated by T drop from $190\,°C$ to $30\,°C$. The scattering maximum appears shortly after the T drop. Then, the maximum intensity increases with time, and the peak position q_m slightly shifts toward the smaller values. For this particular system the intensity increase tends to level off at about 80 s after the T drop. The ordering process typically occurs at the time scale of the order of 100 s. The rate should depend on N and f, as well as temperature, concentration, and type of solvent. A theory of nucleation and growth was presented for the ordering process by Fredrickson and Binder [48].

15A.7   Changes in Properties Accompanied by the Order–Disorder Transition

The flow behavior of block copolymers in bulk [18–21, 23–25] and concentrated solutions [22, 31, 56, 91] have been studied by a number of investigators. They reported the change of the flow behavior from non-Newtonian to Newtonian with increasing temperature and/or decreasing concentration, which was predicted to arise from ODT. In the ordered state the flow is strongly non-Newtonian but in the disordered state the flow is Newtonian below a certain critical shear rate $\dot{\gamma}_c$.

The conclusions drawn from the rheological measurements were first proved, albeit qualitatively, by simultaneous observations of transmitted light intensity (Pico and Williams [22]) and SAXS (Widmaier and Meyer [25]). These conclusions were confirmed more quantitatively by the simultaneous observations of rheology and SAXS by Kraus and Hashimoto [18], and by Hashimoto and Kotaka and their co-workers [31, 91]. Meier [92] proposed a theoretical interpretation on the non-Newtonian flow behavior in the ordered state and explained qualitatively some experimental evidence of

$$\eta \sim \dot{\gamma}^{-1} \tag{15A.65}$$

in terms of an excess energy dissipation associated with thermodynamic energy of mixing of the unlike segments during the flow process, that is, an extra dissipation mechanism absent in single-component and homogeneous systems.

Advanced studies have been reported on the change of rheological behaviors across ODT [12, 77, 93, 94] and on the low-frequency response or the long-time relaxation behavior of block copolymers in the ordered state [12, 77, 95, 96]. Intriguing effects of shear flow on ODT have also been reported recently [97, 98].

References

1. P.J. Flory, *Principles of Polymer Chemistry* (1953) Cornell University Press, Ithaca, NY
2. T. Hashimoto, K. Nagatoshi, A. Todo, H. Hasegawa, and H. Kawai, *Macromolecules 7*, 364 (1974)
3. L. Leibler, *Macromolecules 13*, 1602 (1980)
4. R.J. Roe, M. Fishkis, and C.J. Chang, *Macromolecules 14*, 1091 (1981)
5. T. Hashimoto, M. Shibayama, and H. Kawai, *Polym. Preprints Am. Chem. Soc. Div. Polym. Chem. 23*, 21 (1982)
6. T. Hashimoto, K. Kowsaka, M. Shibayama, and H. Kawai, *Polym. Preprints Am. Chem. Soc. Div. Polym. Chem. 24*, 224 (1983)
7. T. Hashimoto, M. Shibayama, and H. Kawai, *Macromolecules 16*, 1093 (1983)
8. T. Hashimoto, Y. Tsukahara, and H. Kawai, *Polym. J. 15*, 699 (1983)
9. F.S. Bates, *Macromolecules 18*, 525 (1985)
10. T. Hashimoto, In *Thermoplastic Elastomers—A Comprehensive Review.* N.R. Legge, G. Holden, and H.E. Schroeder (Eds.) (1987) Carl Hanser Verlag and Oxford University Press, Munich/New York, Chap. 12, Sect. 3
11. G.H. Fredrickson and E. Helfand, *J. Chem. Phys. 87*, 697 (1987)
12. F.S. Bates and G.H. Fredrickson, *Annu. Rev. Phys. Chem. 41*, 525 (1990)

13. S.L. Aggawal (Ed.), *Block Polymers* 1970 Plenum Press, New York
14. G.E. Molau (Ed.), *Collidal and Morphological Behavior of Block and Graft Copolymers* (1971) Plenum Press, New York
15. N.A. Platzer (Ed.), *Adv. Chem. Series, 142*. American Chemical Society, Washington, D.C., 1975
16. S.L. Cooper and G.M. Estes (Eds.), *Adv. Chem. Series, 176*. American Chemical Society, Washington, D.C. (1979)
17. G. Kraus, In *Block Copolymers: Science and Technology*. D.J. Meier (Ed.) (1983) MMI Press, Gordon & Breach, New York
18. G. Kraus and T. Hashimoto, *J. Appl. Polym. Sci. 27*, 1745 (1982)
19. K.R. Arnold and D.J. Meier, *J. Appl. Polym. Sci. 14*, 427 (1970)
20. C.I. Chung and J.C. Gale, *J. Polym. Sci. Polym. Phys. Ed. 14*, 1149 (1976)
21. E.V. Gouinlock and R.S. Porter, *Polym. Eng. Sci. 17*, 534 (1977)
22. E.R. Pico and M.C. Williams, *Polym. Eng. Sci. 17*, 573 (1977)
23. C.I. Chung and M.I. Lin, *J. Polym. Sci. Polym. Phys. Ed. 16*, 545 (1978)
24. C.I. Chung, H.L. Griesbach, and L. Young, *J. Polym. Sci. Polym. Phys. Ed. 18*, 1237 (1980)
25. J.M. Widmaier and G.C. Meyer, *J. Polym. Sci. Polym. Phys. Ed. 18*, 2217 (1980)
26. See for example, S.K. Ma, *Modern Theory of Critical Phenomena* (1976) Benjamin/Cummings
27. Five fundamental microdomain and morphologies have been found to exist, depending upon volume fraction of $A(f)$ in the two component block polymers: (1) A spheres in B, (2) A cylinders in B, (3) alternating A and B lamellae, (4) B cylinders in A, and (5) B spheres in A.
28. T. Hashimoto, M. Shibayama, M. Fujimura, and H. Kawai, In Ref. [17]
29. M. Shibayama and T. Hashimoto, *Macromolecules 19*, 140 (1986)
30. F.S. Bates, C.V. Berney, and R.E. Cohen, *Macromolecules 16*, 1101 (1983)
31. T. Hashimoto, M. Shibayama, H. Kawai, H. Watanabe, and T. Kotaka, *Macromolecules 16*, 361 (1983)
32. D.J. Meier, *J. Polym. Sci. Pt. C, 26*, 81 (1969); *Preprints Polym. Colloq. Soc. Polym. Sci. Jpn.* Kyoto (1977)
33. Polystyrene and polyisoprene have compressibilities [34] of 220×10^{-6} and 515×10^{-6} MPa^{-1} above T_g, and hence polystyrene–polyisoprene has compressibility of about 3×10^{-4} MPa^{-1}.
34. J. Bandrup and E.H. Immergut (Eds.), *Polymer Handbook* (1975) John Wiley & Sons, New York
35. New intriguing morphologies other than the five fundamental morphologies (A spheres and A cylinders in B-matrix, A/B alternating lamellae, and B cylinders and B spheres in A-matrix) have recently been reported in the ordered state: these are the ordered bicontinuous double diamond structure composed of the tetrapod structural unit [E.L. Thomas, D.B. Alward, D.J. Kinning, D.C. Martin, D.L. Handlin, and L.J. Fetters, *Macromolecules 19*, 2197 (1986); H. Hasegawa, H. Tanaka, K. Yamasaki, and T. Hashimoto, *Macromolecules 20*, 2940 (1987). K.I. Winey, E.L. Thomas, and L.J. Fetters, *Macromolecules 25*, 422 (1992); R.J. Spontak, S.D. Smith, and A. Ashraf, *Macromolecules 26*, 956 (1993)], the lamella catenoid [E.L. Thomas, D.M. Anderson, C.S. Henkee, and D. Hoffman, *Nature 334*, 598 (1988)], the mesh and strut [T. Hashimoto, S. Koizumi, H. Hasegawa, T. Izumitani, and S.T. Hyde, *Macromolecules 25*, 1433 (1992); M.M. Disko, K.S. Liang, S.K. Behal, R.J. Roe, and K.J. Jeon, *Macromolecules 26* 2983 (1993)]
36. E. Helfand and Y. Tagami, *J. Chem. Phys. 56*, 3592 (1972)
37. K.M. Hong and J. Noolandi, *Macromolecules 13*, 964 (1980)
38. M. Benmouna and H. Benoit, *J. Polym. Sci. Polym. Phys. Ed. 21*, 12227 (1983); M. Benmouna, W. Wu, B. Mozer, and A. Lapp, *Macromolecules 18*, 986 (1985)
39. M. Olvera de la Cruz, *J. Chem. Phys. 90*, 1995 (1989)
40. G. Fredrickson and L. Leibler, *Macromolecules 22*, 1238 (1989)
41. T. Hashimoto and K. Mori, *Macromolecules 23*, 5347 (1990)
42. K.M. Hong and J. Noolandi, *Macromolecules 16*, 1083 (1983)
43. R.J. Roe and W.C. Zin, *Macromolecules 17*, 189 (1984)
44. K. Mori, H. Tanaka, and T. Hashimoto, *Macromolecules 80*, (1987)
45. H. Tanaka and T. Hashimoto, *Polym. Commun. 29*, 212 (1988)
46. H. Tanaka and T. Hashimoto, *Macromolecules 24*, 5398 (1991); *24*, 5713 (1991)
47. T. Hashimoto, *Macromolecules 20*, 465 (1987)
48. G.H. Fredrickson and K. Binder, *J. Chem. Phys. 91*, 7265 (1989)
49. See for example, J.D. Gunton, M. San Miguel, and P.S. Sahni, In *Phase Transitions and Critical Phenomena, Vol. 8*. C. Domb and J.L. Lebowitz (Eds.) (1983) Academic Press, New York; K. Binder, In

Materials Science and Technology, Vol. 5. Phase Transformations in Materials. P. Haasen (Vol. Ed.) (1991) VCH, Weinheim, Chap. 7; T. Hashimoto, In *Materials Science and Technology, Vol. 12. Structure and Properties of Polymers.* E.L. Thomas (Vol. Ed.) (1993) VCH, Weinheim, Chap. 6

50. D.J. Meier, Chap. 11 of Ref. [10]
51. E. Helfand and Z.R. Wasserman, *Macromolecules 9*, 897 (1976); *11*, 960 (1978); *13*, 994 (1980)
52. J. Noolandi and K.M. Hong, *Ferroelectrics 30*, 117 (1980)
53. A. Brazovskii, *Sov. Phys. JETP, 41*, 85 (1975)
54. S. Sakurai, H. Kawada, T. Hashimoto, and L.J. Fetters, *Proc. Japan Acad. 69, Ser. B 13* (1993); *Macromolecules 26*, 5796 (1993); D.A. Hajduk, S.M. Gruner, P. Rangarajan, R.A. Register, L.J. Fetters, C. Honeker, R.J. Albalak, and E.L. Thomas, *Macromolecules 27*, 490 (1994)
55. K. Mori, H. Hasegawa, and T. Hashimoto, *Polym. J. 17*, 799 (1985)
56. T. Hashimoto, K. Kowsaka, M. Shibayama, and H. Kawai, *Macromolecules 19*, 754 (1986)
57. Y. Tsukahara, N. Nakamura, T. Hashimoto, H. Kawai, T. Nagaya, Y. Sugimura, and S. Tsuge, *Polym. J. 12*, 455 (1980)
58. T. Hashimoto, Y. Tsukahara, K. Tachi, and H. Kawai, *Macromolecules 16*, 648 (1983)
59. Hereafter we restrict ourselves to the isotropic systems so that $\mathbf{q}$ can be replaced by q.
60. K. Mori, H. Tanaka, H. Hasegawa, and T. Hashimoto, *Polymer 30*, 1389 (1989)
61. S. Sakurai, K. Mori, A. Okawara, K. Kimishima, and T. Hashimoto, *Macromolecules 25*, 2679 (1992), and references cited therein
62. T. Hashimoto, H. Tanaka, and H. Hasegawa, In *Molecular Conformation and Dynamics of Macromolecules ion Condensed Systems.* M. Nagasawa (Ed.) (1988) Elsevier, Amsterdam
63. J.W. Cahn and J.E. Hilliard, *J. Chem. Phys. 28*, 258 (1982)
64. N.A. Rounds and D. McIntyre, cited in E. Helfand and Z.R. Wasserman, *Macromolecules 9*, 879 (1976); R.J. Roe and W.C. Zin, *Macromolecules 13*, 1221 (1980)
65. K. Mori, A. Okawara, and T. Hashimoto, *J. Chem. Phys.* in press
66. T. Hashimoto, M. Shibayama, and H. Kawai, *Macromolecules 13*, 1237 (1980)
67. T. Hashimoto, M. Fujimura, and H. Kawai, *Macromolecules 13*, 1660 (1980)
68. T. Hashimoto, H. Tanaka, and H. Hasegawa, *Macromolecules 18*, 1864 (1985)
69. E.A. DiMarzio, C.M. Guttmann, and J.D. Hoffman, *Macromolecules 13*, 1194 (1980)
70. T. Ohta and K. Kawasaki, *Macromolecules 19*, 2621 (1986); K. Kawasaki, T. Ohta, and M. Kohrogui, *Macromolecules 21*, 2972 (1988)
71. T. Hashimoto, *Macromolecules 15*, 1548 (1982)
72. T. Hashimoto, K. Yamasaki, S. Koizumi, and H. Hasegawa, *Macromolecules 26*, 2895 (1993)
73. It should be noted that the chain dimension in good solvents tends to decrease with increasing Φ_p because of the increasing screening of the excluded volume effects. Thus the domain size should decrease with increasing Φ_p if the excluded-volume effect is a dominant physical factor. However, in reality the excluded volume effect is outweighed by the segregation effect, resulting in the increasing D with increasing Φ_p.
74. K. Mori, H. Hasegawa, and T. Hashimoto, Paper to be submitted.
75. M. Shibayama, T. Hashimoto, and H. Kawai, *Macromolecules 16*, 1434 (1983)
76. K. Mori, H. Hasegawa, and T. Hashimoto, *Polymer 31*, 2368 (1990)
77. F.S. Bates, J.H. Rosedale, and G.H. Fredrickson, *J. Chem. Phys. 92*, 6255 (1990); J.H. Rosedale and F.S. Bates, *Macromolecules 23*, 2329 (1990)
78. T. Hashimoto, T. Ogawa, and C.D. Han, *J. Phys. Soc. Jpn. 63*, 2206 (1994)
79. T. Wolff, C. Burger, and W. Ruland, *Macromolecules 26*, 1707 (1993)
80. T. Hashimoto, Y. Tsukahara, and H. Kawai, *J. Polym. Sci. Polym. Lett. Ed. 18*, 585 (1980); *Macromolecules 14*, 708 (1981)
81. T. Hashimoto, In *Physical Optics of Dynamic Phenomena and Processes in Macromolecular Systems.* B. Sedlacek (Ed.) (1985) Walter de Gruyter, Berlin and New York, 233 pp.
82. T. Hashimoto, K. Kowsaka, M. Shibayama, and S. Suehiro, *Macromolecules 19*, 750 (1986)
83. P.-G. de Gennes, *J. Chem. Phys. 55*, 572 (1971); *J. Phys. (Paris) 36*, 1199 (1975)
84. M. Doi and S. F. Edwards, *J. Chem. Soc. Faraday Trans. 2, 74*, 1789 (1978); *74*, 1802 (1978); *74*, 1818 (1978)
85. J.W. Cahn, *J. Chem. Phys. 42*, 93 (1965)
86. P.-G. de Gennes, *J. Chem. Phys. 72*, 4756 (1980)

87. K. Binder, *J. Chem. Phys. 79*, 6387 (1983)
88. C.R. Harkless, M.A. Singh, S.E. Nagler, G.B. Stephenson, and J.L. Jordan-Sweet, *Phys. Rev. Lett. 64*, 2285 (1990); M.A. Singh, C.R. Harkless, S.E. Nagler, R.F. Shannon, Jr., and S.S. Ghosh, *Phys. Rev. B 47*, 8425 (1993)
89. H.E. Cook, *Acta Metall. 18*, 297 (1970)
90. K. Kawasaki and K. Sekimoto, *Macromolecules 22*, 3063 (1989)
91. H. Watanabe, T. Kotaka, T. Hashimoto, M. Shibayama, and H. Kawai, *J. Rheol. 26*, 153 (1982)
92. D.J. Meier, Invited lecture at 29th Rheology Symp., Soc. Rheology, Japan, October 28–30, 1981, Kyoto, Japan
93. C.D. Han and J. Kim, *J. Polym. Sci. Pt B Polym. Phys. 25*, 1741 (1987); C.D. Han, J. Kim, and J.K. Kim, *Macromolecules 22*, 393 (1989); C.D. Han, D.M. Baek, and J.K. Kim, *Macromolecules 23*, 561 (1990)
94. H.H. Winter, D. Scott, W. Gronski, S. Okamoto, and T. Hashimoto, *Macromolecules 26*, 7236 (1993)
95. K. Kawasaki and A. Onuki, *Phys. Rev. A42*, 3664 (1990)
96. S.T. Milner, *Science 251*, 905 (1991); M. Rubinstein and S.P. Obukhov, *Macromolecules 26*, 1740 (1993)
97. K.A. Koppi, M. Tirrel, and F.S. Bates, *Phys. Rev. Lett. 70*, 1449 (1993)
98. N.P. Balsara and B. Hammouda, *Phys. Rev. Lett. 72*, 360 (1994)
99. N. Sakamoto and T. Hashimoto, *Macromolecules 28*, 6825 (1995)

15B Thermoplastic Elastomers Produced by Bacteria

Karla D. Gagnon

15B.1 Introduction. 466

15B.2 The Structure–Property Relationship 467

15B.3 Polymer Biosynthesis and Characterization 469
    15B.3.1 Biosynthesis with *Pseudomonas oleovorans* 469
    15B.3.2 Biosynthesis with Other Bacteria. 470
    15B.3.3 Biosynthesis Techniques 470
    15B.3.4 Polymer Extraction and Purification 472

15B.4 Characterization 472

15B.5 Morphology. 473

15B.6 Crystallization 475
    15B.6.1 Crystal Structure. 475
    15B.6.2 Crystallization Kinetics. 476
    15B.6.3 Nucleation Study 476
    15B.6.4 Long-Term Crystallization 476

15B.7 Mechanical and Elastic Properties 478
    15B.7.1 Tensile Properties 478
    15B.7.2 Tensile Set 480
    15B.7.3 Hardness 483

15B.8 Biodegradation. 483

15B.9 Summary 484

15B.10 The Future. 485

Acknowledgments 485

References 486

15B.1   Introduction

Concern over the environmental impact of polymers has started an intensive investigation into ways to produce degradable and, more specifically, biodegradable polymers. In this chapter, biodegradable means that through the enzyme-catalyzed biochemical reactions of micro-organisms, the polymer is ultimately converted to carbon dioxide (or methane during anaerobic conditions), water, and nontoxic salts [1, 2]. With the objective of finding potential polymers that could meet these needs, what better place to look than in nature? "What nature makes, nature destroys" is one of the main driving forces behind research into polyesters produced by bacteria. Some of the polyesters produced by bacteria have already been shown to be biodegradable in a variety of ecosystems [3–5].

Besides being biodegradable, the polyesters produced by bacteria have the advantage of arising from renewable resources such as glucose. Sometimes a bacterial food source is found that is considered part of a commercial waste stream. Whey, a waste product in cheese manufacturing, was found to be a good carbon source for growth and polymer production [6].

Polyesters produced by bacteria are generally known as poly(β-hydroxyalkanoates) (PHAs). It has been known since the 1920s that large quantities of polyesters can accumulate in bacteria. As shown in Fig. 15B.1, the polymer accumulates in intracellular inclusion bodies, contributing up to 80% of the total dry weight of the bacteria. Several recent review articles have detailed the many species capable of producing PHAs [7–9].

One such microorganism, *Alcaligenes eutrophus*, produces PHAs that are highly crystalline and exhibit thermoplastic behavior [3]. Poly(β-hydroxybutyrate–co–β-hydroxy-

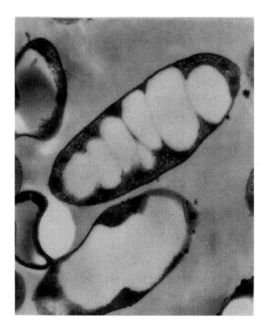

Figure 15B.1   Transmission electron micrograph of a thin section of a PHA-accumulating bacteria at 30,000 × magnification

valerate) (PHB–HV) is a thermoplastic random copolymer that has properties similar to those of polypropylene [3, 10]. This copolymer is commercially available from ICI under the tradename Biopol® and is produced by *A. eutrophus* using a mixture of glucose and propionic acid feed stocks [11].

The existence of elastomers in plants has been known since the 1500s. However, the fact that several polyesters produced by the bacterium *Pseudomonas oleovorans* exhibited rubberlike behavior was reported only in 1990 [12]. One such elastomeric copolymer poly(β-hydroxyoctanoate) (PHO) has been studied extensively and has been shown to have tensile properties similar to those of a commercially available elastomeric alloy type thermoplastic elastomer (TPE) [13, 14]. A brief history of PHAs and particularly the elastomeric polyesters is given in Fig. 15B.2.

This chapter begins by discussing why the polyesters produced by *P. oleovorans* exhibit elastomeric rather than plastic behavior. The results of a comparison between the unique structure of elastomeric microbial polyesters with commercial TPEs will be given. The chapter will then focus on PHO, a well documented microbial TPE, and will include the biosynthesis, extraction, and purification techniques that have been used to maximize polymer production. The physical and morphological characterization data, the crystal structure, the crystallization kinetics, and the results of the mechanical and elastic property evaluations will be reported on this microbial TPE. A summary of these findings along with a look into the future of TPEs produced by bacteria will be presented.

15B.2   The Structure–Property Relationship

The general chemical structure of the most common PHAs produced by bacteria is

$$\left[O-\underset{*}{C}H-CH_2-\overset{O}{\overset{\|}{C}} \right]_x$$

The pendant groups are attached to chiral carbons (*) which are all in the absolute [R] stereoconfiguration [21]. Thus the polymer is 100% isotactic and therefore capable of crystallizing. Stereoregularity of chiral carbons is a common feature of substances produced by living organisms.

Alcaligenes eutrophus produces PHAs with short pendant substituents ($n = 0$ and/or 1) whether grown on short- or long-chain carbon sources. The polymers are highly crystalline and exhibit thermoplastic behavior. Even the random copolymer PHB–HV ($n = 0$ and $n = 1$) produced from a mixture of carbon sources has a high degree of crystallinity because it is isodimorphic [12]. This means that the different repeat units can co-crystallize in each homopolymer's crystal structure, presumably because the incorporation of an additional CH_2 or the lack of the CH_2 in the pendant group is not enough to disrupt the crystal lattice.

P. oleovorans when fed a single carbon source usually produces a random copolymer rather than a homopolymer [32]. The resulting copolymer contains medium-length alkyl

1925	Substance accumulated in bacteria determined to be a polyester, poly(β-hydroxybutyrate)[15].
1941	*Pseudomonas oleovorans* discovered living in and on the cutting oil of a machine shop and given name based on 'oleo' (fat) and 'vorans' (eater)[16].
1962	First patents on PHB production and extraction assigned to W. R. Grace & Co.[17].
1974	Several different repeat units found in microbial polyesters found in sewage sludge[18].
1981	Commercialization of the thermoplastic PHB/HV copolymer by Imperial Chemical Industries under the tradename Biopol[19].
1983	*Pseudomonas oleovorans* found to produce PHO when grown on octane[20].
1986	Investigations into how to trigger polymer production in *P. oleovorans* and what carbon sources will produce PHAs[21].
1988-90	Investigation into water soluble carbon sources and longer chain carbon sources along with characterization results on PHAs produced by *P. oleovorans*[22-25].
1990	Elastomeric material behavior reported for several PHAs produced by *P. oleovorans* [12].
1990	Fed batch biosynthesis of *Pseudomonas oleovorans* developed which improved PHO yield tenfold[26] and dissolved oxygen content of media identified as parameter useful in determining carbon source feeding times[27].
1991	Continuous production of *Pseudomonas oleovorans* investigated[28,29].
1992	Source of elastomeric material behavior for PHO described with mechanical 7and tensile set properties reported and compared to commercial TPEs[13].
1992	Crystallization kinetics and effect of long term crystallization on mechanical properties reported for PHO[30].
1993	26 different bacteria capable of biodegrading PHO reported[4].

Figure 15B.2   A brief history of PHAs with a focus on the discovery and development of TPE PHAs

pendant groups ($n = 3$ to 9) that are a direct result of the longer chain carbon sources required by the bacteria to produce polymer [21–25].

The elastomeric material properties of PHAs produced by *P. oleovorans* are due to the effect of the longer pendant groups in these copolymers. As the pendant groups increase in length, the glass transition temperature falls to a value well below room temperature [12], thereby broadening the temperature range where these materials are in the rubbery regimen. The degree of crystallinity is also significantly reduced in PHAs produced by *P. oleovorans* because there is no evidence of isodimorphism [12]. The intrusion of the different length pendant groups disrupts crystallinity and because of this copolymer effect the overall degree of crystallinity is only about 30%. The combination of a lower T_g and degree of crystallinity results in these materials exhibiting rubberlike elasticity where the crystalline regions act as the physical crosslinks for the otherwise amorphous polymer [13].

How does the structure of microbial TPEs compare to that of commercial TPEs? The most common TPEs are those composed of block copolymers where chemically different segments form a two-phase morphology during solidification. Typically one phase either vitrifies or crystallizes to form a network of hard physical crosslinks, while the soft phase remains amorphous and above its glass transition temperature. Both triblock and multiblock copolymers have been found useful as TPEs.

Microbial TPEs have a somewhat different chemical structure, in that they are stereoregular random copolymers. Like most other TPEs, a two-phase morphology is present. But for the bacterial TPEs, the morphology is comprised of the crystalline and amorphous regions of a chemically homogeneous random copolymer.

15B.3   Polymer Biosynthesis and Characterization

These 100% stereoregular chemical structures have not been successfully synthesized chemically. Producing PHB ($n = 0$) synthetically was attempted using ring opening polymerization of an optically active butyrolactone and special catalyst system [33]. The synthesized polymer was reported to be approximately 73% optically pure, the same purity as the starting butyrolactone. Another attempt using a different catalyst system resulted in PHB with a 59% optical purity [34]. This difference in stereoregularity significantly reduced the degree of crystallinity. On the whole, nature still appears to be the most accurate and least expensive builder of these stereoregular polymers.

15B.3.1   Biosynthesis with *Pseudomonas oleovorans*

Polymer accumulation usually takes place under growth conditions where the environment of the bacteria is limited in some essential nutrient or growth factor but an excess of a carbon food source exists. The carbon source is metabolized and stored as a water-insoluble, osmotically inert polyester [9]. The polymer serves as a reserve carbon and energy source for the microorganisms [31], much like fat is for animals or starch is for plants.

In 1983, while researchers were trying to optimize the production of 1,2-epoxyoctane by *Pseudomonas oleovorans* grown on octane, intracellular inclusions that looked like the commonly known reserve material PHB, were observed [20]. On analysis, the material was found to be a copolymer PHO that was mostly poly(β-hydroxyoctanoate). (See Section 15B.3.)

The PHAs produced by this organism have been investigated when the bacterium was grown on *n*-alkanes, *n*-alkenes, *n*-alkanoic acids, alkenoic acids, sodium salts of *n*-alkanoic acids, and alcohols. A compilation of the results from the different investigations including polymer characterization information is given by Lenz et al. [35].

The flexibility of this organism to grow and produce unusual copolymers on novel substrates or co-substrates has also been demonstrated. Co-substrates were used to coax the bacteria to incorporate repeat units from a non-polymer-producing substrate, if it was mixed with a polymer-producing substrate [35–37]. Copolymers were produced that incorporated the following groups at the terminal position on the pendant chain in a percentage of the repeat units: an unsaturated group [21, 25, 38], a methyl branched group [36], a phenyl group [35, 37], a nitrile group [35], a methyl ester group [35], a benzyl ester group [35], a cyclohexyl group [35], a hydroxyl group [35], a bromo group [35], a chloro group [39], and a fluoro group [39].

The microorganism *P. oleovorans* is flexible in that many different carbon sources can be used to produce unusual PHAs. A constant throughout has been the production of PHAs with medium length pendant groups no matter what carbon source(s) were used. Presumably most of these PHAs would be potential TPEs. The copolymers produced with unusual chemical groups also open the door to subsequent chemical modification to further alter the properties of the PHAs.

15B.3.2   Biosynthesis with Other Bacteria

Pseudomonas oleovorans is not the only organism found to produce PHAs with medium-length pendant groups [18, 40–43]. It seems that the ability to produce PHAs with medium-length pendant groups is a common feature of the fluorescent *Pseudomonads*. Some species even produce the PHAs when relatively short-chain carbon feed sources such as sodium gluconate are used [41, 42].

15B.3.3   Biosynthesis Techniques

There are three main biosynthesis techniques: batch [21, 39], fed batch [23, 26, 27, 44], and continuous [28, 29]. The batch biosynthesis, as the name implies, means all ingredients required for growth and polymer production are mixed in a container, and the container is sterilized and then inoculated with the specific bacteria desired. When the polymer production peaks, the biomass is centrifuged, collected, and lyophilized. The polymer can then be extracted and purified. Several extraction techniques will be discussed later. The optical density (OD), which is a measure of the turbidity of the medium and therefore a reflection of the bacterial concentration, is the primary means of monitoring the culture.

A two-step batch technique has also been used [39]. The bacteria are grown on an inexpensive medium until a high OD is attained. The medium is then centrifuged and the biomass collected and transferred into fresh medium where conditions for polymer production are optimized. This two-step process has the advantage of using the expensive carbon source(s) only for polymer production and not to increase the bacteria population.

The main disadvantage of either batch technique is that the harvest time of the bacteria must coincide with peak polymer accumulation, as seen in Fig. 15B.3. If this is not timed properly, the polymer will quickly be utilized by the bacteria to sustain life in the adverse environmental conditions. In most cultures, the onset of the stationary growth phase of the bacteria (which is induced by a continued hostile environment) corresponds with maximum polymer accumulation. However, the onset of the stationary growth phase can be difficult to determine. Unfortunately, the OD cannot distinguish between bacteria with or without polymer. There is at present no real-time in situ method available to determine when the polymer content has maximized. The result of a poorly timed harvest would be a low polymer yield, even though a high bacteria population could be present. An attempt to increase the concentration of some nutrients and/or carbon sources to improve polymer yield was found to be a problem, since some of the compounds can be toxic to the bacteria at high concentrations [22].

A fed batch biosynthesis technique was developed as a way to increase polymer yield. This negated the effect of harvest timing by precluding polymer utilization by the organisms. This can be accomplished if the carbon source concentration never reaches the limiting concentration that would trigger polymer degradation by the microorganism. This approach requires a knowledge of when essential nutrients or carbon source(s) need replenishing. Bacterial growth should be dramatically increased and polymer utilization prevented, and these effects result in high polymer yields.

The fed batch technique was applied to the production of PHO [23, 26]. By using simple assumptions regarding the carbon content of the biomass, knowledge of the optimum concentration of the carbon source, and noting a linear relationship between the biomass and the OD, the optimum timing of carbon source feedings was estimated. The medium was therefore never deficient in the carbon source so the bacteria did not degrade the accumulated polymer in an attempt to sustain itself. A 10-fold increase in polymer production was reported

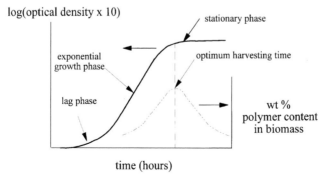

Figure 15B.3 Bacteria growth and polymer production curves typically observed for *Pseudomonas oleovorans* when grown on sodium octanoate

for the same size batch and biosynthesis time [26]. Further refinement of the timing of the carbon source feedings was reported when the dissolved oxygen content of the medium was monitored [27]. The metabolism of the bacteria would dramatically and quickly alter the dissolved oxygen content of the medium if polymer production or polymer degradation processes were occurring.

The continuous culture technique required the removal and replenishing of media at fixed rates. When applied to the biosynthesis of *P. oleovorans*, the times required for polymer accumulation and media replenishing were in conflict, and therefore high polymer yields could not be achieved [28, 29].

15B.3.4   Polymer Extraction and Purification

Polymer extraction and purification have been conducted in three different ways. Commercially, the biomass undergoes a digestive enzyme treatment that removes all biomass, leaving only the polymer. Washing, flocculation, and drying steps are then required to reclaim the polymer [44].

Lab-scale extraction usually relies on solvent extraction with a solvent such as chloroform or acetone. The lyophilized biomass is typically subjected to reflux conditions in the solvent using a Soxhlet extractor. The solvent disrupts the cell walls and the polymer dissolves into the solvent. The polymer solution is collected and reduced by rotary evaporation. The polymer is purified and recovered by precipitation into a nonsolvent.

Another polymer extraction technique relies on the digestion of the cell biomass using sodium hypochlorite, leaving behind only the polymer. Unfortunately, this technique was reported to decrease the molecular weight of the PHA. Recently a study has optimized the conditions that resulted in minimal degradation of the polymer [45].

The rest of this review will focus on PHO, the copolymer produced by *P. oleovorans* when sodium octanoate or octane is used as the sole carbon source. The organism was found to provide a high yield of polymer on these substrates for two reasons. First, the bacteria grew well on the substrates, indicating that a high bacteria concentration could be attained. Second, the bacteria accumulated large quantities of polymer, comprising 25% to 47% of the total dried biomass [22, 23]. Overall polymer yields that averaged 20 g for a 12-liter, 24-h fed batch biosynthesis have been reported [26].

15B.4   Characterization

The chemical structure of PHO is:

The mole percent composition of the different repeat units is indicated.

Physical characterization of PHO was conducted using several techniques. To determine the basic chemical composition, elemental analysis and infrared spectroscopy were used [20]. The copolymer composition was determined on methanolysized polymer using gas chromatography (GC) [13, 14, 22, 23, 30] and gas–liquid chromatography both without (GLC) [20, 21] and with mass spectroscopy (GLC–MS) [24, 40]. Both ^{1}H and ^{13}C nuclear magnetic resonance (NMR) were other techniques cited to help elucidate copolymer composition [12, 25]. Molecular weight descriptions were determined using gel permeation chromatography (GPC) [13, 22, 23, 25, 30, 50]. Thermal transition temperatures and heat of fusion information were determined using differential scanning calorimetry (DSC) [13, 23, 25]. The degree of crystallinity was evaluated using both wide angle X-ray scattering (WAXS) and ^{13}C NMR (^{13}C NMR) [12]. In addition, WAXS studies were done to determine the *d*-spacings [23] and crystal structure [12] of PHO. The onset of mass loss due to thermal decomposition was determined using thermogravimetric analysis (TGA) [13]. The density of PHO was determined using aqueous sucrose solutions by the flotation method [12]. The optical rotation was determined on methanolysized polymer using a polarimeter [21]. All the characterization results are compiled in Table 15B.1.

Thermal decomposition using TGA is one method to describe heat-related degradation of a polymer. A better method is to monitor the changes in molecular weight when the polymer is heated isothermally for long periods of time. This type of experiment indicates that PHB lacks long-term thermal stability above its melting temperature [10, 46, 47]. However, when processed with control of both residence time and temperature, film can be blown from the PHB–HV copolymer. Injection molding is also possible as evidenced by the shampoo bottle presently being sold in Europe. Presumably, the same thermal instability would also be present in the longer side chain PHAs; however, no articles published to date have addressed this concern.

15B.5   Morphology

The molecular weight between crosslinks, M_c, was estimated to better describe the network formed by the physical crosslinks [13, 14]. Stress relaxation experiments were conducted at low strains (<50%), and the equilibrium shear modulus was determined from rubber elasticity theory using Eq. (15B.1).

$$\sigma_{eq} = G_{eq}\left(\lambda - \frac{1}{\lambda^2}\right) \qquad (15B.1)$$

where

G_{eq} = equilibrium shear modulus [MPa];
λ = extension ratio, L/L_0; and
σ_{eq} = equilibrium engineering stress [MPa].

This gave an equilibrium shear modulus of about 2.0 MPa (290 psi). Since the equilibrium shear modulus in the rubber theory is based on point crosslinks rather than crystalline regions, which have a finite size, the Guth–Smallwood equation, [Eq. (15B.2)],

Table 15B.1 Characterization Results for PHO

Parameter	Description	Value	References[a]
Composition (mol%)	C5 (3-hydroxyvalerate)	0–2%	[13, 14]
	C6 (3-hydroxyhexanoate)	6–12%	[20–25]
	C8 (3-hydroxyoctanoate)	75–92%	[30–40]
	C10 (3-hydroxydecanoate)	1–17%	
Molecular weight	M_w (g/mol)	135,000–210,000	[13, 22, 23]
(in THF or CHCl$_3$)	M_n (g/mol)	54,000–99,000	[25, 30, 50]
	PDI	1.6–3.0	
Thermal transitions	T_g (°C)[b]	−35 to −37	[13, 23, 25]
	T_m (°C)[c]	59–61	
	ΔH_m (J/g)	19–34	
Degree of crystallinity	WAXS	25% Crystalline	[12]
	^{13}C NMR	33% Crystalline	[12]
Crystal structure	d-Spacings	d_1–d_9	[23]
	Orthorhombic cell dimensions	$a = 5.12$, $b = 36.05$, $c = 4.55$	[12]
Thermal decomposition[d]	T_{onset} in N$_2$ (°C)	297	[13]
	T_{onset} in air (°C)	293	[13]
Optical rotation	Polarimeter	$-21.9[\alpha]^{20}_{578}$	[21]
Density	Flotation in sucrose solns	1.019 g&so.mL	[12]

[a] References indicated for composition, molecular weight, and thermal transitions are applicable for all values. Other parameter references are specific
[b] Inflection point on second heating
[c] Peak value on first heating
[d] Onset temperature taken as the temperature where the intersection of the baseline with the tangent to the weight loss curve occurs

was used to estimate the effect that filler (in this case the crystalline regions) would have on the modulus [13].

$$\frac{G_f}{G_0} = 1 + 2.5V_f + 14.1V_f^2 \tag{15B.2}$$

where

G = shear modulus,
V_f = volume fraction of filler,
0 subscript refers to unfilled material, and
f subscript refers to filled material [that is, G_{eq} from Eq. (15B.1).

Given that PHO is approximately 30% crystalline by volume [12] then $G_f/G_0 \cong 3$. The unfilled equilibrium shear modulus is therefore about 0.67 MPa (100 psi). Using this value

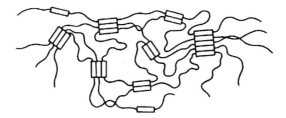

Figure 15B.4   Graphical representation of the network structure of PHO. Each stack of rectangles represents a crystalline region. (Adapted from Ref. [48])

for G_{eq}, the molecular weight between crosslinks, M_c, was then calculated from rubber elasticity theory using Eq. (15B.3).

$$M_c = \frac{\rho RT}{G_{eq}} = \text{approximately } 3600 \text{ g/mol} \qquad (15B.3)$$

where

$\rho = $ polymer density (1.019 g/cm^3) [12],
$R = 8.21 \times 10^6$ cm^3-Pa/mol-K, and
$T = 298$ K.

Using the M_n determined from GPC, the number of physical crosslinks per chain was estimated to be:

$$\frac{M_n}{M_c} = \frac{84,000}{3600} \text{ or approximately } 25 \text{ physical crosslinks per chain}$$

The morphology can be represented by Fig. 15B.4, where the clusters of rectangles represent the crystalline regions. Interestingly, this morphology is very similar to the short block type morphology found for segmented polyurethanes with respect to both M_c and also the number of physical crosslinks per chain.

15B.6   Crystallization

15B.6.1   Crystal Structure

The crystal structure of the medium-length side chain PHAs was determined to be a 2_1 helical structure in an orthorhombic lattice with two molecules per unit cell [12]. In addition, the analysis indicated that longer side chains decreased the fiber period of the helical structure but that the side chains did not interdigitate. It is believed that the side chains, if at least an average of five carbons long, can form ordered sheets as a result of an interchain packing arrangement [12, 49].

15B.6.2   Crystallization Kinetics

Since the crystalline regions are the physical crosslinks for the microbial TPE, the development of crystalline regions and the effect of altering the crystallization conditions were studied and the results of both short- and long-term crystallization studies were reported [30].

Because the TPE PHAs do not exhibit spherulitic texture [12], a relative rate of crystallization was determined by crystallizing samples from the melt for the same time at different temperatures. The development of crystallinity was implied by the values for the heat of fusion observed during the controlled crystallization time.

The goal of the short-term crystallization study was to determine the temperature at which the maximum rate of crystallization would occur. As expected, a Gaussian shaped curve resulted, with the rate reaching a maximum near the median between the glass transition temperature at $-35\,°C$ and the melting temperature at $61\,°C$. The location of the peak indicated that the fastest crystallization rate occurred between $0°$ and $5\,°C$ [30].

The melting temperature varied with crystallization temperature in a manner common for all polymers. The equilibrium melting point of PHO was determined to be approximately $68\,°C$ [30]. As expected, the glass transition temperature, T_g, did not vary significantly with crystallization temperature, T_c.

15B.6.3   Nucleation Study

In an effort to improve the crystallization kinetics, heterogeneous nucleation from the melt was attempted at room temperature [50]. Four different nucleating agents were studied (boron nitride, PHB, talc, and saccharin) at three different loading levels (0.05 wt%, 1 wt%, and 5 wt%). Interestingly, only saccharin, at the 1 wt% loading level, increased the rate of crystallization above that achieved by homogeneous nucleation at the same temperature. However, this increase did not surpass the level achieved at the fastest crystallization rate at approximately $5\,°C$.

15B.6.4   Long-Term Crystallization

The long-term crystallization study conducted at three different crystallization temperatures showed the changes in the melting behavior with time and the effect of temperature on the development of the crystalline regions.

The heat of fusion, ΔH_m, for all three films over a 24-week crystallization time is shown in Fig. 15B.5. Different maximum levels of crystallinity were reached in each film during the 24 weeks of crystallization, attaining constant values after approximately 10 weeks. The maximum heat of fusion obtained was directly proportional to the crystallization temperature; larger heats of fusion were found in films crystallized at higher temperatures [30].

The thermograms in Fig. 15B.6 illustrate the changes that occurred in the shape of the melting endotherm peaks that were typically observed for all films and were due to combined crystallization and annealing effects. Annealing is possible, since the glass transition

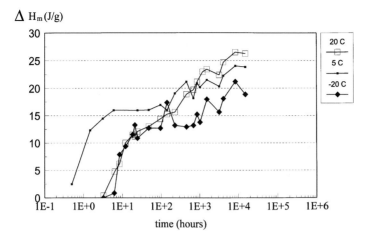

Figure 15B.5   The heat of fusion as a function of log time for PHO crystallized from the melt at the indicated temperatures

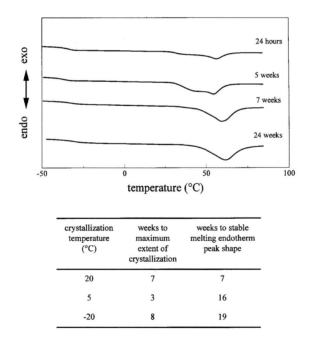

crystallization temperature (°C)	weeks to maximum extent of crystallization	weeks to stable melting endotherm peak shape
20	7	7
5	3	16
-20	8	19

Figure 15B.6   DSC thermograms of PHO crystallized at 20 °C from the melt at various times. The attached table lists the times required for the changes at the different crystallization temperatures

temperature is below all the crystallization temperatures. However, by monitoring the melting temperature, annealing effects were shown to be insignificant, both during the DSC scan and also during the initial 24 h of crystallization.

In general, a broad melting endotherm with an obvious shoulder was observed at first. As crystallization continued, this shoulder became more pronounced and it eventually coalesced into the main peak. The shoulder in the melting endotherm could be explained in terms of imperfect crystals or smaller crystallites, either or both of which would melt at a lower temperature. Over time annealing effects would become more significant, which would allow better organization of the crystallites to occur and the growth of smaller crystallites. The tendency of the crystallites to grow and become more nearly perfect was observed as the disappearance of the shoulder and the narrowing of the endotherm peak over time.

The table in Fig. 15B.6 lists the times required to maximize the extent and stabilize the shape of the melting endotherms at the different temperatures. The time required for these changes depended on the crystallization temperature and crystallization rate.

Overall, the minimum time to stabilize the shape of the endotherm peak shape and to maximize the extent of crystallinity was achieved with the temperature associated with a moderate crystallization rate. This can be considered equivalent to a slow and steady combination of crystallization and annealing.

15B.7   Mechanical and Elastic Properties

15B.7.1   Tensile Properties

Initial mechanical property evaluations of PHO and several PHAs from *P. oleovorans* revealed a stress–strain curve typically observed for elastomers; no yield stress or "knee" appeared in the curve. A tensile modulus of 17 MPa (2,500 psi) and elongation of 250% to 350% were reported although no strain rate was given [12]. The material properties were compared to a poly(*block*-styrene–*block*-butadiene–*block*-styrene) (S–B–S) TPE.

For comparison purposes another mechanical property study conducted on PHO included the test results of samples from different chemical classes of commercially available TPEs [13]. Along with tensile properties and hardness, a quantitative evaluation of the elastic properties of all the TPEs was given. A high tensile set was observed for PHO, and a thermal analysis was conducted on it after deformation to determine the cause of this. The results of the thermal analysis and the effect of thermal history and time on the mechanical properties of PHO are reviewed below.

The stress–strain curve and calculated parameters for PHO crystallized from the melt at room temperature for approximately 3 weeks were determined and are depicted in Fig. 15B.7. The properties of PHO and comparison to commercial TPEs are tabulated in Table 15B.2.

These mechanical property values were found to be dependent on the thermal history of the material and also changed over time. The aging phenomenon occurred because at room temperature, crystallization and annealing continued to take place. The effect of crystallization temperature and time on the mechanical properties is summarized in Table 15B.3.

Table 15B.2 Summary of Tensile Testing[a] and Hardness Results of PHO and the Commercial TPEs

TPE category	Material tradename	E (MPa)	100% Modulus (MPa)	300% Modulus (MPa)	Tensile strength (MPa)	Ultimate elongation (%)	% Tensile set (after 100% elongation)	Hardness (Shore durometer)
Polyester	Hytrel 5556	170	18	25	46	670	56	55D
Amide	Grilon ELX23NZ	71	14	27	42	530	31	47D
Styrenic	Kraton D4141	16	2	4	9	1070	2	52A
Amide	Pebax 2533SN00	11	4	5	21	850	12	22D
Bacterial polyester	PHO	8	2	7	9	380	35	60A
Elastomeric alloy	Santoprene 201-55	7	3	7	7	340	8	58A
Olefinic	TPR 9201-65	7	1	—	2	180	10	64A
Elastomeric alloy	Alcryn 3155-NC	5	4	9	10	370	8	61A
Urethane	Estane 5703-P	5	1	2	16	960	12	56A

[a] Average value for 5 to 10 samples reported. Strain rate of 2 min^{-1} used for all materials
To convert MPa to psi, multiply by 145
Source: Ref. [13]

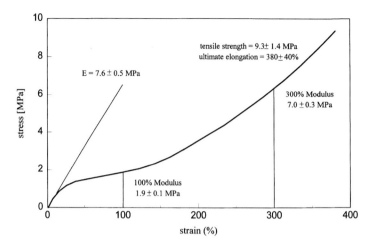

Figure 15B.7   Stress–strain curve for PHO crystallized at room temperature from the melt. Average values and standard deviations from seven samples are given. Strain rate: 2 min^{-1}

After being crystallized at room temperature from the melt and then aged for 2 years, PHO still exhibited a stress–strain curve typical for elastomers [50] even though crystallinity continued to increase (Fig. 15B.5).

15B.7.2   Tensile Set

Tensile set quantifies the deviation of a material from ideal elastic behavior. A high tensile set indicates poor elasticity and is an important consideration for any material considered to be an elastomer. In this study, tensile set was determined by extending the material to a known strain, holding that strain for 10 min, then releasing the contraint and allowing the material to recover for 10 min before final measurement of the gage length.

Table 15B.3 Summary of Mechanical Properties of PHO Crystallized at Different Temperatures and for Different Lengths of Time

Crystallization Temperature	−20 °C		5 °C		20 °C		
Time	1 yr	2 yrs	1 yr	2 yrs	2 wks[a]	1 yr	2 yrs
E (MPa)	3	7.0	6	12	8	9	18
Tensile strength (MPa)	7	13	8	15	9	12	nd[b]
Ultimate elongation (%)	470	500	350	435	380	310	nd
100% Modulus (MPa)	1	2	2	2	2	3	nd
300% Modulus (MPa)	3	5	6	8	7	11	nd

[a] 2 min^{-1} strain rate. All other values for 1 min^{-1} strain rate
[b] nd, no data

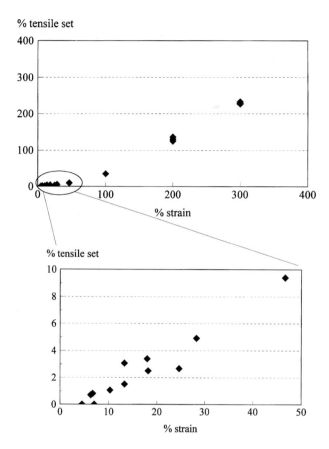

Figure 15B.8   Tensile set of PHO as a function of elongation

The tensile set obtained for PHO at various elongations is depicted in Fig. 15B.8. A substantial increase in tensile set was noted as the elongation was increased. For comparison purposes, the tensile sets of all the TPEs evaluated after 100% elongation were also tested and ranged between 2% for the S–B–S TPE and 56% for the copolyester TPE. PHO had a very substantial tensile set, 35%, compared to the other commercially available TPEs evaluated, only faring better than the copolyester TPE.

Tensile set can result from many sources such as:

- Irreversible orientation or permanent displacement (flow) of the physical crosslinks in the amorphous matrix;
- strain-induced crystallization, which does not melt on release of the deforming stress; and
- deformation induced breakup or rearrangement of the physical crosslinks.

In all cases, a permanent change to the undeformed reference state of the material would occur. Tensile set will be one consequence of this permanent change.

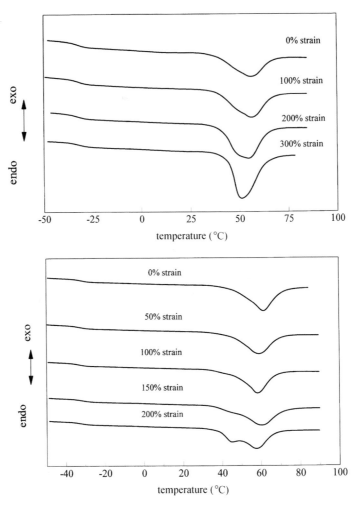

Figure 15B.9   Comparison of PHO stretched to various elongations after both (a) short and (b) long-term crystallization at 20 °C from the melt

Thermal analysis was conducted on stretched samples to investigate the possible sources of tensile set in PHO. It appears all the aforementioned possible sources of tensile set could be occurring in PHO during deformation. At small strains, < 50%, no changes in the transition temperatures or heat of fusion were observed. These results could be explained in terms of physical crosslink orientation or flow in the amorphous matrix. At larger strains, 100% to 300%, a 60% increase in the heat of fusion was observed. This supports the concept of strain-induced crystallization. A decrease in the melting temperature was also noted. One possible explanation is that during deformation the crystal regions break up and become smaller, and therefore melt at lower temperatures. The DSC thermograms are shown in Fig. 15B.9a, which also shows the narrowing of the melting peak as the amount of strain increases. This observation supports the concept that rearrangement of the crystalline regions occurred during deformation.

An experiment was conducted to determine if the material could be induced to recover by slowly and incrementally heating it. Usually, the recovery of block TPEs is possible using this method, since the deformation can induce crystallization in the soft segments. However, since in PHO the physical crosslinks are crystalline regions, strain-induced crystallization directly affects the physically crosslinked network by forming new crosslinks or by altering original ones, or both. Thus in this case, strain-induced crystallization results in a permanent change in the network structure and full recovery does not occur by the application of heat, because heating affects all crystalline regions whether part of the original network or strain-induced.

Figure 15B.9b shows the DSC thermograms of PHO stretched to various elongation after being crystallized from the melt for 24 weeks. Some unusual changes were observed in shape in the melting endotherm as the elongation increased. A shoulder can be seen forming after the 100% and 150% elongations, but after the 200% elongation, a new melting peak emerged at a lower melting point of 45 °C. This new peak may indicate that a different crystal structure was formed with stretching. Another possibility is that chain degradation occurred during the deformation. However, molecular weight data obtained on the PHO after deformation did not show any changes in either the molecular weight or in the molecular weight distribution [50].

15B.7.3 Hardness

The hardness of PHO was determined to be 60 Shore A [13]. PHO was a relatively soft elastomer compared to the commercial TPEs tested (see Table 15B.2). These had hardnesses ranging from Shore 52A to Shore 55D. As the level of crystallization of PHO increased over time, the hardness also increased. For PHO crystallized at room temperature for approximately 6 months, the hardness increased to Shore 79A [30].

15B.8   Biodegradation

Because PHAs are naturally produced and utilized by bacteria in response to environmental changes, the polymer is inherently biodegradable. However, when the material is extracted from the bacteria and processed into useful items, the biodegradability of the polymer must be reevaluated.

Studies have indicated that certain bacteria excrete polymer-degrading enzymes, known as depolymerases, which enable the bacteria to use the processed polymer as their sole exogenous carbon source [51, 52]. Fungi also colonate these materials extensively [10].

The number of studies conducted on the biodegradability PHO to date is rather limited, mainly because of a lack of material availability. An early study showed that PHO was not degraded by the depolymerase known to degrade PHB [39]. A recent study has identified 26 bacteria from various soils, lake water, and activated sludge that used PHO as their sole carbon source [4]. One microorganism from the study was identified as *P. fluorescens* and was found to excrete a PHO depolymerase into the culture media. The extracellular PHO depolymerase was isolated and characterized. Interestingly, this enzyme did not degrade PHB–HV.

If these bacterial TPEs were made part of the consumer market, where would these materials biodegrade? The main disposal system used today, the landfill, has been shown to be lacking in sufficient microbial activity to enable significant degradation of disposed items [53]. Archeological digs of old landfills have found food that is still recognizable and newspapers that are still intact and readable [53]. The landfill system, instead of a way to degrade materials, appears to preserve the waste for significant periods of time.

For the bacterial polyesters, the use of bioreactors [54] may be viable. These give a combination of composting, depolymerization, and recycling. The disposed biomaterials could be used as a feed stock for bacteria that would break down the waste and generate new polymers.

15B.9   Summary

PHO is a TPE unique in its source, bacteria. Because of this natural source, PHO has an unusual chemical structure for a TPE, a stereoregular random copolymer. The polymer is semicrystalline (30%) and in the temperature range between its T_g ($-36\,°C$) and its T_m ($61\,°C$), it exhibits elastomeric behavior. Because of a relatively broad melting peak, however, the maximum temperature at which the material will retain mechanical integrity is approximately $30\,°C$. Typical commercial TPEs are structurally different, being either block copolymers or blended dissimilar polymers.

The statistical rubber elasticity theory, with a correction for the filler effect of the crystalline physical crosslinks, indicated the physically crosslinked network formed when PHO is crystallized from the melt at room temperature is similar to the segmented polyurethanes with a molecular weight between crosslinks of approximately 4000 g/mol and with approximately 25 physical crosslink regions per chain.

The crystallization rate of PHO is very slow. Even at the fastest crystallization temperature several hours are required before the material forms a film with mechanical integrity. An attempt to increase the rate of crystallization through the incorporation of possible nucleating agents was unsuccessful.

The crystallization temperature affected the mechanical properties significantly. Adjustment of the material properties could possibly be achieved through the use of specific crystallization temperatures. It is not known if the material remains elastic over very long times since a combination of crystallization and annealing continues to occur at room temperature. However, the material still exhibited elastomeric properties after being stored at room temperature for over 2 years.

The elastic response of PHO decreased as the amount of strain increased due to deformation induced changes in the crystalline regions and increased overall crystallinity. However, the stress–strain properties, hardness, and tensile set of PHO were found to be in the range of the commercial TPEs.

15B.10 The Future

Biosynthesis, extraction, and purification of polymers from bacteria is an expensive process. Legislation requiring some consumer goods or packaging materials to be compostable may eventually make these biopolymers a more cost-effective alternative to conventional ones.

Ultimately, the use of genetic engineering will be the key to decreasing the cost of these biopolymers. Several researchers have identified the gene sequence that encodes the PHB polymer production machinery in *Alcaligenes eutrophus* [55–60]. This machinery has been transplanted into *E. coli*, which grows faster and with better understood growth conditions [55, 56]. The result was a bacteria that was capable of accumulating polymer up to 98% of its dry weight.

Two groups have placed the polymer-producing genes of PHB from *Alcaligenes eutrophus* into *Pseudomonas oleovorans* [61, 62]. The result was not a copolymer but a blend where different inclusion bodies had accumulated either 100% PHB or a copolymer with medium-length side chains [62]. Researchers have even successfully used recombinant techniques to place the PHB production machinery into plants [63]. Much work is required before there is a harvesting of plastic potatoes [64] but researchers with vision and patience will surely see this come true.

The problem of high tensile set identified for the TPEs produced by bacteria may be overcome by creating a better TPE using different PHAs as building blocks. One possibility is to create a triblock copolymer with the highly crystalline PHB as end blocks and an amorphous PHA as the midblock. *P. oleovorans* when grown on heptane or heptanoate produced a totally amorphous PHA that appears suitable for the midblock [22, 25]. The maximum use temperature would also increase with this type of structure. Of course, the effect on the biodegradability of this engineered microbial TPE would be a real concern. Such engineered materials in which PHAs are used as building blocks have already been reported. Several research groups have described the preparation of PHA-synthetic conjugates as a means to broaden the material properties and overcome the relative thermal instability of the microbial polyesters [65, 66].

The research into polymers produced by bacteria is a prime example of the rich territory awaiting exploration that is found on the border between what are considered disparate research fields. New discoveries can be made when, for example, a microbiologist who keeps referring to a carbon and energy storage material accumulated by bacteria as poly(β-hydroxybutyrate) meets with the polymer scientist who recognizes this storage material is a polyester.

Acknowledgments

I would personally like to acknowledge my co-advisors, Robert W. Lenz and Richard J. Farris, and a much called upon committee member, R. Clinton Fuller, for their encouragement and support while I attended graduate school at the University of Massachusetts, Amherst. I thoroughly enjoyed learning about and working in this fascinating research area. I am grateful to the Office of Naval Research, the Procter & Gamble Company, the National Science

Foundation, and the Focused Giving Program from Johnson & Johnson (R.W.L., R.C.F.) for their financial support.

References

1. R.W. Lenz, *Adv. Polym. Sci. 107*, 1 (1993)
2. G. Swift, *FEMS Microbiol. Rev. 103*, 339 (1992)
3. P.A. Holmes, *Phys. Technol. 16*, 32 (1985)
4. A. Schirmer and D. Jendrossek, *Appl. Environ. Microbiol. 59*, (4), 1220 (1993)
5. H. Brandl and P. Püchner, *Biodegradation 2*, 237 (1992)
6. S. Fidler and D. Dennis, *FEMS Microbiol. Rev. 103*, 231 (1992)
7. A. Steinbuchel, In *Biomaterials: Novel Materials from Biological Sources*. D. Byrom (Ed.) (1991) Stockton Press, New York, pp. 125–213
8. A.J. Anderson and E.A. Dawes, *Microbiol. Rev. 54*, (4), 450 (1990)
9. H. Brandl, R.A. Gross, R.W. Lenz, and R.C. Fuller, *Adv. Biochem. Eng./Biotechnol. 41*, 77 (1990)
10. P.A. Holmes, In *Development in Crystalline Polymers—2*. D.C. Bassett (Ed.) (1988) Elsevier Applied Science, New York, pp. 1–65
11. P.A. Holmes, L.F. Wright, and S.H. Collins, European Patent Appl. EP 52459 (1981), European Patent Appl. EP 69497 (1983)
12. R.H. Marchessault, C.J. Monasterios, F.G. Morin, and P.R. Sundararajan, *Int. J. Biol. Macromol. 12*, 158 (1990)
13. K.D. Gagnon, R.C. Fuller, R.W. Lenz, and R.J. Farris, *Rubber Chem. Technol. 65*, 4 (1992)
14. K.D. Gagnon, R.C. Fuller, R.W. Lenz, and R.J. Farris, *Rubber World 207*, 32, November (1992)
15. M. Lemoigne, *Ann. Inst. Pastuer (Paris) 39*, 144 (1925)
16. M. Lee and A.C. Chandler, *J. Bacteriol. 41*, 373 (1941)
17. J.N. Baptist (to W.R. Grace & Co.), U.S. Patent 3,044,942 91962, U.S. Patent 3,036,959 (1962)
18. L.L. Wallen and W.K. Rohwedder, *Environ. Sci. Technol. 8*, 576 (1974)
19. P.P. King, *J. Chem. Tech. Biotechnol. 32*, 2 (1982)
20. M-J. DeSmet, J. Kingma, H. Wynberg, and B. Witholt, *Enzyme Microb. Technol. 5*, 352 (1983)
21. R.G. Lageveen, G.W. Huisman, H. Preusting, P. Ketelaar, G. Eggink, and B. Witholt, *Appl. Environ. Microbiol. 54* (12), 2924 (1988)
22. H. Brandl, R. Gross, R.W. Lenz, and R.C. Fuller, *Appl. Environ. Microbiol. 54* (8), 1977 (1988)
23. R. Gross, C. DeMello, R.W. Lenz, H. Brandl, and R.C. Fuller, R., *Macromolecules 22*, 1106 (1989)
24. G. Huisman, O. Leeuw, G. Eggink, and B. Witholt, *Appl. Environ. Microbiol. 55*, 8, 1949 (1989)
25. H. Preusting, A. Nijenhuis, and B. Witholt, *Macromolecules 23*, 4220 (1990)
26. K.D. Gagnon, D.B. Bain, R.W. Lenz, and R.C. Fuller, In *Novel Biodegradable Microbial Polymers*. E.A. Dawes (Ed.) (1990) Kluwer, The Netherlands, pp. 449–450
27. E.J. Knee, Jr., M. Wolf, R.W. Lenz, and R.C. Fuller, In *Novel Biodegradable Microbial Polymers*. E.A. Dawes (Ed.), Kluwer, The Netherlands, p. 439.
28. B.A. Ramsay, I. Saracovan, J.A. Ramsay, and R.H. Marchessault, *Appl. Environ. Microbiol. 57* (3), 625 (1991)
29. H. Preusting, J. Kigma, and B. Witholt, *Enzyme Microb. Technol. 13*, 770 (1991)
30. K.D. Gagnon, R.C. Fuller, R.W. Lenz, and R.J. Farris, *Macromolecules 25*, (14), 3723 (1992)
31. E.A. Dawes and P.J. Senior, In *Advances in Microbial Physiology, Vol. 10*. A.H. Rose and D.W. Tempest (Eds) (1973) Academic Press, New York, pp. 135–266
32. A. Ballistreti, G. Montaudo, G. Impallomeni, R.W. Lenz, Y.B. Kim, and R.C. Fuller, *Macromolecules 23*, 5059 (1990)
33. J.R. Shelton, D.E. Agostini, and J.B. Lando, *J. Polym. Sci. Pt. A1 9*, 2789 (1971)
34. N. Tanahashi and Y. Doi, *Macromolecules 24*, 5732 (1991)
35. R.W. Lenz, Y.B. Kim, and R.C. Fuller, *J. Bioactive Comp. Polym. 6*, 392 (1991)

36. K. Fritzsche, R.W. Lenz, and R.C. Fuller, *Int. J. Biol. Macromol. 12*, 92 (1990)
37. Y.B. Kim, R.W. Lenz, and R.C. Fuller, *Macromolecules 24*, 5256 (1991)
38. K. Fritzsche, R.W. Lenz, and R.C. Fuller, *Butterworths-IJBM BEP7 (a)*—Issue LH, April (1990)
39. Y. Doi, Yoshiharu, *Microbial Polyesters* (1990) VCH Publishers, New York
40. G.W. Haywood, A.J. Anderson, and E.A. Dawes, *Biotechnol. Lett. 11*, 471 (1989)
41. G.W. Haywood, A.J. Anderson, D.F. Ewing, and E.A. Dawes, *Appl. Environ. Microbiol. 56*, 3354 (1990)
42. A. Timm and A. Steinbüchel, *Appl. Environ. Microbiol. 56*, 3360 (1990)
43. B.A. Ramsay, I. Saracovan, J.A. Ramsay, and R.H. Marchessault, *Appl. Environ. Microbiol. 58*, 74 (1992)
44. D. Byrom, *Trends Biotechnol. 5*, 246 (1987)
45. E. Berger, B.A. Ramsay, J.A. Ramsay, and C. Charvarie, *Biotech. Technol. 3*, (4), 227 (1989)
46. N. Grassie, E.J. Murray, and P.A. Holmes, *Polym. Degrad. Stab. 6*, 95 (1984)
47. M. Kunioka and Y. Doi, *Macromolecules 23*, 1933 (1990)
48. C.S. Schollenberger and K. Dinbergs, In *Advances in Urethane Science and Technology*. K.C. Frisch and S.L. Reegen (Eds.) (1979) Technomic, CT, p. 3
49. Y. Inoue and N. Yoshie, *Prog. Polym. Sci. 17*, 571 (1992)
50. K.D. Gagnon, Thesis Dissertation, University of Massachusetts, 1993
51. C.J. Lustyand and M. Doudoroff, *Proc. Natl. Acad. Sci. USA 56*, 960 (1966)
52. T. Tanio, T. Fukui, Y. Shirakura, T. Saito, T. Komita, T. Kaiho, and S. Masamune, *Eur. J. Biochem. 124*, 71 (1982)
53. M.M. Nir, *Plastics Eng. 46*, (9), 29 (1990)
54. T. Scherer, Personal communication, University of Massachusetts (1993)
55. S.C. Slater, W.H. Voige, and D.E. Dennis, *J. Bacteriol. 170*, 4431 (1988)
56. P. Schubert, A. Steinbüchel, and H.G. Schlegel, *J. Bacteriol. 170*, 5837 (1988)
57. O. Peoples and A.J. Sinskey, *J. Biol. Chem. 264*, 15298 (1989)
58. A. Steinbüchel and H.G. Schlegel, *Mol. Microbiol. 5*, 535 (1991)
59. O. Peoples and A.J. Sinskey, *J. Biol. Chem. 264*, 15293 (1989)
60. B. Janes, J. Hollar, and D. Dennis, In *Novel Biodegradable Microbial Polyesters*. E.A. Dawes (Ed.) (1990) Kluwer, Dordrecht, p. 175
61. A. Timm, D. Byrom, and A. Steinbüchel, *Appl. Microbiol. Biotechnol. 33*, 296 (1990)
62. H. Preusting, J. Kingma, G. Huisman, A. Steinbüchel, and B. Witholt, *J. Environ. Polym. Degrad. 1*, 11 (1993)
63. Y. Poirier, D.E. Dennis, K. Klomparens, and C. Somerville, *Science 256*, 520 (1992)
64. R. Pool, *Res. News* 1187, 15 September (1989)
65. M. Yalpini, R.H. Marchessault, F.G. Morin, and C.J. Monasterios, *Macromolecules 24* (22), 6046 (1991)
66. M.S. Reeve, S.P. McCarthy, and R.A. Gross, *Macromolecules 26*, 888 (1993)

15C.1   Introduction

Polymer blends or alloys can represent a useful way to combine the desirable features of different materials into a product, and a route that avoids the synthesis of new molecules is attractive. However, most polymer pairs are thermodynamically immiscible and, when mixed, form separate phases comprised essentially of the pure components [1–3]. This situation arises because the interaction between the two types of segments is usually energetically unfavorable and the entropy of mixing such large molecules is so small that there is no net free energy driving force for miscibility. The fact that separate phases are formed is by itself not generally an obstacle to generating useful products from such blends; in fact a multiphase structure is often an asset. However, the nature of the interface between the phases can be quite a problem. As the segmental interaction energy becomes more unfavorable for mixing, the segmental interpenetration between the phases decreases and interfacial tension increases [4]. A large interfacial tension makes it difficult to achieve a fine dispersion of the phases during melt mixing and the morphology attained is somewhat unstable [5–7]. A lack of segmental mixing at the interface translates into relatively poor adhesion between the phases and inferior mechanical behavior of the mixture in the solid state [8, 9]. Thus, the lack of control of morphology and interfacial strength often precludes simple blending of polymer pairs that in principle could lead to interesting property combinations. The use of block copolymers in blends offers unique opportunities to take advantage of the benefits of multiphase polymer systems while solving or avoiding the interfacial problems mentioned above. This is because block copolymers have covalent bonds that bridge the separate phases that each segment type forms. This leads to inherently strong interfaces and to some unique avenues for morphology control.

This chapter reviews two different, but often confused, concepts related to polymer blends containing a block copolymer component. These ideas are compared and contrasted in the schematic illustrations shown in Fig. 15C.1. The one on the left involves what will be referred to here as "solubilization." The idea is to modify a block copolymer, for example, an A–B diblock, by dissolving a homopolymer, H, into the microdomains formed by the segments of A. The schematic on the right illustrates the concept that is referred to here as "compatibilization." In this case a diblock block copolymer is added to a phase-separated

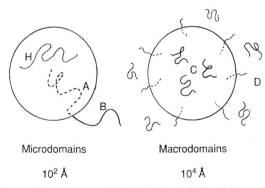

Microdomains   Macrodomains

10^2 Å 10^4 Å

Figure 15C.1   Comparison of the concepts of "solubilization" (left) and "compatibilization" (right)

blend of polymers C and D. If properly designed, the block copolymer will locate at the interface between C and D and solve the "incompatibility" problem mentioned above by lowering interfacial tension and increasing adhesion between the phases. The two concepts differ in many ways, including the size scale of domains. Typically the "macrodomains" of blends are two orders of magnitude larger than block copolymer "microdomains."

The following sections develop in more detail the main concepts of solubilization and compatibilization using as examples studies from the recent literature that concern blends containing styrene–hydrogenated butadiene. A wide variety of such copolymers are commercially available and they are useful for a broad range of applications including blended products.

15C.2 Solubilization

As discussed extensively in previous chapters, the segments of block copolymers segregate into separate phases if the segments are sufficiently long and the energy of mixing the unlike segments is endothermic. The microdomains formed must be comparable in size to the coils of the segments they contain (see left side of Fig. 15C.1). It is possible to dissolve other molecules into these microdomains, causing them to expand. However, the small sizes of these phases and the strong tendency for block copolymer segments to distribute uniformly greatly affects the amount of homopolymer chains that can dissolve in these microdomains. A great deal of research has been concerned with blending homopolymer polystyrene with styrene-based block copolymers and with the question of whether the homopolymer mixes with the polystyrene microdomains or forms separate phases [10–17]. In such blends, solubilization into the microdomain can be driven only by a small gain in combinatorial entropy but is opposed by other physical restraints [18–20]. As a result, it has been found that the homopolymers must generally have molecular weights less than those of the corresponding copolymer blocks to form a mixed polystyrene phase [10–13]. A simple thermodynamic model [20] that accounts for the entropy of (1) mixing the homopolymer and the copolymer styrene segments, (2) stretching the copolymer segments as the microdomain expands, and (3) any conformational rearrangement of the homopolymer coil to allow it to fit into the limited space of the microdomains has been developed. The amount of homopolymer of molecular weight M_H that can be solubilized into the lamellar A domains of an A–B block copolymer when the A segments have molecular weight M_A is shown in Fig. 15C.2. Here $(\phi_H)_s$ is the volume fraction of H that saturates the A domain. The model prediction parallels the experimental observations. The extent of homopolymer solubilization is very small unless the condition $M_H \ll M_A$ is satisfied. The trend predicted by this simple model is very similar to that given by a somewhat more complex model given by Meier [18].

When polystyrene is blended with thermoplastic elastomers (TPEs) of the type S–B–S or S–EB–S, there will be an increase in the hard phase volume where the end blocks reside if solubilization occurs but the properties of this phase, for example, softening point, will not be significantly altered. On the other hand, addition of a different polymer having a higher glass transition temperature, T_g, than polystyrene could raise the softening point of the hard phase and extend the upper end of the temperature range for TPE type behavior. Since poly(2,6-

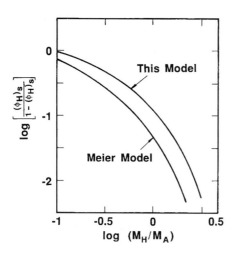

Figure 15C.2   Relative volume of homopolymer (H) solubilized as a function of molecular weight ratio (homopolymer to A block of copolymer) computed from model described in text and from model by Meier. ϕ_H = volume fraction of homopolymer in A domain at thermodynamic saturation. (Reproduced from Ref. [20], with permission of American Chemical Society)

dimethyl-1,4-phenylene oxide) (PPO) is miscible with polystyrene [21], it is a candidate for blending with S–B–S or S–EB–S type copolymers to increase the T_g of the hard phase. The basis for miscibility of polystyrene and PPO is an exothermic heat of mixing [20], which introduces a new driving force for mixing of the styrene end blocks that does not exist in the case of PS homopolymer. Nevertheless the nature of the microdomains can restrict the extent to which PPO homopolymer mixes with the polystyrene end blocks as shown next.

The fact that PPO and polystyrene differ considerably in T_g provides a useful way to determine the extent of solubilization in these mixtures [22, 23]. Figure 15C.3 compares the T_g behavior of blends of two S–EB–S block copolymers with a series of PPO materials [23] having weight average molecular weights ranging from 24,000 to 39,000. The T_g of the hard phase was measured by differential scanning calorimetry (DCS) and is plotted versus the mass fraction of PPO in the blends on a rubber-free basis. On the left, the copolymer has styrene end blocks with molecular weight of 29,000. In this case there is a single hard phase T_g that increases continuously as PPO is added, up to the value for pure PPO. Thus, there is no limit to how much PPO can be solubilized in this case. Of course, the microdomain morphology must change considerably as PPO is added to the hard phase [16, 17, 23]. Figure 15C.4 shows the increase in the size of the hard phase of a similar S–B–S copolymer with added PPO [23]. The line through the data points was calculated assuming the added PPO isotropically swelled the hard phase domain. The right hand side of Fig. 15C.3 shows the T_g for PPO blends with an S–EB–S block copolymer with an end block molecular weight of 10,000. For these smaller microdomains, there is a limit to how much PPO can be solubilized. In this case, the hard phase T_g increases as PPO is added up to the solubilization limit $(\phi_H)_s$ after which two hard phase T_gs are seen. One corresponds to nearly pure PPO, while the other corresponds to the PPO-saturated microdomains.

Figure 15C.5 summarizes the saturation limits found for blends of various PPO materials with a series of styrenic block copolymers [20, 23]. The molecular weights of the PPO or the copolymer midblock do not appear to be significant factors; however, the molecular weight of the polystyrene end block is. When the latter molecular weight is 15,000 or larger there seems

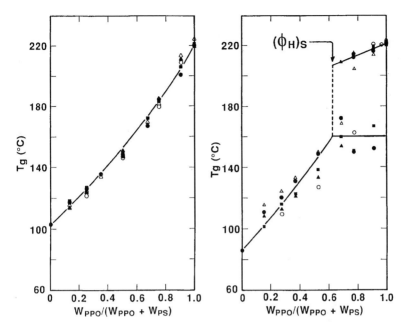

Figure 15C.3   Hard phase T_g by DSC for blends of PPO with S–EB–S block copolymers having polystyrene blocks with molecular weights of 29,000 (left) and 10,000 (right) plotted vs. PPO content on rubber-free basis. Onset of two T_gs defines solubilization limit $(\phi_H)_s$

to be no limit to the extent that PPO can be solubilized but there is a strong decrease in $(\phi_H)_s$ as the end block molecular weight goes below this.

A complete thermodynamic theory for the phase behavior of blends of a homopolymer, H, with block copolymers of the type A–B or A–B–A must describe a number of complex possibilities [24–27]. Various formulations of this problem have appeared in the recent

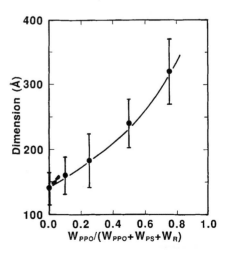

Figure 15C.4   Average minimum characteristic dimension (points) of glassy phase obtained from analysis of electron photomicrographs of PPO blends with a triblock copolymer. Line represents calculated dimension assuming domains swell isotropically. Subscript R denotes rubber phase of midblock segments. (Reproduced from Ref. [23] with permission of American Chemical Society)

literature [18, 19, 24–27]. These focus mainly on the special case where H is chemically identical with one of the blocks of the copolymer, for example, A. The latter restriction amounts to the assumption that H and A mix athermally, and thus ignores the consequences on blend phase behavior of any favorable energetic interaction between them. For the blends of interest here (i.e., H = PPO and A = polystyrene), there is an exothermic interaction between the two types of segments that is an important factor driving miscibility of blends of the corresponding homopolymers. Because of the mathematical complexities of the theoretical formulations mentioned above, it is difficult to gain from them much insight about how a favorable interaction would alter the extent that H would be incorporated into domains of A for phase-separated block copolymers. However, a simple physical picture and model that identifies some of the main factors and gives a semiquantitative assessment of their relative importance has been developed [20]. This approach envisions preexisting domains of A segments and examines some of the thermodynamic consequences of continued addition of H molecules to this phase, including combinatorial entropy of mixing, heat of mixing, conformational extension of A blocks, and conformational compression of H chains. Any contributions from the B phase or the interface (interphase) are ignored; hence, the size of these blocks or their architecture, A–B, A–B–A, etc. has no direct role in the model. The approach used has some similarities with the swelling of polymer networks by solvents. While the model is very simplistic in many respects, it does predict the trends seen experimentally in a semiquantitative manner.

In this model, it is assumed that the heat of mixing of A and H segments (or more precisely the excess free energy of mixing) is given by the simple parabolic form

$$\Delta H_{mix} = B\phi_A \phi_H \qquad (15C.1)$$

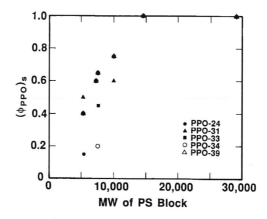

Figure 15C.5 Experimentally determined saturation concentration of PPO in polystyrene microphase as a function of the molecular weight of the polystyrene block of the copolymer. Numbers following PPO in the legend correspond to the nominal weight-average molecular weight of the homopolymer in thousands. Data were determined from glass transition behavior of blends. (Reproduced from Ref. [20], with permission of American Chemical Society)

where $B=$ the interaction energy density and the ϕ_A and ϕ_H are volume fractions of the indicated segments in the mixed domain. The estimates of B for PPO–polystyrene mixtures vary considerably [20] but an appropriate value is probably of the order of -1 cal/cm^3. The volume fraction of homopolymer predicted by this model [20] to be dissolved in the microdomains of A segments at saturation is shown in Fig. 15C.6. The results are shown for the case where the molecular weight of the A segment M_A is increased while the homopolymer molecular weight is fixed ($M_H = 15,000$). The level of solubilization $(\phi_H)_s$ is increased dramatically going from an athermal case ($B=0$ for H = polystyrene and A = polystyrene) to one where there is a exothermic or favorable heat of mixing as in the case H = PPO, A = polystyrene. For a value of B representative of the latter case, the degree of solubilization initially increases rapidly as M_A increases and then seems to level off at high block lengths. This trend is quite similar in shape to what was observed experimentally (see Fig. 15C.5). The experimental results suggest that at high styrene block molecular weights $(\phi_H)_s$ can approach unity. On the other hand, the model predicts for all finite values of B and molecular weights, that $(\phi_H)_s$ is always less than 1, implying that a separate homopolymer phase must form when enough of this component is added to the copolymer. This limit on solubility is actually the result of the assumption imposed in the model that a lamellar morphology exists for all values of ϕ_H. For real blends at high ϕ_H, the mixed phase will become the continuous one with B domains (having convex surfaces) dispersed in it. This will provide significant relaxation of the conformational constraints inherent in the lamellar morphology as ϕ_H approaches unity. Modeling this situation would be very complex and is beyond the scope of the current approach.

The important issue here is the experimental demonstration and theoretical confirmation that the early notions of solubilization in blend copolymer microdomains and the severe limitations of homopolymer molecular weight are applicable only when there is athermal mixing. These rules no longer apply when there is a favorable heat of mixing between H and A. In this case M_H is not a major factor [20]. High levels of solubilization can be obtained,

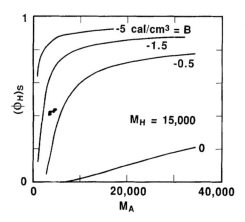

Figure 15C.6   Effect of copolymer segment molecular weight and interaction parameter on saturation concentration when $M_H = 15,000$. (Reproduced from Ref. [20], with permission of American Chemical Society)

particularly as the microdomains get larger (larger M_A). Manifestations of this effect have been reported for other situations [28–31] and systems [32]. This concept can be quite useful for improving the upper use temperature of TPEs.

15C.3   Compatibilization

For some time there has been interest in blend additives that would circumvent the poor mechanical properties that result when most immiscible polymers are blended [33]. Since the goal for such additives is to solve the problem of "mechanical incompatibility," these components are then logically referred to as "compatibilizers" [8, 9, 33]. Ideally compatibilizers should solve the problem of poor interfacial strength and high interfacial tension that exist between domains of polymers that have rather unfavorable energetic interactions between their segments [4].

Early work by Molau [34–37], Riess [12, 38–43], and others [11, 44–46] firmly established the fact that appropriately designed block or graft copolymers act as interfacial agents or emulsifiers for immiscible polymer mixtures. Ideally, a block copolymer will locate at the interface between the two phases as shown on the right side of Fig. 15C.1 with its corresponding segments being associated with the appropriate phase. Since the block copolymer molecule traverses the interface, one can expect greatly improved adhesion between the phases owing to the strength of the covalent chain of this interfacial agent. Graft copolymers might be expected to function similarly. The first comprehensive review of this proposition was published in 1978 [8], and a great deal has been written on the subject since then. The purpose here is to present a focused review of the use of styrene–hydrogenated butadiene block copolymers as the compatibilizing agent. The main body of experimental work considered here was developed in the laboratories of Teyssie [47–60], Heikens [61–65], and the author [66–71] although a few other studies [72–74] presenting similar findings have been published recently. For easy reference, Table 15C.1 summarizes the various block copolymers used in these publications. To facilitate reference to the original work, the copolymers are identified here by the designations used in these papers. Appropriate characterizing information as available has been included in Table 15C.1. The block structures include simple diblocks and triblocks plus some so-called tapered blocks. By control of the microstructure of the butadiene sequences during polymerization one can achieve after hydrogenation olefinlike segments that are essentially comparable to high-density polyethylene (HDPE), low-density polyethylene (LDPE), or ethylene–butene random copolymers. This feature is quantified by the number of ethyl groups per 1000 carbon atoms in the hydrogenated butadiene segment, and controls the level of crystallinity of this phase.

For the most part, we will be concerned with the addition of these block copolymers to mixtures of polystyrene with various polyolefins including LDPE, HDPE, and polypropylene. Of course, the polystyrene segment of these block copolymers should be capable of mixing with the homopolymer polystyrene while the olefinlike segment, depending on its structure, should be at least somewhat compatible with the various homopolymer polyolefins; although, in many cases complete miscibility may not exist [69, 75, 76]. In one example the polystyrene phase also contains a polymer that is closely related to PPO and that forms miscible blends

Table 15C.1 Description of Styrene–Hydrogenated Butadiene Block Copolymers Used as Compatibilizers

Designation	Composition	Block structure	Nature of hydrogenated butadiene block
SE-2 [47, 49] (Teyssie)	Equimolar S–B	S–Bh $M_n = 155,000$	Assumed to be 30 ethyl/1000C
SE-4 [49] (Teyssie)	Equimolar S–B	S–(S–Bh)–Bh tapered block 26,000–22,000–32,000	Assumed to be 30 ethyl/1000C
SE-5 [49] (Teyssie)	Equimolar S–B	S–Bh $M_n = 80,000$	Assumed to be 30 ethyl/1000C
SE-7 [50] (Teyssie)	Equimolar S–B	S–(S–Bh)–Bh tapered block 22,000–22,000–26,000	Assumed to be 30 ethyl/1000C
H-7 [50] (Teyssie)	—	S–Bh 10,400–39,000	5 Ethyl/1000C
None [48] (Teyssie)	—	S–Bh 75,000–80,000	30 Ethyl/1000C
None [65] (Heikens)	—	S–(S–Bh)–Bh tapered block 22,000–22,000–25,000	Unknown
S–EB–S [66–68, 70, 71] (Kraton G1652 Elastomer® from Shell)	29% S by wt.	S–Bh–S triblock 7,200–64,000–7,200	Equivalent to random ethylene butene-copolymer (essentially amorphous)

S, styrene; B, butadiene; Bh, hydrogenated butadiene units

with polystyrene. For comparison, another example is included where the block copolymer segments are not miscible with either of the two homopolymers, HDPE and poly(ethylene terephthalate) (PET), in which case the model shown in Fig. 15C.1 is clearly not applicable.

We begin by briefly reviewing some of the fundamental considerations applicable to the ideal notion embodied in the concept shown on the right side of Fig. 15C.1 and follow this with an extensive presentation of the efficacy of these block copolymers to improve mechanical properties. We conclude by considering the available experimental information related to the effect such additives have on phase morphology, interfacial adhesion, and deformation mechanisms.

15C.3.1 Fundamental Considerations

Optimal design of block copolymers for compatibilizing immiscible polymer blends requires consideration of a variety of issues beyond the intuitive notions embodied in Fig. 15C.1 [8]. Under what circumstances will the block copolymer actually become a surface-active agent? How long must the segments of the copolymer be in order not to simply slip from the homopolymer phases when the interface is stressed? Theories for such complex systems are being developed, but they are not yet able to answer such questions in detail. However, it is possible to understand some of the basic requirements [8].

As demonstrated earlier in this book, the segments of block copolymers will segregate into separate phases if they are sufficiently long and if the energy of mixing the unlike segments is endothermic. For the block copolymer to locate at the blend interface, it should have this propensity to segregate into two phases. Furthermore, the block copolymer as a whole should not be miscible in one of the homopolymer phases. Both these characteristics depend on segmental interactions and molecular weights. The question of anchoring copolymer segments into homopolymer phases is a rheological one and would seem to be assured if the copolymer segments are sufficiently long to be entangled with surrounding chains. Indeed, recent investigations [77–82] using a variety of systems and techniques show this to be the case.

Early considerations of ternary mixtures of homopolymer A–homopolymer B–block copolymer A–B were concerned with the issues of "solubilization" of the homopolymers into the preexisting domains of the block copolymer. That is, given a phase-separated block copolymer, will small amounts of added homopolymers enter these domains or form separate phases? As shown earlier, significant solubilization can be expected only if the homopolymer molecular weight is comparable to or less than that of the corresponding segments of the copolymer [11, 35–38, 40, 44–46, 83–85]. This forecasts a rather pessimistic view for efficient compatibilization of practical blend systems since the segments of block copolymers are usually relatively short (typical segmental molecular weights range from under 10,000 to not more than 100,000, as seen in Table 15C.1). These molecular weights can be compared to those of industrially important homopolymers, for example, typical commercial polystyrenes have molecular weights of the order of several hundred thousands. However, the notions of solubilization of homopolymers into block copolymer domains are not entirely relevant to the question of compatibilization. The former is concerned with the perturbation of a copolymer morphology by adding small amounts of homopolymers, whereas the latter is concerned with the perturbation of a blend of homopolymers by adding small amounts of copolymer. In other words, the real question is not the preservation of a block copolymer domain morphology or the accommodation of homopolymers into this structure, but if the block copolymer will locate at the interface between homopolymer domains when added to this blend. Recent theories [86–92] have addressed this question. The weight of the available theoretical and experimental evidence suggests that regardless of relative molecular weights, in the ideal case block copolymers do generally locate at the interface, lower the interfacial tension, and reduce the size of the homopolymer domains as expected of an emulsifier. Molecular weights do play some role in the efficacy of these processes. Focused theoretical considerations of the general type mentioned above can be extremely helpful in answering important questions about compatibilizer design. However, the results presented subsequently clearly show that the concept shown on the right in Fig. 15C.1 is a very idealized one, and many useful compatibilizers do not function in this way.

15C.3.2 Mechanical Properties

For the reasons outlined earlier, blends of immiscible polymers usually exhibit rather inferior mechanical properties relative to what might be hoped for based on the properties of the individual components comprising the blend. By this, we mean that when plotted versus composition, individual properties of a blend (such as impact strength) fall well below any expected additive behavior such as a simple tie line connecting the values for the pure components. This is especially true for failure properties and particularly those related to ductility of the material (e.g., elongation at break). Small deformation properties such as modulus are not usually very sensitive of the degree to interfacial adhesion between the components of the blend. Thus, in this section we will focus primarily on failure properties for a variety of immiscible blend systems, which in most cases will consist of blends of polystyrene and various polyolefins. These blends have served as useful model systems in many laboratories because the block copolymers required are commercially available or can be synthesized easily. In addition, compatibilized blends of polystyrene and polyethylene have been commercialized in recent years primarily because of their attractive features for certain packaging applications [74].

To judge the efficacy of block copolymers as compatibilizing agents, it is appropriate to examine how adding these materials to a blend of polymers improves the relationships between various mechanical properties and blend composition. In an ideal sense, an effective compatibilizer would raise a particular property from some low value up to one more closely representing an additive value. However, in a practical sense, this type of comparison is complicated by what effects addition of the compatibilizer to the individual blend components will have on their properties. For example, addition of most of the block copolymers given in Table 15C.1 to polystyrene will result in a material that is less stiff and strong but more ductile. Thus, in judging the effects of compatibilization it is necessary to consider these issues, which is usually done by plotting the ternary blend data on the same coordinates as the original binary blend, using a compatibilizer free basis for the composition coordinate and with various property curves for different levels of compatibilizer. In this scheme, zero percent compatibilizer indicates the property curve for the binary mixture.

15C.3.2.1 Polystyrene–Low-Density Polyethylene

Blends of LDPE and polystyrene with various block copolymers have been studied extensively in the laboratories of Teyssie [47, 48] and Heikens [61–64]. Teyssie's work concentrates on failure properties obtained from tensile stress–strain diagrams and will be summarized here; Heikens' deals with a wider range of issues such as adhesion, dilatometric behavior, dynamic mechanical characteristics, and modeling, which will be discussed in part in a later section.

The studies by Teyssie et al. compared effectiveness of the block copolymers identified as SE-2, SE-4, and SE-5 in Table 15C.1 for compatibilizing LDPE with polystyrene. Some key results from this work are reproduced here as Fig. 15C.7. The lower curves on the left and right illustrate the well below additive response of ultimate tensile strength and percent elongation at break for the polystyrene–LDPE binary blends and are typical of other data that have been reported for this system [33, 93–96] and similar immiscible blends [97, 98]. The upper curves show how these block copolymers improve strength and ductility when added to the binary blend. [Note: The original publications show these curves to go through the points for pure polystyrene and pure LDPE. This cannot be the case, as pointed out earlier. For this presentation, the curves have been redrawn accordingly, and, unfortunately, we do not know these properties for polystyrene and LDPE containing 9% block copolymer.] On an absolute scale, the improvements in both strength and ductility caused by adding these copolymers is impressive. The high molecular weight diblock copolymer, SE-2, causes the most improvement in elongation at break but the least improvement in strength. SE-4 and SE-5 have the same molecular weights but the former is a tapered block while the latter is a simple diblock. The tapered block gives slightly better strength and elongation at break than the diblock.

Teyssie et al. [47] conclude that block copolymers are far more effective as compatibilizers for this blend system than graft copolymers that had been used previously [94, 95]. These authors attribute the slightly better performance of the tapered block to a "graded"

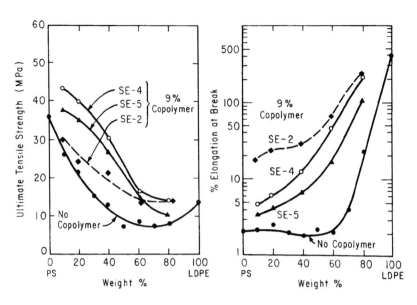

Figure 15C.7   Effect of addition of block copolymers on mechanical properties of polystyrene–low-density polyethylene blends. (Reproduced from Ref. [49], with permission of John Wiley & Sons, Inc.)

modulus profile that it is likely produced at the domain interface. The hydrogenated butadiene segments of their block copolymers apparently have microstructures (see Table 15C.1) quite similar to LDPE. It would be interesting to know about the crystallinity of the hydrogenated butadiene phase and the mechanical properties of the block copolymers to further analyze the interesting data presented.

15C.3.2.2   Polystyrene–High-Density Polyethylene

Blends of HDPE and polystyrene compatibilized with block copolymers have been studied in the laboratories of Teyssie [48, 50] and the author [66]. The former employed the copolymers SE-7, H-7, and the undesignated entry below these in Table 15C.1 while the latter employed a commercial block copolymer, Kraton® G1652 thermoplastic rubber. This is a poly(styrene–block-ethylene–co-butylene–block-styrene) and will be denoted as S–EB–S. It is marketed as part of Shell's family of TPEs. Ideally one could hope to compare the effectiveness of the quite diverse structural characteristics of these copolymers for compatibilization, but differences in the design of experiments in the two laboratories compromise such possibilities. The Teyssie studies were done on blends compounded on a two-roll mill and compression molded into sheet, whereas our blends were compounded in an extruder and injection molded into test bars. These choices reflect, in part, the differences in options when one works with laboratory quantities of specially synthesized compatibilizers versus materials available in commercial quantities. These processing differences sufficiently influence mechanical properties so that direct comparisons of properties cannot be made on an absolute basis. In view of this, our laboratory has subsequently examined the influence of processing methods and conditions on blend properties for some other systems [67, 68] as will be described in part later.

We will present some quantitative results from the rather extensive study by Lindsey in our laboratory [66] and make comparisons and contrasts where possible with the work reported by Teyssie et al. Figure 15C.8 shows the yield strength and modulus for HDPE–polystyrene binary blends and for ternary mixtures containing 20% (based on total blend mass) of the S–EB–S copolymer irrespective of the HDPE/polystyrene ratio. As expected, the yield strength of the binary is below the additivity level defined by a simple tie line, but the departure from additivity is not as severe as seen for roll milled/compression molded mixtures reported by Teyssie [48, 50]. The modulus for these blends shows a simple trend more closely approximating additive behavior. Since S–EB–S has lower strength and modulus than either HDPE or polystyrene, its addition to mixtures of HDPE and polystyrene lowers both properties in a rather systematic manner. In contrast, the diblock copolymers used by Teyssie et al. cause increases in blend strength—no modulus data were reported. This difference is probably due to the nature of the hydrogenated butadiene segments of these materials (which are undoubtedly crystalline) compared to the elastomeric nature of the EB midblock. These mechanical property differences in the block copolymers are translated into the corresponding characteristics of the ternary blends. Figure 15C.9 shows the elongation at break of the compositions shown in Fig. 15C.8. The severe problems of incompatibility of HDPE and polystyrene are quite evident for the binary; however, the presence of the S–EB–S significantly improves this property for all ratios of HDPE to polystyrene. Figure 15C.10 shows for an equal weight ratio of HDPE and polystyrene that ductility continues to improve as the amount of S–EB–S added increases, although a tendency to level off is apparent. The 50 : 50 binary blend prepared by Teyssie et al. has significantly lower elongation at break than those shown in Figs. 15C.8 and 15C.9 owing to differences in processing; however, as a rough guide we can compare the factor by which this property is improved by addition of block copolymers for the two studies. Addition of 9% of the various diblocks resulted in no improvement for H-7, about a factor of 2 improvement for SE-7, and about a factor of 10 improvement for the undesignated diblock. The same amount of S–EB–S increased the elongation at break by about a factor of 4.5 in our studies. The principal advantage of diblocks in this case is the retention of strength whereas S–EB–S causes a loss as seen further in Fig. 15C.11. Because of the many differences among the three diblocks in this example, it is difficult to draw any conclusions about the relative importance of overall molecular weight, segment length, microstructure, or tapering.

A quite serious problem for multiphase blends is weakness at weldlines in moldings, which originates from several mechanisms [98]. Figure 15C.12 illustrates this with strength data for test bars made by injection molding. These were gated at both ends of the bar to give a weldline in the gage section. The lines in this figure were computed from a simple model [98] allowing for the phase mismatch at the weldline. The point here is that addition of a compatibilizer does little to help this problem in the present case.

15C.3.2.3 (Polyether Copolymer–Polystyrene)–High-Density Polyethylene

Schwarz et al. [70, 71] extended the scope of studies on S–EB–S compatibilized blends of polystyrene and HDPE by incorporating an additional polymer, polyether copolymer (PEC). PEC is structurally similar to PPO except for the random incorporation of approximately 5%

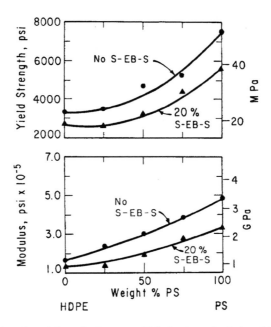

Figure 15C.8   Strength and modulus of polystyrene–high-density polyethylene blends with and without added block copolymer. (Reproduced from Ref. [66], with permission of John Wiley & Sons, Inc.)

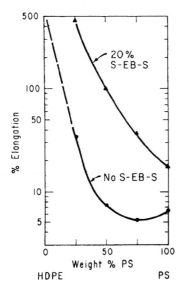

Figure 15C.9   Elongation at break of polystyrene–high-density polyethylene with and without added block copolymer. (Reproduced from Ref. [66], with permission of John Wiley & Sons, Inc.)

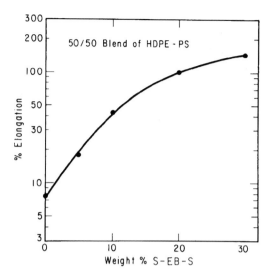

Figure 15C.10   Effect of block copolymer on elongation at break for a 50 : 50 blend of HDPE–polystyrene. (Reproduced from Ref. [66], with permission of John Wiley & Sons, Inc.)

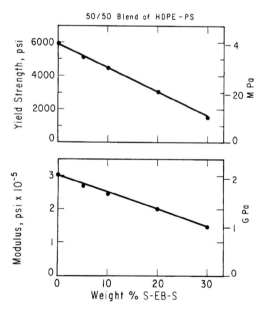

Figure 15C.11   Effect of block copolymer on strength and modulus of a 50 : 50 blend of HDPE–polystyrene. (Reproduced from Ref. [66], with permission of John Wiley & Sons, Inc.)

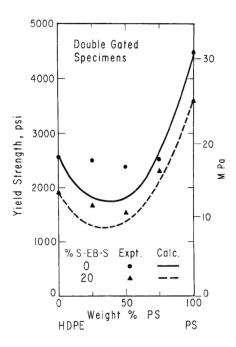

Figure 15C.12   Strength of HDPE–polystyrene blends injection molded to have a weld line. (Reproduced from Ref. [66], with permission of John Wiley & Sons, Inc.)

of trimethyl phenol into the backbone. PEC has virtually the same properties as PPO, including complete miscibility with polystyrene over the full range of compositions. In blends with HDPE, PEC–polystyrene blends form a single glassy phase having a glass transition temperature governed by the relative proportions of PEC and polystyrene. In these blends, the HDPE exists, of course, as a separate phase with a melting point of approximately 145 °C. Because of the high T_g of PEC such blends may have higher use temperatures. An additional and potentially beneficial factor is invoked when the glassy phase of the blend is a mixture of polystyrene and PEC. Any mixing of polystyrene end blocks and polystyrene homopolymer can be driven only by entropic forces since the mixing process is athermal. On the other hand, the miscibility of polystyrene with PPO and PEC is based on an exothermic heat of mixing. This provides an additional driving force for styrene end blocks to mix with a phase containing PEC. As a result, the compatibilizing effect of such copolymers may be better when PEC is present in the glassy phase of these blends rather than pure polystyrene.

The PEC–polystyrene mixtures are designated by the following notation: 80 : 20 PEC–polystyrene = PEC 80 and 60 : 40 PEC–polystyrene = PEC 60. Figure 15C.13 shows the elongation at break for injection molded blends of HDPE with polystyrene, PEC 60, and PEC 80 compatibilized with varying amounts of S–EB–S. Virtually identical trends were observed for Izod impact strength [70, 71].

Some HDPE–polystyrene blends have lower strains at break than pure polystyrene. Addition of S–EB–S increases ductility for both polystyrene and its blends with HDPE. The strain at break exhibited by blends containing 20 pph S–EB–S is approximately an additive function of composition, based on the measured endpoint values as shown. Similar results were obtained by Lindsey et al. [66] for blends of the same components. PEC is significantly more ductile than polystyrene, so the minimum in the strain at break for blends containing

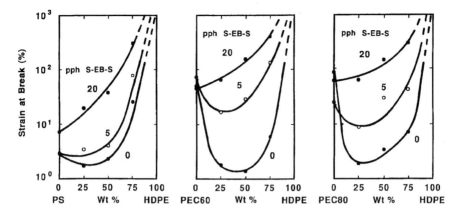

Figure 15C.13   Strain at break for HDPE–polystyrene (left), HDPE–PEC 60 (center), HDPE–PEC 80 (right) blends containing S–EB–S. (Reproduced from Ref. [71], with permission of John Wiley & Sons, Inc.)

either PEC 60 or PEC 80 with HDPE is more dramatic. These results clearly illustrate the inferior properties obtained by combining these two ductile but incompatible polymers. Adding 5 pph S–EB–S has little effect on the strain at break for either PEC 60 or PEC 80 but greatly improves the values for their blends containing HDPE. After adding 20 pph S–EB–S, the minimum in the strain at break is virtually eliminated and the values more nearly approach additivity.

PEC–polystyrene blends containing no HDPE undergo a brittle to ductile transition between 40 and 60 wt% polystyrene. Blends with less than 60% PEC visibly craze and fracture in a brittle mode while PEC 60 and PEC 80 yield and neck throughout the gauge section of the sample prior to fracturing. Yee [99, 100] found that PPO–polystyrene blends also exhibited a similar brittle to ductile transition within the same composition range (from 40% to 60% PPO depending on strain rate). Ductility is enhanced in brittle PEC–polystyrene blends (polystyrene, PEC 20, and PEC 40) by adding S–EB–S.

HDPE–(PEC–polystyrene) blends exhibit low strains at break regardless of the amount of PEC in the glassy phase. The strain at break for these blends increases as a result of adding S–EB–S. The improvement is substantially greater for blends containing PEC-rich glassy phases instead of pure polystyrene. For 75% HDPE blends containing 5 pph S–EB–S, the strain at break increased from about 80% to 140% on changing the glassy phase from pure polystyrene to PEC 60.

These results demonstrate that an S–EB–S triblock copolymer significantly increases the Izod impact strength and strain at break of brittle HDPE–(PEC–polystyrene) blends; however, there is a sacrifice in stiffness and ultimate strength. This compatibilizing effect by S–EB–S is more dramatic when the glassy phase contains PEC rather than pure polystyrene. At least two factors may be at issue. Miscible blends of PEC and polystyrene are considerably more ductile than pure polystyrene. Therefore, the immiscible blends of HDPE with the former involve two relatively ductile phases, but the composite is quite brittle owing to the lack of interfacial adhesion between these phases. Improvement of this adhesion by addition of S–EB–S allows the composite to realize the ductility of both its phases. Thus, the relative change in toughness of these blends is greater than for blends based on polystyrene, which form a brittle phase even after mechanical coupling to HDPE. As described previously, the thermodynamic affinity of PEC for the polystyrene end blocks of S–EB–S may lead to a more effective coupling of HDPE to a PEC–polystyrene phase compared to similar coupling of HDPE to a pure polystyrene phase.

15C.3.2.4 Polystyrene–Polypropylene Blends

Bartlett from our laboratory [67] reported on compatibilization of polypropylene blends with polystyrene using the same S–EB–S mentioned earlier. Figures 15C.14 to 15C.16 show key results for blends made by extruder compounding followed by injection molding. As expected, there is a serious loss in mechanical properties on blending this pair as seen in Fig. 15C.14; however, addition of the S–EB–S greatly improves the ductility (see also Fig. 15C.15). As before, there is a loss in strength and stiffness accompanying this improvement. These improvements in ductility do translate into substantial gains in notched Izod impact strength as seen in Fig. 15C.16. These are mainly caused by the toughening effect of S–EB–S on both polypropylene and polystyrene.

   To illustrate the effect of processing technique and conditions on blend properties, Figs. 15C.17 and 15C.18 show results for materials compounded at two different temperatures in a Brabender batch mixer and then compression molded. As may be seen, the lower mixing temperature gives somewhat poorer values of elongation at break than the higher mixing temperature, probably due to differences in rheological characteristics and opportunities for interfacial knitting. Interestingly, a similar variation in molding temperature had little effect on mechanical properties. In contrast to the previous example involving HDPE, at

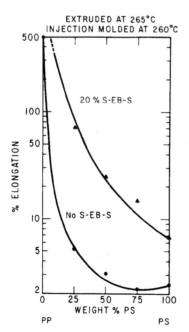

Figure 15C.14   Elongation at break for polypropylene–polystyrene blends with and without added block copolymer. (Reproduced from Ref. [67],with permission of McGraw-Hill, Inc.)

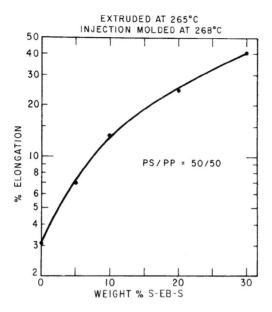

Figure 15C.15   Effect of block copolymer on elongation at break for a 50 : 50 blend of polypropylene–polystyrene blends. (Reproduced from Ref. [67], with permission of McGraw-Hill, Inc.)

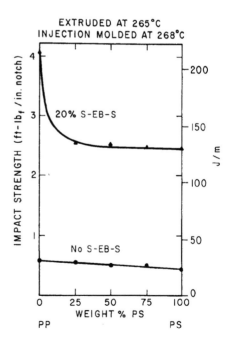

Figure 15C.16   Notched Izod impact strength of polypropylene–polystyrene blends with and without block copolymer. (Reproduced from Ref. [67], with permission of McGraw-Hill, Inc.)

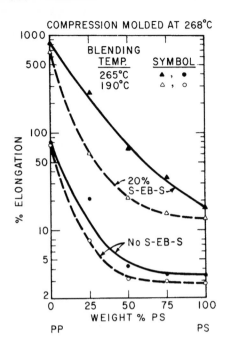

Figure 15C.17   Effect of blending temperature on elongation at break of compression molded polypropylene–polystyrene blends with and without block copolymer. (Reproduced from Ref. [67], with permission of McGraw-Hill, Inc.)

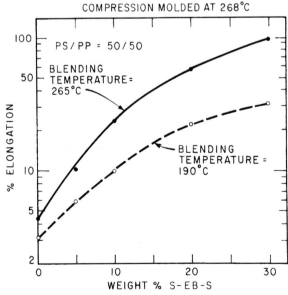

Figure 15C.18   Effect of blending temperature and block copolymer content on elongation at break for a 50:50 polypropylene–polystyrene blend. (Reproduced from Ref. [67], with permission of McGraw-Hill, Inc.)

comparable processing temperatures compression molding gave somewhat higher levels of ductility for polypropylene than did injection molding.

In summary, S–EB–S is an effective compatibilizer for polypropylene–polystyrene blends based on the increases in ductility observed.

15C.3.2.5 Poly(ethylene terephthalate)–High-Density Polyethylene

In the previous examples, the blend components were either structurally identical to one of the block copolymer segments, as in the case of polystyrene, or sufficiently similar, as in the case of the various polyolefins, that one could expect some level of compatibility if not miscibility. Here, we consider the work of Traugott from our laboratory [68] where polystyrene was replaced by poly(ethylene terephthalate) (PET) to see whether the same S–EB–S might produce any beneficial improvements in mechanical behavior in blends with HDPE. Clearly, there is no miscibility between PET and the polystyrene segments of the S–EB–S so the model of Fig. 15C.1 is not applicable.

Binary blends of PET and HDPE exhibit one of the most severe cases of mechanical property deterioration we have ever seen. Figures 15C.19 and 15C.20 show the pertinent values for binary blends plus those to which S–EB–S was added. As before, addition of the S–EB–S reduced strength and stiffness; however, relatively small amounts of S–EB–S caused remarkable improvements in ductility as seen in Fig. 15C.20. Those containing 20% S–EB–S did not break within the full traverse of the Instron tensile tester. The S–EB–S transformed the almost friable PET–HDPE blends into remarkably tough materials. One might ask whether addition of any elastomeric material might accomplish the same thing. To test this notion, blends were prepared using a commercial ethylene–propylene copolymer rubber (EPR), Epcar® 847, instead with the results shown in Fig. 15C.21. These mixtures were substantially no better than the original blend, thus demonstrating the utility of the block copolymer character for this purpose.

Consequently, one must conclude that the beneficial "compatibilizing" effects of this type of block copolymer are not limited to the mechanism implied by Fig. 15C.1. Similar conclusions may be reached from the results presented in Chapter 11 by Gergen.

15C.3.3 Morphology, Adhesion, and Deformation Mechanisms

There is extensive evidence complementing studies such as those described above showing that properly chosen block copolymers significantly alter the morphology of the blends to which they are added [8, 9, 33–65, 68, 93–96]. Basically, the block copolymer appears to emulsify the immiscible mixture, causing significant reductions in domain size. A portion of this response may be attributed to a reduction in interfacial tension; however, it is quite likely that the dominant effect is suppression of coalescence of phases by steric stabilization. During melt processing, domain size is determined by the balance between drop breakup and coalescence. The presence of block copolymers at the interface will help reduce domain size, both by facilitating breakup (lower interfacial tension) and also by inhibiting coalescence (steric stabilization). By decreasing the rate of coalescence, compatibilization leads to more stable morphologies [6, 47–60] which is very important for practical fabrication procedures.

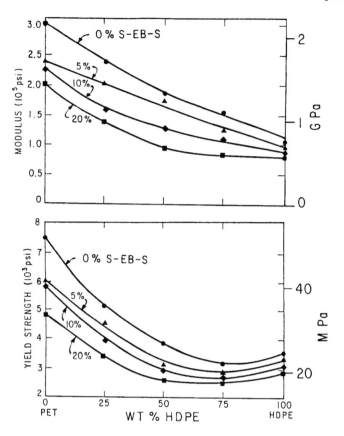

Figure 15C.19   Effect of block copolymer content on modulus and strength of poly(ethylene terephthalate)–high-density polyethylene blends. (Reproduced from Ref. [68], with permission of John Wiley & Sons, Inc.)

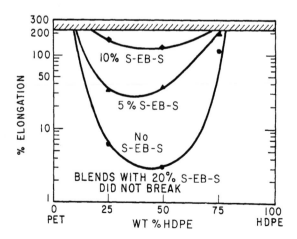

Figure 15C.20   Effect of block copolymer level on percent elongation at break of PET–HDPE blends. (Reproduced from Ref. [68] with permission of John Wiley & Sons, Inc.)

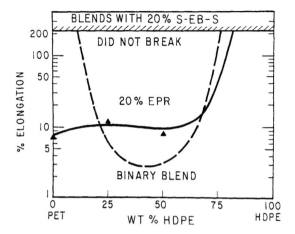

Figure 15C.21   Demonstration that ethylene–propylene elastomer does not compatibilize PET–HDPE blends. (Reproduced from Ref. [68], with permission of John Wiley & Sons, Inc.)

In some cases, the block copolymer appears to induce formation of an interpenetrating network of phases as described more fully in Chapter 11.

Transmission electron microscopy has been used by Teyssie et al. [51] to show that the type of block copolymers described in Table 15C.1 do indeed reside at the interface between polystyrene and polyolefin phases. Observations of fracture surfaces using scanning electron microscopy also strongly support increased interfacial adhesion caused by addition of such block copolymers [8, 47–65, 68, 95, 96]. In summary, these facts leave no doubt that the block copolymer usually plays some interfacial role, although it may, in some cases, have different origins than that implied by Fig. 15C.1.

Clearly interfacial adhesion is one of the key elements in the mechanical compatibilizing effect caused by any additive, and in the remaining part of this section information relevant to this point is reviewed. One approach that has been used to obtain information about the relative adhesion at interfaces between polymer pairs in the blend systems of interest here is the simple lap shear method shown schematically in Fig. 15C.22 [9, 30, 68, 101]. In this method, a three-layer sandwich (A–B–A) is made by laminating together sheets of polymers A and B by compression molding. Notches are cut so that symmetrical lap shear joints are created at the two interfaces when the specimen is pulled in an Instron. Table 15C.2 gives some typical average stresses required to cause adhesive failure at the interfaces between various polymer pairs. As may be expected, adhesion between polystyrene and HDPE, polystyrene and polypropylene, and PET and HDPE is relatively low. On the other hand, adhesion between the S–EB–S and each of these polymers is several-fold larger. That this should be so is easy to understand for polystyrene, HDPE, and polypropylene since the end blocks of the S–EB–S are identical to polystyrene while the midblock of this material should be relatively compatible with HDPE and polypropylene. Addition of PEC to polystyrene greatly improved adhesion to S–EB–S [30] because of the favorable segmental interactions between PPO-type materials with polystyrene as described earlier. The reason for adhesion of S–EB–S to PET is less clear but probably relates to some affinity of the polystyrene blocks with PET. This could be checked by measuring the PET–polystyrene pair, which has not been

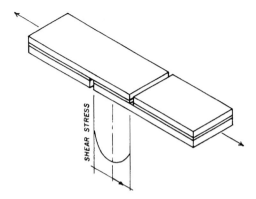

Figure 15C.22   Schematic illustration of lap shear specimen from notched laminate. Variation of stress with position is indicated. (Reproduced from Ref. [101], with permission of John Wiley & Sons, Inc.)

done. In any case, the propensity of S–EB–S to mutually adhere rather well to each of the components in polystyrene–HDPE, polystyrene–polypropylene, and PET–HDPE blends is believed to be a key to the ability of this block copolymer to compatibilize these blends. Most likely, the better affinity of S–EB–S for say PET and HDPE compared to the affinity of PET for HDPE, as indicated by this measure of adhesion, makes contacts between S–EB–S and each of the components during blending more likely than contacts between PET and HDPE. Thus, these results suggest an interfacial role of the S–EB–S in the blend even though the ideal configuration shown in Fig. 15C.1 may not be realizable. The results in Table 15C.2 clearly demonstrate the value of adhesion data for interpreting blend mechanical properties.

There are several problems associated with the type of adhesion information given above. First, the actual values obtained are complex averages since there is a distribution of stresses in the joint formed [9] and so these values depend on the dimensions of the specimen. Second, the values can also depend on the laminating temperature, pressure, and time used to form the specimen. Third, the interfaces are not nascent or virgin ones such as those formed in the blend. Finally, one has no way to relate quantitative values, even if free of the above issues, to conditions of interfacial failure in blends owing to the complex stress conditions that

Table 15C.2 Average Stresses at Failure for Lap Shear Adhesion Test

Polymer pair (A–B)	Average shear stress at failure (KPa)
PS–HDPE	110
PS–PP	248
PET–HDPE	234
S–EB–S/PS	1207
S–EB–S/HDPE	1986
S–EB–S/PP	1227
S–EB–S/PET	593

PS, polystyrene; PP, polypropylene. Other abbreviations are given in the text

develop at domain interfaces during mechanical property determination. A preferred approach would give some indication of in situ interfacial debonding during mechanical testing of the blend itself, since this circumvents all of the problems outlined above. An attractive approach to this is measurement of volume dilation during mechanical stressing of the blend which has been pioneered by Heikens [62, 63, 65] using a relatively simple stress dilatometer.

Some results from the work of Heikens et al. along the lines mentioned above are described next. This group has measured the Poisson ratio, ν, a measure of volume dilation at low strains defined by

$$\nu = \frac{1}{2}\left[1 - \frac{1}{V}\frac{dV}{d\varepsilon}\right]\tag{15C.2}$$

(where V is the sample volume and ε is the uniaxial strain), for blends of polystyrene and LDPE with and without the block copolymer. The results are shown as the next to the last entry in Table 15C.1. The Poisson ratio can reflect events that a simple mechanical characteristic such as modulus cannot. This can be a powerful approach when coupled with modeling of the mechanical behavior in terms of theories for composites such as the Kerner equations [102]. Figure 15C.23 shows the Poisson ratio and modulus for binary blends of LDPE and polystyrene [65] as a function of the volume fraction of polystyrene, ϕ_{PS}, in the blend. The solid line represents the prediction of the Kerner equations while the broken curve is the prediction obtained when the shear and compression moduli of the ductile component are taken as zero. This approximately represents the situation of no interfacial adhesion. As may be seen, the measured Poisson ratio deviates considerably from

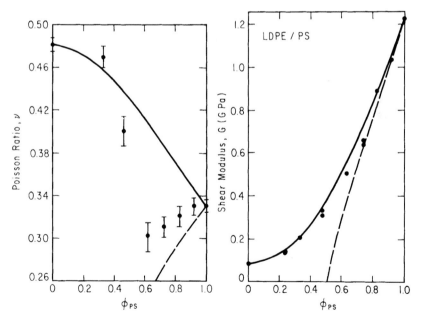

Figure 15C.23   Poisson ratio and modulus of low-density polyethylene–polystyrene binary blends. Curves explained in text. (Reproduced from Ref. [65], with permission of Butterworth and Co. Ltd.)

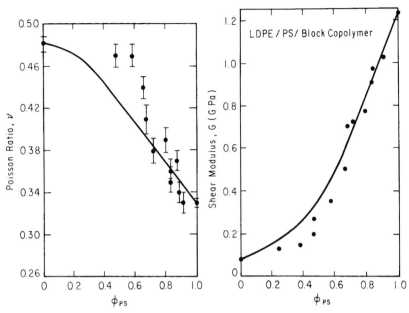

Figure 15C.24   Poisson ratio and modulus of low-density polyethylene–polystyrene blends to which block copolymer has been added. Curves explained in text. (Reproduced from Ref. [65], with permission of Butterworth and Co. Ltd.)

the solid line at high polystyrene contents. These values decrease as more LDPE is added to polystyrene. This trend may be explained by assuming little adhesion between LDPE and polystyrene, since the experimental data deviate in the direction indicated by the dashed curve which represents the extreme limit of no adhesion at all. As may be seen, the modulus is rather insensitive to the extent of adhesion in the region of high polystyrene contents. This approximate analysis does not apply to the LDPE-rich region since the line for no adhesion assumes polystyrene to be the continuous phase. Figure 15C.24 shows the Poisson ratio and modulus for LDPE–polystyrene blends to which various amounts of the block copolymer have been added. Here, ϕ_{PS} includes the homopolymer polystyrene and the polystyrene segments of the block copolymer. The remainder of the material is the LDPE and the hydrogenated butadiene segments of the block copolymer. This division is appropriate since the two segments of the block copolymer have mechanical properties similar to those of polystyrene or LDPE. In this compatibilized case, the Poisson ratio rises when LDPE is added to polystyrene, in more or less good agreement with the Kerner prediction for good adhesion over the range of compositions for which this analysis might be expected to apply. Thus, the Poisson ratio, which reflects volume changes consequent to the complex stress state in the heterogeneous blend, indicates the improved adhesion between LDPE and polystyrene resulting from the presence of the compatibilizing block copolymer.

   A similar stress dilatometer has been used in our laboratory to determine volumetric changes during plastic deformation after the yield point where extensive debonding at the interface may occur. Some typical results are shown in Fig. 15C.25 for a blend containing

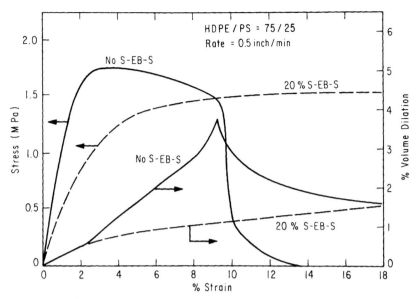

Figure 15C.25   Stress and volume dilation vs. strain for 75 : 25 HDPE–polystyrene blends with and without S–EB–S block copolymer

HDPE–polystyrene in the ratio of 75 : 25 when the ductile HDPE forms the continuous phase. Results are shown for samples with and without S–EB–S as a compatibilizer. For low strains below the yield point, both materials show a small volume increase resulting from the Poisson ratio effect. For the two blends this is essentially indistinguishable on this scale of comparison. However, when irreversible deformations set in beyond yielding (see the stress curves), there is a rapid increase in volume with strain for the blend with no S–EB–S. This increase results from voids formed at the interfaces between domains, owing to the very low interfacial adhesion. When actual failure occurs these voids begin to collapse, as indicated by the volume decrease with further strain. This process is actually a function of time and would occur even if the crosshead travel were stopped. On the other hand, the blend containing S–EB–S does not show this large volume dilation, since apparently such voids are not formed owing to the improved adhesion between phases caused by its presence. It should be pointed out that such measurements can reflect other dilational processes like crazing which would obscure conclusions of this type; however, crazing was not present in the results shown in Fig. 15C.25 since the ductile polyethylene that formed the continuous phase yields by shearing mechanisms that produce insignificant dilation.

Post yield volume dilation has been determined for the same series of HDPE–(PEC–polystyrene) blends described in Fig. 15C.13 [70]. The results are shown in Fig. 15C.26 in terms of the slope of the volume strain curve measured beyond the yield point. It is important to note that for a material whose post yield deformation occurs by crazing, such as high-impact polystyrene, this slope will be approximately unity, while for a material such as polycarbonate that deforms by shear yielding this slope will be essentially zero. Most HDPE–polystyrene compositions do not yield so the only data points possible in Fig. 15C.26 for these blends are those containing 20 pph S–EB–S. In addition to interfacial voiding, the

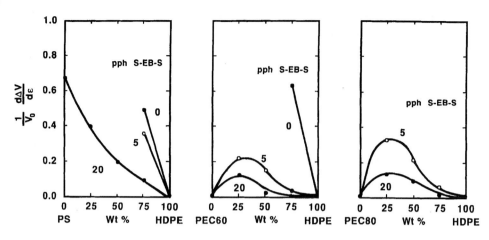

Figure 15C.26   Postyield slope of the volume strain curve for injection-molded HDPE–polystyrene (left), HDPE–PEC 60 (center), and HDPE–PEC 80 (right) containing indicated amounts of S–EB–S. (Reproduced from Ref. [70], with permission of John Wiley & Sons, Inc.)

dilatational response observed in HDPE–polystyrene blends is affected by two mechanisms of post yield deformation (shear yielding and crazing) depending on which component forms the continuous phase. Blends with a polystyrene matrix undergo crazing; hence, reduced dilatation may indicate induced shear yielding in addition to any improved interfacial adhesion in the blend. For blends with a HDPE matrix, dilatation is primarily the result of interfacial void formation since crazing of the dispersed phase is unlikely. Reduction of the volume dilatation on addition of S–EB–S to these blends is due primarily to improved adhesion between the blend components. In all blends, the better interfacial adhesion caused by the S–EB–S compatibilizer shown quantitatively here by stress dilatometry is the primary factor contributing to improved ductility and impact strength described previously which occurs at the expense of stiffness.

PEC 60, PEC 80, and HDPE deform by shear yielding; therefore, volume dilatation in these blends must be the result of interfacial voiding. Reductions in the slope of the volume strain curves shown in Fig. 15C.26 reflect better adhesion on the addition of S–EB–S. The skewed shape of these plots may result from differences in deformability of the individual phases. In blends with a soft matrix (HDPE) and hard dispersed phases (PEC 60 or PEC 80), the soft matrix will deform around these hard particles. Therefore, volume dilation in blends containing 75% HDPE (25% PEC 60 or PEC 80) should be lower than that for blends of the reverse composition (25% HDPE) which consist of a hard matrix containing a soft dispersed phase.

It is interesting to note that the extent of post yield dilatation decreased as the proportion of PEC in the glassy phase was increased to over 60%. At first sight, this might be thought to occur because adding PEC to the glassy phase tends to promote its deformation by shear yielding rather than crazing. However, the possibility of better adhesion of HDPE to PEC-containing phases caused by S–EB–S is another factor to be considered in these blends. This is presumed to be caused by the greater affinity for the polystyrene end blocks of S–EB–S to mix with a phase including PEC rather than polystyrene only. As argued earlier, the driving force for this is the expected exothermic mixing of polystyrene segments with PEC segments, whereas mixing of polystyrene segments from the copolymer with polystyrene homopolymer can be driven only by entropic forces. The results described above suggest this mechanism as an important factor in the decreased dilatation noted. This issue is also believed to contribute to the greater improvements in mechanical properties resulting from the addition of S–EB–S to blends containing PEC. For example, the Izod impact strength of blends of 75% HDPE containing 5 pph S–EB–S is increased from approximately 0.5 ft-lb/in. for the blend with 25% polystyrene to nearly 2.8 ft-lb/in. for the blend containing 25% PEC 80. Of course, the improvement in ductility of the glassy phase when PEC is present contributes to the overall mechanical performance of these blends.

14C.4   Summary

This review has presented selected results that demonstrate that block copolymers containing segments of polystyrene and hydrogenated polybutadiene can cause significant improvements in the mechanical properties of immiscible blends or "compatibilize" them. It seems clear that this is the result of an interfacial role played by the block copolymer in the blend resulting in improved adhesion between the phases. In the most ideal case, this may be the result of the block copolymer acting as a surfactant with its segments penetrating deeply into the homopolymer domains in situations where complete miscibility exists. However, these copolymers function quite effectively also in systems where this idealized monolayer configuration (see left side of Fig. 15C.27) cannot occur. We suggest that in these cases the block copolymer may form an *interphase* between the two components, as suggested on right side of Fig. 15C.27, based on its mutual affinity to wet or adhere to each component which is greater than the affinity of the two components to each other. No doubt many other issues are involved as well.

At this time, it is quite impossible to state conclusively what the "best" block copolymer structure might be for a given application owing to the rather primitive state of the literature on this subject. However, some conclusions can be drawn. There is ample evidence that the block copolymer segments will be firmly anchored into the homopolymer phases when the length of these blocks exceed the critical molecular weight for entanglement coupling [79–

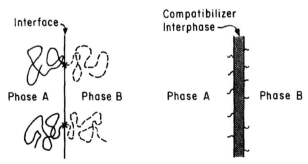

Figure 15C.27   Block copolymer at the interface between polymer phases A and B in ideal configuration (left) and as an interphase (right)

82]. The more stringent thermodynamic requirements on segment molecular weight for "solubilization" do not have to be met since the goals for "compatibilization" are entirely different. However, it is important that the copolymer actually reside at the interface and not form "micelles" in either homopolymer phase. Equilibrium aspects of this issue have been addressed in the recent literature [31]; however, it is not yet clear how this relates to nonequilibrium conditions that certainly prevail in most practical compounding situations. Rheological issues may emerge as important in this case.

This chapter has also reexamined the issue of solubilization of homopolymers in the microdomains of block copolymers for several reasons. First, it is important to point out how this is usually a fundamentally different phenomenon than compatibilization. Second, it was shown that the extent of solubilization can be greatly enhanced when there is a favorable thermodynamic heat of mixing between the homopolymer and the block copolymer segments (i.e., the two polymers are chemically different but miscible). This has practical utility for modifying the properties of the block copolymers and has led to commercial products with higher permissible use temperatures. Finally, there is a useful lesson for the improved design of compatibilizers. Generally one envisions using an A–B type block copolymer to compatibilize a mixture of immiscible A and B homopolymers. However, a better strategy for many purposes is to use the A–B copolymer to compatibilize a mixture of C and D when the A–C and B–D pairs are miscible because of a favorable energetic interaction [103]. These favorable interactions would improve interfacial penetration of segments and adhesion as seen for the (PEC–polystyrene)–S–EB–S system [30]. Compatibilizer systems of this type have been described [60].

Finally, it should be mentioned that a major limitation of using preformed block copolymer of the type described here is the lack of viable and economic synthesis routes to the copolymers needed for compatibilizing commercially important blends. The concept of reactive compatibilization using functionalized additions that react in situ at the interface during processing appears to be a generally more versatile route and is receiving a lot of attention currently [104, 105].

References

1. D. R. Paul and S. Newman (Eds.), *Polymer Blends, Vols. 1 and 2* (1978) Academic Press, New York
2. O. Olabisi, L.M. Robeson, and M.T. Shaw, *Polymer–Polymer Miscibility* (1979) Academic Press, New York
3. D.R. Paul and J.W. Barlow, *J. Macromol. Sci. Rev. Macromol. Chem. C18* 109 (1980)
4. T.A. Callaghan, K. Takakuwa, D.R. Paul, and A.R. Padwa, *Polymer 34*, 3796 (1993)
5. Y. Takeda and D.R. Paul, *J. Polym. Sci. Pt. B Polym. Phys. 30*, 1273 (1992)
6. I. Park, J.W. Barlow, and D.R. Paul, *J. Polym. Sci. Pt. B Polym. Phys. 30*, 1021 (1992)
7. I. Park, H. Keskkula, and D.R. Paul, *J. Appl. Polym. Sci. 45*, 1313 (1992)
8. D.R. Paul, In *Polymer Blends, Vol. 2*. D.R. Paul and S. Newman (Eds.) (1978) Academic Press, New York
9. J.W. Barlow and D.R. Paul, *Polym. Eng. Sci. 24*, 525 (1984)
10. S.L. Aggarwal and R.A. Livigni, *Polym. Eng. Sci. 17*, 498 (1977)
11. T. Inoue, T. Soen, T. Hashimoto, and H. Kawai, *Macromolecules 3*, 87 (1980)

12. G. Reiss, J. Kohler, C. Tournut, and A. Bandaret, *Makromol. Chem. 101*, 58 (1967)
13. A. Skoulios, P. Helffer, Y. Gallot, and J. Selb, *Makromol. Chem. 148*, 305 (1971)
14. J.B. Class, *Rubber Chem. Technol. 58*, 973 (1986)
15. R.J. Roe and W.C. Zin, *Macromolecules 17*, 189 (1984)
16. T. Hashimoto, K. Kimishima, and H. Hasegawa, *Macromolecules 24*, 5704 (1991)
17. H. Tanaka, H. Hasegawa, and T. Hashimoto, *Macromolecules 24*, 240 (1991)
18. D.J. Meier, *Polym. Preprints Am. Chem. Soc. Div. Polym. Chem. 18*, 340 (1977)
19. H. Xie, Y. Liu, M. Jiang, and T. Yu, *Polymer 27*, 1928 (1986)
20. P.S. Tucker and D.R. Paul, *Macromolecules 21*, 2801 (1988)
21. A.R. Shultz and B.M. Gendron, *J. Appl. Polym. Sci. 16*, 461 (1972)
22. A.R. Shultz and B.M. Beach, *J. Appl. Polym. Sci. 21*, 2305 (1977)
23. P.S. Tucker, J.W. Barlow, and D.R. Paul, *Macromolecules 21*, 1678 and 2794 (1988)
24. M.D. Whitmore and J. Noolandi, *Macromolecules 18*, 2486 (1985)
25. K.M. Hong and J. Noolandi, *Macromolecules 16*, 1083 (1983)
26. D. Rigby, J.L. Lim, and R-J. Roe, *Macromolecules 18*, 2269 (1985)
27. M.O. de la Cruz and I.C. Sanchez, *Macromolecules 20*, 440 (1987)
28. H.R. Brown, K. Char, and V.R. Decline, *Macromolecules 23*, 3383 (1990)
29. K. Char, H.R. Brown, and V.R. Decline, *Macromolecules 26*, 4164 (1993)
30. I. Park, J.W. Barlow, and D.R. Paul, *Polymer 31*, 2311 (1990)
31. C. Auschra, R. Stadler, and I.G. Voigt-Martin, *Polymer 34*, 2081 and 2094 (1993)
32. X. Lu and R.A. Weiss, *Macromolecules 26*, 3615 (1993)
33. D.R. Paul, C.E. Vinson, and C.E. Locke, *Polym. Eng. Sci. 12*, 157 (1972) and *13*, 202 (1973)
34. G.E. Molau, In *Block Polymers*. S.L. Aggarwal (Ed.) (1970) Plenum Press, New York, p. 79
35. G.E. Molau, *J. Polym. Sci. Pt. A 3*, 1267, 4235 (1965)
36. G.E. Molau and W.M. Wittbrodt, *Macromolecules 1*, 260 (1968)
37. G.E. Molau, *Kolloid Z.Z. Polym. 238*, 493 (1970)
38. J. Kohler, G. Riess, and A. Bandaret, *Eur. Polym. J. 4*, 173, 187 (1968)
39. G. Riess, J. Periard, and Y. Jolivet, *Angew. Chem. Int. Ed. 11*, 339 (1972)
40. G. Riess and Y. Jolivet, *Adv. Chem. Ser. 142*, 243, American Chemical Society, Washington, D.C. (1975)
41. G. Riess, J. Periard, and A. Bandaret, In *Colloidal and Morphological Behavior of Block and Graft Copolymers*. G.E. Molau (Ed.) (1972) Plenum Press, New York, p. 173
42. P. Gaillard, M. Ossenbach-Santer, and G. Riess, In *Polymer Compatibility and Incompatibility—Principles and Practices*. K. Solc (Ed.) (1982) MMI Press Symp. Series, Vol. 2, Harwood Academic, New York, p. 289
43. G. Hurtrez, D.J. Wilson, and G. Riess, In *Polymer Blends and Mixtures*. D.J. Walsh, J.S. Higgins, and A. Maconnachie (Eds.) (1985) *NATO ASI Series E, VOl. 89*, Martinus Nijhoff, Dordrecht, p. 149
44. M. Moritani, T. Inoue, M. Motegi, and H. Kawai, *Macromolecules 3*, 433 (1970)
45. T. Inoue, T. Soen, T. Hashimoto, and H. Kawai, In *Block Polymers*. S.L. Aggarwal (Ed.) (1970) Plenum Press, New York, p. 53
46. B. Ptaszynski, J. Terrisse, and A. Skoulios, *Makromol. Chem. 176*, 3483 (1975)
47. R. Fayt, R. Jerome, and Ph. Teyssie, *J. Polym. Sci. Polym. Lett. Ed. 19*, 79 (1981)
48. R. Fayt, R. Jerome, and Ph. Teyssie, *J. Polym. Sci. Polym. Phys. Ed. 19*, 1269 (1981)
49. R. Fayt, R. Jerome, and Ph. Teyssie, *J. Polym. Sci. Polym. Phys. Ed. 20*, 2209 (1982)
50. R. Fayt, P. Hadjiandreou, and Ph. Teyssie, *J. Polym. Sci. Polym. Chem. Ed. 23*, 337 (1985)
51. R. Fayt, R. Jerome, and Ph. Teyssie, *J. Polym. Sci. Lett. Ed. 24*, 25 (1986)
52. R. Fayt, R. Jerome, and Ph. Teyssie, *Makromol. Chem. 187*, 837 (1986)
53. R. Fayt, R. Jerome, and Ph. Teyssie, *Polym. Engr. Sci. 27*, 328 (1987)
54. Ph. Teyssie, R. Fayt, and R. Jerome, *Makromol. Chem. Macromol. Symp. 16*, 41 (1988)
55. Ph. Teyssie, *Makromol. Chem. Macromol. Symp. 22*, 83 (1988)
56. R. Fayt, R. Jerome, and Ph. Teyssie, *ACS Symp. Ser. 395*, 38 (1989)
57. R. Fayt, R. Jerome, and Ph. Teyssie, *J. Polym. Sci. Pt. B Polym. Phys. 27*, 775 (1989)
58. R. Fayt and Ph. Teyssie, *Polym. Eng. Sci. 30*, 937 (1990)
59. B. Brahimi, A. Ait-Kadi, A. Ajji, and R. Fayt, *J. Polym. Sci. Pt. B Polym. Phys. 29*, 945 (1991)

15D.1   Introduction

The anionic polymerization of vinylaromatics, 1,3-conjugated dienes, heterocyclics (e.g., oxirane and lactones), and more recently (meth)acrylates has received a great deal of attention since Szwarc's discovery of a possible "living chain" mechanism in 1956 [1]. This remarkable breakthrough has paved the way to the fine tailoring of synthetic polymers. Indeed, in addition to the precise control of molecular weight, molecular weight distribution, molecular architecture, and end-functionality (macromonomers, telechelic polymers) of homopolymers, spectacular advances have also been reported in the tailoring of multiphase block copolymers. Owing to the thermodynamic immiscibility of their constitutive components, block copolymers combine the intrinsic properties of the parent homopolymers along with the additional benefit of some new properties in a strong relation to the phase morphology [2]. These opportunities have been convincingly illustrated by the rapid emergence of thermoplastic elastomers (TPEs) remarkable for a spontaneous and thermo-reversible crosslinking [2], and polymeric emulsifiers of a unique interfacial activity in oil–water and oil–oil emulsions [3] and above all in polymer blends [4].

At the time being, the commercially available TPEs prepared by anionic polymerization are typically of the polystyrene–polydiene–polystyrene type of triblock copolymers (such as S–B–S) synthesized by a two-stage process with monofunctional initiators [5]. As a variant of this method, the styrene–diene diblocks prepared in the first stage can be linked by means of a polyfunctional instead of a difunctional, linking agent with formation of a "star" (or "radial") type of block copolymer instead of a linear triblock copolymer. Hydrogenation of the center polybutadiene block is a second modification of the initially produced S–B–S type of triblock and radial copolymers. The major reasons for these modifications have to be found in the need to improve processability and thermoresistance to degradation of the S–B–S copolymers of the first generation.

Because of the temperature limitations imposed by both the poor thermal and oxidative resistance of polydienes and also the glass transition temperature of the polystyrene end blocks, a serious loss of strength is usually observed at 60 °C to 70 °C. Consequently, there has been much interest in replacing the polystyrene blocks with poly(α-methylstyrene), the T_g of which is 60 °C higher or with a semicrystalline block of a high melting point, and also in replacing the soft polydiene component with polysiloxanes [5, 6]. In a search for novel block copolymers as a source of improved TPEs, this chapter will focus on the synthesis of triblock and star-shaped diblock copolymers containing at least one poly[alkyl(meth)-acrylate] constitutive block. This choice relies on the great range of the properties of this family of polymers, which also have better heat and oxygen resistance than polydienes. Depending on the alkyl substituent, T_g can extend over a very large temperature range, for example, from −60 °C for poly(2-ethylhexylacrylate) up to 130 °C for highly syndiotactic poly(methyl methacrylate) (PMMA). Moreover, very recent progress has been reported on the control of "living" anionic polymerization of alkyl methacrylates and *tert*-butylacrylate, with classical and ligand-modified initiators [7]. These results offer very broad possibilities, especially since the *tert*-butyl acrylate (*t*BA) unit can easily be converted into other acrylates (e.g., of a low T_g) by practically quantitative transalcoholysis. It is also worth pointing out that a difunctional initiator in a polar solvent can be used for the synthesis of the central block, in contrast to polydienes of an essentially 1,4 microstructure which have to be prepared in an apolar medium in which it is difficult to prepare a well-defined soluble difunctional initiator.

15D.2   Synthesis of Poly(MMA–*b*–*t*BA–*b*–MMA) Precursors

Synthesis of fully acrylic triblock copolymers has been considered first. PMMA has been selected as the hard block, because of a T_g close to 130 °C when methyl methacrylate (MMA) is anionically polymerized in tetrahydrofuran (THF) at a low temperature. The choice of poly(*tert*-butylacrylate) (P*t*BA) as the soft block originates from the ability of *tert*-butyl ester to undergo selective acid-catalyzed transalcoholysis by long-chain alcohols (e.g., *n*-butanol and 2-ethylhexanol), leading to low T_g polyacrylates. Three strategies are available for the synthesis of these acrylic triblock copolymers.

1. A two-stage process with a monofunctional organolithium initiator requires the synthesis of a poly(MMA–*b*–*t*BA) diblock, the lithium chain ends of which are then joined by reaction with a coupling agent. It has been shown elsewhere that an organolithium compound complexed with a small molar excess of LiCl initiates the living anionic polymerization of MMA in THF at −78 °C. The sequential addition of *t*BA leads to the expected diblock copolymer [8]. The monomer conversion is complete and the diblock has a narrow molecular weight distribution ($M_w/M_n = 1.15$). Nevertheless, the initiation efficiency drops from ca. 100% for MMA to ca. 80% for the second monomer. The small amount of homo PMMA has been accounted for by a backbiting reaction of the P*t*BA anion onto a MMA subunit [9]. The use of LiCl is essential and the effect of LiCl on the structure and dynamics of the ion pairs has recently been studied by ^{7}Li and ^{13}C NMR spectroscopy [10]. The limitation of this first strategy is the poor efficiency of the coupling reaction, due to the weak nucleophilicity of anions of the acrylate type, even toward activated dihalides, such as α,α′-dibromo *p*-xylene.

2. A three-stage process with monofunctional initiators is a possible strategy because living P*t*BA anions are efficient initiators of the MMA polymerization in the presence of LiCl in THF at −78 °C. Actually, the control of this block copolymerization is still better than in the poly(MMA–*b*–*t*BA) case. The initiation efficiency is high for the two monomers (above 90 to 95%) and no homo P*t*BA is observed in the final product [8]. Since there is no major problem of cross-reactivity with each monomer, this second basic method for the synthesis of poly(MMA–*b*–*t*BA–*b*–MMA) triblocks is quite feasible.

3. A two-stage process with difunctional initiators is, however, much more convenient than the previous three-stage method, since only two monomer additions are required. Moreover, although the order of monomer addition can be reversed, the most favorable situation is the initiation of MMA by living P*t*BA anions. Finally, electron transfer from naphthalene-lithium radical-anions to diphenylethylene or α-methylstyrene (α-MST) forms difunctional anionic initiators in polar solvents [11]. For all these reasons, *t*BA has been first polymerized in THF at −78 °C by a Li dianion derived from α-methylstyrene (naphthalene-lithium/α-MST molar ratio = 1/5) complexed with 10 equivalents of LiCl. MMA polymerization has been carried out under the same experimental conditions. Table 15D.1 reports that a series of triblock copolymers have been prepared in a large range of molecular weight (65,000 to 25,000) and composition. Conversion of both monomers is usually close to

Table 15D.1 Two-Step Synthesis of Poly(MMA–*b*–*t*BA–*b*–MMA)[a]

Sample number	P*t*BA			Poly(MMA–*b*–*t*BA–*b*–MMA)			
	M_n	M_w/M_n	f^b	M_n copol.	M_w/M_n	% Conv.	M_n PMMA
1	20,000	1.20	0.91	70,000	1.15	100%	25,000
2	40,000	1.15	0.90	70,000	1.20	100%	15,000
3	45,000	1.10	0.91	65,000	1.15	100%	10,000
4	50,000	1.15	0.89	70,000	1.15	90%	10,000
5	85,000	1.20	0.93	125,000	1.20	100%	20,000
6	110,000	1.20	0.95	160,000	1.20	100%	25,000
7	120,000	1.15	0.91	190,000	1.25	100%	35,000
8	150,000	1.20	0.92	250,000	1.30	95%	50,000

[a] Experimental conditions. Initiator: α-methylstyryllithium dianion added with 10 equiv. of LiCl·THF, −78 °C

[b] f = initiation efficiency = M_n calc/M_n exp

completion. The polydispersity is rather narrow (M_w/M_n = 1.2) and comparable to that of the first block. The initiation efficiency for the synthesis of the difunctional P*t*BA is high (>90%) and no homo P*t*BA contaminates the final triblock, at least within the limits of GPC detection. In this respect, Figure 15D.1 compares the GPC traces of P*t*BA and the final triblock for the samples 1 and 3 in Table 15D.1.

Poly(MMA–*b*–*t*BA–*b*–MMA) can readily be converted to triblocks containing a central polyacrylate block of a low T_g. Transalcoholysis of P*t*BA by long-chain alcohols can be carried out selectively and quantitatively in the presence of PMMA by acid catalysis. This is a unique approach to the preparation of thermoplastic elastomers with the structure poly(MMA–*b*–*n*- or *s*–alkylacrylate–*b*–MMA). Direct polymerization of similar polymers is not possible because the anionic polymerization of primary and secondary alkyl acrylates cannot be controlled.

15D.3   Derivatization of Poly(MMA–*b*–alkyl acrylate–*b*–MMA)

On the basis of preliminary experiments, the best conditions for the transalcoholysis of P*t*BA consists in dissolving the polymer in an excess of the selected alcohol (*n*-propanol, *n*-butanol, 2-ethylhexanol …) in the presence of *p*-toluenesulfonic acid (PTSA) as a catalyst. After a reflux for several hours, infrared and NMR analysis show that the reaction yield is as high as 97% to 98% and GPC analysis confirms that the molecular weight distribution remains unchanged [12]. No chemical modification is reported when PMMA is treated under the same experimental conditions. Table 15D.2 illustrates that the results obtained for homopolymers can be safely extrapolated to the related poly(MMA–*b*–*t*BA–*b*–MMA) copolymers. Indeed, NMR analysis indicates that transalcoholysis of the P*t*BA central block is close to completion. No detectable *t*BA groups are left, but some of them have been hydrolyzed rather than transalcoholyzed. The carboxylic acid groups have been titrated and found to be in a good agreement with the transalcoholysis yield (Table 15D.2). The absence of chain degradation is

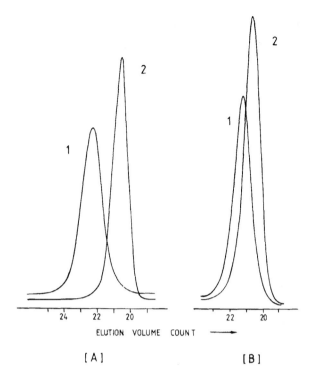

ELUTION VOLUME COUNT ⟶

[A]   [B]

Figure 15D.1   GPC traces of two poly(MMA–b–tBA–b–MMA) samples. [A], sample 1 (Table 15D.1); [B], sample 3 (Table 15D.1). (1) refers to the PtBA block, and (2) refers to the final triblock

clear from the constancy of chain polydispersity. The increase in molecular weight is easily accounted for by the substitution of 2-ethylhexylacrylate (2-EtMA) subunits for tBA subunits. It may thus be concluded that purely acrylic analogues of the S–B–S type of TPEs can be tailored by the sequential living anionic polymerization of tBA and MMA in the presence of a difunctional initiator, followed by the selective transalcoholysis of the PtBA central block with formation of a low T_g poly(alkyl acrylate) inner sequence.

Table 15D.2 Transalcoholysis of poly(MMA–b–tBA–b–MMA) with 2-ethylhexanol[a]

Sample number (Table 15D.1)	M_n ($\times 10^{-3}$) Poly(MMA–b–tBA–b–MMA)	M_w/M_n	Poly(MMA– b–2EtHA–b–MMA)	M_w/M_n	Yield (%) (RMN)	Acid (%)[b] (titr.)
1	25–20–25	1.15	25–28–25	1.15	97	3
2	15–40–15	1.20	15–57–15	1.20	98	2
5	20–85–20	1.20	20–122–20	1.20	98	2

[a] Experimental conditions: 5 wt% copolymer in 2-ethylhexanol + 10 mol% pTSA with respect to tBA units—reflux at 150 °C
[b] Percentage of carboxylic acid subunits formed by hydrolysis

15D.4 Synthesis of Star-Shaped Block Copolymers

The tendency of triblock copolymers to form microdomains in the melt is responsible for a high viscosity. The melt viscosity is indeed usually higher than that of either homopolymer, due to an additional energy required to transport the shorter segments from one microdomain to another one, throughout the immiscible continuous phase. As mentioned in the introduction, a star-branched architecture decreases polymer viscosity at a constant molecular weight, so that the polymer processing is improved [13].

Star-shaped polymers can be prepared by sequential living polymerization in two different ways. The first method, known as the "core-first" method, is based on the synthesis of multifunctional cores that subsequently serve to grow further diblock branches. Polymerization of bis-unsaturated monomers is the usual way to prepare these stable microgel suspensions coated with initiating sites. Alternatively, in the "arm-first" method, either the living diblock precursors initiate the polymerization of a small amount of a bis-unsaturated monomer, or they are added with a mutually reactive multifunctional reagent. The advantage of the "arm-first" method is that each arm can be characterized prior to formation of the more complex star-branched architecture. When the core results from the polymerization of a small amount of a suitable difunctional monomer, no gelation of the reaction medium occurs, due to a shielding effect of the core by the arms, which efficiently prevents "inter-core" reactions from occurring [14, 15]. In this case, there is a random distribution of the number of arms per star-shaped macromolecule. The average number depends on the difunctional monomer/living chain end molar ratio and the chain length of the precursor, which is responsible for a pronounced diffusion control [16, 17].

The "arm-first" method has been considered for the synthesis of star-branched blocks containing at least one acrylic monomer (*t*BA), if not two (*t*BA and MMA). Ethylene glycol bis-methacrylate (EGBMA) has been chosen as the bis-unsaturated monomer because of a reactivity close to that of the (meth)acrylic monomers under consideration. The reaction pathway can be expressed by means of the following equations:

$$n \, M_A + RLi \cdot LiCl \xrightarrow[-78\,°C]{THF} R-(M_A)_{n-1}-M_A^- Li^+ \cdot LiCl \qquad (15D.1)$$
$$(I)$$

$$(I) + m \, M_B \xrightarrow[-78\,°C]{THF} R-(M_A)_n-(M_B)_{m-1}-M_B^- Li^+ \cdot LiCl \qquad (15D.2)$$
$$(II)$$

$$(II) + EGBMA \xrightarrow[(2) \ H^+]{(1) \ THF, -78\,°C} [R-(M_A)_n-(M_B)_m]_p X \qquad (15D.3)$$

where

M_A = styrene (St), MMA;
M_B = *t*BA; and
X = crosslinked core of poly (EGBMA).

A series of representative syntheses of star-branched block copolymers by the "arm-first" method are compiled in Table 15D.3. Diblocks have been prepared in THF at $-78\,°C$, in

Table 15D.3 Synthesis of Star-Branched Block Copolymers by the "Arm-First" Method[a]

Sample number	Macroanions (I)			[EGBMA]/[I] Percent of linking	Star		
	Nature	M_n	M_w/M_n		M_n	M_w/M_n[b]	n (apparent)[c]
1	PSt–b–PtBA	1000–800	1.10	1.0	75	6,000	—	ca. 3.0
2		1000–800	1.10	5.0	98	35,000	1.5	20.0
3		1000–800	1.10	10.0	>99	60,000	1.6	33.0
4	PSt–b–PtBA	5500–6500	1.10	1.0	92	35,000	1.15	3.0
5		5500–6500	1.10	5.0	97	60,000	1.15	5.0
6		5500–6500	1.10	10.0	>99	75,000	1.15	6.0
7	PSt–b–PtBA	15,000–55,000	1.10	10	87	350,000	1.20	6.0
8		20,000–45,000	1.15	5.0	80	225,000	1.25	3.5
9		25,000–65,000	1.10	10	86	410,000	1.20	4.5
10	PMMA–b–PtBA	25,000–95,000	1.10	10	75	630,000	1.25	5.2
11		15,000–40,000	1.10	10	77	250,000	1.25	4.5
12	PαMSt–b–PtBA	25,000–75,000	1.15	5.5	87	360,000	1.20	3.6

[a] Experimental conditions: Initiator = α-MeStLi dianion/10 LiCl in THF at −78 °C. Reaction of EGBMA: 4 h. Yield = 99% to 100%

[b] M_w/M_n calculated from the elution peak of the coupled species

[c] $n = M_n$ Star/M_n arm

the presence of a molar excess of LiCl as a ligand of the organolithium active species. EGBMA was added to the living diblocks in various molar ratios and the coupling efficiency analyzed.

Figure 15D.2 shows that living polystyryllithium chains quantitatively initiate the anionic polymerization of *t*BA in THF at −78 °C. The efficiency of EGBMA in forming a crosslinked core onto which all the linear diblocks are attached clearly depends on the molar excess of the bis-unsaturated monomer. A fivefold molar excess gives rise to a coupling efficiency of 97% to 98%, when the molecular weight of the precursors is not too high (e.g., 12,000). A slight improvement is reported when a 10-fold molar excess of EGBMA is used, whereas a further increase has no additional effect. This observation appears to be valid even though the molecular weight of the precursor is increased. In this case, the linking efficiency

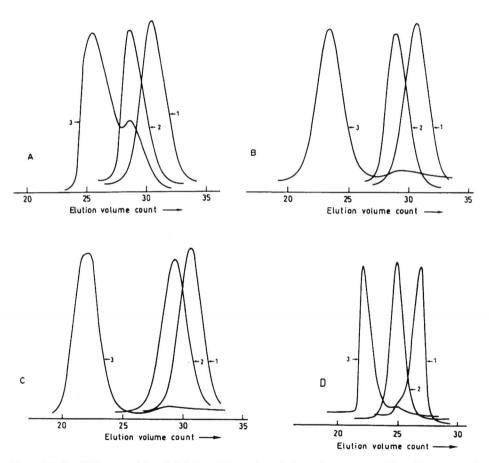

Figure 15D.2   GPC traces of four (PS–*b*–P*t*BA)$_n$X star-branched copolymers reported in Table 15D.3. [A], sample 1; [B], sample 2; [C], sample 3; [D], sample 4. GPC trace of the PS block (1), the (PS–*b*–P*t*BA) diblock (2), and the star-branched diblock (3)

drops below 90%, in an apparent dependence on the block associated to the living PtBA component. Comparison of samples 7 and 11 would indeed indicate that substitution of PMMA for polystyrene (PSt) decreases the linking efficiency at a constant molar excess of EGBMA and a comparable molecular weight. Except for the very short length poly(St–b–tBA) ($M_n = 1800$), the molecular weight distribution of these star-branched blocks was essentially constant ($M_w/M_n =$ ca. 1.20) and not very much broader than that of the linear precursors (1.10). There is an apparent parallelism in the M_w/M_n values for the star-branched blocks and the average number of arms per star (n). For values of n between 3 and 6 M_w/M_n is constant (ca. 1.20). For higher values of n (22 and 30) M_w/M_n increases to 1.5 and 1.6 (samples 2 and 3, Table 15D.3). It is obvious that a large number of branches are attached to the poly EGBMA core only when the chain length of the linear diblocks is very short (samples 2 and 3; $M_n = 1800$). An increase in molecular weight (samples 4 to 6; $M_n = 12,000$) results in a drastic decrease in n, although the linking efficiency remains unaffected. When the diblock molecular weight exceeds a few tens of thousand, then the linking is no longer quantitative but n is constant. These observations are a clear consequence of a diffusion-controlled process, as previously reported by Roovers and Bywater for the synthesis of comb-shaped macromolecules by grafting [18].

It is worth mentioning that the value of n has been approximated from the (M_n star/M_n arm) ratio, in which M_n star results from GPC measurements, although no universal calibration is available. Recently, a series of poly(MMA–b–tBA) star-branched blocks was synthesized using the same techniques as reported here. These polymers were analyzed with a multidetection GPC, that is, an instrument that combines refractive index and viscometric detections together with an on-line light scattering detector. Molecular weight of the diblocks ranged from 20,000 to 35,000 and composition from 20 to 70 wt% PMMA. As a rule, n is in the same range as reported in Table 15D.3 ($3.5 < n < 10$) [19].

Although outside the scope of this chapter, it is worth pointing out that star-branched polymers consisting of at least two different arms have been prepared according to the same reaction pathway. The key point is to prepare the living precursors independently [e.g., PMMA and PtBA, PtBA, and poly(St–b–tBA), polystyrene and poly(St–b–tBA)] and to mix them together before the addition of EGBMA. Linking efficiencies of 85% to 90% have been reported.

Transalcoholysis of PtBA containing star-branched diblocks has been carried out as previously reported and found to be of the same efficiency, that is, complete disappearance of the $tert$-butyl groups and formation of 2% to 3% of residual carboxylic acids.

15D.5   Mechanical Properties of Triblock and Branched Block Copolymers

15D.5.1   Poly(MMA–b–2EtHA–b–MMA) Triblock Copolymers

The ultimate mechanical properties of two poly(MMA–b–2EtHA–b–MMA) triblocks, of the same composition (ca. 30 wt% PMMA) but of a very different molecular weight (70,000 vs.

Table 15D.4 Effect of Molecular Weight on the Mechanical Properties of Poly(MMA–*b*–2EtHA–*b*–MMA)

Sample number	M_n ($\times 10^{-3}$) Poly(MMA–*b*–2EtHA–*b*–MMA)	σ_B (MPa)	ε_B (%)
1	10–50–10	1.2	103
2	35–170–35	1.5	180

To convert MPa to psi, multiply by 145

240,000), were measured at a crosshead speed of 50 cm/min at 25 °C. These samples were roll-milled at 170 °C for 5 min and then compression molded under the same conditions of time and temperature. Table 15D.4 shows that the ultimate tensile strength (σ_B) was very low (≤ 1.5 MPa or 220 psi) whatever the molecular weight, in contrast to the elongation at break (ε_B) which moderately increased with molecular weight. No significant improvement resulted from the esterification of the residual carboxylic acid groups in the central block with diazomethane. The mechanical performances of the fully acrylic triblocks are thus very disappointing compared to the traditional S–B–S TPEs, which have tensile strengths of about 30 MPa (4,300 psi) and elongations of up to 800% [2].

Although the origin of the poor tensile properties of the acrylic triblocks under consideration is still an unresolved question, it is interesting to point out that tensile strengths of S–B–S polymers of a constant polystyrene content are independent of molecular weight, so long as the polystyrene length is high enough to cause the formation of strong, well-separated domains under the testing conditions [2]. This general observation addresses the question of the phase separation in the poly(MMA–*b*–2EtHA–*b*–MMA) copolymers. Is it reduced because of the poor thermal stability of the copolymers under the processing conditions? If not, are PMMA and poly(2EtHA) partially miscible?

It was shown that these triblocks were not degraded when processed at 170 °C, although this temperature is close to the upper limit for the processing. For instance, a decrease in molecular weight (ca. 30%) and a dramatic increase in polydispersity (M_w/M_n) from 1.15 to 2.65 was observed at a processing temperature of 200 °C. Phase separation in all the investigated triblocks was confirmed by a series of techniques, such as transmission electron microscopy, solid-state NMR, differential scanning calorimetry (DSC), and dynamic mechanical measurements. Figure 15D.3 is a typical example of the temperature dependence of the storage modulus (E') and tan δ of the poly(MMA–*b*–2EtHA–*b*–MMA) copolymers reported in Table 15D.4. For the high molecular weight sample (Fig. 15D.3B) glass transitions of the soft phase and the hard phase were clearly observed. However, in the temperature range between the maxima, tan δ does not decrease as much as expected, which may be indicative of an intermediate transition. In the lower molecular weight sample (Fig. 15D.3A) there was a dramatic decrease in both the intensity and temperature of the transition assigned to the hard phase, which suggests a partial miscibility of the two blocks in this polymer. Thermal analysis (DSC) supports this assumption, since a transition at ca. 40 °C is observed in addition to T_gs at −55 °C and 125 °C, respectively, for the high molecular weight triblock (Fig. 15D.4). A diffuse interphase was identified by solid-state NMR (broad line ^{1}H NMR and high-resolution ^{13}C NMR). Above ca. 30 °C, three spin–spin relaxation times (^{1}H NMR) were clearly observed and assigned to the hard phase (short T_2), the soft phase (long T_2), and the

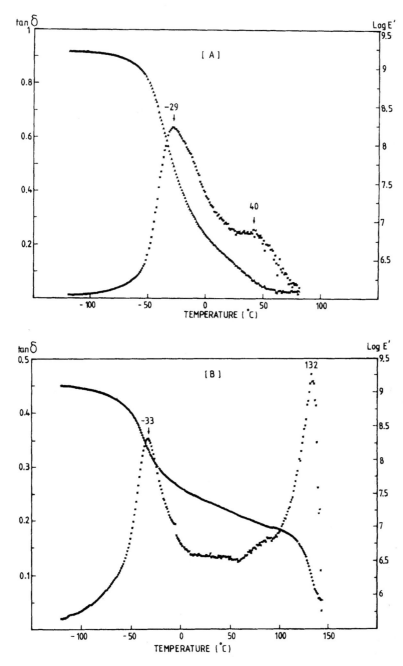

Figure 15D.3   Dynamic mechanical analysis (1 Hz, 5 °C/min) of the poly(MMA–*b*–2EtHA–*b*–MMA) copolymers reported in Table 15D.4. [A], sample 1; [B], sample 2 (*E*′ in Pa)

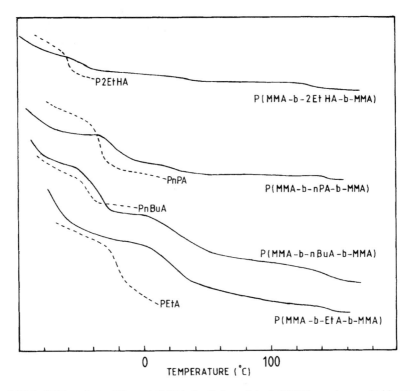

Figure 15D.4   DSC analysis of the poly(MMA–*b*–alkyl acrylate–*b*–MMA) copolymers (Table 15D.5) and the parent homopoly(alkyl acrylates)

interphase (intermediate T_2), respectively. Composition of these phases and their relative proportions were measured in relation to the molecular characteristics of the triblocks. These data will be detailed elsewhere.

For the sake of comparison, the poly(MMA–*b*–*t*BA–*b*–MMA)s used as the precursor of sample 2 in Table 15D.4 were transalcoholyzed with *n*-butanol, *n*-propanol, and ethanol, respectively. Figure 15D.4 compares the DSC traces of the four triblocks, which differ from each other only by the nature of the central block. Although not very pronounced, the T_g of the PMMA block was observed close to the expected value (130 °C). In the case of the poly(*n*-butyl acrylate) and poly(*n*-propyl acrylate) central blocks, T_g was increased by several degrees compared to the parent homopolymers, and a very broad intermediate transition was observed. For the polymer with a poly(ethyl acrylate) center block, a new transition at $+5\,°C$ was observed instead of the usual T_g at $-27\,°C$. These experimental data have been confirmed by dynamic mechanical measurements. Table 15D.5 shows that the tensile properties were poor whatever the central polyacrylate block. Two observations must, however, be emphasized. First, the sample preparation might have an effect on the tensile properties, as illustrated by a very significant improvement when poly(MMA–*b*–*n*BA–*b*–MMA) was cast from a toluene solution at 25 °C rather than processed at 170 °C. The slow evaporation of a toluene solution may improve the phase separation process, since toluene is a better solvent for PMMA than for PnBA. Second, the highest ultimate tensile strength was

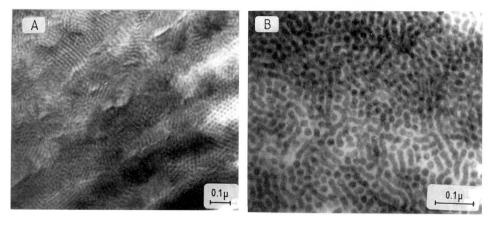

Figure 15D.5   Transmission electron micrographs of a (PS–b–PnBA)$_n$X star-branched copolymer (sample 1; Table 15D.6) [A], and the (PS–b–PtBA)$_n$X precursor [B]

observed for the block copolymer with a poly(ethyl acrylate) center block even though it showed partial miscibility of the two blocks.

15D.5.2   Star-Branched Block Copolymers

It is worth comparing the mechanical properties of star-branched blocks and the related triblocks, since the core in star-branched copolymers should improve the elastomeric behavior of the soft phase. Moreover, a higher fluidity in the melt state may improve the phase separation and thus the tensile properties of the star-shaped blocks. Finally, polystyrene was substituted for PMMA, so as to avoid the problem of the partial miscibility of the hard block and the soft block.

Table 15D.6 compares two star-shaped blocks of comparable molecular weight, composition, and average number of arms. These samples were tested as films cast from a solution in various solvents. The choice of solvent obviously affected the mechanical properties of the star-branched polymer with PMMA end segments, in contrast to similar

Table 15D.5   Mechanical Properties of Poly(MMA–b–AlkA–b–MMA) Prepared by Transalcoholysis of a Common Poly(MMA–b–tBA–b–MMA) Precursor

Sample number	M_n ($\times 10^{-3}$) (PMMA–b–PAlkA–b–PMMA)	Alkyl group	σ_B (MPa)	ε_B (%)
1	35–170–35	2-Ethylhexyl	1.5	180
2	35–115–35	n-Butyl	3.0	78
3[a]	35–115–35	n-Butyl	7.0	325
4	35–90–35	Ethyl	15	140

[a] Sample prepared by solvent-casting of a toluene solution, instead of by roll-milling and compression molding, at 170 °C for 5 min, respectively

To convert MPa to psi, multiply by 145

Table 15D.6 Mechanical Properties of (PSt–*b*–P*n*BA)_nX and (PMMA–*b*–P*n*BA)_nX Star-Branched Block Copolymers. Films Cast from Various Solvents

Sample number	$M_n \times 10^{-3}$ (GPC value)	Solvent	Composition (wt%)		Ultimate tensile strength σ_B (MPa)	Elongation at break ε_B (%)
			PS	PnBA		
1	(PS–*b*–P*n*BA)_nX	Toluene	20	80	11.0	580
2	10–40	CCl₄	20	80	12.5	530
	(n = 4.2)					
3	(PMMA–*b*–P*n*BA)_nX	Toluene	27	73	12.5	460
4	15–40	CCl₄	27	73	7.5	270
	(n = 4.5)					

To convert MPa to psi, multiply by 145

polymers with polystyrene end segments. This observation more likely indicates a stronger driving force to phase separation when the central poly(*n*-butyl acrylate) block (PnBA) is polymerized with polystyrene rather than with PMMA. In this respect, Fig. 15D.5 illustrates the well-defined two-phase morphology of sample 1 (Table 15D.6) in comparison with the *t*BA containing precursor. When poly(St–*b*–*n*BA)_nX was processed in the melt instead of being solvent-cast, the ultimate tensile strength was preserved but the elongation at break was reduced (samples 1 and 2 in Table 15D.6, compared to sample 1 in Table 15D.7). This may indicate some reduction in the phase separation due to a limited chain mobility in the melt. Nevertheless, the two-phase structure was well defined when observed by dynamic mechanical analysis (Fig. 15D.6). Values of σ_B and ε_B remained essentially unchanged when the molecular weight of the arm was increased from 50,000 to 95,000 and

Table 15D.7 Mechanical Properties of (PSt–*b*–P*n*BA)_nX and (PSt–*b*–PEA)_nX Star-Branched Blocks

Sample number	$M_n \times 10^{-3}$ (GPC value)	n^b (GPC value)	Composition (wt%)		Ultimate tensile strength σ_B (MPa)	Elongation at break ε_B (%)
			PS	PnBA		
	(PS–*b*–P*n*BA)_nX					
1	10–40	4.2	20	80	11	380
2	15–60	5.0	20	80	11	375
3	20–50	3.5	30	70	10	415
4	30–65	4.5	30	70	9.0	375
	(PSt–*b*–PEtA)_nX					
5	15–50	5.0	23	77	15.5	350

[a] Samples were prepared by roll-milling at 100 °C for 10 min, followed by compression-molding at 180 °C for 10 min. They were tested at a crosshead speed of 2 cm min⁻¹ at room temperature
[b] Average number of arms per star
To convert MPa to psi, multiply by 145

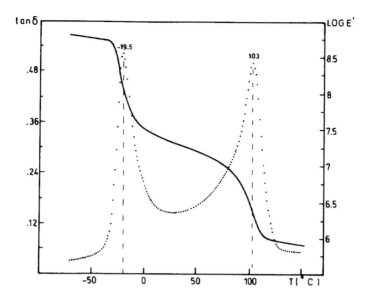

Figure 15D.6 Dynamic mechanical analysis (1 Hz, 5 °C/min) of a (PSt–*b*–P*n*BA)$_n$X star-branched copolymer. (Sample 4; Table 15D.7) (*E'* in Pa)

polystyrene/PnBA wt ratio was varied from 20 : 80 to 30 : 70 (samples 1 to 4, in Table 15D.7). Substitution of poly(ethyl acrylate) for PnBA improved σ_B although ε_B remained constant (samples 2 and 5 in Table 15D.7). The properties of the samples reported in Tables 15D.6 and 15D.7 were very poor. σ_B and ε_B were approximately half the values expected for the S–B–S type of TPE. Finally, the two main modifications considered in this section, that is, star-branching of the acrylic blocks (sample 3a in Table 15D.5 and sample 3 in Table 15D.6) and substitution of a nonacrylic hard block (polystyrene) for PMMA (Table 15D.6) did not substantially improve the performances of the acrylic triblocks described in this study.

15D.6   Conclusions

The ligated anionic polymerization of MMA and *t*BA has proved to be a very powerful tool for the macromolecular engineering of poly[alkyl(meth)acrylate]-based materials. Indeed, triblocks of the poly(MMA–*b*–*t*BA–*b*–MMA) type have been prepared and the related poly(MMA–*b*–*t*BA)$_n$X star-branched equivalents as well. Molecular weight and composition can be changed at will and fitted to values characteristic of the well-known S–B–S type of TPE. In this respect, the acid-catalyzed transalcoholysis of the P*t*BA block with long alkyl chain alcohols has been carried out selectively with formation of low T_g central blocks. Surprisingly, tensile properties of the related triblock and star-shaped copolymers are disappointing. Although this may be due to partial miscibility of the hard PMMA block and the soft poly(alkyl acrylate) block, the reason for the poor mechanical properties of both the fully acrylic triblock and star-branched copolymers, and also the similar poly(S–*b*–

nBA)$_n$X stars is still undetermined. Since replacement of PMMA by a nonacrylic hard block (polystyrene) has had no effect on the mechanical behavior, the poor elastomeric properties of low T_g poly(alkyl acrylates) may be related to the apparent increase in the chain cross-section produced by the bulky alkyl ester pendant groups. In this respect, the mechanical testing of covalently crosslinked poly(2EtHA) and poly(nBA) should provide helpful information.

Acknowledgments

The authors are very much indebted to ATOCHEM-ELF (France) for scientific and financial support. They are also grateful to the "Services Fédéraux des Affaires Scientifiques, Techniques et Culturelles" for general support to the laboratory in the frame of the "Poles d'Attraction Interuniversitaires: Polymères." They express their consideration to Dr. W. Demarteau (UCB, Drogenbos) for the dynamic mechanical measurements and to Mrs. M.C. Guesse for her technical assistance.

References

1. M. Szwarc, M. Levy, and R. Milkovich, *J. Am. Chem. Soc. 78*, 2656 (1956)
2. G. Holden and N.R. Legge, In *Thermoplastic Elastomers—A Comprehensive Review*. N.R. Legge, G. Holden, and H.E. Schroeder (Eds.) (1987) Hanser, Munich, Chap. 3 (See also Chap. 3 in this edition)
3. G. Riess, In *Thermoplastic Elastomers—A Comprehensive Review*. N.R. Legge, G. Holden, and H.E. Schroeder (Eds.) (1987) Hanser, Munich, Chap. 12 (2)
4. R. Fayt, R. Jérôme, and Ph. Teyssié, In *ACS Symposium Series 395*. L.A. Utracki and R.A. Weiss (Eds.) (1989) American Chemical Society, Washington, D.C., Chap. 2
5. M. Morton, In *Thermoplastic Elastomers—A Comprehensive Review*. N.R. Legge, G. Holden, and H.E. Schroeder (Eds.) (1987) Hanser, Chap. 4 (See also Chap. 4 in this edition)
6. R. Jérôme, R. Fayt, and Ph. Teyssié, *Thermoplastic Elastomers—A Comprehensive Review*. N.R. Legge, G. Holden, and H.E. Schroeder (Eds.) (1987) Hanser, Munich, Chap. 12 (7)
7. Ph. Teyssié, R. Fayt, J.P. Hautekeer, C. Jacobs, R. Jérôme, L. Leemans, and S.K. Varshney, *Makromol. Chem. Macromol. Symp. 32*, 61 (1990)
8. S.K. Varshney, C. Jacobs, J.P. Hautekeer, Ph. Bayard, R. Jérôme, and Ph. Teyssié, *Macromolecules 24*, 4997 (1991)
9. C. Jacobs, S.K. Varshney, J.P. Hautekeer, R. Fayt, R. Jérôme, and Ph. Teyssié, *Macromolecules 23*, 4024 (1990)
10. J.S. Wang, R. Warin, R. Jérôme, and Ph. Teyssié, *Macromolecules 26*, 6776 (1993)
11. D. Lipkin, D.E. Paul, J. Towsend, and S.I. Weissman, *Science 177*, 534 (1953)
12. C. Jacobs, Ph.D. Thesis, University of Liège (Belgium), 1992
13. W.W. Graesley, T. Masuda, J. Roovers, and N. Hadjichristidis, *Macromolecules 9*, 127 (1976); W.W. Graesley and J. Roovers, *Macromolecules 12*, 959 (1979)
14. W. Burchard and H. Eschweg, *Polymer 16*, 180 (1975)
15. W. Funke and O. Okay, *Macromolecules 23*, 2623 (1990)
16. D.J. Worsfold, J.G. Zilliox, and P. Rempp, *Can. J. Chem. 47*, 3379 (1969)
17. F. Afshar-Taromi, Y. Gallot, and P. Rempp, *Eur. Polym. J. 25*, 1153 (1989)
18. J.E.L. Roovers and S. Bywater, *Macromolecules 4*, 443 (1974)
19. J. Lesec, M. Millequant, M. Patin and Ph. Teyssié, In *Chromatographic Characterization of Polymers*, Adv. Chem. Series 247, *Th. Provder, H.G. Barth and M.W. Urban (Eds.) ACS, Washington, DC 1995, ch. 13*

15E Novel Optical and Mechanical Properties of Diacetylene-Containing Segmented Polyurethanes

Paula T. Hammond and Michael F. Rubner

15E.1 Introduction 538
    15E.1.1 Introductory Remarks. 538
    15E.1.2 Overview of Diacetylenes and Diacetylene Macromonomers 539
    15E.1.3 Segmented Polyurethane–Diacetylene Macromonomers 541

15E.2 Characterization of Polyurethane–Diacetylene Segmented Elastomers 543
    15E.2.1 WAXD and Thermal Characterization of Polyurethane–Diacetylenes 543
    15E.2.2 Studies of Domain Morphology 545

15E.3 Mechanical Properties of Polyurethane–Diacetylenes 547
    15E.3.1 Effects of Molecular Weight and Annealing on Un-Cross-Polymerized
            Elastomers 547
    15E.3.2 Effects of Diacetylene Cross-Polymerization on Thermal and Mechanical
            Properties 549

15E.4 Linear Optical Properties of Segmented Polyurethane–Diacetylenes 553
    15E.4.1 Thermochromism 553
    15E.4.2 Mechanochromism 557
    15E.4.3 Morphological Model of Deformation 564

15E.5 More Recent and Future Developments 566

15E.6 Conclusions 569

Acknowledgments 570

References 570

15E.1   Introduction

15E.1.1   Introductory Remarks

The extremely versatile class of thermoplastic elastomers (TPEs) known as segmented polyurethanes (SPUs) has expanded from its roots as processable rubbers to include materials for high-technology industry. Today, SPUs are found in biomaterials, optical adhesive and waveguide materials, and protective coatings for microelectronics packaging. Functionalization of a segmented polyurethane with linear or nonlinear optical moieties presents the opportunity to develop TPEs that play an active, rather than a passive, role in sensors, optical switches, and other devices. The ability to couple the optical and mechanical behavior of a TPE presents a world of opportunities, which include the ability to process highly oriented, optically anisotropic materials, and the possibility of an extensible mechanooptic sensor that may be coated onto a surface or molded into any number of desirable shapes.

The development of a new series of diacetylene-containing elastomers has successfully combined the elastomeric properties of SPUs with the optical properties of polydiacetylenes [1–3]. This novel class of materials has ushered thermoplastic elastomers into the arena of optical applications. In the polyurethane–diacetylene (PU–DA) copolymers, on irradiation in the solid state, a polydiacetylene network is produced in the hard domains. The resulting cross-polymerized polyurethanes undergo color changes that are inherently coupled to elastomeric strain (mechanochromism), as well as to temperature changes (thermochromism). The mechanochromic behavior in the diacetylene SPUs has been observed up to 700% strain, and involves dramatic color changes from blue to red or yellow. The chromic responses are generally reversible within a given strain or temperature range.The chemical structure of PU–DAs may be altered to vary the mechanical and elastomeric properties from very stiff, small-strain materials to highly viscoelastic, soft rubbers. Such alterations change the range of mechanochromic response as well as the potential applications of these materials.

As mentioned above, the PU–DAs are candidates for mechanooptic stress or pressure sensors, and the excellent coating properties of polyurethanes make sensory coatings an interesting application. In addition, the polydiacetylene backbone of the polyurethanes provides a molecular scale probe of the level of stress within the hard (or soft) segments. The visible absorption spectra of the highly anisotropic PDA chain can be used with visible dichroism to gain information on the orientation of polymer segments on deformation, processing, or heating. The PU–DAs discussed in this chapter also provide new opportunities in the area of post-processable thermoplastics; the solid state topochemical polymerization of the diacetylene groups produces a covalently bonded network of conjugated polydiacetylene chains. Following processing of the elastomer into a specific form, such as a film or fiber, ultraviolet light or other types of irradiation can be used to convert the thermoplastic into a thermoset. The degree of effective crosslinking imposed on the polyurethane can be controlled to induce conversion of an initially low-modulus elastomer to a strong, tough rubber, and ultimately to a high-modulus, brittle material by simply adjusting the radiation dosage.

Since the initial synthesis of the polyurethane elastomers, considerable effort has gone toward understanding their mechanical, thermal, and optical behavior. This chapter describes some of the most important findings of these studies. The importance of this new class of materials in the areas of sensor technology, optical spectroscopy of polymers, and materials post-processing has been noted; however, the diacetylene-containing elastomers also offer examples of the interplay between conventional polymer science and optically responsive organic materials.

15E.1.2   Overview of Diacetylenes and Diacetylene Macromonomers

The polymerization of diacetylene groups is a solid-state, topochemical process that was first characterized by G. Wegner [4, 5]. Diacetylene monomer, arranged in a crystal lattice, undergoes a 1,4 addition with its nearest neighbor to yield a conjugated polymer. The diffusionless process involves rotation of the monomer units within the lattice. Generally, the substituents remain in approximately the same positions throughout the reaction as the diacetylene groups shift in position to attain contact with adjacent groups. For this reason, changes in crystallographic spacing and symmetry following topochemical polymerization are relatively small. For a large number of monomers, this process occurs in one solid phase, thus yielding perfect polymer crystals [6]. Methods such as optical spectroscopy and gel permeation chromatography have been used to observe high degrees of polymerization in single crystals [7]. The reactivity of diacetylene monomers is determined in large part by the packing of the monomer unit; the topochemical polymerization can take place only in an ordered lattice of monomer molecules stacked such that specific geometric constraints on parameters such as stacking distance (d) and angle between the polydiacetylene axis and the diacetylene monomer (γ) are satisfied, as shown in Fig. 15E.1a. Diacetylene reactivity is also affected by the mobility of the side group. Even with the appropriate lattice parameters d and γ, some diacetylenes may not be reactive due to the lack of flexibility of the substituent R group.

The linear optical properties of polydiacetylenes are highly dependent on the effective conjugation length of the main chain. The effective conjugation length is loosely defined as the distance in number of p orbitals along which electrons can travel; higher conjugation lengths result in lower energies required for electronic processes in the polymer. Changes in the effective conjugation length result in subsequent changes in the wavelength of absorption in the visible spectra. Other effects, such as direct tension or compression effects on the en–yne alternating backbone, and electronic interactions with the substituent side chain groups, can also affect the energy at which a polydiacetylene chain absorbs in the visible region. These effects often give a color change with changes in temperature or an application of stress, and are known as thermo- or mechanochromism, respectively. Chromism may also be observed on solvation or extraction of polydiacetylenes. In studies of thermochromic behavior in soluble polydiacetylenes, the changes in the visible absorption spectra have been correlated to changes in the polymer conjugation length due to a loss of planarity along the polymer backbone during the solvation process. This information, followed by FTIR and NMR studies of substituted polydiacetylenes, led to the realization that chromism is driven primarily by conformational changes of the polydiacetylene chain, rather than changes in its electronic bonding state [10–12].

The conjugated, one-dimensional backbone of polydiacetylene chains and their highly ordered states are the source of many interesting optical and electrical properties in addition

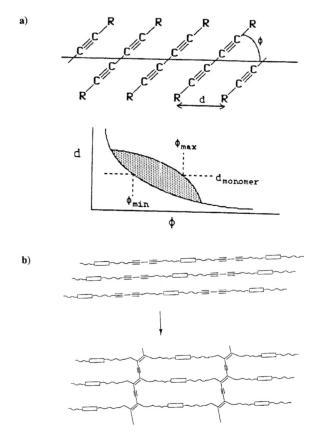

Figure 15E.1   (a) Topochemical polymerization of diacetylenes. (b) Cross-polymerization of macromonomer chains illustrated for a semirigid main chain polymer. (Rectangular areas indicate structurally rigid portions of the polymer chain)

to chromism. Although polydiacetylenes cannot be doped to obtain metal-like conductivity, they have very large electronic carrier mobilities along the chain direction ($>10^3$ cm^2 v^{-1} s^{-1}), making these materials of interest for electronic devices [12]. It has been found that ion implantation of some polydiacetylene crystals using ion beam irradiation increases their conductivity by 15 orders of magnitude [13]. The optical properties of polydiacetylene also include a large nonlinear optical susceptibility coefficient. This has sparked interest in the use of diacetylene-containing materials in electrooptical and optical devices. The polydiacetylene backbone vibrations are also Raman active; therefore, Raman spectroscopy has been used to characterize the vibrational modes in polydiacetylene crystals [14, 15]. Because polydiacetylene vibrations change frequency with the application of stress, large single crystal "fibers" have been used to determine strain distributions of fibers embedded in a composite matrix [16, 18]. The stress state of the polydiacetylene backbone can be measured accurately even in thick or opaque samples using Raman spectroscopy. The resolution of frequency vs. strain plots is extremely high for polydiacetylenes, and researchers have used polydiacetylene crystals as molecular "strain gauges" in composites.

15E.1.3   Segmented Polyurethane–Diacetylene Macromonomers

Linear polymers containing diacetylene groups within the repeat unit also have been synthesized and characterized. Such polymers are often referred to as *macromonomers*; when exposed to heat or light radiation, the diacetylene units react with neighboring macromonomers to produce a conjugated backbone transverse to the chain direction of the macromonomer [4]. For the topochemical reaction to occur, the geometrical packing criteria discussed earlier must be met; therefore, well-ordered semicrystalline polymer systems are often more reactive. However, amorphous polymers have also been found to react, particularly if some type of paracrystalline or semiordered state exists. The macromonomer reaction process, termed cross-polymerization to distinguish it from the more typical random polymer crosslinking, is shown in Fig. 15E.1b; in the schematic, a polymer with alternating rigid and flexible groups is shown to undergo cross-polymerization.

On the premise that an ideal material would have the tough elastomeric properties of a TPE and the chromic properties of polydiacetylene chains, the Rubner research group developed a series of segmented copolymers that contain diacetylene groups incorporated in the hard segments [1]. On exposure to suitable radiation or thermal annealing, the diacetylenes undergo solid-state topochemical reaction to form a conjugated polydiacetylene network connecting the host polyurethane segments. This cross-polymerization process imparts large differences in the mechanical and optical properties of the polyurethanes, with only small morphological changes.

The structures of the polyurethanes designed and synthesized by Rubner and associates are shown in Fig. 15E.2. The polymers consist of urethane–diacetylene hard segments for which the average number of diisocyanate monomer units incorporated per segment is two. The hard segments alternate with poly(tetramethylene oxide) (PTMO) soft segments with molecular weights of 1000, 2000, and 3000. The hard segment structures are varied to obtain a range of morphologies and mechanical characteristics. Hard segments based on 4-4'-methylenebis(phenyl isocyanate) (MDI) contain two aromatic rings between the polyurethane linkages to provide a semirigid hard segment structure. Hexamethylene diisocyanate (HDI) is used to obtain a highly regular, aliphatic hard segment. In all cases, the diol chain extender contains the diacetylene group. The length of the chain extender is either one methylene group (2,4-diacetylene diol) or four methylene groups (5,7-diacetylene diol). The nomenclature used for the series of PU–DAs investigated by the Rubner group is as follows: hard segment type-chain extender–PTMO molecular weight (e.g., HDI-2,4-1000). This nomenclature will be used throughout this review to refer to this set of polyurethanes.

The polyurethanes are synthesized using a two-step solution polycondensation method. First, dihydroxyl terminated PTMO is end-capped with diisocyanate using a 2 : 1 ratio of isocyanate to PTMO. Next, the diacetylene diol chain extender is added in a 1 : 1 ratio with the PTMO prepolymer. This second step results in the formation of polyurethane linkages as the molecular weight of the polymer increases. The weight average molecular weights of these materials range from 30,000 to 100,000 or greater. The use of the two-step solution polymerization differs from many industrial methods for the production of polyurethanes, such as one-shot bulk polymerizations or reaction injection molding, which generally result in a very large distribution of hard segment sizes. The size distribution of hard segments is expected to be much narrower for the PU–DAs. The well-defined chemical structure of these materials make them ideally suited for morphological and mechanical studies.

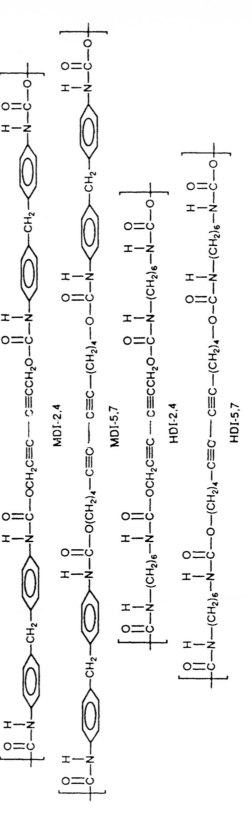

Figure 15E.2 Chemical structures of hard segments for PU–DAs

Segmented polyurethanes with diacetylene groups were also introduced by Liang and Reiser at approximately the same time as Rubner's materials [19]. The intended application of the Reiser polyurethanes was for use as deep UV photoresistant materials. The hard segments were based on hexamethylene diisocyanate or on cyclohexanediyl diisocyanate, using 2,4-hexadiyn-1,6-diol as a chain extender. The soft segments used included PTMO, polybuta-diene, polycaprolactone, and poly(dimethylsiloxane-ethylene oxide) dicarbinol. In their investigations, the authors exposed select areas of PU–DA films to UV light using a photolithographic mask, and then used a variety of solvents to wash away the unexposed areas. The photoreactivity of each of the various polymers has been investigated, and a detailed study has been completed on the effects of segmented polyurethane morphology on UV reactivity [20].

More recently, other researchers have also begun to synthesize and examine the properties of the PU–DAs. These groups include the work of Young and associates [21–23], who have developed polyurethanes of similar structure to those of Rubner, but using a single-step bulk polymerization method that results in different polydispersities and film morphologies. A primary focus of this work has been to examine the optomechanical properties of the polymers using Resonance Raman spectroscopy to measure stress in the polydiacetylene backbone during deformation. Huang and Edelman [24] have recently produced polyetherurethaneureas that contain diacetylene groups in the hard segments. These materials are synthesized from hexamethylene diisocyanate, 2,4-hexadiyn-1,6-diol, and an amine-terminated oligoether. A range of morphologies were obtained by varying the amount of hard segment; the diacetylene reactivities were seen to vary with morphology as well.

15E.2 Characterization of Polyurethane–Diacetylene Segmented Elastomers

15E.2.1 WAXD and Thermal Characterization of Polyurethane–Diacetylenes

Segmented polyurethanes phase segregate to form thermodynamically favored "hard" and "soft" domains. For the PU–DAs the reactive diacetylene groups reside in the hard domains. It is within these domains that cross-polymerization takes place. The rate and extent of diacetylene polymerization is affected by the order and packing characteristics of segments in the hard domain. As with most segmented polyurethanes, the hard domain order is dependent on the history of the sample. The morphology of the segmented PU–DAs also affects their mechanical properties, deformation behavior of the films, and mechanochromic response. Given the importance of the phase morphology, the thermodynamic behavior, and the ordering of the PU–DAs, this section will address these basic characteristics of these polymers.

The thermal transitions of some of the PU–DAs, as determined by differential scanning calorimetry (DSC), are shown in Table 15E.1 [25]. The samples used were cast from a toluene–tetrahydrofuran solvent combination that has been shown to best promote phase separation and more ordered hard domains. Endothermic melt transitions at approximately 100 °C to 120 °C have been observed for each of the elastomers, with the exception of MDI-2,4-1000. These transitions are due to melting of the crystalline portions of the polyurethane hard domains, as indicated by DSC, thermomechanical analysis (TMA) and visual observations of melt behavior [3]. The most pronounced melt endotherms are those of HDI-5,7-1000 and -2000, indicating that the HDI polyurethanes are considerably more crystalline. On the other hand, MDI-2,4-1000, which contains 32% hard segment by weight, is noncrystalline; no high-temperature peaks appear for this material (see Table 15E.1). Wide angle X-ray diffusion (WAXD) results verify these findings, indicating sharp, distinct Bragg reflections for the HDI polymers, but no crystalline refections for the MDI polymers. Reasons for the low crystallinity of MDI-2,4-1000 include the lack of flexibility needed for the somewhat bulky MDI unit to achieve crystallinity with an average hard segment size of only two MDI units. Increasing the length of the chain extender apparently allows the additional flexibility needed for a small amount of crystalline ordering to occur, and a small melt endotherm is observed for MDI-5,7-1000.

Glass transition temperatures of each of the polyurethanes are also listed in Table 15E.1. The HDI materials have lower glass transition temperatures than the MDI systems of the same PTMO molecular weight, an indication that the degree of phase separation is greater in the aliphatic HDI polyurethanes. The aliphatic, crystalline nature of the HDI hard segments appears to make the HDI systems the most highly phase separated of this series. The effect of doubling the PTMO molecular weight is to increase the degree of phase separation, as seen by the reduction of -20 °C in the glass transition temperature. Soft segment crystallization is also observed in the PTMO 2000 polymers, as shown by the low-temperature melt endotherms present in those elastomers. Infrared spectral analysis of the PU–DAs was used to attain a better sense of the phase segregation in these copolymers. A convenient measure of phase separation that has been used in other polyurethane studies is a measurement of the fraction of inter-urethane-bonded carbonyl [26]. The ratio of hydrogen bonded carbonyl absorbance versus total carbonyl absorbance is determined and used as a comparative measure. It was found that cast films of the PU–DAs exhibited a wide range of inter-urethane bonding. HDI-5,7-1000 had the highest percentage of inter-urethane bonding both before and after annealing, indicating the largest degree of phase segregation. HDI-2,4-1000 has the next highest fraction, followed by the MDI elastomers, both of which exhibit the lowest degrees of phase segregation.

Table 15E.1 Thermal Transitions of the Polyurethane–Diacetylene Elastomers

Material	Annealing treatment	Low-temperature DSC (T_g, T_{exo}, T_{endo}) (°C)	Endotherms (°C)	Exotherms (°C)
MDI-5,7-1000-(D)	90 °C, 45 min	−54, none, none	none, 103 (122)[a]	321
MDI-5,7-2000	90 °C, 45 min	−72, −28, 11	64, 107	315
HDI-5,7-1000	90 °C, 45 min	−73, none, none	46, 107	319
HDI-5,7-2000-(C)	90 °C, 45 min	−76, none, 13	none, 103	331
HDI-2,4-1000	90 °C, 45 min	−73, none, none	45, 94	219

[a] Represents weak high-temperature shoulder

All of the PU–DAs exhibit exotherms at 200–300 °C or greater. This exotherm has been attributed to a combination of thermally induced, liquid state cross-polymerization, and thermal degradation. Thermal studies have shown that the size of these exotherms generally decrease with irradiation, indicating that some of the enthalpy is probably due to polymerization of the diacetylene groups in the liquid state; thermal gravimetric analysis indicates that degradative weight loss is also observed at these temperatures. This phenomenon has been observed with a large number of diacetylene monomers and macromonomers.

15E.2.2   Studies of Domain Morphology

Small angle X-ray scattering (SAXS) has also been used to study the morphology of the SPUs [27]. The averaged spacing obtained from these studies gave a general idea of the characteristic domain size for each polymer. HDI-5,7-1000 and -2000 exhibit interdomain spacings of 130 Å and 140 Å, respectively; these values are roughly twice the length of the HDI hard segment (70 Å), and are close to the expected average length obtained from the two-step polymerization process. MDI-2,4-1000 and MDI-5,7-1000 films have long spacings of 82 Å and 91 Å; the length of the MDI 2,4 hard segment is 30 Å. A SAXS Porod analysis confirmed that the phase interface is more diffuse for MDI elastomers than for HDI-based copolymers. The relative degree of phase separation, as determined by the electron density variance and Porod's invariant, was also lower for the MDI as compared to the HDI elastomers. Finally, it was found that cross-polymerization had little or no effect on any of these parameters, confirming the idea that morphology is not disrupted by cross-polymerization.

To further develop models of this unique series of PU–DAs and their mode of deformation, an electron microscopy study of their morphology and microstructure also has been performed [28, 29]. This study has provided important information on both the superstructure and microstructure of the MDI and HDI polyurethanes, as well as some verification of preliminary models derived from thermal analysis and X-ray scattering. Transmission electron microscopy (TEM) was used to examine the thick (0.125 mm or 5 mil) solution cast, electron beam irradiated films used for mechanical and mechanooptical testing, as well as extremely thin films of polymer cast from solution directly onto TEM copper grids, and then irradiated with UV light. These thin films were illustrative of the representative morphologies of each of the polyurethane types. Figure 15E.3a contains a TEM micrograph of HDI-5,7-1000; the darker regions contain the osmium-stained hard domains, and the light regions are soft segment rich domains. A prespherulitic (hederitic) superstructure is evident throughout this film, with electron-dense centers of approximately 0.8 to 1.0 μm (8000 to 10,000 Å) in size. The branches split into many fine hairlike fibrils that are approximately 100 Å to 200 Å wide. In general, the hard and soft domains form phases that are well defined, with sharp phase boundaries and an organized interconnected superstructure. In contrast, MDI-2,4-1000 exhibits an ill-defined superstructure, as shown in Fig. 15E.3b). The macrostructure consists of hard segment rich globules dispersed in a continuous soft segment phase. Both the sizes and shapes of these structures are irregular; the diameters of the dark domains range from 0.1 to 6 μm (1000 to 60,000 Å) in diameter. The phase boundaries between hard and soft domains appear to be more diffuse than that of HDI-5,7-1000, which suggests a lower degree of phase separation in this polymer. Unlike the HDI-

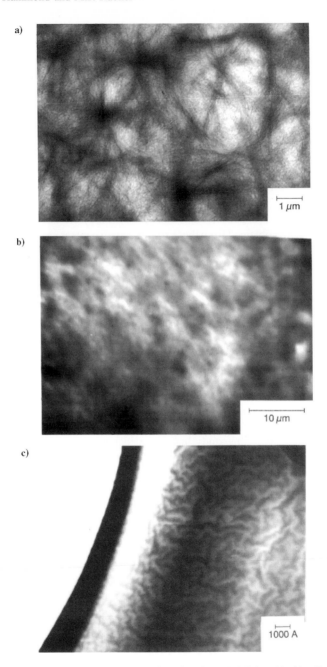

Figure 15E.3   Transmission electron micrographs of various PU–DAs; (a) thin film of HDI-5,7-1000 cast from solution onto copper grid; (b) thin film of MDI-2,4-1000 cast from solution onto grid; (c) ultracryomicrotomed cross-section of a slowly evaporated solvent cast film of HDI-5,7-2000. In all cases, dark regions represent hard domains

5,7-1000 films, the hard and soft domains do not appear to be highly interdispersed, and there is no apparent connectivity between the isolated hard domain superstructures.

The mechanical and mechanochromic properties of these materials were measured on much thicker films cast during a long-term (2- or 3-day) solvent evaporation period under a steady nitrogen flow. TEM revealed that the superstructures observed in the slow evaporated thin films had coalesced into larger two-dimensional structures in the thicker films. Spherulites became flattened disks in the plane of the film, and cross-sections perpendicular to the film plane revealed lamellar domains with much greater thicknesses of 300 Å to 400 Å, thought to be the result of coalescence of two or more hard domains during the solvent evaporation process. Figure 15E.3c shows a cross-section of an HDI-5,7-2000 slow-evaporated film. It is clear that the major aspects of the polyurethanes' morphology—well phase-separated and highly interconnected domains—are visible in these samples, as well as in the thin films.

15E.3   Mechanical Properties of Polyurethane–Diacetylenes

The tough elastomeric properties of SPUs have been attributed to the dissipation of stress through the deformation and reorganization of the hard domains. Therefore, it is important to understand specifically the effect of hard domain cohesiveness and rigidity on polyurethane mechanical behavior. The polydiacetylene backbone can be formed in both highly organized, interconnected crystalline PU–DA hard domains and in poorly ordered, noncontinuous paracrystalline domains. The reactivity, extent of polymerization, and therefore the effective increase in rigidity of the hard domains and change in tensile properties differ for each of these cases. The examination of the mechanical properties of these two types of polyurethane materials presents the opportunity to gain a deeper understanding of deformation for two morphological extremes.

15E.3.1   Effects of Molecular Weight and Annealing on Un-Cross-Polymerized Elastomers

The mechanical properties of a select number of PU–DAs of varying molecular weight were determined by Nallicheri and Rubner [25a], and are shown in Table 15E.2 and Fig. 15E.4. Each of the elastomers exhibits a relatively low initial modulus at very low strains of 2% to 5%, followed by strain softening as the polymer is extended to strains of 200% and higher; in some cases, the modulus increases at higher strains due to strain-induced crystallization of the soft segments. The moduli, ultimate tensile strengths, and ultimate strains observed for these polyurethanes are close to those of other SPUs with similar composition and molecular weights [30]. In comparing the mechanical properties of these polymers to common industrial polyurethanes, it should be kept in mind that the percentage of hard segment in the PU–DAs is relatively low, ranging from 22% to 35%.

Increased molecular weights result in higher ultimate tensile strengths (UTS), but little or no change in the ultimate strain or the initial modulus. These increases in UTS are similar to

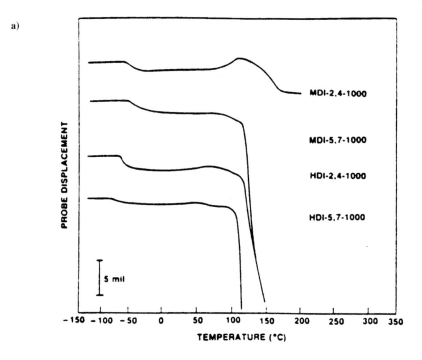

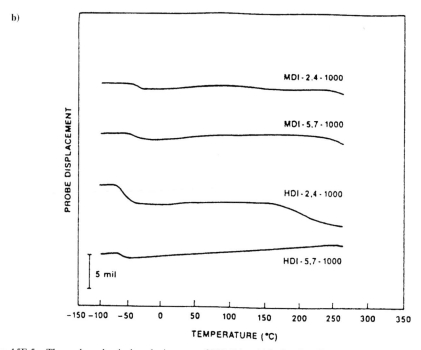

Figure 15E.5   Thermal mechanical analysis scans of PU–DAs: (a) before irradiation and (b) after 2 weeks of gamma irradiation

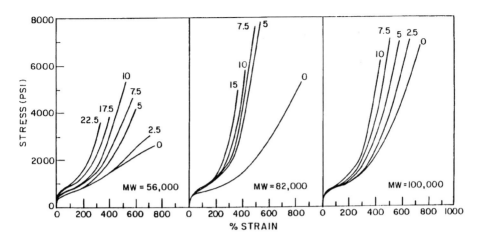

Figure 15E.6   Stress–strain curves of MDI-5,7-1000 of different molecular weights as a function of cross-polymerization. Numbers indicate radiation dosage in megarads

At very high irradiation levels, the hard domains become crosslinked to the point that plastic deformation can no longer take place at high stress. The result of this embrittlement is that the UTS decreases as the hard domains become less efficient at warding off the propagation of cracks. Thus, there exists an optimal level of irradiation at which the UTS reaches its maximum value.

When the molecular weight is increased, the effects of cross-polymerization become less significant. The higher molecular weight samples have more fully established virtual crosslinks due to the larger number of hard segments per polymer chain. The effects of chain pullout and slippage are reduced for the high molecular weight polymers. For this reason that effects of cross-polymerization are not as strongly felt in these cases. This appears to be true in general for any form of improved hard domain reinforcement. For example, well-annealed samples also show less sensitivity to the effects of cross-polymerization. Generally, molecular weight or annealing effects that improve the UTS by increasing the cohesion of the hard domain structure render further improvements obtained from crosslinking less noticeable.

The stress–strain curves of HDI-5,7-1000 and HDI-2,4-1000 elastomers as a function of radiation dosage are shown in Fig. 15E.7. The phase morphology of these elastomers is highly interconnected and well phase separated; the hard domains are known to be highly crystalline as well. The application of strain to the HDI PU–DAs results in simultaneous deformation of the hard and the soft domains. Cross-polymerization increases the rigidity of the hard domains, rendering them less deformable. The UTS is increased as the hard domains are reinforced, and the loss of plasticity of the hard segments gives lower ultimate strains. There is no strain amplification effect because the soft segments are not able to become fully extended in the HDI morphology. However, the interconnected nature of the domains means that modifications to hard domain rigidity greatly affect modulus. The 100% strain modulus, in fact, increases by a factor of 2 when exposed to 0.56 Mrad.

The mechanical behavior of the HDI-based PU–DAs is generally less responsive to radiation-induced changes. This is consistent with the fact that the hard domains are well

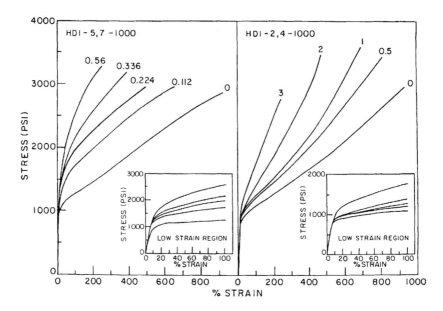

Figure 15E.7   Stress–strain curves of HDI-5,7-1000 and HDI-2,4-1000 as a function of radiation dosage. Numbers indicate radiation dosage in megarads

phase segregated and highly crystalline. The well-ordered domains do not experience as dramatic an improvement in ultimate tensile strength. The sensitivity of the cross-polymerization effects to molecular weight and annealing has not been measured at this time. Preliminary experiments suggest that molecular weight is not as dominant a variable as it is with the MDI-based equivalents.

The changes in the amount of hysteresis in the PU–DAs with increased degrees of cross-polymerization support the models of mechanical deformation described above [32]. Cross-polymerization of MDI-5,7-1000 decreased hysteresis; this is due to the reinforcement of hard domain cohesion. The hard domains were less likely to break up prematurely when crosslinked, thus reducing energy losses and permanent set. Cross-polymerization of HDI-5,7-1000 increased hysteresis. In this case, deformation involves the simultaneous disruption of highly interconnected hard and soft domains. Crosslinking the hard domains only serves to increase the energy losses experienced with strain, as more irreversible plastic deformation is required to extend the polyurethane.

Cross-polymerization also affects the thermal and swelling characteristics of PU–DAs. The melt/disordering endotherms of the moderately crosslinked elastomers used for the mechanical studies exhibit broadening, decreased enthalpies, and higher temperatures when compared to nonirradiated equivalents. The fact that the melt endotherm is still present indicates incomplete conversion, and therefore some mobility of the polymer chains. Highly cross-polymerized samples exhibit no endotherms, and have very poor mechanical strength. Lightly cross-polymerized PU–DAs swell in solvents such as toluene and tetrahydrofuran (THF). In fact, the degree of swelling can be controlled by varying the degree of cross-polymerization.

15E.4   Linear Optical Properties of Segmented Polyurethane–Diacetylenes

Each of the MDI and HDI series of PU–DAs can be cross-polymerized using UV or electron beam irradiation to give highly colored elastomeric films. The color obtained, generally blue, purple or red, reflects the degree of planarity and the static strain effects of the resulting network of conjugated backbones. The optical absorption spectra of polydiacetylene chains are expressions of the relative amounts of order, packing, and degree of strain experienced by the conjugated backbone in its host environment. It is therefore possible to use the polydiacetylene chain as a visual monitor of changes that take place within the host monomer or polymer matrix. Thus it indicates whether that matrix is an amorphous or highly ordered crystalline environment, and whether it is in the solid, liquid, or solvated states. The varied morphologies of the polyurethanes provide an opportunity to examine the relationships between the absorption spectra or color and the status of the host system.

15E.4.1   Thermochromism

A comprehensive examination of the thermochromic behavior of cross-polymerized PU–DAs illustrates the different types of chromism that can be observed in these polymers [2]. Specifically, the optical changes observed for HDI-5,7-1000 were found to be quite different in nature from those of HDI-2,4-1000 and the MDI systems. In every case, the spectral shifts and variations in linewidth could be correlated to molecular processes taking place within the hard domains of the polyurethanes as the temperature was increased. For the visible absorption studies discussed below, the hard domains of all polyurethane films were lightly cross-polymerized (<5% crosslinked) by exposing cast films to UV or electron beam irradiation. At such low levels of cross-polymerization, the films remained transparent, although deeply colored, allowing the visible spectroscopy to be done in transmission mode. Under these circumstances, the predominant host environment of the polydiacetylene chains is that of the original polyurethane host hard segments to which the conjugated backbone is covalently bonded.

Figure 15E.8 shows the visible absorption spectra of cross-polymerized HDI-2,4-1000 taken at increasing temperatures. At room temperature, an excitonic peak at 570 nm is visible, along with a broad second band representative of a large distribution of conjugation lengths present in the polydiacetylene. The vibronic sidebands of the conjugated bonds in the hard domain are convoluted with this second peak. On heating, the excitonic peak gradually decreases in intensity as the absorption band shifts to higher energies; these effects occur as the polymer undergoes thermal expansion, and finally begins to melt at higher temperatures. The peaks become broader with temperature, indicating a wide range of poorly organized polydiacetylene conformations present. The gradual shift to lower wavelengths and the loss of order observed in the spectra illustrate a clear example of an order–disorder transition. This transition is reversible up to 100 °C, which is close to the melting temperature of the un-cross-polymerized polyurethane hard segments. Beyond this point, the disordering of the hard domains that is characteristic of the melting process induces irreversible changes in the

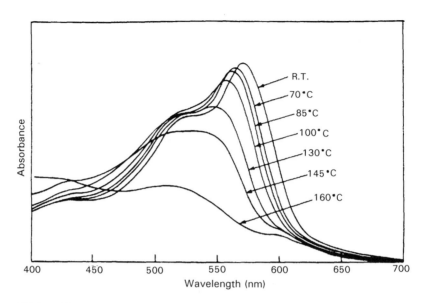

Figure 15E.8   Visible absorption spectra of cross-polymerized HDI-2,4-1000 as a function of temperature (spectra have been arbitrarily offset for clarity)

conjugation length of the polydiacetylene backbone. As the melting range is spanned, the excitonic transition peak decreases in intensity until it is completely lost in a featureless band. The thermochromism is affected by the melting of the unreacted macromonomer lattice; in fact, fully cross-polymerized HDI-2,4-1000 does not undergo any melting behavior or thermochromic transition at all. The presence of residual diacetylene macromonomer in the lattice affects the conformation of the conjugated backbone.

At room temperature, the elastomeric film is deep red in color. As it is heated, the color changes from red to yellow. It has also been found that when cooling the film to liquid nitrogen temperatures, the film becomes blue. This color change is due to thermal contraction of the polymer lattice, which causes a gradual shift of the excitonic transition toward higher wavelengths, and is a direct result of the high coefficient of thermal expansion of this lightly cross-polymerized PU–DA. In summary, the HDI-2,4-1000 polymer undergoes chromic transitions that directly correspond to the thermodynamic changes in the host polymer lattice.

The nature of the thermochromic transition of HDI-5,7-1000 is quite different from that of the HDI-2,4 elastomer. Instead of a gradual order–disorder transition in which the features of the absorption curves become broadened and poorly defined, the spectra undergo dramatic changes while retaining a high degree of order. The visible absorption spectra taken during a series of heat/cool thermal cycles of cross-polymerized HDI-5,7-1000 are shown in Figs. 15E.9a and b. The first excitonic transition slowly moves to lower wavelengths due to thermal expansion of the polyurethane matrix. An entirely new peak appears at 90 °C and continues to increase in size with increasing temperature, at the expense of the original 650 nm peak. At 160 °C, the original peak is completely replaced by the new one. The population of polydiacetylene chains then exists completely in this second, well-ordered phase with a peak at 511 nm.

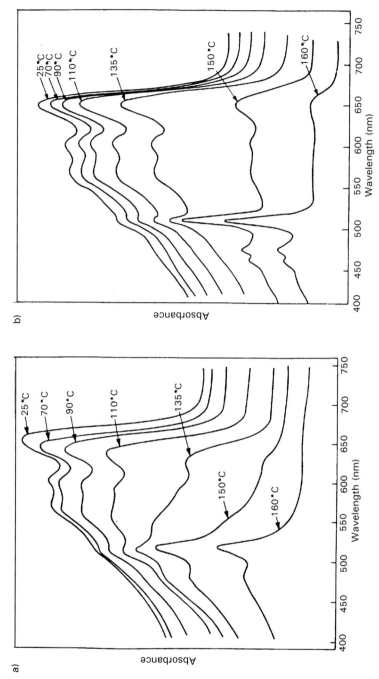

Figure 15E.9   Visible absorption spectra of cross-polymerized HDI-5,7-1000 as a function of temperature: (a) heated state and (b) cooled to room temperature after heating to the indicated temperatures (spectra have been arbitrarily offset for clarity)

The color changes observed during this thermochromic transition extend from a blue film corresponding to the original excitonic peak at 650 nm, to a yellow color characteristic of the new absorption curve. The thermochromism is completely reversible up to approximately 130 °C. Above this temperature, the yellow phase remains highly visible in the spectrum on cooling to room temperature, but the size of the blue phase is diminished. The blue phase does remain present when the temperature is returned to 25 °C, however, even after cooling from 160 °C. The coexistence of the two primary peaks suggest that they represent two thermodynamic phases in which the hard domains may exist, each with its own characteristic average polydiacetylene conjugation length. The yellow phase is at least as highly ordered as the blue phase, judging from the clear excitonic peak, and the well-defined vibronic sidebands that are visible in the room temperature spectra of Fig. 15E.9b. These sidebands, which appear at 475 and 462 nm, correspond to the double and triple bond vibrations of the conjugated polydiacetylene backbone; the definition of these peaks at higher temperatures was compromised by thermally induced line broadening. The presence of such a highly ordered visible absorption spectrum at such low wavelengths is quite rare for polydiacety-lene-containing materials. The implication is that a well-ordered, regular arrangement of polydiacetylene chains exists that has a low effective conjugation length.

FTIR studies of the inter-urethane hydrogen bonding of the cross-polymerized poly-urethanes indicate that cross-polymerization results in the retention of hydrogen bonding well above the original polymer melting points. However, a change in the slope of the percent inter-urethane bonding versus temperature curve is found for HDI-5,7-1000, as opposed to the constant slope derived for HDI-2,4-1000 [3]. This slope change indicates a change in the thermal expansion coefficient of HDI-5,7-1000 at 90 °C, and corresponds directly to the onset of the thermochromic transition observed optically. These findings imply that a crystal-lographic or conformational transition occurs at this temperature that involves a volume change of the original polymer lattice. In fact, to further understand what is taking place on a molecular level, Rubner [10] performed the FTIR study of poly-ETCD mentioned earlier; ETCD is a urethane-substituted diacetylene monomer with a structure similar to that of the HDI 5,7 hard segments. It was found that the thermochromic transition of poly-ETCD involves a localized conformational rearrangement of the methylene spacer groups at the thermochromic temperature, which results in the relaxation of the conjugated polydiacetylene chain to a conformation less conducive to electron delocalization. NMR studies on poly-ETCD and on diacetylene-polyamides suggest that these conformational changes are *trans*-gauche transitions of the alkyl side groups. It is believed that HDI-5,7-1000 undergoes a similar conformational rearrangement to form the yellow phase. In both cases, it is the retention of hydrogen bonding that prevents complete relaxation and disordering of the polydiacetylene backbone. HDI-2,4-1000 undergoes thermal expansion and melting, and hydrogen bonding is ultimately lost in this system, as seen in an order–disorder thermo-chromic transition.

The visible absorption spectra of cross-polymerized MDI-2,4-1000 were taken at room temperature, and also after heating to 120 °C and cooling back to 25 °C. A primary absorption peak at 630 nm was observed for this PU–DA. The lack of definition of the absorption band is due in part to the low degree of ordering and phase segregation in the polyurethane film. After heating MDI-2,4-1000, permanent changes are seen in the absorption curve. The absorption band has shifted to higher energies, and an expansive, completely featureless curve at 500 nm remains. The gradual decrease in wavelength and broadening of the original bands that occurs on increasing temperature is characteristic of an order–disorder transition, the result of the gradual disordering of the hard domains as the material is heated. Interestingly, the absorption spectral changes are analogous to those observed during the dissolution of soluble hydrogen-bonded polydiacetylenes. The paracrystalline hard domains of the segmented polyurethane undergo a gradual breakup and disordering as the polymer is heated; these changes are irreversible, and the color change from purple to red becomes permanent upon heating.

15E.4.2 Mechanochromism

Just as the visible absorption spectrum of the conjugated polydiacetylene backbone is sensitive to changes in temperature, it is also responsive to mechanical stress and deformation. In fact, the polydiacetylene chain is a highly sensitive molecular probe that can be used to gain insight into the state of stress of its host lattice. The mechanochromic properties of the segmented polyurethanes can be used to observe the relative amount of stress experienced by the hard domains.

An additional aspect of the polydiacetylene backbone that can be exploited in mechanochromic studies is the anisotropy of the conjugated polymer chain. The optical absorption coefficients of polydiacetylene chains are highly directional, with the predominant component existing along the conjugated backbone. Dichroism measurements using polarized light can be used to assess the degree and direction of orientation of the polydiacetylene backbone within the hard domains. This technique, combined with the change in the shape of the absorption spectra with mechanical stress, provides the opportunity to closely examine the deformation process of the segmented polyurethanes with varying chemical structure, morphology, and hard domain rigidity. Several studies have been completed on the mechanochromism of the PU–DA segmented block copolymers [2, 33–35].

A simple mechanooptical measuring technique has been used to examine the chromic transitions and strain-induced orientation in the hard domains of the polyurethanes. Thin films cast from a THF–toluene solvent mixture are irradiated to achieve low levels of cross-polymerization. The visible absorption spectra are measured during tensile deformation of the films. The sample is subjected to loading and unloading cycles at increasing levels of strain. For each strain level, horizontally and vertically polarized visible spectra are recorded while the sample is held in the strained state; the sample is then relaxed to zero load, and polarized spectra are measured in the relaxed state. The cycle is repeated to higher strain levels until the sample fails mechanically. Results for MDI-2,4-1000 are shown in Figs. 15E.10 and 15E.11 for the stretched and relaxed state, respectively.

The visible spectra recorded at 0% strain represents the initial state of the polydiacetylene backbone; the lowest energy peak appearing at 625 nm is the excitonic peak typically seen in polydiacetylenes and it gives rise to the blue color seen prior to straining the MDI-2,4-1000 film. Figure 15E.10a presents data obtained from the sample in the vertically polarized stretched state; as the level of strain increases, the peaks gradually broaden and shift to higher energies (i.e., lower wavelengths). This is seen as a blue to red color change in the film. As stress is applied to the elastomer, some of it is transferred to the stiff hard segments, which contain conjugated polydiacetylene backbones. As these backbones become distorted due to

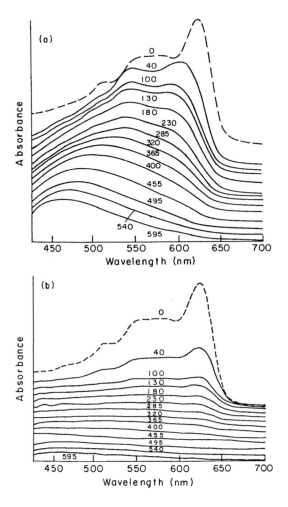

Figure 15E.10   Visible absorption spectra of MDI-2,4-1000 recorded in the stretched state as a function of increasing strain: (a) vertically polarized light and (b) horizontally polarized light. Numbers indicate strain levels (spectra have been arbitrarily offset for clarity)

the stress, co-planarity of the pi orbitals is lost and the conjugation length decreases. The broadening of the peaks is indicative of the widening distribution of conjugation lengths present in the sample. At high strains, only a broad, featureless peak remains, and at 500% strain and higher (not shown), strain-induced crystallization is observed as a change from a transparent to a translucent film.

The absorption spectra in horizontally polarized light (Fig. 15E.10b) indicates that there is very little shift in energy for conjugated backbones oriented lateral to the strain direction; thus the level of stress is relatively low in this direction. Also note that the area under the curves decreases rapidly with increasing strain, indicating orientation of the polydiacetylene

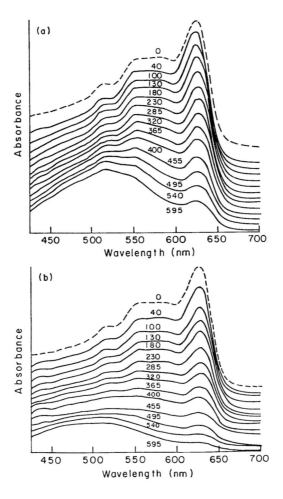

Figure 15E.11   Visible absorption spectra of MDI-2,4-1000 recorded after relaxation from the indicated strain levels as a function of increasing strain: (a) vertically polarized light and (b) horizontally polarized light. Numbers indicate strain levels (spectra have been arbitrarily offset for clarity)

chains along the direction of stress during tensile elongation. Because the polydiacetylene backbone is approximately perpendicular to the host polyurethane hard segments, we can conclude that the hard segments connected by polydiacetylene chains orient perpendicular to the stress direction during the deformation process. It is apparent from vertically polarized spectra of the polymer in its relaxed state, shown in Fig. 15E.11, that the mechanochromic effect is highly reversible up to 350% strain. At higher strains, some permanent changes are noted; the relative intensity of the excitonic transition at 615 nm decreases with respect to the broad peaks at lower wavelengths, and a high energy tail is formed at very high strain levels. These changes are evidence of the gradual permanent disruption of hard domain ordering that occurs at high strains. Similar observations are made for the visible spectra taken with horizontal polarization. As in the thermochromic behavior of MDI-2,4-1000, this transition appears to be to an order–disorder transformation. However, unlike the case of thermochromism, the transition is quite reversible to fairly high levels of strain.

The visible absorption spectra of HDI-5,7-1000 in the stretched state is shown in Fig. 15E.12. The excitonic transition appears at 615 nm, and the secondary, broad absorption band at lower wavelengths is due to disordered polydiacetylene chains with a wide distribution of effective conjugation lengths that probably exist in less ordered areas of the hard domains. Figure 15E.12a presents data from the vertically polarized stretched state. At low levels of strain, a new peak appears at 495 nm; this peak is representative of a new phase with a lower average conjugation length. Just as was observed in the thermochromic behavior of HDI 5,7 elastomers, the hard domains undergo a pronounced phase transition that results in the appearance of a second, well-ordered phase. This is the first reversible mechanically induced chromic transition reported, and it illustrates the concept of a well-ordered, controlled mechanochromic response. In this case two different, ordered molecular environments exist for the polydiacetylene backbone; on stretching, the yellow phase with a peak at 495 nm increases in proportion, as the blue phase at 615 nm is decreased. The appearance of this transition at strains as low as 15% is thought to be the result of a highly interconnected phase morphology, in which stress is transferred to the hard domains early in the deformation process. Both peaks gradually broaden and shift to higher energies. At high strains, only a broad, featureless peak at 450 nm remains. On relaxation, the absorption spectra (not shown) illustrate reversibility up to 300% strain. After stretching to higher strain levels, the yellow phase remains present in the spectra, although the blue phase continues to dominate the relaxed state absorption spectra even when stretched to 700% strain. This phase transition is very similar to the thermochromic transition observed for HDI-5,7-1000. It is probable that the same sort of *trans*-gauche conformational rearrangement of the methylene spacer groups occurs on stretching the elastomer as upon heating it.

The absorption spectra in horizontally polarized light (Fig. 15E.13b) indicates that there is very little shift in energy among conjugated backbones oriented lateral to the strain direction, thus the level of stress is relatively low in this direction. Also note that the area under the curves decreases rapidly with increasing strain, indicating orientation of the polydiacetylene chains along the direction of stress.

Dichroic ratios from visible absorption studies of MDI-2,4-1000, HDI-5,7-1000, and HDI-5,7-2000 were calculated by taking the ratio of vertical and horizontal polarized spectra at each strain level. These results are shown in Fig. 15E.13 [35]. It is apparent that in the stretched state a great deal of orientation takes place in the polyurethanes. This suggests that initially, the hard segments are stacked with their molecular axes perpendicular to the long axis of the hard domains. The domains orient parallel to the stress direction during the deformation process, thus causing the conjugated backbone connecting the hard segments to also line up along the stress direction. There is a maximum in the dichroic ratio for each of the polyurethanes shown. For the film samples shown, the maximum occurs at approximately 300% strain; as will be discussed shortly, the point at which the maximum appears, and even the appearance of a maximum at all, depends on the degree of cross-polymerization and the thermal history of the polymer samples. The observed decrease or leveling of the vertical orientation with increased strain may be due to the physical break up of the original hard domains, and subsequent reordering of the hard segments along the direction of stress. This type of hard segment orientation has been suggested by Bonart [36], Seymour et al. [26], and

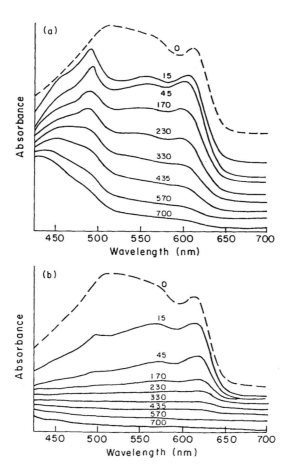

Figure 15E.12   Visible absorption spectra of HDI-5,7-1000 recorded in the stretched state as a function of increasing strain: (a) vertically polarized light and (b) horizontally polarized light. Numbers indicate strain levels (spectra have been arbitrarily offset for clarity)

Kimura et al. [37]. It is important to note that because the visible spectra data only include information on hard segments linked by polydiacetylene chains, and the levels of cross-polymerization are low in these materials, it is not clear from this technique alone whether the average, noncrosslinked hard segment orients transverse or parallel to the stress direction. Infrared dichroism would provide this information.

The dichroic ratios of the HDI elastomers indicate greater degrees of orientation in the hard domains than in the MDI polyurethane. This is consistent with the fact that the HDI hard domains are highly crystalline, and are more likely to exhibit higher order parameters in the stretched state than the paracrystalline MDI polymer. On the other hand, the dichroic ratios calculated for the relaxed states of the polyurethanes indicate that HDI-5,7-1000 experienced the greatest levels of residual orientation, an indication of permanent set and deformation of the hard domains, due to the fact that the hard and soft domains tend to deform simultaneously in

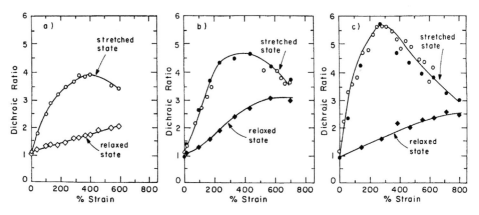

Figure 15E.13   Dichroic ratios of (a) MDI-2,4-1000, (b) HDI-5,7-1000 and (c) HDI-5,7-2000 in the stretched and relaxed states as a function of increasing strain level. For the HDI-based elastomers, the closed data points represent hysteresis runs, whereas the open points represent continual stretching runs

the interconnected HDI phase morphology, and that the hard domains experience a great deal of permanent deformation. The domains of HDI-5,7-2000 are thought to be highly intercon-nected, as well; however the longer soft segment molecular weight allows for a greater amount of overall extensibility, and therefore, less permanent set. Finally, MDI-2,4-1000 does not have an interconnected morphology. This segmented copolymer consists of more isolated hard domains in a continuous soft segment matrix, thus allowing the soft segments to deform initially, followed by hard segment deformation and breakup at higher strains.

Investigations on the effect of increased cross-polymerization on the polyurethanes have recently been completed, and indicate that at low conversion levels, the deformation orientation behavior of the hard domains is a function of the degree of conversion [28]. For example, the visible absorption dichroic ratios of HDI-5,7-2000 films irradiated at three different levels are shown in Fig. 15E.14. The levels of cross-polymerization were too low to note any change in the stress–strain behavior of the polyurethane samples. Despite this fact, even slight changes in the degree of cross-polymerization can make large differences in the degree of orientation and the permanent deformation experienced by the hard domains. At 0.02 Mrad, the dichroic ratio does not get much higher than 5, whereas a dichroic ratio of 9 is achieved in the 0.08 Mrad sample. The amount of permanent set is also considerably higher in the 0.08 than in the 0.02 Mrad sample, particularly at strains below 500% to 600%, as indicated by the much lower dichroic ratios in the relaxed state of the lower irradiation sample. Finally, at very low conversions, a maximum is not present in the plot of dichroic ratio versus strain level; instead, the vertical orientation of the hard domains increases monotonically up to 650% strain. On the contrary, the 0.08 Mrad sample clearly undergoes a maximum at 350% strain, followed by a decrease in vertical orientation. As discussed above, the presence of a maximum indicates the breakup and realignment of the original hard domains.

These differences can be understood if one considers the fact that cross-polymerization effectively increases the rigidity and cohesiveness of the hard domains, and that the HDI morphology consists of considerably interconnected domains. As the hard domains become

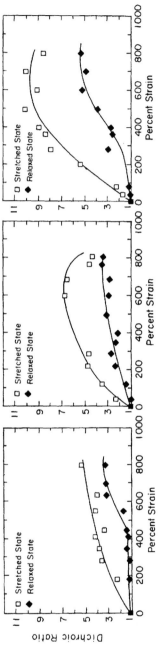

Figure 15E.14   Dichroic ratios of HDI-5,7-2000 as a function of increasing electron beam irradiation levels

more cohesive, they will become more oriented in a stress field, as they begin to bear more load relative to the highly extensible soft segments. The extensibility of the hard domains is effectively decreased, however, resulting in the early rupture and reordering of the hard segments. Increased rigidity is also evidenced in the actual absorption spectra obtained from the samples. The onset of the mechanochromic transition actually takes place at slightly higher strains. Apparently, greater strain levels are needed to induce the conformational transition in the more highly cross-polymerized samples. The fact that these differences were detectable even at very small degrees of cross-polymerization indicates the large effect of cross-polymerization on the hard domain rigidity. The insignificant differences found in the mechanical stress–strain curves of the samples, on the other hand, suggest that these changes in rigidity must be occurring only in very localized areas. Similar observations have been made for MDI-2,4-1000. Visible absorption and dichroic observations made in a separate study on samples that were more highly crosslinked revealed that the hard domains remain intact to larger strains; for example, the dichroic ratio does not reach a maximum within the strain limits of the elastomer. These observations are in keeping with the proposed mode of deformation for the MDI series, in which isolated hard domains are reinforced by cross-polymerization, and are therefore less likely to be permanently deformed at low or moderate strains. In short, just as the effects of cross-polymerization on hysteresis are reversed for the MDI versus the HDI polyurethanes, so the effects of irradiation on the visible dichroism results differ, and for similar reasons. Remarkably, the polydiacetylene cross-polymerization process appears to be capable of micromanipulating deformation behavior on a very localized level at low dosages, and of making significant changes in the overall mechanical properties at high radiation levels.

15E.4.3   Morphological Model of Deformation

A model describing the deformation of the cross-polymerized PU–DAs can be constructed from the information obtained from mechanical stress–strain data, mechanochromic and visible dichroism measurements, and the TEM and SAXS results. Figure 15E.15 is a schematic of the proposed deformation process for an HDI-based PU–DA. Randomly oriented lamellar or cylindrical domains are shown, based on the lamellar regions seen in HDI-5,7 elastomers. In the model shown, it is assumed that these regions consist of single stacks of hard segments oriented perpendicular to the long axis of the domains. The variation in lamellar thickness is due to the range of hard segment lengths present in the polyurethane. For these polyurethanes, the average hard segment length should be two repeat units, but some variation in size about a standard distribution is to be expected. It has been mentioned that the lamellar regions in the micrographs may be superstructure composites of two or more stacks of hard segment; this is not illustrated in the model shown here, but the hard segment orientation and deformation behavior for this case should be quite similar. Soft segments exist as randomly coiled, entangled chains connecting the hard domains. Branching, as seen in the TEM micrographs, is shown here as the result of smaller hard segments separating to accommodate soft segments, as suggested by Fridman and Thomas [38] in their spherulitic model.

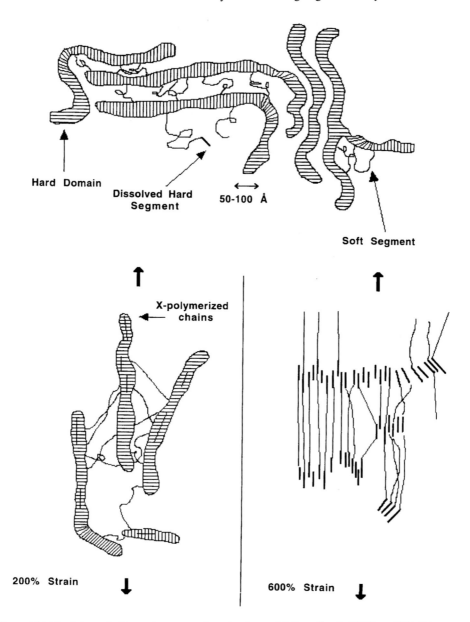

Figure 15E.15   Schematic illustrating proposed mechanisms of deformation in HDI-based PU–DAs

At low levels of strain, the hard domains are shown orienting along the strain direction. The polydiacetylene backbone is shown connecting cross-polymerized hard segments within the domains. The vertical orientation seen in visible dichroism is explained by the transverse orientation of the hard segments. The soft segments begin to orient parallel to the direction of stress, as hard domains are sheared past each other. At strains greater than 300%, the hard segments are pulled out of their original domains, and begin to reorient along the stress direction. This causes the dichroic ratio to reach a maximum value and then decrease as the average direction of the polydiacetylene backbones is altered. The high levels of stress and distortion are also seen in the decrease of average conjugation length.

The data obtained from visible dichroism yields information only on the hard segments connected by the polydiacetylene backbone. At low levels of cross-polymerization, a small minority of hard segments will have reacted to form the polydiacetylene chain. The non-cross-polymerized hard segments may undergo very different deformation behavior; without the added cohesiveness of the cross-polymerized network, the hard domains may break up at much lower strain levels, or orient so that the hard segments are parallel, rather than perpendicular, to the direction of tensile stress. Many studies, including those of Seymour, Allegrezza, and Cooper [26], and Bonart [36], have revealed parallel orientation of hard segments on deformation, sometimes from the point at which stress is applied. As mentioned earlier, infrared dichroism studies would clarify this issue. Visible light dichroism studies also give an accurate picture of the effects of increased rigidity in the hard domains, and the state of mechanical stress of this unique class of segmented copolymers.

15E.5   More Recent and Future Developments

The next frontier for the segmented elastomer diacetylenes lies in the use of molecular scale design of this unique class of materials to achieve new levels of versatility and control in their thermo- and mechanochromic behavior. This may be accomplished by further functionaliza-tion of the polyurethanes, and systematic variation of the chemical structure of both hard and soft segments to achieve the desired range or ranges of chromic response. The development of more highly functionalized elastomers is currently being accomplished by the incorpora-tion of the diacetylene group into the soft segments, as well as the hard segments, of the polyurethanes [39]. The thermochromic response of the polydiacetylene macromonomer systems is often tied to the melting transitions of the host polymer. Because the melting range of the soft segment is generally 30 °C to 100 °C below that of the hard segment, it is actually possible to observe thermochromic transitions from each phase. The lower melting temperatures of the soft segments result in chromic changes from 30 °C to 70 °C, while the more rigid hard segments undergo thermochromic transitions from 90 °C to 120 °C. Variation of the chemical structure of the hard and soft segments allows the temperature or strain level at which chromism is observed to be varied, thus resulting in new multifunctional elastomeric materials whose response is controlled by design. Potential applications and uses for such materials include the use of polydiacetylene as a molecular probe in mechanical studies. Lightly cross-polymerized samples would allow the separate observation of deformation and orientation behavior in the hard and soft segments using visible absorption experiments. The ability to crosslink the soft as well as the hard segments also provides the opportunity to further enhance and vary the mechanical properties in the solid state.

The structures and maximum melt transitions of the polyester–diacetylene soft segments that have been synthesized so far are shown in Fig. 15E.16. Their molecular weights range from 600 to 3000 g/mol, and they contain hydroxyl end groups to allow their incorporation into polyurethanes via conventional methods. These materials exhibit melting behavior at or near room temperature or body temperature, making them interesting for temperature sensors in the packaging industry, the medical industry, and for use as novelty or display items. The structures were varied to achieve desired melt ranges, and to obtain the desired level of crystallinity in the soft domains. Soft segments which are highly crystalline result in nonelastomeric mechanical properties; however, some degree of crystalline ordering is desirable to provide the ordered environment required for solid state topochemical poly-merization of the diacetylenes. The ability to easily vary the melt temperature, and therefore the temperature and/or strain level of the chromic response, by straightforward chemical modifications is a key concept in the issue of molecular design of these polyurethane systems.

The polyesters undergo a large shift in their absorption band when heated beyond the melt temperature. For example, the original excitonic transition of poly-10,12-docosadiyne-1,22 malonate is at 650 nm at room temperature. When heated beyond $30\,^\circ$C, a sharp decrease in the original peak is observed, and a new broad, featureless peak appears at

Series		$T_m(^\circ C)$
1	$-[O(CH_2)_9-C\equiv C-C\equiv C-(CH_2)_9O-\overset{O}{\overset{\|}{C}}(CH_2)_1\overset{O}{\overset{\|}{C}}]_n-$	35
	$-[O(CH_2)_9-C\equiv C-C\equiv C-(CH_2)_9O-\overset{O}{\overset{\|}{C}}(CH_2)_3\overset{O}{\overset{\|}{C}}]_n-$	50
	$-[O(CH_2)_9-C\equiv C-C\equiv C-(CH_2)_9O-\overset{O}{\overset{\|}{C}}(CH_2)_4\overset{O}{\overset{\|}{C}}]_n-$	59
2	$-[O(CH_2)_9-C\equiv C-C\equiv C-(CH_2)_9O-\overset{O}{\overset{\|}{C}}CH_2CH_2\overset{CH_3}{\overset{\|}{CH}}CH_2\overset{O}{\overset{\|}{C}}]_n-$	22
	$-[O(CH_2)_9-C\equiv C-C\equiv C-(CH_2)_9O-\overset{O}{\overset{\|}{C}}$ (benzene ring) $\overset{O}{\overset{\|}{C}}]_n-$	34
3	$-[O(CH_2)_9-C\equiv C-C\equiv C-(CH_2)_9O\left\{\begin{array}{l}30\%\ \ \overset{O}{\overset{\|}{C}}CH_2CH_2\overset{CH_3}{\overset{\|}{CH}}CH_2\overset{O}{\overset{\|}{C}}\\ 60\%\ \ \overset{O}{\overset{\|}{C}}(CH_2)_4\overset{O}{\overset{\|}{C}}\end{array}\right\}]_n-$	48
	$-[\left\{\begin{array}{l}30\%\ O(CH_2)_8O\\ 70\%\ O(CH_2)_9-C\equiv C-C\equiv C-(CH_2)_9O\end{array}\right\}\overset{O}{\overset{\|}{C}}(CH_2)_4\overset{O}{\overset{\|}{C}}]_n-$	46

Figure 15E.16   Diacetylene-containing polyester soft segment structures

475 nm. This particular polyester is interesting in that the thermochromic transition is triggered at body temperature. Polyurethanes have been synthesized using the soft segments shown in Fig. 15E.16 by the two-step method using hexamethylene diisocyanate and 5,7-dodecadiyne-1,12 diol as the chain extender. Preliminary results indicate that the resulting polymers exhibit two separate thermochromic transitions; detailed studies of these multifunctional materials will be addressed in later work. Once characterized, this new family of PU–DAs may be tailored by chemical modifications or processing conditions to achieve polymeric materials sensitive to a number of temperature ranges or mechanical strain levels.

These new materials may also undergo selective cross-polymerization, in which only the hard domains undergo solid state cross-polymerization, or in which both hard and soft segments are crosslinked. This is done by controlling the thermal conditions under which UV irradiation induces cross-polymerization. When the polymer film is heated to a temperature above the melt point of the soft segments, but below the melt temperature of the hard domains, the polyester soft segments in the melt state are too disorganized to undergo topochemical cross-polymerization. In this case, only the hard domains may be cross-polymerized. If irradiation takes place at room temperature, on the other hand, both the hard and soft segments are in their semicrystalline states, and are ordered enough to undergo diacetylene crosslinking.

The first set of curves shown in Fig. 15E.17 are the visible absorption spectra of an HDI based polyurethane with diacetylene groups present solely in the hard domains. It is clear that the absorption spectra are the same whether diacetylene polymerization occurs at room temperature, or between the soft and hard segment melt temperatures. Polyurethanes containing diacetylenes solely in the soft segment exhibit visible absorption spectra only if cross-polymerized below the soft segment melt temperature, as shown by the second set of curves. Above $T_{m,s}$, no cross-polymerization takes place, as evidenced by the flat absorption curve shown in Fig. 15E.17. When a polyurethane containing diacetylenic hard and soft segments is irradiated at room temperature, as shown in the third set of curves, the absorption spectra of hard and soft domains are superimposed. Irradiation between the two melt temperatures results in only the absorption spectrum characteristic of the hard domains. The ability to selectively crosslink the hard versus the soft segments means that the effects of crosslinking separate environments may be studied using optical studies, and the organization and mechanical responses of the two domains may be examined independently.

A different approach to multifunctional thermochromic materials has recently been described using the mesogenic transitions of liquid crystalline polymers or host monomers as a means of inducing two or more first-order thermochromic transitions [40]. Aromatic liquid crystalline diacid and diol diacetylene monomers, and aromatic semirigid main chain liquid crystalline polyesters containing the diacetylene group in the repeat unit, have been designed, once polymerized or cross-polymerized, to exhibit chromic transitions at the melting and clearing points of the corresponding host environment. The concept of controlled chromic behavior via the engineering of liquid crystalline diacetylenes may be extended to the hard or soft segments of a segmented polyurethane. The future development of segmented PU–DAs may well include liquid crystalline polyurethanes, using liquid crystalline polyester–diacetylenes, or diacid or diol mesogenic monomers as a basis for new polymers. The range of thermo- and mechanochromic behavior could then be expanded to include the multiphasic behavior of mesogenic polymers, and the high level of orientation possible in such systems might yield highly anisotropic optical properties that could prove interesting in

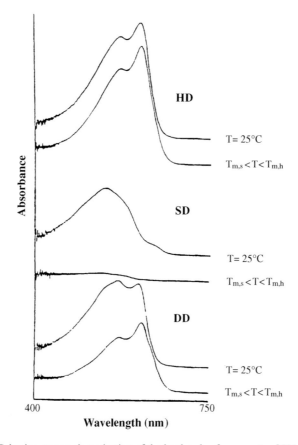

Figure 15E.17   Selective cross-polymerization of the hard and soft segments of PU–DA

a number of linear optical applications, including the possibility of new mechanooptical effects.

15E.6   Conclusions

With the development of diacetylene-containing segmented polyurethanes, the behavior of thermoplastic elastomers has been expanded to include the mechanooptic and chromic properties characteristic of polydiacetylenes. The presence of the diacetylene group in this new class of polymers also provides a means to topochemically crosslink TPEs in the solid state using irradiation techniques, and thus obtain large variations in the polymer's mechanical properties without altering its original morphology. The possibilities for applications of PU–DAs are numerous. The diacetylene cross-polymerization could be used to form highly crosslinked, rigid fibers or films with enhanced mechanical properties. The thermochromic and mechanochromic properties could be used to design optical sensors for stress or temperature, or could be further developed as molecular probes to better the understanding of the deformation process in more conventional polyurethanes. The future of these materials will be directed by the further development and evaluation of these materials in practical applications. As further functionalization and molecular design of PU–DAs is achieved, and as the ability to control and manipulate the optical and mechanical properties of these systems on higher levels is mastered, these polymers will reach their potential, and new applications will evolve.

Acknowledgments

The authors gratefully acknowledge the Office of Naval Research for their continued support of this research.

References

1. M.F. Rubner, *Polym. Mater. Sci. Eng. 53*, 683 (1985)
2. M.F. Rubner, *Macromolecules 19*, 2129 (1986)
3. M.F. Rubner, *Macromolecules 19*, 2114 (1986)
4. G. Wegner, *Die Makromolekulare Chemie 134*, 219 (1970)
5. G. Wegner, *Z. Naturforsch 246*, 824 (1969)
6. V. Enkelmann, In *Polydiacetylenes: Advances in Polymer Science, Vol. 63* (1984) Springer-Verlag, New York
7. H. Sixl, *Polydiacetylenes: Advances in Polymer Science, Vol. 63* (1984) Springer-Verlag, New York
8. D. Bloor, In *Crystallographically Ordered Polymers*. D.J. Sandman (Ed.) (1987) American Chemical Society, Washington, D.C.
9. G.N. Patl, R.R. Chance, and J.D. Witt, *J. Chem. Phys. 70*, 4387 (1979)
10. M.F. Rubner, D.J. Sandman, and C. Velazquez, *Macromolecules 20*, 1296 (1987)
11. H. Tanaka, M.A. Gomez, A.E. Tonelli, and M. Thakur, *Macromolecules 22*, 1208 (1989)
12. G. Wegner, In *Polydiacetylenes: Advances in Polymer Science, Vol. 63* (1984) Springer-Verlag, New York
13. B.S. Elman, D.J. Sandman, and M.A. Newkirk, *Appl. Phys. Lett. 46*, 100 (1985)
14. M.E. Morrow, R.C. Dye, and C.J. Eckhardt, *Chem. Phys. Lett. 158*, 499 (1989)
15. M.E. Morrow, K.M. White, C.J. Eckhardt, and D.J. Sandman, *Chem. Phys. Lett. 140*, 263 (1987)
16. D. Galiotis, R.J. Young, P.H. Yeung, and D.N. Batchelder, *J. Mater. Sci. 19*, 3640 (1984)
17. C.F. Fan and S.L. Hsu, *J. Polym. Sci. Polym. Phys. Ed. 27*, 337 (1989)
18. C.F. Fan and S.L. Hsu, *Macromolecules 22*, 1474 (1989)
19. R.C. Liang and A. Reiser, *Polymer Preprints Am. Chem. Soc. Div. Polym. Chem. 26*, 327 (1985)
20. R.C. Liang, W-Y.F. Lai, and A. Reiser, *Macromolecules 19*, 1685 (1986)
21. R.J. Day, X. Hu, J.L. Stanford, and R.J. Young, *Polym. Bull. 27*, 353 (1991)
22. X. Hu, J.L. Stanford, R.J. Day, and R.J. Young, *Macromolecules 25*, 672 (1992)
23. X. Hu, J.L. Stanford, R.J. Day, and R.J. Young, *Macromolecules 25*, 684 (1992)
24. S.J. Huang and P.G. Edelman, *J. Appl. Polym. Sci. 41*, 3 (1990)
25. (a) R.A. Nallicheri and M.F. Rubner, *Macromolecules 23*, 1005 (1990); (b) R.A. Nallicheri and M.F. Rubner, *Macromolecules 23*, 1017 (1990)

26. R.W. Seymour, A.E. Allegrezza, and S.L. Cooper, *Macromolecules 6*, 897 (1973)
27. L.J. Buckley, P.T. Hammond, and M.F. Rubner, *Macromolecules 26*, 2380 (1993)
28. P.T. Hammond and M.F. Rubner, Presentation at the ACS Rubber Division, *Frontiers in Polymer Science*, Louisville, 1992
29. P.T. Hammond, 1991, Morphological Studies of Diacetylene-Containing Polyurethanes, Massachusetts Institute of Technology, Dept. Mat. Sci. Eng., unpublished report
30. C.G. Seefried, J.V. Koleske, and F.E. Critchfield, *J. Appl. Polym. Sci. 19*, 2493, 2503 (1975)
31. C.S. Schollenberger and K. Dinsbergs, *J. Elastom. Plast. 11*, 58 (1979)
32. R.A. Nallicheri and M.F. Rubner, *Macromolecules 24*, 526 (1991)
33. P.T. Hammond and R.A. Nallicheri, *Mater. Sci. Eng. A126*, 281 (1990)
34. R.A. Nallicheri and M.F. Rubner, In *Materials Research Society Symposium* (1990), p. 577
35. R.A. Nallicheri and M.F. Rubner, *Macromolecules 24*, 517 (1991)
36. R. Bonart, *J. Macromol. Sci. Phys. B2*, 115 (1968)
37. I. Kimura, H. Ishihira, H. Ono, N. Yoshihara, S. Nomura and H. Kawai, *Macromolecules 7*, 355 (1974)
38. I.D. Fridman and E.L. Thomas, *Polymer 21*, 388 (1980)
39. K.D. Zemach and M.F. Rubner, Presentation at the Materials Research Society Annual Fall Meeting, Boston, Massachusetts, 1993
40. P.T. Hammond and M.F. Rubner, *Macromolecules* (1994)

16   Applications of Thermoplastic Elastomers

G. Holden

16.1 Introduction. 574

16.2 Composition 576
    16.2.1 Phase Structure 576
    16.2.2 Molecular Structure 576
    16.2.3 Phase properties 576
        16.2.3.1 Hard Phase. 578
        16.2.3.2 Elastomer Phase. 578
        16.2.3.3 Hard Phase/Soft Phase Ratio 579

16.3 Commercial End-Uses of Thermoplastic Elastomers. 580
    16.3.1 Polystyrene–Elastomer Block Copolymers 580
        16.3.1.1 Replacements for Vulcanized Rubber 581
        16.3.1.2 Adhesives, Sealants, and Coatings 585
            16.3.1.2.1 Pressure–Sensitive Adhesives 586
            16.3.1.2.2 Assembly Adhesives. 587
            16.3.1.2.3 Sealants 587
            16.3.1.2.4 Coatings. 587
            16.3.1.2.5 Oil Gels 587
        16.3.1.3 Blends with Thermoplastics or Other Polymeric Materials. 588
            16.3.1.3.1 Thermoplastics 588
            16.3.1.3.2 Thermosets 589
            16.3.1.3.3 Asphalt Blends. 589
            16.3.1.3.4 Wax Blends. 589
    16.3.2 Multiblock Copolymers 590
        16.3.2.1 Replacements for Vulcanized Rubber 593
        16.3.2.2 Adhesives, Sealants, and Coatings 594
        16.3.2.3 Polymer Blends 594
    16.3.3 Hard Polymer–Elastomer Combinations 595
        16.3.3.1 Replacements for Vulcanized Rubber 598
        16.3.3.2 Polymer Blends 598

16.4 Economics and Summary 599

References. 600

16.1   Introduction

Thermoplastic elastomers (TPEs) are materials that combine the processing characteristics of thermoplastics with the physical properties of vulcanized rubbers. Those based on segmented polyurethanes were first introduced in the 1950s [1, 2]. In 1965, the announcement [3] and commercial introduction of products of this type based on styrenic block copolymers also generated much interest in the rubber industry. Since then, TPEs have gained considerable importance. Their first applications spanned two industries—rubber and thermoplastics. At first the growth of these new products and of the pioneering thermoplastic polyurethane elastomers was slow. At that time the rubber industry had little thermoplastic know-how or processing equipment. The thermoplastics industry had this know-how and equipment but lacked knowledge of which rubber products to aim at or of how to sell into the rubber market. However, the unique feature of TPEs—their ability to provide products with most of the physical properties of conventional vulcanized rubbers, but without going through the process of vulcanization—is so attractive that they became commercially successful.

Because of their excellent impact strength, some TPEs were used to replace thermoplastics. However, the first area in which TPEs became commercially important was as replacements for vulcanized rubbers. In this application, the economic advantages of eliminating the compounding of rubbers with fillers, plasticizers, and vulcanizing agents, as well as avoiding the slow and costly process of vulcanization, led to a rapid growth. The diversification of the major rubber (i.e., tire) companies into plastics ventures such as poly(vinyl chloride) (PVC) aided considerably in this growth. TPEs are now estimated to have about 5% of the total rubber market and about 10% of the nontire rubber market. These market shares are expected to grow to about 6% and 12% by 1996 [4]. Worldwide annual consumption should be over 1,000,000 metric tons by the end of the century.

The unique feature of TPEs can be best appreciated by comparing their properties to those of other commercial polymers, as in Fig. 16.1. In this figure, polymers are compared using two criteria—their mechanical properties at room temperature (either hard, flexible, or rubbery) and the means by which they are formed into the final product (thermoset or thermoplastic). Six classes result—hard thermosets, flexible thermosets, rubbery thermosets, hard thermoplastics, flexible thermoplastics, and rubbery thermoplastics. The first five classes have been known for many years and the introduction of the TPEs completed the picture. The fact that the TPEs do not require vulcanization is of course the key point. In the terminology of the plastics industry, vulcanization is a thermosetting process. As such, it is slow, irreversible, and takes place on heating. In contrast, in TPEs, the change from a fluid, processable melt to a solid rubbery article is fast and reversible. This change takes place on cooling or in some cases, on the addition and removal of a solvent. These transitions are shown diagrammatically in Fig. 16.2.

This ability of TPEs to become fluid on heating and then solidify on cooling gives manufacturers the ability to produce rubberlike articles using the fast processing equipment (injection molders, blow molders, extruders, etc.) that have been developed for the plastics industry. The intensive (and expensive) compounding and vulcanization steps of conventional rubber processing are eliminated, as are the vulcanization residues in the final product. Scrap can usually be reground and recycled. Output is greatly increased and manpower reduced. However, because these materials are TPEs, they have some deficiencies. Especially

	Thermosetting	Thermoplastic
Rigid	Epoxies Phenol–formaldehyde Urea–formaldehyde	Polystyrene Polypropylene Poly(vinyl chloride) High-density polyethylene
Flexible	Highly filled and/or highly vulcanized rubbers	Low-density polyethylene EVA Plasticized PVC
Rubbery	Vulcanized Rubbers (NR, SBR, IR etc.)	Thermoplastic Elastomers

Figure 16.1   Comparison of TPEs with conventional plastics and rubbers

in the softer grades, such properties as compression set (particularly at elevated temperatures), upper service temperature and resistance to solvents and oils are often not as good as with conventional vulcanized elastomers. Thus soft TPEs have not found many applications as replacements for vulcanized rubbers in such areas as automobile tires, fan belts, or radiator hose. However, they have found many other such applications in areas where these properties are less important (e.g., footwear, wire insulation, and milk tubing). They have also found many applications in such rapidly growing markets as hot melt adhesives, sealants, polymer modification, and asphalt blending. The harder products, particularly those based on polyurethanes, polyesters, and polyamides (see later) usually have much better resistance to oils and solvents and also to compression set. Thus they are used in such applications as brake hose, timing belts, and automobile grease boots for steering or drive train linkages.

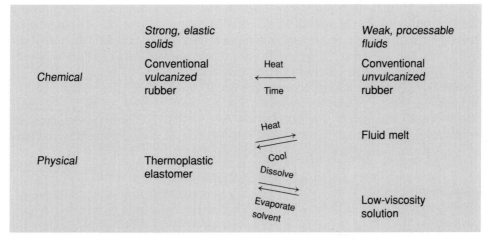

Figure 16.2 Chemical and physical changes

16.2   Composition

The applications of TPEs have been described in a recent book [5]. They are of course dependent on the properties of the individual types, which in turn are determined by their compositions. Three classes of TPEs are commercially important. They are:

- Polystyrene/elastomer block copolymers,
- multiblock copolymers, and
- hard polymer/elastomer combinations.

16.2.1   Phase Structure

Almost all commercial TPEs have one feature in common—they are phase separated systems in which one phase is hard and solid at room temperature while another phase is an elastomer. In many cases the phases are chemically bonded by block or graft copolymerization. In others, a fine dispersion is apparently sufficient. The hard phase gives these TPEs their strength. Without it, the elastomer phase would be free to flow under stress and the polymers would be unusable. When the hard phase is melted, or dissolved in a solvent, flow can take place and so the TPEs can be processed. On cooling or evaporation of the solvent, the hard phase solidifies and the TPEs regain their strength. Thus, in a sense, the hard phase in a TPE acts in a similar manner to the sulfur crosslinks in conventional vulcanized rubbers and the process by which it does so is often referred to as physical crosslinking.

16.2.2   Molecular Structure

Although the first two classes of TPEs listed above are block copolymers, they have important structural differences. Most of the polystyrene–elastomer class have the general formula S–E–S, where S represents a polystyrene block and E an elastomer block. Others have a branched structure with the general formula $(S-E)_n x$, where x represents a multifunctional junction point. In contrast, the multiblock copolymers have the general formula H–E–H–E–H–E ... or $(H-E)_n$, where H represents a hard thermoplastic block, which is usually crystalline at service temperatures. In these multiblock copolymers, the molecular weight distributions of both the individual blocks and the polymer as a whole are very broad. The segmental molecular weights are relatively low compared to those of the polystyrene–elastomer block copolymers. The last class, the hard polymer–elastomer combinations, are usually intimate mixtures of the two phases, although in some cases grafting of one polymer onto the other can take place.

16.2.3   Phase Properties

Since most TPEs are phase separated systems, they show many of the characteristics of the individual polymers that constitute the phases. For example, each phase has its own glass transition temperature (T_g) [or crystal melting point (T_m), if it is crystalline], and these in turn determine the temperatures at which a particular TPE goes through transitions in its physical properties. Thus, when the modulus of a TPE is measured over a range of temperatures, there are three distinct regions (see Fig. 16.3). At very low temperatures, both phases are hard and so the material is stiff and brittle. At a somewhat higher temperature the elastomer phase becomes soft and the TPE now resembles a conventional vulcanizate. As the temperature is further increased, the modulus stays relatively constant (a region often described as the "rubbery plateau") until finally the hard phase softens. At this point, the TPE becomes fluid. Thus, TPEs have two service temperatures. The lower service temperature depends on the T_g of the elastomer phase while the upper service temperature depends on the T_g or T_m of the hard phase. Values of T_g and T_m for the various phases in some commercially important TPEs are given in Table 16.1. Note that since the hard phase in the TPE begins to soften below its T_g or T_m, practical upper service temperatures are somewhat below the values given in Table 16.1. They also depend on the stress applied. An unstressed part (e.g., one undergoing heat sterilization) will have a higher upper service temperature than one that must support a load. Similarly, practical lower service temperatures are somewhat above the T_g of the elastomer phase. The exact value depends on the extent of hardening that can be tolerated in the final product.

Other effects of the properties of the individual phases on the properties of the TPEs are as described in the following sections.

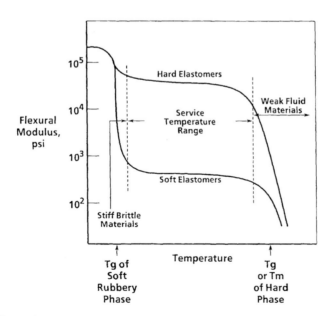

Figure 16.3   Stiffness of typical thermoplastic elastomers at various temperatures

Table 16.1 Glass Transition and Crystal Melting Temperatures[a]

Thermoplastic elastomer type	Soft, rubbery phase T_g (°C)	Hard phase T_g or T_m (°C)
Polystyrene–elastomer block copolymers		
S–B–S	−90	95 (T_g)
S–I–S	−60	95 (T_g)
S–EB–S	−60	95 (T_g) and 165 (T_m)[b]
Multiblock copolymers		
Polyurethane–elastomer block copolymers	−40 to −60[c]	190 (T_m)
Polyester–elastomer block copolymers	−40 to −60[c]	185–220 (T_m)
Polyamide–elastomer block copolymers	−40 to −60[c]	220–275 (T_m)
Polyethylene–poly(α-olefin) block copolymers	−50	70 (T_m)
Poly(ether-imide)–polysiloxane block copolymers	−50	200 (T_g)
Hard polymer–elastomer combinations[d]	−60	165 (T_m)

[a] Measured by differential scanning calorimetry
[b] In compounds containing polypropylene
[c] The values are for polyesters and polyethers, respectively
[d] The values are for polypropylene–EPDM or EPR combinations

16.2.3.1   Hard Phase

The choice of polymer in the hard phase strongly affects the oil and solvent resistance of the TPEs. Even if the elastomer phase is resistant to a particular oil or solvent, if this oil or solvent swells the hard phase, all the useful physical properties of the TPE will be lost. Thus, the polystyrene–elastomer block copolymers have little or no resistance to most organic solvents unless they are blended with solvent resistant polymers such as polypropylene or polyethylene. However, this lack of solvent resistance also allows polystyrene–elastomer block copolymers to be applied from solution, an important feature in many of their applications (see later). In the multiblock copolymers, there are five types of hard segments [polyurethane, polyester, polyamide, polyethylene, and poly(ether-imide)]. All but the last are crystalline and so have some resistance to oils and solvents. Solvent resistance is especially good for the first three types, in which the hard segments are polar as well as crystalline. Most materials in the last class (the hard polymer–elastomer combinations) also have a crystalline polymer as the hard phase. This is often polypropylene or a propylene copolymer, although PVC is now becoming more important. This crystalline hard phase also gives the hard polymer–elastomer combinations some resistance to oils and solvents.

16.2.3.2   Elastomer Phase

In the polystyrene–elastomer block copolymers, the elastomer phase controls both the stability and the stiffness of the products. Three elastomers are commonly used in the commercial versions of these materials—polybutadiene, polyisoprene, and poly(ethylene-co-butylene). The corresponding block copolymers are denoted S–B–S, S–I–S, and S–EB–S. Poly(ethylene-co-propylene) or EP is also used in a few materials. The properties of S–EP–S block copolymers are very similar to those of S–EB–S analogues and in this chapter, descriptions of the properties of S–EB–S block polymers apply very closely to S–EP–S block copolymers of similar molecular structure. The elastomer segments in S–B–S and S–I–S block copolymers are unsaturated and contain one double bond per original monomer unit. Typical S–B–S and S–I–S block copolymers have one thousand or more double bonds in each elastomer chain. These double bonds are an obvious source of instability and so limit the use of the polymers in applications where they are exposed to high temperatures, ultraviolet light, or ozone. In contrast, S–EB–S block copolymers are saturated and thus much more stable. If polymer degradation takes place, S–B–S block copolymers crosslink and eventually become hard, insoluble, and infusible. Under similar conditions, S–I–S block copolymers undergo chain scission and so become softer and weaker. As noted, S–EB–S block copolymers are much more resistant to degradations, but if it does occur, they also undergo chain scission. Another important point is the difference in stiffness between block copolymers based on these three elastomers. Because of differences in the molecular weights between entanglements in the elastomer phase (see Chapter 3), S–I–S block copolymers are softer than corresponding S–B–S block copolymers and these in turn are softer than corresponding S–EB–S block copolymers. All these elastomers are readily swollen by hydrocarbon oils. In compounding the styrenic block copolymers, the oils used are chosen so as to have low aromatic content. Thus they are not compatible with the polystyrene segments and so the elastomer network remains effective. Another important difference is in cost. S–B–S copolymers are the least expensive. Prices of S–I–S equivalents are about 20% higher. The S–EB–S block copolymers are about double the price of the similar S–B–S block polymers.

The multiblock copolymers with polyurethane, polyester, and polyamide hard segments can be considered together. All can be made with polyether elastomer segments. In some cases polyesters (including polycaprolactones) can also be used. The polymers having polyester elastomer segments are tougher and more resistant to oils, solvents, and thermal degradation whereas analogues with polyether elastomer segments have better hydrolytic stability and are more flexible at low temperatures. The polyurethane–elastomer block copolymers with polycaprolactone elastomer segments have better resistance to hydrolysis than those with the other polyester segments and are considered to be premium products. All three types have good electrical properties [6]. The polymers with polyethylene hard segments have ethylene–α-olefin copolymers as the elastomers. These are thermally stable but are swollen by oils and solvents. In the poly(ether-imide)–polysiloxane multiblock copolymers, the polysiloxane segments give outstanding thermal stability.

In the last class (the hard polymer–elastomer combinations) the elastomer is often an ethylene–propylene random copolymer (EPR) or a similar material with a small amount of out-of-chain unsaturation (EPDM). Butyl and natural rubbers are also used. These elastomers are readily swollen by oils and solvents. The elastomers can be crosslinked during mixing (a process referred to as dynamic vulcanization; see Chapter 7). This improves both solvent resistance and compression set. Products in which the elastomer is either a nitrile rubber or an ethylene interpolymer have excellent oil resistance.

16.2.3.3 Hard Phase/Soft Phase Ratio

Not surprisingly, TPEs with a high proportion of hard phase are themselves relatively hard and stiff. As the proportion of hard phase is further increased, the products lose their elastomeric character and instead become leathery. Finally, with still further increases in the proportion of hard phase, they become hard thermoplastics, usually with high impact resistance. Polystyrene–polybutadiene block copolymers with high styrene content have been commercialized under the trade name K-Resin[®] (Phillips Petroleum Co.).

16.3   Commercial End-Uses of Thermoplastic Elastomers

Since their commercial introduction, many types (and grades within types) of TPEs have been produced to meet specific end use requirements. At least 40 manufacturers have entered the field.

16.3.1   Polystyrene/Elastomer Block Copolymers

Like most conventional vulcanized rubbers—and unlike most thermoplastics—the polystyrene–elastomer block copolymers are never used commercially as pure materials. To achieve the particular requirements for each end-use, they are compounded with other polymers, oils, resins, fillers, etc. In almost all cases, the final products contain less than 50% of the block copolymer. Thus a study of their end-uses is in effect a study of how they are blended to achieve the properties needed for the particular application. Trade names of some of the commercial products are listed in Table 16.2.

Before discussing the end-uses in detail, it is important to consider how the added materials are distributed with respect to the two phases in the block co-polymer. For any additive there are four possibilities.

1. *It can mix with the polystyrene phase.* In this case the additive increases the relative volume of the polystyrene phase and so makes the product harder. The glass transition temperature of the additive should be similar to or greater than that of polystyrene (100 °C); otherwise its addition will reduce the high temperature performance of the final product.

2. *It can mix with the elastomer phase.* Conversely, in this case the additive decreases the relative volume of the polystyrene phase and so makes the product softer. The addition also changes the glass transition temperature of the elastomer phase. This in turn affects such end-use properties as tack and low-temperature flexibility.

3. *It can form a separate phase.* Unless the molecular weight of the additive is substantially less than that of either type of segment in the block copolymer, this is the most likely outcome (see Chapters 3, 12 and 15C). Thus only low molecular weight resins and oils are compatible with either of the existing two phases—polymeric materials tend to form a separate third phase. This polymeric third phase is usually co-continuous with the block copolymer and so confers some of its own characteristic properties on those of the final blend.

4. *It can mix with both phases.* This is usually avoided because it reduces the degree of separation of the two phases and so weakens the product.

This ability to be blended with so many different materials gives the polystyrene–elastomer block copolymers an exceptionally wide range of end-uses. Some examples are:

Table 16.2 Some Trade Names of Thermoplastic Elastomers Based on Styrenic Block Copolymers

Trade name (Mfr)	Type	Elastomer segment	Notes
Kraton® D and Cariflex TR (Shell)	Linear and branched	B or I	General-purpose, soluble. Also compounded products
Vector® (Dexco)[a]	Linear	B or I	
Solprene® (Phillips)[b]	Branched	B	
Finaprene® (Fina)	Linear	B	
Tufprene™ and Asaprene™ (Asahi)	Linear	B	General-purpose, soluble.
Coperbo™ (Petroflex)	Linear	B	Not available as
Calprene™ (Repsol)	Linear and branched	B	compounded products
Europrene Sol T® (Enichem)	Linear and branched	B or I	
Quintac® (Nippon Zeon)	Linear	I	
Stearon® (Firestone)	Linear	B	High polystyrene content
K-Resin® (Phillips)	Branched	B	Very high polystyrene content. Hard and rigid
Kraton® G (Shell)	Linear	EB or EP	Improved stability, soluble when uncompounded
Septon™ (Kuraray)	Linear	EP	Elastomer block is hydrogenated polyisoprene
Dynaflex® (GLS)	Linear	B or EB	
Multi-flex™ (Multibase)	Linear	EB	
Hercuprene® (J-VON)[c]	Linear	B or EB	Only compounded products
Flexprene® (Teknor Apex)	Linear	B	
Tekron® (Teknor Apex)	Linear	EB	
Elexar® (Teknor Apex)[d]	Linear	EB	Wire and Cable compounds
C-Flex® (Concept)[e]	Linear	EB	Medical applications. Contains silicone oil

[a] Joint venture of Dow and Exxon
[b] No longer made in U.S.A.
[c] Formerly J-Plast
[d] Formerly produced by Shell
[e] Now Consolidated Polymer Technologies Inc.

16.3.1.1   Replacements for Vulcanized Rubber

Many of these applications have been described in recent articles [7, 8]. In this end-use the products are usually manufactured by machinery originally developed to process conventional thermoplastics. Examples are injection molding, blow molding, and blown film and profile extrusions. S–I–S block copolymers have not been used very much in this application but there are many applications for compounded products based on S–B–S and S–EB–S block copolymers. A list of some of the compounding ingredients and their effects on the properties of the compounds is given in Table 16.3. Products can be produced that are as soft as 5 Shore A [9] to as hard as 55 Shore D (see Table 16.4). Quite large amounts of these compounding ingredients can be added and the final products often contain only about 25% of the S–B–S or S–EB–S block copolymers. From an economic point of view, this is most important. For example, it enables compounds based on S–EB–S block copolymers to

Table 16.3 Compounding Styrenic Block Copolymers

	Oils	Polystyrene	Polyethylene	Polypropylene	Fillers
Hardness	Decreases	Increases	Increases	Increases	Small increase
Processability	Improves	Improves	Improves	Improves, especially with S–EB–S	Improves
Effect on oil resistance	None	None	Improves	Improves	None
Effect on ozone resistance	None	Some improvement	Improves	Improves	None
Cost	Decreases	Decreases	Decreases	Decreases	Decreases
Other	Decreases U.V. resistance	—	Often gives satin finish	Improves high-temperature properties	Often improves surface appearance

compete with those based on polypropylene–EPDM or EPR blends, even though the S–EB–S alone is several times more expensive than the polypropylene, the EPDM, or the EPR.

Polystyrene is often used as a compounding ingredient for S–B–S block copolymers. It acts as a processing aid and makes the products stiffer. Mineral oils are also processing aids but make the products softer. Naphthenic oils are preferred. Oils with high aromatic contents should be avoided, since they plasticize the polystyrene domains. Inert fillers such as whiting, talc, and clays can be used in these compounds. They have only a relatively small effect on physical properties but reduce cost. Frequently, up to 200 parts (per 100 parts of the block copolymer) are used in compounds intended for footwear applications. Reinforcing fillers such as carbon black are not required and, in fact, large quantities of such fillers make the final product stiff and difficult to process.

S–EB–S block copolymers can be compounded in a similar manner to their S–B–S analogues. One important difference is that for S–EB–S block copolymers, polypropylene is a preferred additive, acting in two different ways to improve the properties of the compounds. First, it gives the compounds better processability. Second, when the compounds are processed under high shear and then quickly cooled (as, for example, in injection molding or extrusion), the polypropylene and the S–EB–S/oil mixture form two continuous phases. Polypropylene is insoluble and has a relatively high crystal melting point (about 165 °C). This continuous polypropylene phase significantly improves both the solvent resistance and upper service temperature of these compounds. Another advantage of S–EB–S polymers is that because of their lower midsegment solubility parameter, they are very compatible with mineral oils. Large amounts of these oils can be added without bleedout and this allows soft compounds to be produced. Paraffinic oils are preferred, because they are more compatible with the EB center segments than the naphthenic oils used with similar S–B–S based compounds. Again, oils with high aromatic contents should be avoided. Blends of S–EB–S block copolymers with mineral oils and polypropylene are transparent. This is probably due to the fact that the refractive index of an S–EB–S/oil mixture almost exactly matches that of

Table 16.4 Properties of Compounded Styrenic Block Copolymers

Product	Kraton® D2109	Kraton G2705	Dynaflex® G2706[a]	Dynaflex D3202[b]	Multi-Flex® RP 6568	Multi-Flex A 8832	Multi-Flex A 4001 LC	Kraton G7720	Dynaflex GX-6705	Dynaflex GX-7722	Elexar 8431	Elexar 8451
Base polymer	S–B–S	S–EB–S	S–EB–S	S–B–S	S–EB–S	S–EB–S	S–EB–S	S–EB–S	S–EB–S	S–EB–S	S–EB–S	S–EB–S
Application	Milk tubing and film	Medical	Medical	General purpose	Automotive air bag covers	Automotive, sound barrier	Automotive, low compression set	Automotive, general purpose	Automotive, Extra soft	Automotive, low compression set	Wire and cable	Wire and cable
Properties												
Hardness												
Shore A or D	48A[c]	55A	28A	50A	57A	88A	40A	60A	5A	57A	64A	82A /34D
Tensile strength												
MPa	11[c]	6.0	7.6	5.6	4.3	4.3	4.5	5.0	1.7	5.0	9.6[c]	12[c]
(psi)	(1600[c])	(850)	(1100)	(800)	(620)	(620)	(660)	(750)	(240)	(750)	(1400[c])	(1700[c])
300% Modulus												
MPa	2.0[c]	2.8	1.4	4.3	1.4	3.8	3.1	2.5	0.3	2.5	3.6[c]	4.4[c]
(psi)	(300[c])	(400)	(200)	(600)	(200)	(550)	(450)	(350)	(45)	(350)	(500[c])	(650[c])
Elongation												
(%)	850[c]	700	750	400	1400	430	900	700	1840	700	650[c]	650[c]
Specific gravity	0.94	0.90	0.89	1.0	0.94	1.88	1.08	1.19	0.90	1.19	0.92	1.01

[a] Formerly Kraton G2706
[b] Formerly Kraton D3202
[c] Measured on extruded samples; all others measured on injection molded samples

crystalline polypropylene. Surprisingly, even oils that themselves are stable to UV radiation reduce the stability of the blends, but the effects can be minimized by the use of UV stabilizers and absorptive or reflective pigments. Blends with silicone oils are used in some medical applications [10]. The same inert fillers used in the S–B–S-based compounds can also be used with the S–EB–S analogues. In addition, barium or strontium sulfate fillers can be used to produce very high density compounds used for sound deadening applications. In another development, fire retardants can be added to produce compounds that will qualify under many of the current regulations.

Compounds based on both S–B–S and S–EB–S TPEs require some protection against oxidative degradation and in some cases, against sunlight also, depending on their end-uses. Hindered phenols are effective antioxidants and are often used in combination with thiodipropionate synergists. Benzotriazoles are effective UV stabilizers and they are often used in combination with hindered amines. If the product does not have to be clear, titanium dioxide or carbon black pigments give very effective protection against sunlight.

Compounding techniques are relatively simple and standard. There is one important generalization—the processing equipment should be heated to a temperature at least 20 °C above the upper service temperature of the block copolymer (see Table 16.1) or the melting point of the polymeric additive, whichever is greater. The use of cold mills, etc. can result in polymer breakdown, which is not only unnecessary but also detrimental to the properties of the final product.

Unfilled or lightly filled compounds can be made on a single screw extruder fitted with a mixing screw. The length/diameter ratio should be at least 24 : 1. If large amounts of fillers or fire retardants are to be added, these can be dispersed on either a twin screw extruder or a closed intensive mixer that discharges into an extruder. After extrusion, the compound is fed to a pelletizer. This can be either a strand cutting or an underwater face cutting system. If the first type is used, it is important to remember that rubbery compounds must be cut rather than shattered. Thus, the blades of the cutter should be sharp and the clearance between the fixed and the rotating blades should be minimized. With this type of pelletizer, the strands must be thoroughly cooled before they enter the cutter. A chilled water bath can be used to increase production rates.

Trial batches or small-scale production runs can be made using a batch type closed intensive mixer. Mixing times of about 5 min are usually adequate. After this mixing, the hot product is passed to a heated two-roll mill. When it has banded, it is cut off, allowed to cool, and finally granulated. In this case also, the granulator blades must be sharp and clearances minimized.

S–B–S and S–EB–S block copolymers with high molecular weight polystyrene segments and/or high styrene contents are very difficult to process as pure materials. Oiled versions with oil contents from about 25% to about 50% are commercially available. These process much more easily and of course more oil can be added during mixing.

Many end-users prefer to buy precompounded products and numerous grades have been developed for the various specialized end-uses. Representative examples and some details of their properties are given in Table 16.4. The end-uses covered include milk tubing, shoe soles, sound deadening materials, wire insulation, and flexible automotive parts. After priming, the parts can be coated with paints that are also flexible. If a compound contains a particular homopolymer (e.g., polystyrene or polypropylene), it will adhere well when insert molded or extruded against the same pure homopolymer, or against other compounds containing it. This allows the production of parts having a rigid framework supporting a soft, flexible outer surface [8].

The physical properties of compounded TPEs based on both S–B–S and S–EB–S block copolymers are sensitive both to processing conditions and to processing equipment. Thus, it is most important to make test samples under conditions and on equipment similar to those that will be used in production. Misleading results will be obtained if, for example, prototype parts or test pieces are compression molded when the actual products will be made by extrusion or injection molding.

Conditions for processing these compounds by injection molding, blow molding, extrusion, etc. have been discussed in some detail in a recent publication [11]. As a generalization, compounds based on S–B–S block copolymers are processed under conditions suitable for polystyrene whereas those based on S–EB–S block copolymers are processed under conditions suitable for polypropylene. These compounds usually have relatively high surface friction. Ejection of the molded parts can be difficult, especially with softer products. Use of a release agent or a suitable coating on the mold makes for easier ejection and if possible the mold should be designed so that the ejection can be air assisted. Tapering the sides of the mold is also helpful, as is the use of stripper rings. The use of small-diameter ejection pins should be avoided, since they tend to deform the molded part rather than eject it.

Ground scrap from molding is reusable and is usually blended with virgin product. Grinding is relatively easy if the conditions required for successful pelletization are met, that is, if the grinder blades are kept sharp and the clearances minimized.

Various compounds have been developed to allow the production of blown and extruded (slot cast) film, including heat shrinkable films [12]. These are based on both S–B–S and S–EB–S block copolymers. These can be very soft and flexible. They also exhibit low hysteresis and low tensile set. A significant advantage is that they can be used in contact with skin and in contact with certain foods.

One unusual application is the use of solutions of S–EB–S/oil blends to replace natural rubber latex in the manufacture of articles such as surgeon's gloves [13]. These blends have two advantages. First, they are more resistant to attack by oxygen or ozone. Second, natural rubber latex contains proteins that produce dangerous allergic reactions in some patients. The saturated elastomer phase and extreme purity of the S–EB–S avoids both these problems.

16.3.1.2   Adhesives, Sealants, and Coatings

These are very important applications for polystyrene–elastomer block copolymers and probably the fastest growing. Again, the products are always compounded and the subject has been extensively covered [14, 15]. As previously mentioned, the effects of the various compounding ingredients depend on the region of the phase structure with which they associate. Since there are three different elastomers used in these block copolymers, each has particular resins and/or oils with which it is most compatible. Table 16.5 gives details of the various resins and oils suitable for use with each of these elastomers and also with the polystyrene phase. Ingredients that go into both phases are usually avoided, since they make the phases more compatible with each other, and thus make the product weaker. Polymers that go into neither phase (e.g., polypropylene) are used in some hot melt applications (see later),

16.3.1.3   Blends with Thermoplastics or Other Polymeric Materials

The polystyrene–elastomer block copolymers are technologically compatible with a surprisingly wide range of other polymeric materials, that is, the blends show improved properties when compared to the original polymers. Impact strength usually is the most obvious improvement, but others include tear strength, stress crack resistance, low-temperature flexibility, and elongation. Thermoplastic and thermoset polymers can be modified in this way, as can asphalts and waxes.

16.3.1.3.1 Thermoplastics   TPEs have several advantages in this application. The other elastomers that can be blended with thermoplastics to improve their impact resistance (e.g., SBR, EPDM, and EPR) can normally be used only in the unvulcanized state—the vulcanized products usually cannot be dispersed (an exception is the process of vulcanization during dispersion; see Chapter 7). Since these unvulcanized elastomers are soft and weak, they reduce the strength of the blends and so only limited amounts can be added. However, TPEs are much stronger, even though they are unvulcanized, and so there is less limitation on the amounts that can be added. Blending is usually carried out in the processing equipment (injection molders, extruders, etc.), especially if a TPE with low viscosity is used. The thermoplastic elastomers form a separate phase and so do not change the T_g or T_m of the thermoplastic into which they are blended. Thus, these blends maintain the upper service temperature of the original thermoplastic. Three large-volume thermoplastics—polystyrene, polypropylene, and polyethylene (both high and low density)—can be modified using polystyrene–elastomer block copolymers [26, 27]. Because of their lower price, S–B–S block copolymers are most commonly used with these thermoplastics. In polystyrene there are two important applications—upgrading high-impact polystyrene to a super high impact product and restoring the impact resistance that is lost when flame retardants are mixed into high-impact polystyrene. Polypropylene has very poor impact resistance at low temperatures. This also can be improved by adding styrene–elastomer block copolymers. Impact improvement of any polymer is usually accompanied by a loss in clarity, because the added polymer forms a separate phase with a different refractive index. However, blends of S–EB–S with polypropylene are about as transparent as the pure polypropylene, probably because of a match in refractive indices. Linear low density polyethylene LLDPE/S–EB–S mixtures can be used instead of the pure S–EB–S and blends of these two polymers with polypropylene also retain the clarity of the pure polypropylene [28]. Blends with polyethylene are mostly used to make blown film, where they have improved impact resistance and tear strength, especially in the seal area. Polystyrene–elastomer block copolymers are also blended with engineering thermoplastics such as poly(phenylene oxide) and polycarbonate. S–EB–S block copolymers are often used in these blends because of their better stability. In more polar thermoplastics such as polyamides and polyesters, maleated S–EB–S polymers are recommended as impact modifiers [27].

A slightly different application is the use of polystyrene–elastomer block copolymers to make useful blends from otherwise incompatible thermoplastics (see Chapter 15C). Polystyrene, for example, is completely incompatible with polyethylene or polypropylene and blends of this type form a two-phase system with virtually no adhesion between the phases. Thus when articles made from them are stressed, cracks easily develop along the phase boundaries and the products fail at low elongations. Addition of a low molecular weight S–B–S or S–EB–S block copolymer changes this behavior and converts the blends to more ductile materials [27]. Similar results were reported on blends of poly(phenylene oxide) with polypropylene compatibilized with S–EB–S and S–EP and on polyamides compatibilized with polyolefins using functionalized S–EB–S copolymers. In another example, a higher molecular weight S–EB–S was used in a blend with a polycarbonate and a polypropylene. Compared to the pure polycarbonate, the blend had almost the same upper service temperature combined with improved solvent resistance, lower cost, and lower density [29].

16.3.1.3.2 Thermosets   Sheet molding compounds (SMCs) are thermoset compositions containing unsaturated polyesters, styrene monomer, chopped fiber glass, and fillers. They are cured to give rigid parts that are often used in automobile exteriors. Special types of polystyrene–elastomer block copolymers have been developed as modifiers for these compositions and give the final products improved surface appearance and better impact resistance [30, 31].

In an entirely different application, TPEs are blended with silicone rubbers containing either vinyl or silicon hydride functional groups. The silicone rubbers containing the vinyl groups are pelletized separately from those containing the silicon hydride groups. When melted and mixed together in the processing equipment, the two groups react under the influence of a platinum catalyst to form an interpenetrating network of the silicone rubber and the TPE. The products are useful in medical applications. This end-use was originally developed using polyurethane–elastomer block copolymers. It has now been extended to other TPEs including S–EB–S and polyester–elastomer block copolymers [32].

16.3.1.3.3 Asphalt Blends   The block copolymer content of these blends is usually less than 20% and even as little as 3% can significantly change the properties of asphalts. The polystyrene–elastomer block copolymers make the blends more flexible (especially at low temperatures) and increase their softening point. They decrease the penetration and reduce the tendency to flow at high service temperatures, such as those encountered in roofing and paving applications. They also increase the stiffness, tensile strength, ductility, and elastic recovery of the final products. Melt viscosities remain relatively low and so the blends are still easy to apply. As the polymer concentration is increased to about 5%, an interconnected polymer network is formed. At this point the nature of the blend changes from an asphalt modified by a polymer to a polymer extended with an asphalt. It is important to choose the correct grade of asphalt; those with low asphaltenes content and/or high aromaticity in the maltene fraction usually give the best results [33]. Applications include road surface dressings such as chip seals (these are applied to hold the aggregate in place when a road is resurfaced), slurry seals, asphalt concrete (this is a mixture of asphalt and aggregate used in road surfaces), road crack sealants, roofing, and other waterproofing applications [34–36]. Because of their lower cost, S–B–S block copolymers are usually chosen for this application, but in roofing and paving applications the S–EB–S block copolymers are also used because of their better UV, oxidative, and thermal stability.

16.3.1.3.4 Wax Blends   S–I–S and S–B–S block copolymers can be used in blends with waxes. Because of their limited compatibility, they usually require the addition of resins. S–EB–S block copolymers are more compatible. The products are used to give flexible coatings for paper products and can be applied by curtain coaters [37].

16.3.2   Multiblock Copolymers

The structure of these block copolymers is an alternating series of polymeric segments. One type of segment forms a hard phase (usually crystalline) while the other forms an amorphous phase that gives the elastomeric properties. Those in which the hard phase is crystalline can be subdivided into four types (polyurethane–elastomer, polyester–elastomer, polyamide–elastomer, and polyethylene–poly(α-olefin) block copolymers). The polyurethane–elastomer block copolymers were developed first and the polyester–elastomer block copolymers somewhat later. The polyamide–elastomer block copolymers are more recent. The properties of all these three types complement those of the polystyrene–elastomer block copolymers in that they are generally harder, tougher, more expensive, and more resistant to oils and solvents. The fourth type (the polyethylene–poly(α-olefin) block copolymers) are a newer development and are generally softer and lower in cost. A discussion of the manufacture, structure, and properties is given in Chapters 2, 5, 8, and 9 and in a recent handbook [5]. Another type of multiblock copolymer is based on alternating poly(ether-imide) hard segments and polysiloxane elastomer segments. It should be noted that in this case the hard phase is amorphous. Some trade names of commercial products of all five types are listed in Table 16.6.

In the polyurethane–elastomer block copolymers the crystalline hard segments are polyureas or polyurethanes. The amorphous elastomeric segments are either polyethers or polyesters (usually adipates, but polycaprolactones can also be used). The polymers with adipate polyester elastomer segments are tougher and have better resistance to abrasion, swelling by oils, and oxidative degradation. Corresponding polymers with polyether elastomer segments have much better hydrolytic stability (this is probably the main weakness of the polyester-based materials) and are more resistant to fungus growth. They are also more flexible at low temperatures and more resilient. Thus they give less heat increase under repeated cyclic stress. Those with polycaprolactone elastomer segments are premium grades with a combination of superior oil and solvent resistance and quite good hydrolytic stability. Properties of some typical grades of polyurethane–elastomer block copolymers are given in Table 16.7.

The polyester–elastomer block copolymers share many of their characteristics of polyurethane–elastomer block copolymers. The main difference is that the hard segments are crystalline polyesters. Like the polyurethane–elastomer block copolymers, they are tough, oil-resistant materials with high tear strength. They have excellent resistance to compression set and flex cracking. At low strain levels, they show very little hysteresis, that is, they behave almost like a perfect spring. They are generally somewhat harder than the corresponding polyurethane–elastomer block copolymers. Compared to polyurethane–elastomer block copolymers of the same hardness, they have higher modulus (i.e., they are stiffer), better resistance to creep, and show less variation of modulus with temperature. This is claimed to give them a wider service temperature range. Some work has been reported on the use of plasticizers in these materials [38]. Those tested were suitable for use with PVC and were most compatible with the softer grades of the polyester–elastomer block copolymers. At least 50 parts of some of these plasticizers could be added to 100 parts of the polyester–elastomer block copolymer. The products were softer and more processable. Flame retardants can also be mixed with these polyester–elastomer block copolymers and some flame-retardant grades

Table 16.6 Some Trade Names of Thermoplastic Elastomers Based on Multiblock Copolymers

Trade name (Mfr)	Hard segment	Elastomer segment	Notes
Estane® (B.F. Goodrich) Q-Thane®a (Morton International) Pellethane®a (Dow) Desmopan™ and Texin™ (Bayer)[b] Elastollan® (BASF)	Polyurethane	Polyether or amorphous polyester	Hard and tough. Abrasion and oil resistant. Good tear strength. Fairly high priced
Urafil (Akzo) Hytrel® (DuPont) Lomod® (GE) Ecdel® (Eastman) Riteflex® (Hoechst Celanese) Arnitel® (DSM)	Polyester	Polyether or amorphous polyester	Similar to polyurethanes and more flexible at low temperatures
Pebax® (Atochem) Vestenamer® (Huls) Grilamid® (Emser) Montac® (Monsanto)[c] Orevac® (Atochem)[c]	Polyamide	Polyether or amorphous polyester	Similar to polyurethanes but can be softer. Very good at low temperatures
Engage® (Dow) Flexomer (Union Carbide) Exact® (Exxon)	Polyethylene	Poly(α-olefins)	Flexible and low cost. Very good at low temperatures
Siltem® (GE)	Poly(ether-imide)[d]	Polysiloxane	Fire retardant, used in wire and cable insulation

[a] Including some with polycaprolactone segments
[b] Formerly marketed by Mobay and Miles
[c] For hot melt adhesives
[d] Amorphous. All the other hard segments are crystalline

Table 16.7 Properties of Polyurethane–Elastomer Block Copolymers

Product	Estane 58133	Estane 58311	Q-Thane PC86	Texin 480A	Pellethane 2102-90A	Pellethane 2103-70A	Ellastalon C-60AW[a]
Type	Polyester	Polyether	Polyester	Polyester	Polycaprolactone	Polyether	Polyester
Hardness Shore A or D	55D	85D	50D	86A	90A	70A	60A
Tensile strength MPa	40	47	52	41	48	24	26
(psi)	(5800)	(6800)	(7500)	(6000)	(7000)	(3500)	(3800)
100% Modulus MPa	14	6.2	8.3	5.1	10.7	3.5	2.8
(psi)	(2100)	(900)	(1200)	(750)	(1500)	(500)	(400)
Elongation (%)	500	530	475	500	500	500	870
Specific gravity	1.22	1.11	—	1.20	1.20	1.06	1.18

[a] Contains proprietary plasticizer

Table 16.8 Properties of Polyester–Elastomer Block Copolymers

Product	Hytrel 4056	Hytrel 5526	Hytrel 6356	Hytrel 7246	Lomod ST3090A	Lomod TE3055A
Hardness	40D	55D	62D	72D	90A	54D
Shore A or D	92A					
Tensile strength						
MPa	26	40	41	46	14	23
(psi)	(4000)	(5800)	(6000)	(6600)	(2000)	(3300)
100% Modulus						
MPa	9.5	19	20[a]	25[a]	—	—
(psi)	(1100)	(2000)	(2500)[a]	(3800)[a]	—	—
Elongation (%)	550	500	420	350	250	290
Specific gravity	1.17	1.20	1.22	1.25	1.13	1.22

[a] After yielding at lower elongation

are commercially available. The properties of some representative grades of polyester–elastomer block copolymers are given in Table 16.8.

The polyamide–elastomer block copolymers are one of the newer developments in TPEs. They are also block copolymers with alternating hard and soft segments. The hard crystalline segments are polyamides while the soft segments are either polyethers or polyesters. One notable feature is that there are many different polyamides from which to make a choice. These include nylon 6, nylon 6,6, nylon 11, and nylon 12 and also polyamides containing aromatic groups. These all have different melting points and degrees of crystallinity and so quite a wide range of property variations are available [39]. When these are combined with the various polyethers and polyesters, the range of property variations becomes even wider. Quite soft materials can be produced (as low as 65 Shore A hardness), as can harder counterparts (up to 70 Shore D). Those with polyester soft segments have excellent resistance to thermal and oxidative degradation whereas those with polyether soft segments are more resistant to hydrolysis and more flexible at lower temperatures. Moisture uptake is normally quite low, but hydrophillic products that are very permeable to water vapor have been described [40]. As far as resistance to deformation is concerned, the upper service temperature depends on the choice of polyamide segment and can be as high as 200 °C. Resistance to oils and solvents is also good.

The polyethylene–poly(α-olefin) block copolymers are newer, lower cost materials. They are very flexible at low temperatures but their upper service temperature is rather low. They have some resistance to oils and solvents. The poly(ether-imide)/polysiloxane alternating block copolymers [41] are hard (about 70 on the Shore D scale) and differ from the previous four types in that the hard phase is amorphous. Thus they do not have such good resistance to oils and solvents. However, they have excellent thermal and hydrolytic stability.

The properties of some representative grades of polyamide–elastomer and poly(ether-imide)–polysiloxane block copolymers are given in Table 16.9. Properties of a representative polyethylene–poly(α-olefin) block copolymer are given in Table 5.1.

Table 16.9 Typical Properties of Polyamide–Elastomer and Poly(ether-imide)–Polysiloxane Block Copolymers

Product	Pebax 2533	Pebax 4033	Pebax 6333	Grilamid Ely 60	Siltem[a] STM1500
Hardness	25D	40D	63D	62D	69D
Shore A or D	75A	—	—	—	—
Tensile strength					
MPa	29	33	49	36	25
(psi)	(4200)	(4800)	(7100)	(5100)	(3600)
100% Modulus					
MPa	4.3	10	19	—	—
(psi)	(630)	(1450)	(2700)	—	—
Elongation (%)	350	620	680	300	105
Specific gravity	1.01	1.01	1.01	1.01	1.18

[a] Fire retardant grade—intended for wire insulation

16.3.2.1   Replacements for Vulcanized Rubber

This is the most important end-use of the polyurethane–elastomer, polyester–elastomer, and polyamide–elastomer block copolymers. They are fabricated into the final article by the typical processing techniques developed for conventional thermoplastics, that is, injection molding, blow molding, extrusion, etc. Their crystalline hard segments make them insoluble in most liquids. Thus they are often used as replacements for oil-resistant rubbers such as neoprene because they have better physical properties (e.g., tear and tensile strengths) at temperatures up to about 150 °C. Applications include flexible couplings, seal rings, automotive steering boots, gears, footwear (including ski boots), wheels, timing and drive belts, tire chains, industrial hose, and outer coverings for wire and for optical fiber cables. The polyurethane–elastomer block copolymers with polyether elastomer segments have significant medical applications, including their use in body implants [42]. Polyester–elastomer block copolymers have been used to make transparent replacements for glass bottles to be used in medical applications [43]. All three types of these elastomers are usually used uncompounded, although calcium carbonates and radiopaque fillers can be added [44]. The final articles can be metallized or painted. Special elastomeric paints have been developed that match the appearance of painted automotive sheet metal. Parts painted in this way are used in car bodies. After processing has been completed, the finished parts may be annealed by overnight heating at about 115 °C. This allows the crystallization of the hard phase to become more complete and this in turn improves the physical properties of the product.

It is most important that these block copolymers are dried before processing. If they are not, water will be released when the hot, molten polymer is being processed and the final article will contain bubbles and will possibly be degraded. Drying times from 1 to 2 h at temperatures of about 100 °C are recommended and reground polymer or color concentrates should be dried also. During processing, hot, dry air should be circulated through the hopper in which the dry polymer is held, to prevent water being reabsorbed.

For injection molding applications, mold release can be a problem, as it is with most TPEs. The possible solutions are covered in the similar section dealing with the applications of the polystyrene–elastomer block copolymers (see above). Recommended conditions for extrusion include the use of relatively high-compression screws (about $3:1$ compression ratio) with length-to-diameter ratios of at least $24:1$. Profile, blown film [45] and sheet extrusion can all be used with these polymers. They can also be laminated to fabric backings.

In both molding and extrusion of polyurethane–elastomer block copolymers the melt temperature should be about 200 °C. Low screw speeds should be used to avoid degradation and the machine should be purged before shutdown [46].

Processing of polyester–elastomer block copolymers is generally easier than with the polyurethane–elastomer block copolymers because thermal degradation is less of a problem and this gives the polyester–elastomer block copolymers a wider processing window. One important point is that the recommended processing temperature of the polyester–elastomer block copolymers varies with the hardness of the material [47]. Thus the melt temperature during processing of softer (Shore D 40) grades should be about 185 °C whereas for harder (Shore D 72) grades it should be about 240 °C.

The recommended processing temperatures for polyamide–elastomer block copolymers are about 60 °C above the melting points of the polyamide segments [39] and processing techniques are similar to those used with other polyamides, such as nylon 6.

Suggested applications of polyethylene–poly(α-olefin) block copolymers include molded and extruded goods, such as shoe soles, weatherstrips, tubing, and wire insulation. They can be compounded with fillers and both naphthenic and paraffinic oils. Processing is similar to that of polyethylene and thermal degradation is not usually a problem.

The main application of the poly(ether-imide)–polysiloxane block copolymers is in flame-resistant wire and cable covering [41], where they combine very low flammability with a low level of toxic products in the smoke. This unusual and vital combination of properties justifies their relatively high price of about $17/lb.

16.3.2.2   Adhesives, Sealants, and Coatings

This is a relatively small application for these polymers. They are usually applied as hot melts [48], although some grades can be applied from solutions in such polar solvents as methyl ethyl ketone or dimethyl formamide. They have some uses in footwear, including attaching the shoe soles to the uppers, and also in coextrusion, where they act as adhesive interlayers between dissimilar polymers.

16.3.2.3   Polymer Blends

Polyurethane–elastomer block copolymers have been blended with a wide range of other thermoplastics and elastomers [49] (see also Chapter 2). Nonpolar polymers such as polyolefins were about the only materials tested that showed little or no compatibility. Probably the most important blending material is plasticized PVC (see Chapter 6), although other elastomers (including S–B–S and S–I–S TPEs) or polar thermoplastics such as polycarbonates can also be used. Blends of plasticized PVC with polyurethane–elastomer block copolymers have good flex life, oil resistance, abrasion resistance, and low-temperature flexibility and are used in footwear.

Like S–EB–S block copolymers, polyurethane–elastomer block copolymers can be blended with reactive silicone rubbers to give products intended for medical applications [32]. In fact, they are probably the most important TPE in this end-use.

Polyester–elastomer block copolymers can be blended with PVC (see Chapter 6). Both plasticized and unplasticized PVCs have been used and also chlorinated PVC (CPVC) [50]. Compared to the plasticized PVC alone, the blends have better flexibility (particularly at low temperatures) and improved abrasion resistance. In unplasticized PVC and CPVC, the polyester–elastomer block copolymers act as polymeric plasticizers. They can also be blended with thermoplastic polyesters such as poly(butyleneterephthalate) to give relatively hard, impact-resistant products.

The polyamide–elastomer block copolymers can be blended with a number of other polymers, including polystyrene, nylon, polyacrylates, nitrile rubber, and polycarbonates (see Refs. [46–55] of Chapter 9).

The polyethylene–poly(α-olefin) block copolymers are recommended for blending with polypropylene or polyethylene to give hard polymer–elastomer combinations, similar to those described in the next section. The products are claimed to have better properties than the corresponding polypropylene–EPDM or polypropylene–EPR blends.

16.3.3   Hard Polymer–Elastomer Combinations

A very large number of these hard polymer–elastomer combinations have been investigated [51] and are described in more detail in Chapters 5, 6 and 7. Trade names of some commercial products are given in Table 16.10. These combinations are usually produced by intensively mixing the elastomer and the hard polymer together at a temperature above that at which the hard polymer becomes fluid, or less frequently, by polymerizing both polymers in the same reactor (see Chapter 5).

Morphology is a critical factor and in most cases a continuous interpenetrating network of the two components is formed, rather than a dispersion of one in the other [52]. In some cases, crosslinking agents are added so that the elastomer phase is vulcanized under intensive shear to give a fine dispersion of the crosslinked elastomer in the hard polymer. This process is known as "dynamic vulcanization" [53] and is described in Chapter 7. It gives the products better resistance to solvents and to compression set. Combinations of polypropylene with EPDM and EPR (both mixtures and dynamic vulcanizates) were the first and are still the largest commercial application for these products. They have been available for about 20 years and are believed to have the second largest share of the market for polymer elastomers, after the polystyrene–elastomer block copolymers [4]. The rubber phase can be extended by the addition of mineral oils, giving the products improved processability [52]. However, the amount that can be added is limited by the fact that the rubber phase is already relatively weak (especially if it is not vulcanized) and so further significant loss of strength cannot be tolerated. High-density polyethylene improves impact resistance at low temperatures [54]. Judging from values given for their densities [54, 55], most products are not highly filled, although some grades intended for sound deadening applications are claimed to contain up to 65% by weight of filler [44]. Dynamic vulcanizates based on blends of polypropylene with butyl rubber [56, 57], natural rubber [58], and nitrile rubber [55] have also been described. The last type requires the use of a polypropylene–nitrile rubber graft copolymer as a compatibilizer.

When suitably pigmented and stabilized, the hard polymer–elastomer combinations based on polypropylene and either EPR, EPDM, natural rubber, or butyl rubber have good

Table 16.10 Some Trade Names of Thermoplastic Elastomers Based on Hard Polymer–Elastomer Combinations

Trade name (Mfr)	Type	Hard polymer	Elastomer	Notes
Ren-Flex™ (Dexter) Hifax® (Himont) Polytrope® (Schulman) Telcar® (Teknor Apex) Ferroflex® (Ferro) Flexothene® (Quantum)[a]	Blend	Polypropylene	EPDM or EPR	Relatively hard, low density, not highly filled
Santoprene® (AES)[b] Sarlink® 3000 and Sarlink® 4000 (Novacor)[c] Uniprene® (Teknor Apex)	DV[d]	Polypropylene	EPDM	Better oil resistance, low compression set, softer
Hifax® XL (Himont)				
Trefsin® (AES) Sarlink 2000 (Novacor)[c]	DV	Polypropylene	Butyl rubber	Low permeability, high damping
Vyram® (AES)	DV	Polypropylene	Natural rubber	Low cost
Geolast® (AES)	DV	Polypropylene	Nitrile rubber	Oil resistant
Alcryn® (DuPont)	Blend	Chlorinated poly-olefin	Ethylene interpolymer	Single phase, soft, oil-resistant
Sarlink® 1000 (Novacor)[c] Chemigum® (Goodyear) Apex® N (Teknor Apex) Elastar® (Nippon Zeon)	DV Blend Blend —[e]	PVC	Nitrile rubber	Oil-resistant
Rimplast® (Petrarch Systems)	Blend	Blend of other TPEs with silicone rubbers		Medical applications

[a] Blend of PP and EPR produced in the polymerization reactor
[b] Advanced Elastomer Systems—a joint venture between Monsanto and Exxon
[c] Now a part of DSM
[d] Dynamic Vulcanizate—a composition in which the soft phase has been dynamically vulcanized, that is, crosslinked during mixing (see Chapter 7)
[e] Elastomer phase reported to be ionically crosslinked [67]

resistance to oxidative and hydrolytic degradation and to weathering. The products can vary quite widely in hardness, from 35 Shore A up to 55 Shore D. Details of the properties of some of these thermoplastic rubbers based on combinations of polypropylene with these relatively nonpolar rubbers are given in Table 16.11. Other commercially important types of hard polymer–elastomer combinations include those based on chlorinated polyolefins with ethylene interpolymers (these are claimed to be single-phase systems) [59] and combinations of PVC with nitrile rubbers [60–65]. Further details of both types are given in Chapter 6. Types of nitrile rubber have been developed that are specifically intended for blending with

Table 16.11 Properties of Hard Polymer–Elastomer Combinations Based on Nonpolar Elastomers

	Ferroflex		Sarlink		Telcar	Santoprene		Trefsin
Product	FF-100	FF-1900	3140	4380[a]	302	101-73 201-73	103-40 203-40	3201-60
Type	Blend	Blend	DV[b]	DV	Blend	DV	DV	DV
Hardness								
Shore A or D	68A	75D	40A	81	75A	73A	40D	60A
Tensile strength								
MPa	4.5	28	4.6	6.1	11	8.3	19	4.6
(psi)	(600)	(4000)	(670)	(885)	(1600)	(1200)	(2750)	665
100% Modulus								
MPa	—	—	—	0.8	3.2	3.2	8.6	1.9
(psi)	—	—	—	115	465	(470)	(1250)	(275)
Elongation (%)	1300	—	650	400	900	375	600	355
Specific gravity	0.90	0.94	0.94	1.31	0.88	0.98	0.95	0.97

[a] Fire retardant grade—intended for wire insulation
[b] Dynamic Vulcanizate—a composition in which the soft phase has been dynamically vulcanized, that is, crosslinked during mixing (see Chapter 7)

PVC. If the final product is intended for extrusion applications, a precrosslinked nitrile rubber (i.e., one crosslinked during the process of its manufacture) is often preferred, but if it is intended for injection molding, uncrosslinked versions may give better results [63, 64]. In some cases the nitrile rubber is dynamically vulcanized [65]. Published compositions often contain fillers and a liquid ester plasticizer such as dioctyl phthalate [60, 61, 63, 64]. All these TPEs based on chlorinated polyolefins or PVC as the hard polymer have good oil resistance, although this resistance is reduced if the rubber phase contains too much plasticizer [64]. It is also affected by the type of plasticizer used. The combinations based on a dynamically vulcanized nitrile rubbers are claimed to have especially good oil resistance [55, 65].

Properties of some typical grades of hard polymer–elastomer combinations based on these more polar rubbers are given in Table 16.12.

16.3.3.1 Replacements for Vulcanized Rubber

This is by far the most significant end use for these materials. Much effort has been devoted to developing specialized grades for various applications, including flame-retardant products. They have numerous applications, particularly in the automobile and appliance industry, in weather stripping, and in wire insulation. One significant advantage is the opportunities that they give for part consolidation [66], that is, they allow assemblies of several pieces of rubber and plastic, often held together with metal clamps, to be replaced by a single molded part.

The polypropylene–EPDM or EPR combinations are especially useful as replacements for general-purpose vulcanized rubbers such as natural rubber, SBR, and EPDM. The products are generally precompounded by the manufacturer to meet the requirements of the end-user. Processing conditions are similar to those used for polypropylene and since the products are relatively stable, degradation during processing is rarely a problem. Scrap can

Table 16.12 Properties of Hard Polymer–Elastomer Combinations Based on Polar Elastomers

	Geolast		Alcryn MPR			Sarlink	Chemigum	
Product	701-70	701-87	2060 BK	2070 NC	2080 BK	1570 UV	03050/ 03150	03070/ 03170
Type	DV[a]	DV	Blend	Blend	Blend	DV	Blend	Blend
Hardness								
Shore A or D	70A	87A	60A	70A	80A	71A	50A	70A
Tensile strength								
MPa	6.2	12	7.5	8.3	11	12	12	17
(psi)	(900)	(1750)	(1100)	(1200)	(1650)	(1700)	(1700)	(2500)
100% Modulus								
MPa	3.3	6.0	2.9	3.9	6.4	3.9	2.9	5.9
(psi)	(480)	(875)	(430)	(580)	(935)	(565)	(420)	(850)
Elongation (%)	265	380	400	375	285	450	470	400
Specific gravity	1.00	0.98	1.10	1.20	1.17	1.20	1.20	1.23

[a] Dynamic Vulcanizate—a composition in which the soft phase has been dynamically vulcanized, that is, crosslinked during mixing (see Chapter 7)

easily be reground and reworked (see the suggestions for grinding soft products previously given in the section on polystyrene–elastomer block copolymers). Predrying is often advantageous [55]. If the products are to be painted, an undercoat is normally needed, although some directly paintable grades have been developed. Ejection of injection molded parts can be a problem, particularly with the softer grades (again, see the suggestions given in the section of polystyrene–elastomer block copolymers). A significant advantage is that these products can be coextruded with polypropylene or insert molded against it [66]. This allows the production of parts such as profile extrusions with regions of different hardness. These are used in weather-strips, gaskets, and similar applications.

Polypropylene–natural rubber combinations will probably compete in the lower cost end of the market, whereas polypropylene–butyl rubber combinations will be especially useful in applications where low gas permeability is important [65].

The materials based on the more polar elastomers, that is, the combinations of either polypropylene or PVC with nitrile rubbers or the combinations of chlorinated polyolefins with ethylene interpolymers, have many applications, especially in areas that take advantage of their excellent oil resistance. All the usual thermoplastic processing techniques (injection molding, blow molding, profile and sheet extrusion) can be used to make end products. Some examples are wire insulation (particularly under the hood), fuel, and hydraulic hose [55, 59, 65]. The simple blends of PVC and nitrile rubbers should compete effectively in low-cost applications, particularly if oil resistance is important. One such application is in injection molded shoe soles. These blends are also reported to have made considerable inroads in automotive applications, especially in Japan. All these TPEs will adhere well when melt processed in conjunction with the appropriate hard polymer. Thus those based on polypropylene adhere to polypropylene and those based on PVC or chlorinated polyolefins adhere to PVC. This allows them to be insert molded or coextruded against the hard polymer to give products with regions of different hardnesses.

16.3.3.2   Polymer Blends

This is not a large application for these TPEs although the use of polypropylene–EPDM or EPR combinations to upgrade scrap polypropylene has been suggested [10].

16.4   Economics and Summary

The economic aspect of TPEs is not simply a function of their price. If it was, they would have achieved little commercial success, since their raw material cost is significantly above that of conventional vulcanized rubbers. Equally (and perhaps more) important are the cost savings they bring because of fast processing, scrap recycle, etc. Another significant factor is the new processing techniques that they have introduced to the rubber industry. These include blow molding, hot melt coating of pressure-sensitive adhesives, and direct injection molding of footwear. Thus rather that a simple cost comparison based on raw material prices, the "value in use" of TPEs should be considered when they are being evaluated as possible replacements for more conventional materials. Bearing this caveat in mind, the price ranges of the various types of TPEs are summarized in Table 16.13. This table also includes data on the specific gravity and hardness of the products. The former is particularly important when used in conjunction with the price, since the multiple of the two gives the relative price per unit of volume, and this (rather than the price per unit of mass) is the quantity that determines the cost of the end product.

Table 16.13 Approximate Price and Property Ranges for Thermoplastic Elastomers[a]

	Price range (cents/lb)	Specific gravity	Hardness range
Polystyrene–elastomer block copolymers			
S–B–S (pure)	85–130	0.94	65A–75A
S–I–S (pure)	100–130	0.92	32A–37A
S–EB–S (pure)	185–280	0.91	65A–75A
S–B–S (compounds)	90–150	0.9–1.1	40A–45D
S–EB–S (compounds)	125–225	0.9–1.2	5A–60D
Polyurethane–elastomer block copolymers	225–350	1.05–1.25	70A[b]–75D
Polyester–elastomer block copolymers	235–375	1.15–1.40	35D–80D
Polyamide–elastomer block copolymers	340–400	1.0–1.15	60A–65D
Polyethylene–poly(α-olefin) block copolymers	80–90	0.85–0.90	65A–85A
Poly(ether-imide)–polysiloxane block copolymers	1700	1.18	70D
Polypropylene–EPDM or EPR blends	90–150	0.9–1.05	50A–65D
Polypropylene–EPDM dynamic vulcanizates	165–300	0.95–1.0	35A–50D
Polypropylene–butyl rubber dynamic vulcanizates	210–360	0.95–1.0	40A–80D
Polypropylene–natural rubber dynamic vulcanizates	140–160	1.0–1.05	60A–45D
Polypropylene–nitrile rubber dynamic vulcanizates	200–250	1.0–1.1	70A–50D
PVC–nitrile rubber blends	130–140	1.10–1.33	50A–90A
Chlorinated polyolefin–ethylene interpolymer blends	220–250	1.10–1.25	60A–80A

[a] These price and property ranges do not include fire retardant grades or highly filled materials for sound deadening
[b] As low as 60A when plasticized

References

1. C.S. Schollenberger, H. Scott, and G.R. Moore, *Rubber World 137*, 549 (1958)
2. W.H. Charch and J.C. Shivers, *Tex. Res. J. 29* (1959)
3. J.T. Bailey, E.T. Bishop, W.R. Hendricks, G. Holden, and N.R. Legge, Presented at the ACS Rubber Division Meeting, Philadelphia, PA, October 1965. *Rubber Age 98* (10), 69 (1966)
4. M.S. Reisch, *C and E News*, p. 30 (May 4, 1992). See also Ref. [131] of Chap. 8
5. B.M. Walker and C.E. Rader (Eds.), *Handbook of Thermoplastic Elastomers*, 2nd edit. (1988) Van Nostrand Reinhold, New York
6. V.M. Kotian, In Ref. [5], Chap. 13
7. G. Holden and K.H. Speer, *Automot. Polym. Design 8* (3), 15 (1988)
8. G. Holden and X-Y. Sun, Presented at the 4th International Conference on Thermoplastic Elastomer Markets and Products sponsored by Schotland Business Research, Orlando, FL, February 13–15, 1991
9. R.J. Deisler, Presented at the 4th International Conference on Thermoplastic Elastomer Markets and Products sponsored by Schotland Business Research, Orlando, FL, February 13–15, 1991
10. *Modern Plastics 60*(12), 42 (1983)
11. Technical Bulletin SC:445-90, Shell Chemical Co., Houston, TX (1990) (A revised version is in press)
12. Technical Bulletin SC:1105-90, Shell Chemical Co., Houston, TX (1990)
13. D.A. Buddenhagen, N.R. Legge, and G. Zscheuschler (to Tactyl Technologies, Inc.), U.S. Patent 5,112,900 (March 12, 1992)
14. Technical Bulletin SC:198-92, Shell Chemical Co., Houston, TX (1992)
15. D.J. St. Clair, *Rubber Chem. Technol. 55*, 208 (1982)
16. Technical Bulletin SC:72-85, Shell Chemical Co., Houston, TX (1985)
17. G. Kraus, K.W. Rollman, and R.A. Gray, *J. Adhesion 10*, 221 (1979)
18. S.G. Chu and J. Class, *J. Appl. Polym. Sci. 30*, 805 (1985)
19. W.M. Halper and G. Holden, in Ref. [5], Chap. 2
20. J.R. Erikson, *Adhesives Age 29*(4), 22 (1986). Reprinted as Technical Bulletin SC:889-86, Shell Chemical Co., Houston, TX (1986)
21. G. Holden and S.S. Chin, Presented at the Adhesives and Sealants Conference, Washington D.C., March 1986
22. G. Holden, Presented at the Adhesives and Sealants Council Seminar, Chicago IL, October 1982
23. D.M. Mitchel and R. Sabia, In Proceeding of the 29th International Wire and Cable Symposium, November 1980, p. 15
24. Technical Bulletin SC:1102-89, Shell Chemical Co., Houston, TX (1989)
25. J.Y. Chen (to Applied Elastomerics, Inc.), U.S. Patent 4,369,284, (January 18, 1983) and U.S. Patent 4,618,213 (October 21, 1986)
26. A.L. Bull and G. Holden, *J. Elastom. Plast. 9*, 281 (1977)
27. Technical Bulletin SC:165-93, Shell Chemical Co., Houston, TX (1993)
28. G. Holden and D.R. Hansen (to Shell Oil C.), U.S. Patent 4,904,731 (February 27, 1990)
29. G. Holden, *J. Elastom. Plast. 14*, 148 (1982)
30. C.L. Willis, W.M. Halper, and D.L. Handlin Jr., *Polym. Plast. Technol. Eng. 23*(2), 207 (1984)
31. Technical Bulletin SC:1216-91, Shell Chemical Co., Houston, TX (1991)
32. B.C. Arkles, *Med. Device Diagnost. Indust. 5*(11), 66 (1983); see also B.C. Arkles (to Petrarch Systems), U.S. Patent 4,500,688 (February 19, 1985)
33. E.J. van Beem and P. Brasser, *J. Inst. Petroleum 59*, 91 (1973)
34. S. Piazza, A. Arcozzi, and C. Verga, *Rubber Chem. Technol. 53*, 994 (1980)
35. J.L. Goodrich, *Asphalt Paving Technol. Proc. AAPT 57* (1988)
36. M.G. Bouldin, J.H. Collins, and A. Berker, *Rubber Chem. Technol. 64*, 577 (1991)
37. Technical Bulletin SC:1043-90, Shell Chemical Co., Houston, TX (1990)
38. Hytrel Bulletin HYT-302(R1), E.I. DuPont de Nemours & Co. (1981)
39. G. Deleens, In *Thermoplastic Elastomers—A Comprehensive Review*, 1st edit. (N.R. Legge, G. Holden, and H.E. Schroeder (Eds.) (1987) Hanser and Oxford University Press, Munich and New York
40. D.G. Davis and J.B. Conkey, Presented at the 6th International Conference on Thermoplastic Elastomer

Markets and Products sponsored by Schotland Business Research, Orlando, FL, January 15–17, 1992

41. J. Mihalich, Presented at the 2nd International Conference on Thermoplastic Elastomer Markets and Products sponsored by Schotland Business Research, Orlando, FL, March 15–17, 1989

42. M. Szycher, V.L. Poirier, and D. Demsey, *Elastomerics 115*(30), 11 (1983)

43. R. School, in Ref. [5], Chap. 9

44. R. School, *Rubber Plast. News 49* (May 7, 1984)

45. Hytrel Bulletin HYT-452, E.I. DuPont de Nemours & Co. (1981)

46. E.C. Ma, in Ref. [5], Chap. 7

47. T.W. Sheridan, in Ref. [5], Chap. 6

48. F.A. Quinn, V. Kapasi, and R. Mattern, Presented at the 6th International Conference on Thermoplastic Elastomer Markets and Products sponsored by Schotland Business Research, Orlando, FL, January 15–17, 1992

49. H.W. Bonk, R. Drzal, C. Georgacopoulos, and T.M. Shah, Presented at the 43rd Annual Technical Conference of the Society of Plastics Engineers, Washington, D.C., 1985

50. Hytrel Bulletin IF-HYT-370,025. E.I. DuPont de Nemours & Co. (1976); see also R.W. Crawford and W.J. Witsiepe (to DuPont), U.S. Patent 3,718,715 (February 27, 1973)

51. A.Y. Coran, R. Patel, and D. Williams, *Rubber Chem. Technol. 55*, 116 (1982)

52. E.N. Kresge, *Rubber Chem. Technol. 64*, 469 (1991)

53. A.M. Gessler (to Esso Research and Engineering Co.), U.S. Patent 3,037,954 (June 5, 1962)

54. C.D. Shedd, in Ref. [5], Chap. 3

55. C.P. Rader, in Ref. [5], Chap. 4

56. R.C. Puydak and D.R. Hazelton, Presented at the 2nd International Conference on Thermoplastic Elastomer Markets and Products sponsored by Schotland Business Research, Orlando, FL, March 15–17, 1989

57. R.C. Puydak and D.R. Hazelton, Presented at the Symposium on Thermoplastic Elastomers sponsored by the ACS Rubber Division, Nashville, TN, November 3–6, 1992

58. A.J. Tinker, Presented at the Symposium on Thermoplastic Elastomers sponsored by the ACS Rubber Division, Cincinnati, OH, October 18–21, 1988

59. J.G. Wallace, in Ref. [5], Chap. 5

60. M. Stockdale, Presented at the Symposium on Thermoplastic Elastomers sponsored by the ACS Rubber Division, Cincinnati, OH, October 18–21, 1988

61. P. Tandon, Presented at the 3rd International Conference on Thermoplastic Elastomer Markets and Products sponsored by Schotland Business Research, Dearborn, MI, March 28–30, 1990

62. P. Tandon and M. Stockdale, Presented at the 4th International Conference on Thermoplastic Elastomer Markets and Products sponsored by Schotland Business Research, Orlando, FL, February 13–15, 1991

63. L.B. Kliever and R. DeMarco, Presented at the Symposium on Thermoplastic Elastomers sponsored by the ACS Rubber Division, Nashville, TN, November 3–6, 1992

64. L.B. Kliever, Presented at the 7th International Conference on Thermoplastic Elastomer Markets and Products sponsored by Schotland Business Research, Orlando, FL, February 11–12, 1993

65. M.J. Thompson, Presented at the 6th International Conference on Thermoplastic Elastomer Markets and Products sponsored by Schotland Business Research, Orlando, FL, January 15–17, 1992

66. R.B. Mattix, Presented at the Symposium on Thermoplastic Elastomers sponsored by the ACS Rubber Division, Nashville, TN, November 3–6, 1992

67. *Chemical Week*, p. 28 (September 25, 1985)

17 Future Trends

G. Holden, N.R. Legge, H.E. Schroeder, and R.P. Quirk

"Unreal, unduly optimistic forecasts of future research and development are all too prevalent, especially in chapters like this where the writers do not have to pay the piper—but who could look at thermoplastic elastomers without adopting a rosy view of the future?" This was the opening sentence in the corresponding chapter of the first edition of this work. Now, almost 10 years later, we have the advantage of hindsight. With this advantage, our opinion has not changed. It is even clearer today that this field of work presents an unusual combination of great academic and industrial interest, with much to be learned, and an eagerly receptive market awaiting the fruits of discovery.

For over 50 years new thermoplastic elastomer (TPE) materials have emerged, (many of them as surprises), well before the physical science had advanced paradigms that could explain and predict their properties. The same situation also applies to many new polymerization systems. However, science has advanced and a better understanding of both the polymer properties and the processes of their production has been achieved. Our first edition appeared at an opportune time, when the first stages of this process were well underway. With the realization that almost all TPEs owe their essential properties to the presence of at least two polymeric phases, one fluid and the other solid (in the normal operating temperature range), plus the realization that these phases can be combined in many ways, both physical and chemical, the possibilities for new useful TPE structures now appear almost boundless.

To create these new TPEs we have at our disposal a very broad choice of polymerization reactions and catalysts that can be used to enchain the most diverse of monomers, polar and nonpolar, in almost any ratio. We can control monomer ordering and even chirality. Control of molecular weight and of molecular weight distribution is no longer a mystery. There is also a wealth of fluid or elastomeric structures and an even larger variety of glassy or crystalline polymers to combine with them. The rapidly developing science of the rheology of these systems allows us to understand polymer flow properties and the relationships between processing characteristics and chain structure, chain length, branching, and molecular weight distribution. Improved analytical techniques, such as the recent advances in gel permeation chromatography, have allowed us to define these molecular parameters. Morphological research on these fascinating products is showing us the bases for physical and mechanical properties. Present knowledge of the chemical structures can explain solubility, stability to solvents and chemicals, and resistance to degradation by heat, light, and oxidation. With such assets at our disposal it is now possible to undertake material research problems with the expectation that particular property goals can be approximated. Although surprising results still occur and form the basis for exciting new discoveries, we are often able to predict polymer properties in advance. So why should we not be positive in assessing this future?

The content of the preceding chapters supports this thesis. The extent of our under-standing is remarkable, yet in no case does it appear that the subject is exhausted. Rather, the explications of the authors all seem to presage an interesting future. The real question is not whether, but where, the new TPE products will appear. Of course, this leaves open the question of economics insofar as practical products are concerned. Here again, experience has shown innumerable compounding modifications, or selection of use areas, that can provide success. Against this background we propose to look at the future by touching first on the properties that might appear desirable, then on the possibilities of achieving them, and finally on the supporting sciences.

The rapid and easy processing of TPEs have made them welcome in the marketplace. Because of these processing advantages, even though their raw material prices may be relatively high compared to those of conventional elastomers, the cost of the final product is often less. They have quickly filled many product requirements formerly held by conventionally cured rubbers. They have been particularly welcome where their elastomeric physical properties at normal use temperatures are adequate and the retention of properties at elevated temperatures, conveyed by a chemical (covalent) crosslink, is not needed. Some examples are in wire coatings, automotive parts, shoe soles, and adhesives. We expect that with some limitations we shall see the TPEs fill out the entire continuum of properties now provided by the conventional cured rubbers. This continuum now goes from the extremely soft silicone rubbers to the ultrastable perfluoroelastomers and includes products with extraordinary chemical and solvent resistance. In addition, we expect the TPEs to offer properties that cannot be achieved with the usual cured rubbers. They fill out the transition zone between the hardest conventional rubbers and the high-impact plastics. This area has been partially occupied already by the so-called "engineering" high-modulus TPEs such as the polyurethanes, copolyesters, and copolyamides. Where performance requirements are less critical, TPEs based on polyolefins have emerged as a major factor in this market. We believe that the TPEs will meet the developing needs of the hard, resilient materials more easily and more economically than the cured rubbers.

At the other end of the hardness range, TPEs can be compounded to give extremely soft products, even gels.

Certain characteristics are associated with the families of cured rubbers. These include:

- *Elasticity*—long-range extension and retraction, snap, softness, resilience—the domain of natural rubber and the general purpose synthetics [poly(styren-co-butadiene) random copolymer (SBR), polyisoprene (IR), and ethylene–propylene–diene (EPDM)], plus polychloroprene and the silicones
- *Solvent resistance* plus most of the above characteristics—the realm of almost all of the "general purpose" specialty rubbers ranging from butadiene–acrylonitrile rubbers (NBR) to the more oxygen- or heat-stable hydrogenated versions (HNBR), polychloroprene, chlorosulfonated polyethylene and related polyethylene derivatives, and also polyacrylates
- *Thermal and oxidative stability*—typified by the ethylene copolymers, such as EPDM elastomers, HNBR, the silicones, polyacrylates, and for more extreme demands—the various fluoroelastomers
- *Chemical resistance*—the hydrofluoroelastomers, fluorosilicones, fluorophosphazenes, and the perfluorocarbon elastomers
- *Abrasion resistance*, toughness—found in most cured rubbers to varying degrees but outstanding in the crosslinked polyurethanes
- *Resistance to creep, permanent set, compression set*—load bearing capacity, the hallmark of all cured elastomers

- *Stability to deformation at elevated temperatures*—characteristics of all cured rubbers up to the limit of the thermal stability of the crosslinks or the polymer chains.

Clues to discovery of TPEs to match properties of cured rubbers lie in the very structures that convey the desired properties. Elastomer type molecules quite naturally will be first choice for the fluid matrix (elastomeric phase) but many other fluids with low T_g are quite suitable. Similarly, selection of the hard or physical crosslinking segment will depend on a fit between properties sought and the hard polymer structure known to convey them. In these two categories there are hundreds of available polymers and more keep coming. The problems, of course, are in the reactions to form the desired segmented copolymer molecules, for example, triblock copolymers, randomly segmented block copolymers, ionomers, polymer blends, etc.

For snappy, elastomeric characteristics, experience suggests that the best prospects are in structures where the fluid phase has a very low T_g, and is present in high proportion relative to the hard phase. The latter should be separate and discontinuous. It should be almost insoluble in the fluid phase because mutual solubility will raise the T_g of this phase.

Triblock copolymers with shorter, higher melting, glassy or crystalline hard blocks offer one approach to improved resilience. Crystalline hard blocks will improve solvent resistance. Higher melting hard blocks should also raise the heat distortion temperature and give increased resistance to creep and set. Where anionic or carbocationic polymerizations do not yield the desired structure, capping reactions followed by coupling or further polymerization are alternatives. In addition, there is always the possibility of new methods of enchainment. Stability in the elastomeric phase will be obtained by further use of saturated elastomers based on polyolefins as replacements for the simple polydienes. Silicones and other new chains of diverse sorts with heteroatoms such as phosphazines offer similar opportunities to improve low- and high-temperature performance.

These same comments could also apply to the multisegmented copolymers $(A–B)_n$, such as the polyurethanes, copolyetheresters, copolyamides, etc. Because of phase incompatibility, it is not easy to take this specific approach with dynamically vulcanized melt mixed blends. However, steps in this direction can be taken by incorporation of suitable A–B diblock copolymers as interfacial adjuncts to convey compatibility and processability as well as enhanced physical properties.

Ionomers seem to offer a unique opportunity to embed very small "hard" ionic clusters in an elastomer matrix and thus achieve a very soft rubber with good properties. This is probably the closest we can come to the model of the cured elastomer with one covalent link per 100 to 200 chain length atoms. Ionomers based on the fusible Na, K, NH$_4$, etc., salts unfortunately show excessive creep, probably because of a very low energy barrier to the dislocation of some ions in the cluster. Higher melting or infusible salt combinations, such as the Pb, Zn, or Ca salts used in the Surlyn development, may be combined with a temporary fluxing agent or salt to enhance processability. On a more speculative level is the possibility of using thermally dissociating bonds that recombine as the melt is cooled, for example, diene adducts, or groupings that on heating dissociate to relatively unreactive free radicals that recombine on cooling.

Resistance to swelling by solvents is not easily achieved with the copolymers accessible through the relatively inexpensive anionic polymerizations used in triblock copolymers. The polyurethane, copolyester, and copolyamide TPEs all show outstanding resistance to nonpolar and many polar solvents because the crystalline hard segments are insoluble and their elastomeric segments are polar. However the raw materials for their production are relatively costly. As shown by Alcryn and other blends, the goal of solvent resistance is probably attainable at lower cost by melt blending hard and soft polymers, with or without dynamic vulcanization. This approach may be considered as technical compatibilization. It is particularly attractive to industrial managers because so many commercial polymers with desired properties are available for blending. Capital investment in a new plant is minimal. In contrast, commercial production of a new polymer requires a long and costly development and large amounts of capital. The rapidly increasing knowledge of the rheology and morphology of polymer blends as well as the availability of polymeric (A–B) dispersants and compatibilizers further expands the product possibilities.

Chemical, solvent, and high-temperature resistance of the highest order is likely not a practical goal for TPEs. First, the concept of thermoplastic processing is antithetical to stability at very high temperatures. For example, a silicone or hydrofluoroelastomer is stable for long periods at 225 °C and resists short excursions to over 275 °C. Appropriate hard segments are not easily found. An aromatic polyamide, ether, or ester melting above this point would present some very interesting processing problems, somewhat like those involved in perfluorocarbon resins. Even if temporary fluxing or plasticizing agents were found, the aromatic polyamides, esters, and ethers do not possess the stability of a fluorocarbon chain when exposed at high temperatures to corrosive chemicals, water, and acids.

On the other hand, chemical and temperature resistance in the intermediate range should be an attractive goal for TPEs. There are many polycondensation reactions available and new ones are often discovered. The condensation polymers can be designed to possess many of the desired properties without greatly increasing cost. In addition, new intermediates for condensation reactions could be created from substituted polyolefins by equipping them with end groups for attachment of hard segments, thus forming various sorts of segmented copolymers. Group transfer polymerization (GTP) [1–9] is described in Chapter 14. It brings many acrylic monomers into the picture for the creation of tri- and polyblock copolymers. Straight-chain, star-shaped, and branched structures are all possibilities. GTP would be particularly useful since it has many characteristics of a living polymer system. This enables the preparation of monodisperse copolymer blocks and chain-end functionalized polymers. To illustrate, a difunctional monomer such as ethylene dimethacrylate could be converted via the silyloxy reagent to difunctional initiator. This in turn, on reaction with methyl or butyl acrylate, could yield a monodisperse, elastomeric polyacrylate segment. Further treatment with an acrylic monomer, such as methyl methacrylate or acrylonitrile, could give higher melting segments. As research on GTP continues there will undoubtedly be many more monomers that will work. It is to be expected that, in addition to new elastomers, novel glassy and crystalline hard segments will be developed.

One of the most exciting advances in polymer science has been the development of methods of living polymerization. These allow the controlled polymerization of an ever-widening range of monomer types proceeding via a wide range of chain polymerization mechanisms [10] (see also Chapter 14). Thus, in addition to GTP and the traditional living anionic polymerization methodology, which resulted in the commercial development of styrenic thermoplastic elastomers (see Chapter 3), methods have been developed for living and/or controlled cationic [11] (see Chapter 13), Ziegler–Natta [12] (see Chapter 5), ring-opening methathesis [13], coordination [14], and even radical [15–17] systems that have many of the characteristics of living polymerizations. Obviously, the wide range of monomers that can be polymerized by these mechanically diverse methods allows the preparation of many new block, segmented, functionalized and branched polymers for use as TPEs.

TPEs usually show adequate abrasion resistance and toughness and so setting these properties as new product objectives will not be a hindrance. The suitability of S–B–S triblock copolymers for shoe soles, and utility of the polyurethane and copolyester TPEs in industrial wheels, gears, and uninflated tires attests to their quality in this respect. Of course, they are not suitable for applications where they are exposed to temperatures at which the hard segments melt. For example, polyurethane TPEs will melt or even depolymerize if, in tire applications, they are subjected to panic stop or skids that produce high local temperatures from frictional effects.

Creep resistance, permanent set, and compression set resistance of the TPEs of all classes are usually inferior to those of the corresponding cured rubbers over the normal service temperature range of the latter. When TPEs are subjected to excessive tensile or compressive forces, slippage of the polymer molecules in the hard segment phase occurs. Amorphous hard segments creep while crystalline ones often draw or orient to form more perfect crystals. Prospects for improving these characteristics are reasonably good. Polyurethane, copolyester, and copolyamide TPEs with a high proportion of hard segments have very good resistance to creep or set. We believe that these traits can be improved and even extended somewhat to softer TPEs by improving the crystallinity. This means using either more crystallizable (higher melting) hard segments or monodisperse hard segments that are able to form more perfect crystals. To illustrate, in a copolyetherester, polybutylene terephthalate crystallizes more quickly and more perfectly than it does in polybutylene terephthalate homopolymer. Also when a copolyetherester is annealed under tension it draws, with increase in crystallinity and corresponding improvement in resistance to set and creep.

There appears to be little hope that TPEs can match the resistance to deformation at elevated temperatures that is shown by cured rubbers. Yet by the use of higher melting hard segments, remarkable properties are within reach. For example, a copolyester with 75% butylene terephthalate hard segments has better mechanical properties at 150 °C than almost all cured rubbers. This property should be greatly improved by the use of short, high melting aromatic polymer segments. Processability requirements will, however, dictate very careful study of the soft segment accompanying structures. In this area of polymer properties it is interesting to note [18] that the comb-graft copolymer of short grafts of polypivalolactone (PPVL) on an EPDM backbone showed mechanical properties that in many respects closely resembled those of a chemically vulcanized rubber. The PPVL appeared to exist in discrete crystalline domains of sizes in the range of 10 to 100 nm (100 to 1,000 Å). The graft copolymers were very strong and resistant to compression set.

Future research must also consider the potential of compounding new products to produce the desired properties and also to broaden the economic range of these products. There are many examples in the preceding chapters, such as the effects of oils and plasticizers in adjusting the phase volumes in melt mixed blends of EPDM and polypropylene, both with and without dynamic vulcanization (Chapters 7 and 5); the addition of isotactic polypropylene to semicrystalline propylene–α-olefin copolymers to provide improved properties via co-crystallization (Chapter 5); the blending of polymers, resins, and oils with the styrenic TPEs to broaden their range of physical properties (Chapters 3 and 16). On the other end of the spectrum, we have seen the significant effects of 20% of S–EB–S on the physical properties of blends of two thermoplastics (Chapter 15C), and the effect of less than 5% of S–B–S or S–EB–S on the properties of asphalt (Chapter 16).

There are other new preparative techniques we wish to draw to your attention. Four of them are described in this book. The availability of soluble metallocene catalysts that can effect controlled polymerization of a variety of alkenes holds promise of many new TPEs based on alternating crystalline and amorphous segments. This is one system that is now yielding commercial TPEs such as the ethylene–α-olefin multiblock copolymers described in Chapter 5. Extension of this system to produce TPEs based on multiblock copolymers of isotactic polypropylene and atactic polypropylene has been recently reported [19, 20]. The ideal low-cost TPE is probably a multiblock copolymer of isotactic polypropylene and an ethylene–α-olefin random copolymer, but so far this Holy Grail seems to have eluded the searchers. Chapter 13 covers controlled carbocationic polymerizations, including living polymerizations. These inifer systems, which proceed by reversible termination steps, can yield well-defined, living linear or three-arm star polyisobutylenes (PIB) from which one can proceed to macromers, ionomers, block copolymers, and telechelics. The use of macro-monomers to make TPEs based on a wide variety of starting monomers and/or polymers is described in Chapter 14 and TPEs produced by bacteria are described in Chapter 15B.

The final method is entirely different, as claimed by Falk and Van Beck [21]. The product is a TPE prepared by a sequential emulsion polymerization to give a core-shell system. The rigid thermoplastic core, for example, a lightly crosslinked styrene–acrylonitrile copolymer, is polymerized as a latex. A transition layer is then formed by simultaneously adding the monomers of the core and the shell (butyl acrylate), to the emulsion polymerization, thus surrounding and encapsulating the core. Finally, the monomers of the shell are added and polymerized to form the encapsulating outer shell. Small amounts (about 5%) of functionally reactive monomers are added to the core (acrylic acid), and to the shell (2-hydroxyethyl methacrylate) with the intent that these will link the core and shell lightly during melt mixing. The approximate proportions of the components (by weight of the total polymer) are: core 20%, shell 30%, transition 50% . The presence of the core, transition layer, and shell is essential for TPE properties. Omitting the transition layer, or the functional linking monomers, results in a soft, low strength gum. Compounding the product with 0.1% to 1% of a metal oxide, such as zinc oxide, improves extrusion and molding operations. The core-shell polymer is processable in normal thermoplastic operations. Although the physical properties of these polymers are not equal to those of many of the current TPE base copolymers, they are, nevertheless, those of a TPE. If this approach can be optimized to provide improved physical properties, it will have a major impact on the TPE area.

Of course, all the above possibilities would be much less interesting if it were not for the commercial uses of TPEs. They have made a significant impact on the polymer industry and continue to show a healthy growth rate. Table 17.1 illustrates this point. Over the 10-year period from 1990 to 2000, annual worldwide consumption is predicted to grow from almost 1.4 billion lbs to over 2.4 billion lbs [22]. Considering that the average price is at least \$1/lb, this is of great commercial importance.

To sum up—there are many possible structures for new TPEs. The essential principle, now well known, is the requirement for at least two interdispersed polymeric phases. At normal operating temperatures, one is fluid (above its T_g) and the other solid (below its T_g or T_m), with some interaction between them. A good guide for future R&D would be to avoid a prior mind-set on a given polymer structure, for example $(A–B)_n$, or $(A–B–A)$, but to select

Table 17.1 Estimated Worldwide Consumption of Thermoplastic Elastomers (Millions of pounds per year)

	1990	1995	2000
Styrenic block copolymers	625	805	990
Polyurethane or polyester–elastomer block copolymers	245	335	445
Polyolefin blends and thermoplastic vulcanizates	390	570	785
All others	120	165	190
Totals	1380	1875	2410

Source: Ref. [22]

polymer segments to achieve the properties desired. In reviewing elastomeric property ranges, one may conclude that the trend toward TPEs with stability at higher temperatures will continue. Another area of significant effort will be aimed at chemical, oil, and temperature resistance, with emphasis on the melt-mixing of blends of polymers. In the triblock copolymer TPEs, new polymerization techniques can lead to products with polyolefin center blocks and crystalline end blocks. New TPEs will begin to fill in the entire range of conventional crosslinked rubbers. Undoubtedly there will be limitations, which we have discussed here; however, we view the scope and utility of TPEs as being much broader than generally considered today. In the preceding chapters we have seen a wealth of information on the TPE systems which we believe will be applied to expand this product area. Everything points to a research area that is fast moving and rewarding—and which will become more so!

References

1. O.W. Webster, et al., *Polymer Preprints 24* (2), 52 (1983); *J. Am. Chem. Soc. 105*, 5706 (1983), U.S. Patent 4,544,724, October 1, 1985 (to DuPont)
2. D.V. Sogah and O.W. Webster, *Polym. Preprints 24* (2), 54 (1983)
3. D.V. Sogah and O.W. Webster, *Macromolecules 19*, 1775 (1986)
4. O.W. Webster, *Polym. Preprints 27* (1), 161 (1986)
5. D.Y. Sogah, *Polym. Preprints 17* (1), 163 (1986)
6. W.R. Hertler, *Polym. Preprints 27* (1), 165 (1986)
7. W.B. Farnham and D.Y. Sogah, *Polym. Preprints 27* (1), 167 (1986)
8. F. Banderman, H.D. Sitz, and H.D. Speikamp, *Polym. Preprints 27* (1), 169 (1986)
9. W.J. Brittain, *Rubber Chem. Technol. 65*, 580 (1992)
10. R.P. Quirk and J. Kim, *Rubber Chem. Technol. 64*, 450 (1991)
11. J.P. Kennedy and B. Ivan *Designed Polymers by Carbocationic Macromolecular Engineering. Theory and Practice* (1992) Hanser, Munich and New York
12. T. Keii, Y. Doi, and K. Soga, In *Encyclopedia of Polymer Science and Engineering*, 2nd edit. J.I. Kroschwitz (Ed.) (1989) John Wiley & Sons, New York, Supplemental Volume, p. 436
13. B.M. Novak, W. Risse, and R.H. Grubbs, *Adv. Polym. Sci. 102*, 47 (1992)
14. S. Inoue and T. Aida, In *Encyclopedia of Polymer Science and Engineering, Vol. 7*, 2nd edit. J.I. Kroschwitz (Ed.) (1987) John Wiley & Sons, New York, p. 412

15. M.K. Georges, R.P.N. Veregin, P.M. Kazmaier, and G.K. Hamer, *Macromolecules 26*, 2987 (1993)
16. R.P.N. Veregin, M.K. Georges, P.M. Kazmaier, and G.K. Hamer, *Macromolecules 26*, 5316 (1993)
17. P.M. Kazmaier, K.A. Moffat, M.K. Georges, R.P.N. Veregin, and G.K. Hamer, *Macromolecules 28*, 2987 (1995); G.Y-S. Lo, E.W. Otterbacher, R.G. Pews, and L.H. Tung, *Macromolecules 27*, 2241 (1994)
18. C-K. Shih and A.C.L. Su, In *Thermoplastic Elastomers—A Comprehensive Review.* N.R. Legge, G. Holden, and H.E. Schroeder (Eds.) (1987) Hanser and Oxford University Press, Munich and New York
19. W.J. Gauthier, J.F. Corrigan, N.J. Taylor, and S. Collins, *Macromolecules 28*, 3771 (1995)
20. W.J. Gauthier and S. Collins, *Macromolecules 28*, 3779 (1995)
21. J.C. Falk and D.A. Van Beck, (to Borg-Warner Chemicals Inc.), U.S. Patent 4,473,679 (September 25, 1984), filed December 12, 1983
22. H.R. Blum, Personal communication, Chem Systems Inc., Tarrytown, NY

Index

Abbreviations (*see* Nomenclature)
Abrasion Resistance *33, 39, 135, 137, 143, 145, 146, 148, 150, 151, 249, 250, 604, 607*
Adhesion *251, 511, 512*
Adhesive Applications *292, 575, 585–587, 594*
Adipic Acid *232*
Alcaligenes eutrophus 466
Aliphatic Amides *230, 235, 246*
Alkylaluminum Coinitiators *367*
Alkyllithium Initiators *4–6, 48, 49, 51, 52, 74–76, 91, 398, 400, 402, 404, 405*
Alkylsodium and Alkylpotassium Initiators *5*
Alpha-Omega Polymers *396, 397, 403, 406*
Anionic Polymerization *3–6, 51–54, 74–78, 87, 89, 91, 97, 298, 336, 337, 380, 381, 398–406, 410, 413, 414, 522, 525, 605, 606*
Annealing *20, 26, 28, 29, 37, 85, 193, 201–204, 209, 547, 548, 551*
Antioxidants *24, 147, 148, 195, 212–215, 246, 584*
Applications, General *39–41, 49, 61, 124, 125, 142, 147, 150, 151, 185–189, 211, 214, 221, 222, 254, 255, 263, 538, 567, 569, 574–599*
Applications of Ionomers *291, 292*
Asphalt Blend Applications *292, 293, 589*
Automotive Applications *40, 124, 125, 142, 150, 188, 221, 254, 574, 575, 589, 593, 598*
Azeleic Acid *232*

Bacterial Polymerization *10*
Bicontinuous (Gyroid) Structure *359–361*
Bigrafts *377–379*
Biodegradation *467, 483, 484*
Biosynthesis *469–472*
Blends *2, 6–8, 10, 37–39, 102, 103, 106, 119–125, 142, 148, 150, 154, 219, 220, 254, 269, 288–290, 314–331, 425, 490–518, 588, 589, 594, 595, 598, 605, 606, 609*
Block Copolymer Ionomers *284*
Block Copolymers *48–68, 72–99, 112, 115, 192–213, 298–332, 336–364, 380–389, 430–460, 490, 491, 497, 522–524, 526,*

530, 532, 534, 535, 576, 578–590, 605–609
Block Length Distribution in Polyester Ethers *194–198*
Blow Molding *141, 148, 196, 211, 212, 581, 585, 593, 598*
Blown Film Applications *39, 222, 581, 585, 588, 594*
Branched Polymers (*see also* Star and Radial Polymers) *50, 53, 67, 68, 78, 80, 81, 86, 197, 336, 341, 415, 416*
Brittle Point (*see* Service Temperature)
Bromobutyl Rubbers (*see* Halobutyl Rubbers)
B-S-B Block Copolymers *50, 67, 68*
Butyl Rubber (*see* Polyisobutylene and Butyl Rubber)

Calendering *140*
Carbocationic Polymerization *9, 54, 113, 365–391, 605, 608*
Carboxylated Ionomers *273, 278–280, 283–285, 287, 288*
Carboxylic Acids *195, 230–234, 246*
Cast Films (*see* Solution Cast Films)
Cation Effects in Ionomers *259, 260, 262, 265, 266, 287, 288, 290*
Chain Entanglements *55–57, 59, 73, 82, 94, 155, 176–178, 285, 517*
Chain Extenders and Extension *16, 21, 22, 26, 27, 34*
Chain-end Transformation *388*
Characterization of PHO *467, 474*
Chemical Milling Applications *587*
Chlorinated Polyolefins *7, 130, 596–598*
Chlorobutyl Rubbers (*see* Halobutyl Rubbers)
Chlorosulfonated Polyethylene (Hypalon) *375–377*
Clarity *221, 258, 262, 267, 290, 425, 582, 588, 593*
Clash-Berg Temperature (*see* Service Temperature)
Classification *576*
Coextrusion (*see* Extrusion and Coextrusion)
Compatibility and Compatibilizers *10, 54, 59, 61, 123, 155, 178–184, 192, 195, 266, 288–290, 490, 496–498, 517, 518, 579, 580, 582, 585–588, 594*

Compositional and Sequence Distribution *18–20, 24, 29–31, 115*

Compounding *291, 574, 579, 580–587, 589, 594–596, 598*

Compression Set *135, 143, 146, 147, 151, 154, 155, 157, 249, 250*

Concentration Fluctuations *442*

Condensation Polymerization *2, 3, 7, 194–197, 396, 606*

Conjugated Polymers *538, 539*

Consumption *192, 608, 609*

Controlled Initiation *367*

Cooperative Phenomena *430*

Coordination Polymerization *606*

Copolyester/PVC Blends (*see* PVC/ Copolyester Blends)

Costs (*see* Prices)

Coupling Agents *52, 53, 75, 78, 89, 91, 380, 416, 523*

Critical Molecular Weights for Domain Formation *59, 65, 66, 80, 83, 414*

Cross-linking by Radiation *538, 539, 541, 551–553, 557, 562, 563, 587*

Cross-linking following Processing *538, 539, 557, 587*

Cross-links *2, 5, 6, 8, 17, 24, 31, 48, 53, 83, 84, 130, 144, 145, 154, 159, 167, 258, 273, 540, 541, 545, 549–554, 562, 568–570, 576, 579, 587, 595, 596*

Cross-polymerization *540, 541, 545, 549–554, 562, 568–570*

Crystal Melting Temperature (Tm) *18, 19, 23, 26–30, 34, 103, 104, 170, 193, 194, 198–205, 208, 214, 215, 228–230, 233, 234, 240, 302, 303, 474, 476, 482, 484, 544, 566–569, 577, 578, 582, 588, 592, 606–608*

Crystallinity and Crystalline Segments *17, 25–27, 82, 96–99, 111, 117, 155, 169, 174, 175, 177, 178, 189, 197–205, 208, 209, 217, 219–221, 230, 233, 255, 268, 300, 301, 304, 467, 469, 474–478, 484, 576–578, 590–593, 605–607, 609*

Curatives *170–173*

Cyclized Polyisoprene Segments (cyPIP) *388*

Degradation (*see* Stability, Thermal, Ozone and UV)

Diacetylene *10*

Diblock Copolymers *75, 89, 336, 338, 344–350, 352, 357, 359, 360, 400, 410, 420, 421*

Dicarboxylic Acids *192, 213–217, 231–234*

Differential Scanning Calorimetry (DSC) (*see* Thermal Analysis (including DSC, DTA and TMA))

Differential Thermal Analysis (DTA) (*see* Thermal Analysis (including DSC, DTA and TMA))

Diffusion (*see* Permeability)

Diffusivity *458*

Diisocyanates *16–18, 21, 23, 232, 233, 541, 568*

Dilatometry *513*

Disordered State *337, 349, 360, 431, 436–439, 443–446, 450, 451, 457*

Di- and Multifunctional Initiators *48, 53, 76–78, 87, 380, 381, 384, 523, 526*

Domain Theory *4, 6, 48–51, 83–89*

Double Diamond Structure (*see* Ordered Bicontinuous Double Diamond (OBDD)

Dynamic Vulcanization *7, 103, 104, 133, 154–189, 220, 579, 595–597, 599, 605, 607*

Economics *574, 599, 604, 608*

Eisenberg-Hird-Moore (EHM) Model *273, 277, 280, 285*

Elastic Fibers *2, 3, 7, 16, 25, 26, 215, 216,*

Elastomers, Conventional *4, 5, 48, 49, 88, 89, 158, 167, 186, 574, 575, 588, 596, 598, 604, 605, 607, 608*

Electrical Properties *136, 148, 251, 252*

Electron Microscopy *58, 63–65, 80, 81, 91, 92, 144, 159, 163, 164, 166, 168, 206, 268, 278, 279, 339–341, 353–355, 385, 446, 511, 533, 545–547, 564, 566*

Elongation (*see* Physical Properties)

Emulsion Polymerization *410, 417, 608*

EPDM and EPR Rubbers *6, 102, 103, 119, 122–125, 157–162, 173, 174, 183, 184, 298, 300, 369–371, 377–379, 391, 509, 579, 582, 588, 595, 596, 598, 599, 607*

Esterification *195–197, 230, 232–235*

Extraction and Purification of Biopolymers *472*

Extrusion and Coextrusion *17, 22, 24, 37, 40,*

134, 138–140, 143, 144, 187, 188, 263, 581, 584, 585, 593, 594, 596, 598
E-EB-E and E-EP-E Block Copolymers 66, 97–99, 102, 112, 113, 115–117

Fibers (see Elastic Fibers)
Fillers 55, 56, 58, 73, 82, 130, 134, 143, 144, 162, 165, 170, 580, 582, 584, 586, 587, 589, 593, 594
Flame Resistance 134, 136, 386
Flex Properties 250, 251
Flory-Huggins Interaction Parameter 301, 303
Fluctuation Effects 347–350, 352–354
Footwear Applications 33, 39, 147, 151, 254, 575, 582, 584, 593, 594, 598
FOR (see SEE)
Free Radical Polymerization 409, 410, 416–425, 606
Friction 133
Friedel-Crafts Acids 367, 368, 376
Functional End Groups 396, 399–410, 413, 415–420, 425
Functionalized Block Copolymers 588, 589
Functionalized Polyolefins 178–181, 183, 184

Gel Permeation Chromatography 74–77, 89–92, 528, 529
Genetic Engineering 485
Ginzburg-Landau Formalism 459
Glass Transition Temperature (Tg) 19, 30, 49, 50, 54, 57, 59, 61, 65, 74, 86–91, 103, 104, 130, 131, 144, 192, 193, 196, 201–205, 208, 209, 212, 228–230, 234, 250, 274, 280, 285, 286, 290, 298–304, 308–311, 325, 354, 378, 381, 383–387, 397, 404, 408, 422, 423, 445, 469, 474, 484, 491, 492, 522–525, 530, 576–578, 580, 588, 608
Graft Copolymers 1, 4, 9, 102, 117, 118, 123, 178–181, 264, 269, 366–368, 376–378, 391, 396–398, 414–425, 496, 499, 544, 549, 595, 607
Grignard Reagent Initiators 405
Group Transfer Polymerization (GTP) 389, 407, 408, 411, 606

Halato Telechelic Ionomers 280, 281, 286
Halobutyl Rubbers 373–379
Hardness (see Physical Properties)
Hard Segment Modification in Polyester Ethers 193–198, 213–216, 221

Harvest Timing for Biopolymers 471
Hetroarm Block Copolymers 86, 87, 89, 94, 95
Hexagonally Modulated Lamellae (HML) Structure 359
Hexagonally Perforated Layer (HPL) Structure 359
Hexamethylene Adipate and Carbonate Glycols 233
Historical Review 2–6, 16, 31, 48, 49, 192–194, 221, 468
Hydrogen Bonding 25, 26, 29–31
Hydrogenation 8, 10, 54, 97, 102, 105, 298, 299, 303, 304, 380, 522
Hydrolytic Stability 35, 36, 149, 151, 213, 221, 240, 246–248, 252, 579, 591, 596
Hysterysis (see Stress Softening – Mullins Effect)
Hy (B–I–B) (see E–EB–E and E–EP–E Block Copolymers)

Impact Modifier Applications 38, 292, 293, 588, 589, 598
Infrared Dichroism 238
Infra-red Spectroscopy 539, 556
Injection Molding 134, 138, 144, 187, 188, 221, 222, 581, 584, 585, 588, 593, 596, 598, 599
Interface 302, 338, 340, 343, 349, 359, 432, 490
Interfascial Tension (see Surface Energies)
Interpenetrating Polymer Networks (IPNs) 298, 312–332
Interphase Adhesion 92, 95, 96
Intracellular Inclusion Bodies in Bacteria 466
Ionic Clusters 268, 269, 389, 390
Ionic Elastomers 272
Ionomer Blends 291
Ionomers 8, 38, 258–270, 272, 389, 390, 605, 608
Ionomers, Cement Filled 264, 265
Ionomers, Diamene 267
Ionomers, Terpolymer 266
Isocyanate 2, 3, 8, 230, 231, 233
Isoviscous Mixing 315, 320, 321, 330

Kinetics (see Reaction Kinetics)

Light Scattering 66, 67, 460

Lithium Metal Initiators *4, 5, 48*
Lower Service Temperature (*see* Service Temperature)

Macromonomers *8, 9, 370, 371, 396–426, 522, 539–543, 608*
Mechanical Properties (*see* Physical Properties)
Mechanochrosism *538, 547, 557–564*
Melt Viscosity (*see* Viscosity and Viscoelastic Properties)
Melting Temperature (*see* Crystal Melting Temperature, Service Temperature)
Metallocene Catalysis *105, 109, 608*
Metathesis (Ring opening) *606*
Michael Insertion Polymerization *408*
Microdomains *337, 431*
Microscopy (*see* Electron Microscopy, Optical Microscopy)
Miscibility *65–67, 314, 340, 434, 491, 530, 580*
Mixing (*see* Blends)
Moisture Stability (*see* Hydrolytic Stability)
Molecular Weight between Chain Entanglements (*see* Chain Entanglements)
Morphology *25–27, 31, 41, 63–65, 72, 80, 89, 117, 120, 154, 157, 162–164, 166, 168, 197, 198, 201, 204–220, 229, 236–239, 273, 279, 305, 312–319, 339–343, 349, 357, 359–361, 397, 424, 425, 469, 473, 498, 509, 522, 534, 543, 545–547, 549, 562, 564–566, 606*
Multiblock Copolymers *2–4, 6–8, 17–41, 50, 67, 105–125, 192–213, 228, 230, 246, 538–570, 576, 578, 579, 590–595, 599*
Multifunctional Initiators (*see* Di- and Multifunctional Initiators)

Napthalenedicarboxylic acid esters *210, 214*
Natural Rubber *2, 4, 10, 162, 165, 172–175, 579, 585, 595, 596, 598, 599*
NBR/PVC Blends (*see* PVC/NBR Blends)
Neoprene (*see* Polychloroprene (CR))
Neutralization *258*
Neutralization, Partial *259, 263*
Neutron Scattering (*see also* Small Angle Neutron Scattering) *207, 277, 278*
Nitrile Rubber (*see* Poly(acrylonitrile-co-butadiene) (NBR))
Nomenclature *1, 366, 578*
Nucleating Agents *192, 476*

Nylon (*see also* Polyamide) *8, 133, 234, 235*

Oil and Solvent Resistance *33–34, 36, 133, 135, 137, 143, 145, 147, 148, 151, 154, 155, 168, 170, 182, 195, 217, 240, 246, 249, 254, 575, 578, 579, 582, 587, 590–598, 603–606, 609*
Oil Gel Applications *587*
Oils, Compounding *119, 155, 162, 165, 579, 580–582, 584–587, 594, 595*
Opacity (*see* Clarity)
Optical Microscopy *97, 157, 206*
Optical Properties *10, 63, 65–67, 474, 538–540, 547, 553–568*
Optical Spectroscopy *539, 553–561, 564, 566–569*
Ordered Bicontinuous Double Diamond (OBDD) *340–342, 344, 357, 360*
Ordered State *337, 349, 430–433, 442, 446, 451, 457*
Order-Disorder Transition Temperature *351–354, 356, 358, 359, 431, 439–443, 445, 452, 454–456, 460*
Order-Disorder Transition (ODT) *9, 337–339, 346–354, 357–361, 430–460, 554*
Oxidation (*see* Stability, Thermal, Ozone and UV)

Particle Size *159, 160*
Permeability *162, 166, 167*
Phase Boundry (*see* Interface)
Phase Growth *316, 317, 320, 321*
Phase Separation *4, 17, 29, 30, 50, 54, 59, 61, 62, 65–67, 72, 81–83, 86, 91–95, 195, 204–206, 209, 219, 228, 301–304, 311, 337, 342, 346, 368, 382, 388, 397, 413, 417–422, 530, 532*
Physical Crosslinks *6, 17, 50, 102, 117, 192, 229, 285, 313, 397, 469, 473, 576, 605*
Physical Properties *2, 5, 7, 18, 20, 30–35, 49, 54–58, 73–75, 79, 82–93, 95–98, 105, 106, 110, 114–116, 121–123, 131–135, 137, 143, 145–148, 151, 154, 158–165, 167–175, 177, 181–184, 186, 194, 200, 201, 203, 210–218, 236, 241–244, 247, 249, 260, 261, 265–267, 272, 283, 285–288, 305–308, 322–331, 372–379, 382–388, 390, 422–425, 478–483, 485, 498,*

502, 503, 507, 508, 510, 513–516, 529, 530, 533–535, 547–553, 583, 589, 591–593, 597

Plastics (*see* Thermoplastics)

Plasticizers 24, 130, 143, 144, 147, 148, 169–171, 288, 291, 582, 591, 594, 595–597, 599

Plastomers 105

Polyacrylate Segments 3, 4, 10, 318, 388, 422–425, 522–525, 532, 606

Polyamide Segments 7, 8, 230–236, 240–255, 578, 579, 590, 592–595, 599, 604–607

Polyamides (Nylon) 166–171, 172–175, 183, 184, 316–318, 320, 330, 331, 588, 589, 594, 595

Polybutadiene 4, 5, 49, 55, 57, 59, 61

Polybutadiene, Hydrogenated Segments (*see also* E-EB-E and E-EP-E) 97, 98, 115–117, 304, 352, 496, 497

Polybutadiene Segments (*see also* S-B-S, S-B), 3, 5, 6, 61, 65, 68, 85–89, 300, 350, 352, 371, 377, 403, 413, 420, 421, 578, 580, 581, 586

Polybutylene 316, 330, 331

Polycarbonate 8, 38, 171–175, 220, 322, 323, 325, 327–331

Polycarbonate Segments 51

Polycarbonate-esteramide (PCEA) 228, 230–236, 239, 241–243, 246

Polychloroprene (CR) 169, 172–175, 375, 376

Polydiacetylene 538–540, 567, 569

Polyester 2, 8, 10, 38, 172–175, 193, 219, 325–327, 330, 331, 497, 509–512, 588, 589

Polyester Biopolymers 466–485

Polyester/PVC Blends (*see* PVC/Copolyester Blends)

Polyester Segments 2, 3, 7, 8, 18, 19, 28, 34, 186, 194, 213–216, 229–233, 245, 246, 578, 579, 589–595, 599, 604–607

Polyesteramide (PEA) 228, 230–233, 236, 238, 239, 241–243, 245–247, 250–252, 254

Polyester-Diacetylene Segments 567–569

Polyether Ester 192–222

Polyether Segments 2, 3, 7, 8, 18, 19, 30, 34, 186, 192, 217–219, 229–236, 245, 246, 578, 579, 590, 591

Polyetheresteramide (PEEA) 228, 230–233, 236, 238, 239, 241–243, 246, 252

Polyether-b-amide (PE-b-A) 228, 230, 233, 236, 238, 239, 241–243, 245–247, 251–254

Polyethylene 10, 102, 103, 122, 158, 172–175, 274, 291, 496, 499–506, 510–517, 575, 578, 582, 588, 595

Polyethylene, Carboxylated 258

Polyethylene Segments (*see also* E-EB-E and E-EP-E) 6, 61, 97, 98, 304, 352–354, 356, 578, 579, 590, 592, 594, 595, 599, 608

Polyindene Segments (*see* Substituted Polystyrene Segments)

Polyisobutylene and Butyl Rubber 57, 102, 120, 162, 164, 173, 174, 579, 595–599

Polyisobutylene Segments (*see also* S-IB-S) 9, 48, 54, 377, 378, 381–391, 608

Polyisoprene 4, 5, 48, 49, 56, 57, 59

Polyisoprene Segments (*see also* S-I-S, S-I) 5, 6, 54, 75, 77, 78, 82–85, 94, 299, 371, 400, 401, 419, 421, 578, 581, 586

Polymer Blends (*see* Blends)

Poly(alpha-methylstyrene) Segments (*see* Substituted Polystyrene Segments)

Poly(butylene terephthalate) (PBT) (*see* Poly(tetramethylene terephthalate), Polyester)

Poly(ethylene oxide) (PEO) (*see* Polyoxyethylene Glycol)

Poly(para-chloromethylstyrene) Segments (*see* Substituted Polystyrene Segments)

Poly(propylene oxide) (PPO) (*see* Polyoxypropylene Glycol)

Poly(tert-butylstyrene) Segments (*see* Substituted Polystyrene Segments)

Poly(tetramethylene oxide) (PTMO) (*see* Polyoxytetramethylene Glycol)

Polyolefins 102, 154–166

Polyols 17, 30, 34, 35, 192–197, 212–217, 230, 232–234, 240, 254, 541, 543, 544, 549, 567, 568

Polyoxyalkylene Glycol 192–197, 212–217, 233–235, 254

Polyoxyethylene Glycol, Poly(ethylene oxide) (PEO) 192, 217, 233, 234, 240

Polyoxypropylene Glycol, Poly(propylene oxide) (PPO) 197, 210, 217, 218, 233, 234, 254

Polyoxytetramethylene Glycol, Poly(tetramethylene oxide) (PTMO) 192–

199, 212–217, 233, 234, 541, 544, 549
Polypivalolactone Segments 102, 118
Polypropylene 6, 8, 10, 106, 107–112, 114,
    119–125, 155–166, 172–175, 178–184,
    220, 291, 316–318, 320, 328–331, 496,
    506–509, 511, 512, 575, 578, 582, 584,
    585, 588, 589, 595, 596, 598, 599, 607
Polypropylene Segments 6, 113–115, 118, 608
Polysiloxane Segments 51, 91–94, 355, 522,
    578, 579, 590, 593, 594, 599
Polystyrene 5, 10, 49, 54, 66, 67, 172–175,
    425, 491, 492, 499–509, 511–517, 575,
    580, 582, 584, 585, 588
Polystyrene Segments (see also S-B-S, S-B, S-
    I-S, S-I, S-EB-S, S-EB, S-IB-S) 1, 3, 5, 6,
    48–51, 53, 54, 59, 62, 65–68, 74–96, 298,
    350, 356, 369–375, 378, 379, 381–382,
    398–404, 409–411, 413–425, 491–494,
    499–509, 511, 512, 515–517, 522, 529,
    533, 534, 576, 578–582, 584–589, 606, 607
Polyurethane 1, 2, 7, 183, 184
Polyurethane Ionomers 284, 285
Polyurethane Segments 2, 3, 7, 8, 10, 17, 21–
    30, 51, 89, 103, 538, 549, 574, 578, 579,
    589–591, 593, 594, 599, 604–607
Polyurethane-Diacetylene (PU-DA)
    Segments 538, 541–570
Polyvinylchloride (PVC) 1, 2, 7, 130, 325,
    326, 574, 575, 578, 591, 594–599
Poly(2-vinylpyridine) Segments 406
Poly(acrylonitrile-co-butadiene) (NBR) 155,
    162, 166–171, 173–174, 178–184, 579,
    593, 596, 597, 599
Poly(alpha-olefin) Segments 6, 7, 578, 579,
    590, 592, 594, 595, 599, 607, 608
Poly(beta-hydroxyalkanoates) (PHA) 466,
    467–470, 472, 485
Poly(beta-hydroxybutyrates) (PHB) 469, 470,
    485
Poly(beta-hydroxyoctanoates) (PHO) 467,
    471–484
Poly(cyclohexlene terephthalate) 214
Poly(epichlorohydrin) 183, 184
Poly(etherimide) Segments 578, 579, 590, 593,
    594, 599
Poly(ethylene terephthalate) (PET) (2GT) (see
    also Polyester) 215
Poly(ethyleneoxide) Segments 3, 217, 355
Poly(ethylene-co-alphaolefin) 102, 105
Poly(ethylene-co-butylene) 57

Poly(ethylene-co-butylene) Segments (see also
    S-EB-S, S-EB, E-EB-E) 8, 10, 54, 97,
    298, 300, 578, 581, 586
Poly(ethylene-co-methacrylic acid) 258
Poly(ethylene-co-propylene) (see also EPR and
    EPDM) 57
Poly(ethylene-co-propylene) Segments (see
    also S-EP-S and E-EP-E) 54, 299, 352–
    354, 356
Poly(ethylene-co-vinyl acetate) 123
Poly(iso-alkylacrylate) Segments 405, 522–
    524
Poly(methyl methacrylate) Segments 78, 350,
    355, 356, 404, 405, 407–411, 423–425,
    522–526, 529, 530, 532–535
Poly(mycrene) Segments 75, 76
Poly(n-alkylacrylate) Segments 405, 522, 524,
    532–535
Poly(phenylene oxide), Poly(phenylene
    ether) 67, 491–496, 501, 504–506, 515–
    517
Poly(propylene sulfide) Segments 76, 89
Poly(propylene-co-hexene-1) 106
Poly(tert-butylacrylate) Segments 404, 420,
    522–526, 528, 529, 532–535
Poly(tetramethylene isophthalate) (4GI) 194,
    201, 209
Poly(tetramethylene phthalate) (4GP) 195
Poly(tetramethylene terephthalate) (4GT) 192–
    195, 211–216
Poly(trimethylene bibenzoate) 215
Prepolymers, Chlorinated 373–376
Prices 579, 582, 594, 599
Processability and Processing Conditions 37,
    95, 96, 120, 125, 138–142, 144, 149–151,
    156, 170, 187, 188, 252–254, 259, 262,
    311, 424, 425, 530, 538, 584, 585, 593,
    594, 598, 603–605, 607, 608
Production Amounts (see Consumption)
Proton Traps 391
Pseudomonas olevorans 467, 469–472, 485
PVC/Copolyester Blends 148–150, 220
PVC/NBR Blends 2, 7, 130, 143–150
PVC/Polyurethane Blends 150, 151

Radial Polymers (see also Branched and Star
    Polymers) 381, 382, 384–386, 390, 522,
    526–529, 533, 534
Random Copolymers 103, 467
Reaction Kinetics 74, 411–414, 417, 418

Reinforcing Fillers (*see* Fillers)

Rheology (*see* Viscosity and Viscoelastic Properties)

Residual Solvent Effects *79, 80*

Resins *580, 585–587, 589*

Sample Preparation *75, 89, 90*

Scanning Electron Microscopy (SEM) (*see* Electron Microscopy)

Scattering—Structure Factor Relationship *435, 436*

Scrap Recycle Applications *41, 574, 585, 593, 598, 599*

Sealant Applications *575, 586, 587, 589*

Second Order Transition Temperature (Tg) (*see* Glass Transition Temperature (Tg))

Semi-Aromatic Amides Segments *230, 232, 246*

Semi-Crystalline Polymers *122*

Sequential Polymerization *51–53, 74, 381, 382, 523*

Service Temperature *34, 131, 135, 137, 145, 146, 148, 154, 157, 182, 189, 195, 214–218, 222, 228, 230, 240, 242, 254, 255, 575, 577, 578, 582, 586, 588, 589, 591, 592, 604–607, 609*

Sheet Molding Compound Applications (SMC) *589*

Shell-Core Model *274, 276, 277*

Shoes (*see* Footwear Applications)

Silicone Rubber *589, 594, 596*

Size Exclusion Chromatography (*see* Gel Permeation Chromatography (GPC)

Small Angle Neutron Scattering (SANS) *340, 342, 343, 347, 351–354, 358*

Small Angle X-ray Scattering (SAXS) (*see* X-ray Scattering and Diffraction)

Soft Segment Modification in Polyester Ethers *193–195, 217–219*

Solid State Polymerization *539*

Solubility Parameter *62, 63, 85, 86, 94, 176, 301–304, 317, 321*

Solubilization *90, 491, 518*

Solution Cast Films *79, 91*

Solution Polymerization *3–6, 51–54, 74–78, 87, 89, 91, 97, 233, 398–424, 441*

Solution Properties *62, 63, 66, 233, 235, 439–441, 439, 449–453, 459, 560, 575, 586, 594*

Solution Viscosity (*see* Solution Properties)

Solvent Resistance (*see* Oil and Solvent Resistance)

Solvents *50–53, 57, 59, 62–64, 66, 68, 79, 80, 83, 84, 233, 433, 439, 447*

Spectroscopy (*see* Optical Spectroscopy)

Stability, Thermal, Ozone and UV *34, 36, 54, 99, 133–135, 137, 143–149, 151, 162, 165, 212–214, 230, 240, 243, 245, 251, 268, 284, 298, 366, 385, 473, 474, 485, 522, 579, 582, 584, 585, 589, 591, 592, 596, 598, 603, 605, 606*

Star Polymers (*see also* Branched and Radial Polymers) *78, 86, 87, 94, 341, 345, 522, 526–529, 533–536*

Stereocomplexes *388*

Stereoregular Polymers *467*

Stereoregular Random Copolymers *469, 484*

Stereo-Block Copolymers *107*

Stress Induced Crystallization *210, 239*

Stress Softening—Mullins Effect *58, 65, 84, 98*

Stress-Optical Properties *65, 538, 547, 557–564*

Strong Segregation Limit (SSL)—Experiment *338–343, 358*

Strong Segregation Limit (SSL)—Theory *343–345, 358*

Structure *48–51, 67, 68, 197, 198, 239, 467, 469*

Structure—Property Relationships *57–60, 78–99, 576–580*

Substituted Polystyrene Segments *51, 54, 76, 86–93, 303, 350, 370, 373, 375, 378, 379, 381–387, 391, 401, 413, 522*

Sulfonated EPDM Ionomers *278, 280, 284, 285*

Sulfonated Ionomers *275, 277, 278, 280–292, 389, 390*

Sulfonated Polypenteneomer Ionomers *273, 278, 282, 285–287*

Surface Behavior—Experiment *355, 356, 358*

Surface Behavior—Theory *357*

Surface Energies *95, 96, 155, 175–178*

Surlyn *258, 264, 267–269*

Swelling in Solvents *57, 59, 83, 84, 133, 135, 137, 143, 147, 151, 159, 552*

Synthesis *24, 51–54, 74–78, 87, 89, 91, 97, 103, 105–115, 195, 196, 230–235, 278, 280, 380–390, 398–422, 523, 524, 526–*

529, 541–543, 608
S-B Block Copolymers *5, 49, 58, 67, 340, 343, 380, 414, 421, 430, 441, 459, 586*
S-B-S Block Copolymers *5, 6, 10, 49–51, 53, 55, 57–63, 65–68, 72, 73, 75, 85–88, 94, 99, 102, 112, 298, 303–309, 311, 312, 329, 330, 340, 350, 366, 380, 397, 414, 439, 491, 522, 530, 578, 579, 581–589, 599, 607, 608*
S-EB Block Copolymers *586, 587*
S-EB-S Block Copolymers *8, 54, 57–59, 66, 298, 304–331, 380, 491, 497, 500–509, 515–517, 578, 579, 581–589, 599, 607, 608*
S-EP-S Block Copolymers *54, 303, 578, 579*
S-I Block Copolymers *5, 49–51, 58, 340–342, 355, 359, 430, 432, 434, 438, 441–443, 446, 449, 586*
S-IB-S Block Copolymers *9, 54, 57, 366, 381–384*
S-I-S Block Copolymers *5, 6, 49–51, 53, 56–59, 63, 75, 79–85, 88, 94, 95, 99, 112, 298, 350, 354, 366, 380, 578, 579, 581, 586, 589, 599*

Tackifiers (*see* Resins)
Tear Strength (*see* Physical Properties)
Telechelic Polymers *389, 390, 396, 522*
Tensile Properties (*see* Physical Properties)
Terephthalate Esters *193, 194, 215, 216*
Tetramethylene Azealate Glycol *233*
Theoretical Aspects of Block Copolymers *4, 6, 8–10, 336–364, 430–460*
Thermal Analysis (including DSC, TGA, TMA) *23, 26–30, 99, 104, 106, 147, 201, 205, 212, 219, 220, 237, 239, 283, 284, 304, 473, 477, 482, 483, 530, 532, 543–545, 549, 550*
Thermochromism *539, 553–560, 566, 568*
Thermodynamics *8, 9, 65, 66, 301, 336–364, 431, 448, 456, 491–495*
Thermogravimetric Analysis (TGA) (*see* Thermal Analysis (including DSC, DTA and TMA))
Thermoplastics *1, 2, 5, 6, 8, 10, 49, 154, 175, 192, 574, 575, 580, 588, 594*
Thermoplastic Vulcanizate (*see* Dynamic Vulcanization)
Thermoset Polymers *538, 574, 575, 589*

Theta Solvents *62*
Topochemical Cross-linking and Polymerization *538–541, 570*
Tradenames *125, 154, 221, 273, 581, 590, 596*
Transalcoholysis *522–525, 529, 532, 533, 535*
Transmission Electron Microscopy (TEM) (*see* Electron Microscopy)
Transparency (*see* Clarity)

Upper Service Temperature (*see* Service Temperature)
UV Stabilizers *24, 212, 584*

Viscous and Viscoelastic Properties *49, 59–61, 66–68, 75, 80, 82, 88, 93–95, 99, 131, 134, 136, 145, 161, 174, 175, 185–187, 205, 211, 212, 219, 253, 259–261, 265–267, 287, 290, 298, 304–312, 325–327, 331, 350, 351, 358, 385, 430, 460, 531, 534, 535, 575, 586, 587, 589, 594, 606*
Vulcanized Rubbers *2, 5, 49–51, 54, 55, 59, 65, 73, 88, 89, 92, 96, 130, 133, 155, 166, 185, 308, 311, 325, 326*
Vulcanizing Agents (*see* Curatives)

Water Absorbtion *194, 218, 222, 260, 262*
Wax Blend Applications *589*
Weak Segregation Limit (WSL)— Experiment *350–354, 358, 359*
Weak Segregation Limit (WSL)—Theory *338, 345–350, 358, 360, 361*
Wide Angle X-ray Scattering (WAXS) (*see* X-ray Scattering and Diffraction)
Wire Covering Applications *33, 39, 124, 125, 136, 147, 148, 150, 188, 251, 252, 254, 581, 583, 584, 590, 593, 594, 597, 598*
WLF Approach *61*

X-ray Diffraction (*see* X-ray Scatttering and Diffraction)
X-ray Scattering and Diffraction *25–27, 62, 65, 80, 193, 203, 206–209, 268, 274–276, 340, 341, 347, 351, 352, 358, 359, 437–440, 444–445, 449–451, 459, 473, 474, 543–545, 564*

Yarusso-Cooper (YC) Model *275, 277*

Ziegler-Natta Catalysts *4, 103, 105, 113–115, 300, 371, 606*

Biographies

Geoffrey Holden was born and educated in England. After obtaining his Ph.D. from the University of Manchester, he came to the United States in 1958. He worked at the Torrance, CA Technical Center of the Shell Development Company until 1974, when he spent a year at the Shell Laboratory in Delft, the Netherlands. In 1975 he returned to this country and worked at Westhollow Research Center of the Shell Development Company in Houston, Texas, until his retirement in 1992.

Most of his work has been in the area of thermoplastic elastomers. He was one of the original inventors of the styrenic thermoplastic elastomers and has published several papers and articles that deal with the relationship of their properties to their structure. He is the co-editor of a book "Thermoplastic Elastomers – A Comprehensive Review" that covers the whole field of these materials. He was the Chairman of the 1991 Gordon Conference on Elastomers and is one of the original committee members of the Thermoplastic Elastomers Group of the ACS. In 1993 he received the ACS Rubber Division Award for his work in the chemistry of thermoplastic elastomers. He is now a consultant in this field.

N. R. Legge obtained the degrees of B.Sc. (Honors Chem.) and M.Sc. (Phys. Chem.) from the University of Alberta and did his doctoral research at McGill University.

He entered the elastomers research field at Polymer Corp., Sarnia, Ontario, 1945. There he led research projects on oil extended SBR and cold (5 °C) polymerized acrylonitrilebutadiene (NBR) rubber. In 1951 he was appointed Director of Research and Development for Kentucky Synthetic Rubber Corp. in Louisville, and directed research projects sponsored by the Office of Rubber Reserve.

Dr. Legge joined Shell Development Company, Emeryville, California, Laboratory in 1955 and spent the last 23 years of his career with the Shell Companies. He led research and development on high cis-polyisoprene (IR), high cis-polybutadiene (BR), ethylene-propylene (EPR) rubbers. He is best known as leader of the Shell team which discovered and developed, in the years 1961 to 1972, the triblock styrene-diene thermoplastic elastomers.

After retiring from Shell he has been active as a consultant in polymers. In 1987 he received from the Rubber Division of the American Chemical Society the Charles Goodyear Medal for his research in thermoplastic elastomers.

Roderic P. Quirk is Kumho Professor of Polymer Science and Chair of the Department of Polymer Science at the University of Akron, Ohio. The author of over 100 professional publications, he was a Fellow of the Japan Society for Promotion of Science in 1990 and currently serves on the advisory boards of the *Journal of Macromolecular Science-Reviews, Polymer International* and the *Chinese Journal of Polymer Science*. He has held appointments in the Department of Chemistry at the University of Arkansas, Fayetteville (1969–1978), at Michigan Molecular Institute (1979–1983) and as Visiting Professor at the Institut Charles Sadron, CNRS, Strasbourg, France (1991), Tokyo Institute of Technology (1990) and the University of Akron (1976); he gained industrial experience at Ethyl Corporation (1963), 3M Company (1964) and Phillips Petroleum Company (1974). Dr. Quirk received the B.S. degree (1963) in chemistry from Rensselaer Polytechnic Institute, Troy, New York and the M.S. (1965) and Ph.D. (1967) degrees in organic chemistry from the University of Illinois, Champaign-Urbana.

Following completion of his undergraduate (A.B. Summa cum laude) and graduate (A.M, PhD) studies at Harvard University in 1938, **Herman E. Schroeder** began a 42 year career with the DuPont

Company that culminated in 17 years as Director of Research and Development. Before his retirement in 1980 Dr. Schroeder contributed to the development of a variety of specialty elastomers, including: the first good (vinyl pyridine copolymer) adhesive for bonding nylon to rubber, DuPont's pioneering work on polyetherurethanes and the thermoplastic polyetherester Hytrel, the fluoroelastomers Viton and Kalrez, the ethylene/acrylic rubber Vamac and Nordel, the first sulfur-curable ethylene propylene-diene rubber.

Since retiring he has been active as an industrial consultant and lecturer and has served as consultant on material science for art conservation at several museums. Dr. Schroeder was honored by the International Institute of Synthetic Rubber Producers for his many contributions to the rubber industry (1979). In 1984 he received the Charles Goodyear Medal from the American Chemical Society in recognition of contributions in specialty elastomers, and in 1990 the Lavoisier medal from DuPont in recognition of his many commercially successful innovations.